Diagnosis, Characterization and Treatment of Emerging Pathogens

Diagnosis, Characterization and Treatment of Emerging Pathogens

Editors

Shengxi Chen

Fabio Zicker

MDPI

Basel • Beijing • Wuhan • Barcelona • Belgrade • Novi Sad • Cluj • Manchester

Editors
Shengxi Chen
Biodesign Center for
BioEnergetics
Arizona State University
Tempe
USA

Fabio Zicker
Center for Technological
Development in Health
Oswaldo Cruz Foundation
- Fiocruz
Rio de Janeiro
Brazil

Editorial Office
MDPI
St. Alban-Anlage 66
4052 Basel, Switzerland

This is a reprint of articles from the Special Issue published online in the open access journal *Microorganisms* (ISSN 2076-2607) (available at: https://www.mdpi.com/journal/microorganisms/special_issues/O45NXJID7H).

For citation purposes, cite each article independently as indicated on the article page online and as indicated below:

Lastname, A.A.; Lastname, B.B. Article Title. *Journal Name* **Year**, *Volume Number*, Page Range.

ISBN 978-3-7258-0668-3 (Hbk)
ISBN 978-3-7258-0667-6 (PDF)
doi.org/10.3390/books978-3-7258-0667-6

Contents

About the Editors vii

Preface ix

Shengxi Chen
Diagnosis, Characterization and Treatment of Emerging Pathogens
Reprinted from: *Microorganisms* **2023**, *11*, 2032, doi:10.3390/microorganisms11082032 1

Derich Shalbie Simon, Chee-Wei Yew and Vijay Subbiah Kumar
Multiplexed Reverse Transcription Loop-Mediated Isothermal Amplification Coupled with a Nucleic Acid-Based Lateral Flow Dipstick as a Rapid Diagnostic Method to Detect SARS-CoV-2
Reprinted from: *Microorganisms* **2023**, *11*, 1233, doi:10.3390/microorganisms11051233 6

José Manuel Reyes-Ruiz, Omar García-Hernández, Gustavo Martínez-Mier, Juan Fidel Osuna-Ramos, Luis Adrián De Jesús-González, Carlos Noe Farfan-Morales, et al.
The Role of Aspartate Aminotransferase-to-Lymphocyte Ratio Index (ALRI) in Predicting Mortality in SARS-CoV-2 Infection
Reprinted from: *Microorganisms* **2023**, *11*, 2894, doi:10.3390/microorganisms11122894 25

Elena V. Bogoslovskaya, Galina M. Tsyganova, Anastasiia O. Nosova and German A. Shipulin
Analysis of the Frequency of Mutations at Diagnostic Oligonucleotide Sites and Their Impact on the Efficiency of PCR for HIV-1
Reprinted from: *Microorganisms* **2023**, *11*, 2838, doi:10.3390/microorganisms11122838 39

Juthamas Phadungsombat, Huong Thi Thu Vu, Quynh Thi Nguyen, Ha Thi Van Nguyen, Ha Thi Nhu Nguyen, Bich Thi Dang, et al.
Molecular Characterization of Dengue Virus Strains from the 2019–2020 Epidemic in Hanoi, Vietnam
Reprinted from: *Microorganisms* **2023**, *11*, 1267, doi:10.3390/microorganisms11051267 51

Olga Vladimirovna Orlova, Dina Viktorovna Glazkova, Ramil Rafailovich Mintaev, Galina Mikhailovna Tsyganova, Felix Anatolevich Urusov, German Alexandrovich Shipulin and Elena Vladimirovna Bogoslovskaya
Comparative Evaluation of the Activity of Various Lentiviral Vectors Containing Three Anti-HIV Genes
Reprinted from: *Microorganisms* **2023**, *11*, 1053, doi:10.3390/microorganisms11041053 67

Qianniu Li, Xiaoyu Tang, Ling Zhou, Xiaocheng Lv, Long Gao, Tian Lan, et al.
Combined Analysis of the Whole Transcriptome of Piglets Infected with SADS−CoV Virulent and Avirulent Strains
Reprinted from: *Microorganisms* **2023**, *11*, 409, doi:10.3390/microorganisms11020409 80

André Valencio, Miriam Aparecida da Silva, Fernanda Fernandes Santos, Juliana Moutinho Polatto, Marcelo Marcondes Ferreira Machado, Roxane Maria Fontes Piazza and Ana Cristina Gales
Capture ELISA for KPC Detection in Gram-Negative Bacilli: Development and Standardisation
Reprinted from: *Microorganisms* **2023**, *11*, 1052, doi:10.3390/microorganisms11041052 99

María José Cárdenas Espinosa, Tabea Schmidgall, Jessica Pohl, Georg Wagner, Benedikt Wynands, Nick Wierckx, et al.
Assessment of New and Genome-Reduced *Pseudomonas* Strains Regarding Their Robustness as *Chassis* in Biotechnological Applications
Reprinted from: *Microorganisms* **2023**, *11*, 837, doi:10.3390/microorganisms11040837 113

José R. Aguirre-Sánchez, Beatriz Quiñones, José A. Ortiz-Muñoz, Rogelio Prieto-Alvarado, Inés F. Vega-López, Jaime Martínez-Urtaza, et al.
Comparative Genomic Analyses of Virulence and Antimicrobial Resistance in *Citrobacter werkmanii*, an Emerging Opportunistic Pathogen
Reprinted from: *Microorganisms* **2023**, *11*, 2114, doi:10.3390/ microorganisms11082114 . . . . . . **124**

Pilar Marcos, Aoife Doyle, Paul Whyte, Thomas R. Rogers, Máire McElroy, Seamus Fanning, et al.
Characterization of Food Chain *Clostridioides difficile* Isolates in Terms of Ribotype and Antimicrobial Resistance
Reprinted from: *Microorganisms* **2023**, *11*, 1296, doi:10.3390/ microorganisms11051296 . . . . . . **141**

Andris Dishlers, Ivars Petrovskis, Dace Skrastina, Ieva Zarina, Ilva Lieknina, Juris Jansons, et al.
PreS1 Containing HBc VLPs for the Development of a Combined Therapeutic/Prophylactic Hepatitis B Vaccine
Reprinted from: *Microorganisms* **2023**, *11*, 972, doi:10.3390/microorganisms11040972 . . . . **153**

Ruyu Zhao, Xiaoxue Tang, Huiyao Lin, Chen Xing, Na Xu, Bingxin Dai, et al.
Knocking Down Gm16685 Decreases Liver Granuloma in Murine Schistosomiasis Japonica
Reprinted from: *Microorganisms* **2023**, *11*, 796, doi:10.3390/microorganisms11030796 . . . . **170**

Jiajia Zeng, Shuaini Yang, Yuqing Tuo, Xiaoyu Zha, Ruoyuan Sun, Tingsha Lu, et al.
IL-27 Signaling Promotes Th1 Response by Downregulating IL-10 Production in DCs during Chlamydial Respiratory Infection
Reprinted from: *Microorganisms* **2023**, *11*, 604, doi:10.3390/microorganisms11030604 . . . . **184**

Xiaoyu Tang, Ruiyu Zhang, Long Gao, Xiaocheng Lv, Yuan Sun and Jingyun Ma
LncRNA 8244-ssc-miR-320-CCR7 Regulates IFN-β during SVA Infecting PK-15 Cells
Reprinted from: *Microorganisms* **2023**, *11*, 688, doi:10.3390/microorganisms11030688 . . . . **200**

Haoran Zhang, Zuyu Sun, Yi Liu, Rongrong Wei and Nanying Che
Biomarkers Correlated with Tuberculosis Preventive Treatment Response: A Systematic Review and Meta-Analysis
Reprinted from: *Microorganisms* **2023**, *11*, 743, doi:10.3390/microorganisms11030743 . . . . **214**

Jie-Li Hu and Ai-Long Huang
Dynamics of Hepatitis B Virus Covalently Closed Circular DNA: A Mini-Review
Reprinted from: *Microorganisms* **2023**, *11*, 600, doi:10.3390/microorganisms11030600 . . . . **230**

Dequan Pan, Wei Wang and Tong Cheng
Current Methods for the Detection of Antibodies of Varicella-Zoster Virus: A Review
Reprinted from: *Microorganisms* **2023**, *11*, 519, doi:10.3390/microorganisms11020519 . . . . **249**

Giuseppe A. Ramirez, Marco Ripa, Samuele Burastero, Giovanni Benanti, Diego Bagnasco, Serena Nannipieri, et al.
Drug Reaction with Eosinophilia and Systemic Symptoms (DRESS): Focus on the Pathophysiological and Diagnostic Role of Viruses
Reprinted from: *Microorganisms* **2023**, *11*, 346, doi:10.3390/microorganisms11020346 . . . . **266**

About the Editors

Shengxi Chen

Dr. Shengxi Chen is an Associate Professor in the Biodesign Center for BioEnergetics at Arizona State University. He has made significant progress in the development of novel protein models to study the function and dynamics of proteins. Previous to his present position, Dr. Chen worked at the University of Virginia. In his time there, he contributed primarily to research that concerned the incorporation of natural and unnatural amino acids into several kinds of protein and to the study of their activities. Recently, Dr. Chen has developed a series of DNA-affibody nanoparticles potentially useful for inhibiting the growth of breast cancer cells overexpressing the HER receptors. As PI of the PRMRP F15 Discovery Award, Dr. Chen developed tandem toehold-mediated displacement reactions carried out on a DNA tetrahedron structure for amplifying a fluorescent signal. Combined with a single molecule technique, as few as six copies of dengue RNA in a sample could be detected.

Fabio Zicker

Fabio Zicker, MD, PhD is a Specialist in Science, Technology and Innovation at the Center for Technological Development in Health (CDTS) of the Oswaldo Cruz Foundation (Fiocruz). He has a scientific career in epidemiology and previously served as a senior staff member of the UNICEF/UNDP/World Bank/WHO Special Programme for Research and Training in Tropical Diseases (TDR) at Geneva headquarters and the Pan American Health Organization (PAHO) in Venezuela and Washington, DC. Dr. Zicker has extensive international experience in managing technical and scientific collaborations, networks and projects with a focus on strengthening individual and institutional research capabilities in Latin America, Africa and Asia.

Preface

Emerging infectious diseases have shown the potential to rapidly spread in the population. Recently, SARS-COV-2 infected billions of people and caused more than 6 million deaths worldwide. At the same time, there are many other different emerging pathogens, such as dengue, Ebola, and diarrheagenic E. coli, threatening the health of billions of populations worldwide. To provide help for fighting these emerging infectious diseases, this reprint collected 19 publications focus on the development of novel detection methods against these pathogens, including SARS-COV-2, HIV, dengue virus, and bacteria. New protein and RNA markers are explored. Finally, drug treatments and their side effects are summarized in this reprint.

Shengxi Chen and Fabio Zicker
Editors

microorganisms

MDPI

Editorial

Diagnosis, Characterization and Treatment of Emerging Pathogens

Shengxi Chen

Biodesign Center for Bioenergetics, Arizona State University, Tempe, AZ 85287, USA; shengxi.chen.1@asu.edu; Tel.: +1-480-965-5969

Emerging infectious diseases are perhaps the most rapidly spreading diseases. COVID-19 is an infectious disease caused by severe acute respiratory syndrome coronavirus 2 (SARS-CoV-2). The outbreak of COVID-19 started in December 2019 and has infected billions of people, causing about 7 million deaths worldwide to date [1]. Multiple SARS-CoV-2 variants emerged globally, with some demonstrating increased transmissibility or potential resistance to antibodies. Variants of concern, such as the Alpha, Beta, Gamma, Delta, and Omicron variants, caused more fatal syndromes [2]. To date, the world is still suffering from this virus. Another example is Dengue, whose clinical syndrome caused by the dengue virus, presents with a range of manifestations ranging from self-limited febrile illness to hemorrhagic fever and shock, resulting in death [3]. The global incidence of dengue virus infections has increased 30-fold during the last 50 years, with an estimated 50–100 million symptomatic infections occurring each year and over half of the world's population living at risk of infection [3]. At the same time, there are many other different emerging pathogens, such as malaria, monkeypox, Zika, Ebola, West Nile, and diarrheagenic E. coli, threatening the health of billions of people worldwide.

A comprehensive understanding of the structures of these emerging pathogens and their interactions with hosts is important to develop effective therapeutic and preventive approaches. SARS-CoV-2 is formed by a single-strand positive-sense RNA (~29.9 kb), 16 non-structural proteins (nsp1—16), and four structural proteins—nucleocapsid protein (N), envelope protein (E), membrane protein (M), and spike protein (S) [4]. The nucleocapsid protein forms the capsid outside the SARS-CoV-2 RNA genome, and the RNA genome is further packed by envelope proteins, membrane proteins, and spike proteins. The non-structural proteins mediate SARS-CoV-2 RNA transcription, protein translation, protein splicing, and viral replication. Among them, the proteins nsp7, nsp8, and nsp12 form the RNA-dependent RNA polymerase (RdRp) to transcript viral RNA [4]. SARS-CoV-2 is an airborne-transmission virus. The trimers of the S protein of SARS-CoV-2 bind to the angiotensin converting-enzyme 2 (ACE2) receptor of host cells to initiate the infection [5]. The S protein undergoes proteolytic cleavage by several host proteases, such as cathepsin L, furin, and TMPRSS2, resulting in virus–host membrane fusion. The viral RNA is released into the host cytoplasm to replicate its genetic material and assemble new viral particles utilizing the host and its own machinery.

The early and accurate diagnosis of COVID-19 allows for appropriate treatment and timely isolation of infected individuals, reducing the risk of further transmission. The diagnosis of COVID-19 typically involves a combination of clinical syndromes, diagnostic imaging (e.g., chest X-rays or computed tomography scans), and laboratory testing [6]. The primary laboratory methods for detecting COVID-19 are through the detection of the RNA and protein antigens (N, S proteins) of the SARS-CoV-2 virus [7]. The detection of human antibodies (IgA, IgG, IgM) is used to evaluate the infection history of the SARS-CoV-2 virus. Reverse transcription polymerase chain reaction (RT-PCR) tests are the gold standard for detecting the presence of the SARS-CoV-2 virus. These RT-PCR tests detect the specific sequences of SARS-CoV-2 RNA in respiratory samples, such as nasopharyngeal

Citation: Chen, S. Diagnosis, Characterization and Treatment of Emerging Pathogens. *Microorganisms* **2023**, *11*, 2032. https://doi.org/10.3390/microorganisms11082032

Received: 6 July 2023
Revised: 3 August 2023
Accepted: 4 August 2023
Published: 8 August 2023

swabs or throat swabs. The process involves amplifying and analyzing the conserved viral RNA sequences to confirm the presence of the SARS-CoV-2 virus. The RT-PCR tests are highly accurate but require specialized laboratory equipment and may take several hours to produce results [8]. Isothermal amplification techniques have been developed to detect SARS-CoV-2 RNAs as alternative methods to RT-PCR. Reverse transcription loop-mediated isothermal amplification (RT-LAMP), nucleic acid sequence-based amplification (NASBA), rolling circle amplification (RCA), recombinase polymerase amplification (RPA), and transcription-mediated amplification (TMA) are used for isothermal-based SARS-CoV-2 RNA tests. These isothermal detection methods are coupled to a variety of portable devices for readouts, making them more accessible and user-friendly. Next-generation sequencing (NGS) technology is capable of comprehensively sequencing the whole SARS-CoV-2 genome and identifying newly emerging variants, and this technology has been used to discover SARS-CoV-2 and its variants. All the RNA diagnosis methods mentioned above require high-cost materials and highly skilled bioinformatics staff, thus restricting their broad use. Comparatively, rapid antigen/antibody tests are designed to detect specific viral proteins or human antibodies associated with the SARS-CoV-2 virus. These tests provide relatively faster results (usually within 15–30 min) and are less expensive compared to RT-PCR tests [8]. In addition, the rapid antigen tests are suitable to be used at home by any individual. Lab diagnosis of the SARS-CoV-2 virus is crucial for providing appropriate treatments and interventions to infected individuals, as well as tracking the spread of COVID-19 on a local, national, and global scale. It is also useful to assess the prevalence of the virus, identify hotspots, and control the spread of the virus. However, there are several unmet clinical needs for the diagnosis of COVID-19. RT-PCR and its alternative methods for RNA detection require several hours to produce results, thus delaying the decisions for timely treatment and suitable isolation of SARS-CoV-2-infected patients. It is an urgent need to develop a faster RNA test that could provide results within minutes, allowing for more immediate decisions in the future. Antigen/antibody tests are more rapid alternatives to PCR tests, but they are less sensitive and may have a higher rate of false-negative results, especially in asymptomatic individuals or those with low viral loads. Thus, improvements in the sensitivity and specificity of antigen/antibody tests in the future are needed to enhance their utility in diagnosing COVID-19.

Efforts to control the SARS-CoV-2 virus include a combination of vaccination campaigns, viral testing, contact tracing, mask wearing, social distancing, and adapting public health measures as new information emerges. However, these efforts cannot thoroughly prevent the spread of SARS-CoV-2 infection. Numerous patients still require drug treatment for SARS-CoV-2 infection. Remdesivir is an antiviral drug that was initially developed to treat Ebola. It targets the RNA-dependent RNA polymerase (RdRp) of the SARS-CoV-2 virus to inhibit viral replication [9]. It has been approved in several countries for the treatment of COVID-19 in certain patient populations, aged 12 and older. The FDA has authorized molnupiravir for emergency use as a drug to treat COVID-19 patients, which is another RdRp inhibitor. Molnupiravir is used by RdRp to transcript viral RNA, thus increasing the frequency of viral RNA mutations and impairing SARS-CoV-2 replication in humans [10]. Another anti-SARS-CoV-2 drug, Paxlovid (a combination of two drugs—nirmatrelvir and ritonavir) has also been authorized for emergency use by the FDA. Nirmatrelvir is a peptidomimetic inhibitor of SARS-CoV-2's main protease (Mpro). It covalently binds to the catalytic cysteine (Cys145) residue of Mpro and prevents the virus from processing the polyprotein precursors that are required for viral replication [11]. Ritonavir is an HIV protease inhibitor, which inhibits the metabolizing enzyme cytochrome P450 3A (CYP3A) to lengthen the half-life of nirmatrelvir. In Paxlovid, rintonavir acts as a pharmacological enhancer for nirmatrelvir [12]. To date, five anti-SARS-CoV-2 mAb products (bebtelovimab, sotrovimab, bamlanivimab + etesevimab, casirivimab + imdevimab, and tixagevimab + cilgavimab) have received Emergency Use Authorizations (EUA) from the FDA to treat outpatients with mild to moderate COVID-19 [13]. These antibodies bind to nonoverlapping epitopes of the S protein of SARS-CoV-2. Nevertheless, they are not currently authorized for use because the dominant Omicron subvariants are not susceptible to these products. COVID-19 convalescent plasma has also received EUA from the FDA for the treatment of hospitalized patients with COVID-19. In addition, three immunosuppressive drugs, baricitinib, dexamethasone, and tocilizumab, have been used in some cases of severe COVID-19 to control the excessive immune response [14]. To date, ongoing research and clinical trials continue to explore new drugs and therapeutic strategies for COVID-19.

Dengue virus is a member of the family of flaviviruses and is related to yellow fever, West Nile, Zika, and Japanese encephalitis. It is a rapidly spreading endemic viral disease in subtropical and tropical regions of the world [3]. Dengue virus is a mosquito-borne pathogen frequently transmitted by *Aedes aegypti*, *Aedes albopictus*, and *Aedes polynesiensis*. Currently, this disease threatens more than 2.5 billion people in more than 100 countries including the Americas, the Western Pacific, Southeast Asia, the Eastern Mediterranean, and Africa. It is estimated that 50–100 million dengue infections cause 24,000 deaths annually.

Dengue virus infection can cause severe systemic diseases, such as high-grade fever, bleeding, and shock, which results in a fatality rate as high as 10% without proper treatment in one week [3]. However, the early recognition of dengue and proper treatment reduce the fatality rate to <1%. To date, no vaccine or specific drug against dengue virus has been successfully developed. Thus, the early detection of dengue infection is extremely important to cure patients and prevent the spread of the epidemic. Dengue virus is a single-stranded encapsulated RNA virus containing a positive-sense RNA of 11 kb in length, multiple copies of three structural proteins (the capsid protein (C), membrane protein (M), and the envelope protein (E)), and seven nonstructural proteins (NS1, NS2A, NS2B, NS3, NS4A, NS4B, and NS5) [15]. Infection with dengue virus causes a broad spectrum of clinical symptoms, many of which are similar to other infections such as flu virus. Thus, a diagnosis of dengue virus infection based only on clinical syndromes is not reliable. An early laboratory diagnosis is required to confirm the dengue virus infection since more than 50% of patients have a mild non-specific fever or are asymptomatic at the early stage of the disease. During 1–5 days after the onset of symptoms, the dengue virus can be isolated from patients to diagnose the infection and the serotypes. However, this technique requires tissue culture assays, which require 1–2 weeks for detection. Another effective diagnostic method is to detect antibodies and/or antigens of dengue in blood samples. Recently, rapid diagnostic tests (RDT) that simultaneously detect the NS1 antigen and IgM antibody of the dengue virus have been reported. Unfortunately, the cross-reactivity between flaviviruses limits the sensitivity and specificity of these RDT methods [16,17]. To date, the diagnosis of dengue viral RNA in patient's blood, serum, or plasma using a reverse transcriptase PCR (RT-PCR) and real-time RT-PCR (rRT-PCR) have been shown to have the highest sensitivity and specificity of all early-detection methods [18,19]. RT-PCR and rRT-PCR methods require multiple thermocycling steps, which is time-consuming and requires special thermocycling equipment. To avoid these multiple thermocycling steps, isothermal amplification methods (RT-LAMP, RT-RPA, and NASBA) that require a single-step isothermal RNA amplification temperature have been used to detect four dengue virus serotypes [20]. Even though these isothermal amplification methods require only a single thermocycling step (a single melting step at 60–65 °C and an amplification reaction at 37–42 °C), they still require special temperature control equipment. Recently, tandem toehold-mediated displacement reactions (tTMDR) were developed to amplify the dengue RNA signal at room temperature without any additional equipment [21,22]. This method has been used to detect four serotypes of dengue virus with a limit of as low as six copies of RNA per sample. However, these RNA molecular assays at the point of care have been hindered by expensive reagents and equipment, high test complexity, and the extensive laboratory infrastructure required for performance. In addition, dengue is prevalent in many low-resource settings, where access to sophisticated diagnostic tools may be limited. Therefore, it is required to develop rapid, accurate, and economical molecular tests that improve clinical care and inform targeted public health interventions in the future, especially in remote or underserved areas.

Currently, there is no specific antiviral drug available to treat dengue fever. Treatment for dengue mainly focuses on supportive care to manage symptoms and prevent complications. Research is ongoing to develop specific antiviral drugs for dengue. Several potential inhibitors targeting NS3, NS5, and RdRp of the dengue virus are being investigated [23]. However, no specific antiviral drug had been approved for the treatment of dengue to date. In addition, there is no approved specific antiviral drug to treat other flaviviruses either. More efforts are required to develop effective drugs to inhibit these flaviviruses in the future.

In summary, emerging infectious diseases have caused serious damage to human health and continue to threaten the health of billions of people worldwide. Research on the characterization of specific genes or protein targets for the detection or treatment of emerging pathogens, novel methods and strategies for early detection of emerging pathogens, and novel agents to inhibit emerging pathogens are required to fight against these emerging infectious diseases.

Funding: This research was funded by Atlanta Center for Microsystems Engineered Point-of-Care Technologies (ACME-POCT), grant number A473111.

Conflicts of Interest: The author declares no conflict of interest.

References

1. World Health Organization (WHO). *WHO Coronavirus (COVID-19) Dashboard*; World Health Organization: Geneva, Switzerland, 2023. Available online: https://covid19.who.int/ (accessed on 5 July 2023).
2. Sah, R.; Rais, M.A.; Mohanty, A.; Chopra, H.; Chandran, D.; Bin Emran, T.; Dhama, K. Omicron (B.1.1.529) variant and its subvariants and lineages may lead to another COVID-19 wave in the world? An overview of current evidence and counteracting strategies. *Int. J. Surg. Open.* **2023**, *55*, 100625. [CrossRef] [PubMed]
3. World Health Organization (WHO). *Dengue: Guidelines for Diagnosis, Treatment, Prevention and Control*; World Health Organization: Geneva, Switzerland, 2009. Available online: http://www.who.int/tdr/publications/documents/dengue-diagnosis.pdf (accessed on 6 July 2023).
4. Wang, M.Y.; Zhao, R.; Gao, L.J.; Gao, X.F.; Wang, D.P.; Cao, J.M. SARS-CoV-2: Structure, biology, and structure-based therapeutics development. *Front. Cell. Infect. Microbiol.* **2020**, *10*, 587269. [CrossRef] [PubMed]
5. Hoffmann, M.; Kleine-Weber, H.; Schroeder, S.; Krüger, N.; Herrler, T.; Erichsen, S.; Schiergens, T.S.; Herrler, G.; Wu, N.H.; Nitsche, A.; et al. SARS-CoV-2 cell entry depends on ACE2 and TMPRSS2 and is blocked by a clinically proven protease inhibitor. *Cell* **2020**, *181*, 271–280.e8. [CrossRef] [PubMed]
6. Umakanthan, S.; Sahu, P.; Ranade, A.V.; Bukelo, M.M.; Rao, J.S.; Abrahao-Machado, L.F.; Dahal, S.; Kumar, H.; Kv, D. Origin, transmission, diagnosis and management of coronavirus disease 2019 (COVID-19). *Postgrad. Med. J.* **2020**, *96*, 753–758.
7. Fernandes, Q.; Inchakalody, V.P.; Merhi, M.; Mestiri, S.; Taib, N.; Moustafa Abo El-Ella, D.; Bedhiafi, T.; Raza, A.; Al-Zaidan, L.; Mohsen, M.O.; et al. Emerging COVID-19 variants and their impact on SARS-CoV-2 diagnosis, therapeutics and vaccines. *Ann. Med.* **2022**, *54*, 524–540. [CrossRef]
8. Sharma, A.; Balda, S.; Apreja, M.; Kataria, K.; Capalash, N.; Sharma, P. COVID-19 diagnosis: Current and future techniques. *Int. J. Biol. Macromol.* **2021**, *193*, 1835–1844. [CrossRef]
9. Kokic, G.; Hillen, H.S.; Tegunov, D.; Dienemann, C.; Seitz, F.; Schmitzova, J.; Farnung, L.; Siewert, A.; Höbartner, C.; Cramer, P. Mechanism of SARS-CoV-2 polymerase stalling by remdesivir. *Nat. Commun.* **2021**, *12*, 279. [CrossRef]
10. Kabinger, F.; Stiller, C.; Schmitzová, J.; Dienemann, C.; Kokic, G.; Hillen, H.S.; Höbartner, C.; Cramer, P. Mechanism of molnupiravir-induced SARS-CoV-2 mutagenesis. *Nat. Struct. Mol. Biol.* **2021**, *28*, 740–746. [CrossRef]
11. Marzi, M.; Vakil, M.K.; Bahmanyar, M.; Zarenezhad, E. Paxlovid: Mechanism of action, synthesis, and in silico study. *Biomed. Res. Int.* **2022**, *2022*, 7341493. [CrossRef]
12. Hashemian, S.M.R.; Sheida, A.; Taghizadieh, M.; Memar, M.Y.; Hamblin, M.R.; Bannazadeh Baghi, H.; Sadri Nahand, J.; Asemi, Z.; Mirzaei, H. Paxlovid (Nirmatrelvir/Ritonavir): A new approach to COVID-19 therapy? *Biomed. Pharmacother.* **2023**, *162*, 114367. [CrossRef]
13. Widyasari, K.; Kim, J. A review of the currently available antibody therapy for the treatment of coronavirus disease 2019 (COVID-19). *Antibodies* **2023**, *12*, 5. [CrossRef]
14. Albuquerque, A.M.; Eckert, I.; Tramujas, L.; Butler-Laporte, G.; McDonald, E.G.; Brophy, J.M.; Lee, T.C. Effect of tocilizumab, sarilumab, and baricitinib on mortality among patients hospitalized for COVID-19 treated with corticosteroids: A systematic review and meta-analysis. *Clin. Microbiol. Infect.* **2023**, *29*, 13–21. [CrossRef]
15. Perera, R.; Kuhn, R.J. Structural proteomics of dengue virus. *Curr. Opin. Microbiol.* **2008**, *11*, 369–377. [CrossRef]
16. Yow, K.S.; Aik, J.; Tan, E.Y.; Ng, L.C.; Lai, Y.L. Rapid diagnostic tests for the detection of recent dengue infections: An evaluation of six kits on clinical specimens. *PLoS ONE* **2021**, *16*, e0249602. [CrossRef]

17. Luvira, V.; Thawornkuno, C.; Lawpoolsri, S.; Thippornchai, N.; Duangdee, C.; Ngamprasertchai, T.; Leaungwutiwong, P. Diagnostic performance of dengue NS1 and antibodies by serum concentration technique. *Trop. Med. Infect. Dis.* **2023**, *8*, 117. [CrossRef]
18. Warrilow, D.; Northill, J.A.; Pyke, A.; Smith, G.A. Single rapid TaqMan fluorogenic probe based PCR assay that detects all four dengue serotypes. *J. Med. Virol.* **2002**, *66*, 524–528. [CrossRef]
19. Tsai, J.-J.; Liu, W.-L.; Lin, P.-C.; Huang, B.-Y.; Tsai, C.-Y.; Lee, P.-Y.A.; Tsai, Y.-L.; Chou, P.-H.; Chung, S.; Liu, L.-T.; et al. A fully automated sample-to-answer PCR system for easy and sensitive detection of dengue virus in human serum and mosquitos. *PLoS ONE* **2019**, *14*, e0218139. [CrossRef]
20. Kabir, M.A.; Zilouchian, H.; Younas, M.A.; Asghar, W. Dengue detection: Advances in diagnostic tools from conventional technology to point of care. *Biosensors* **2021**, *11*, 206. [CrossRef]
21. Gao, M.; Daniel, D.; Zou, H.; Jiang, S.; Lin, S.; Huang, C.; Hecht, S.M.; Chen, S. Rapid detection of a dengue virus RNA sequence with single molecule sensitivity using tandem toehold-mediated displacement reactions. *Chem. Commun.* **2018**, *54*, 968–971. [CrossRef]
22. Gao, M.; Waggoner, J.J.; Hecht, S.M.; Chen, S. Selective detection of dengue virus serotypes using tandem toehold-mediated displacement reactions. *ACS Infect. Dis.* **2019**, *5*, 1907–1914. [CrossRef]
23. Obi, J.O.; Gutiérrez-Barbosa, H.; Chua, J.V.; Deredge, D.J. Current trends and limitations in dengue antiviral research. *Trop. Med. Infect. Dis.* **2021**, *6*, 180. [CrossRef] [PubMed]

microorganisms

Article

Multiplexed Reverse Transcription Loop-Mediated Isothermal Amplification Coupled with a Nucleic Acid-Based Lateral Flow Dipstick as a Rapid Diagnostic Method to Detect SARS-CoV-2

Derich Shalbie Simon, Chee-Wei Yew and Vijay Subbiah Kumar *

Biotechnology Research Institute, Universiti Malaysia Sabah, Kota Kinabalu 88400, Malaysia; derich_shalbie_simon_mz20@iluv.ums.edu.my (D.S.S.); cheewei.yew@ums.edu.my (C.-W.Y.)
* Correspondence: vijay@ums.edu.my

Abstract: Due to the high reproduction rate of COVID-19, it is important to identify and isolate infected patients at the early stages of infection. The limitations of current diagnostic methods are speed, cost, and accuracy. Furthermore, new viral variants have emerged with higher rates of infectivity and mortality, many with mutations at various primer binding sites, which may evade detection via conventional PCR kits. Therefore, a rapid method that is sensitive, specific, and cost-effective is needed for a point-of-care molecular test. Accordingly, we developed a rapid molecular SARS-CoV-2 detection kit with high specificity and sensitivity, RT-PCR, taking advantage of the loop-mediated isothermal amplification (LAMP) technique. Four sets of six primers were designed based on conserved regions of the SARS-CoV-2 genome: two outer, two inner and two loop primers. Using the optimized protocol, SARS-CoV-2 genes were detected as quickly as 10 min but were most sensitive at 30 min, detecting as little as 100 copies of template DNA. We then coupled the RT-LAMP with a lateral flow dipstick (LFD) for multiplex detection. The LFD could detect two genic amplifications on a single strip, making it suitable for multiplexed detection. The development of a multiplexed RT-LAMP-LFD reaction on crude VTM samples would be suitable for the point-of-care diagnosis of COVID-19 in diagnostic laboratories as well as in private homes.

Keywords: SARS-CoV-2; COVID-19; diagnosis; loop-mediated isothermal amplification; lateral flow dipstick; nucleocapsid gene; membrane gene; envelope gene

Citation: Simon, D.S.; Yew, C.-W.; Kumar, V.S. Multiplexed Reverse Transcription Loop-Mediated Isothermal Amplification Coupled with a Nucleic Acid-Based Lateral Flow Dipstick as a Rapid Diagnostic Method to Detect SARS-CoV-2. *Microorganisms* **2023**, *11*, 1233. https://doi.org/10.3390/microorganisms11051233

Academic Editors: Shengxi Chen and Fabio Zicker

Received: 14 April 2023
Revised: 30 April 2023
Accepted: 3 May 2023
Published: 7 May 2023

1. Introduction

Severe acute respiratory syndrome coronavirus 2 (SARS-CoV-2) was first identified in December 2019 during an outbreak originating from a market in Wuhan, China [1]. This novel coronavirus is the cause of coronavirus disease 2019 (COVID-19) and is only the seventh known coronavirus that has been infectious to humans [2,3]. The rapid spread of COVID-19 has resulted in the World Health Organization (WHO) declaring a global pandemic. The rising number of infected individuals has caused the collapse of hospital systems as they cannot be accommodated [4,5]. This led to countries deciding to close their borders to manage the spread of the disease. However, with many countries adopting national vaccination policies, international travel opened again. Therefore, variants from all around the world were allowed to spread worldwide [6,7]. However, due to the rapidly mutating nature of the viral RNA, many new variants emerged with the ability to evade vaccine-induced immunity. Mutations on common PCR primer binding sites cause the misdiagnosis of COVID-19 via RT-PCR tests [8].

The spread of viral disease can be controlled by quickly identifying and isolating infected individuals. Therefore, it is important to have rapid and accurate diagnostic assays. Throughout the pandemic, various techniques were described and employed for the mass screening of COVID-19. This included real-time reverse transcription polymerase chain reaction (real-time RT-PCR) detection, droplet digital PCR (ddPCR) [9,10], Clustered Regularly Interspaced Palindromic Repeats (CRISPR) [11], nanomaterial-based techniques [12], antigen rapid tests (RTK-Ag), antibody rapid tests (RTK-Ab), cell culture, electron microscopy to the chest and CT-scans [13]. However, these techniques were limited by the lack of facilities, trained personnel, and accuracy.

Currently, the common methods used are RT-PCR detection and RTK-Ag. RTK-Ag is widely used for its rapidness and ease of use but is not as accurate as RT-PCR. RTK-Ag detects the viral proteins that are already present in the patient samples. Meanwhile, RT-PCR is regarded as the gold standard in SARS-CoV-2 detection as it can amplify low amounts of viral genetic material to a detectable amount [14,15]. Therefore, this test is highly specific and sensitive but requires trained personnel and advanced facilities, is expensive and requires a long time. As an alternative, isothermal nucleic acid amplification methods such as Loop-mediated isothermal amplification (LAMP) have been established as a rapid, specific, sensitive, and robust diagnostic method suitable for high-throughput screening [16,17].

LAMP is a method to exponentially amplify a specific nucleic acid region at isothermal conditions [18,19]. Amplification can be observed within 15–60 min at 60 °C to 65 °C. The target regions can be amplified with high efficiency by only using a heating block or water bath, solving the temperature dependency of PCR. Therefore, this method does not require expensive equipment such as a thermocycler or real-time PCR machine and can be performed by individuals without prior training. This method is suitable for clinical diagnostics in a resource-poor environment. LAMP (and RT-LAMP) are commonly used in diagnostic microbiological fields to detect pathogens such as viruses (HIV [20], SARS-CoV-1 [21] and MERS-CoV [22]), bacteria (Tuberculosis [23] and *Salmonella* [24]), nosocomial bacteria (*Acinetobacter baumannii*) [25], fungal pathogens (*Pneumocystis jirovecii*) [26] and parasites (*Ortleppascaris sinensis* [27] and *Phytophthora ramorum* [28]), as well as in the detection of antibiotic-resistant genes (β-lactamases genes [29]).

LAMP products can be visualized either by agarose electrophoresis or through colorimetry. These visualization techniques come with their own advantages and disadvantages. However, in a multiplexed system, these methods cannot be used to differentiate the amplified targets. Therefore, the use of a Lateral Flow Dipstick (LFD) is best suited to detect and differentiate target genes when performing multiplex amplifications. Here, each set of primers could be modified with specific antigen labels to enable rapid detection with the LFD.

In this study, novel LAMP primers were designed to detect SARS-CoV-2 Nucleocapsid (N), Membrane (M) and Envelope (E) genes. These primers are designed on conserved regions of the SARS-CoV-2 genes and aligned against closely related coronaviruses. Using a strand displacing DNA polymerase with reverse transcription activity, a single-enzyme RT-LAMP reaction was achieved. This reaction could detect 100 copies of the control plasmids in 30 min, which could be observed through the formation of bands on a lateral flow dipstick or color changes by SYBR Green Staining. We believe this method will be useful as an alternative to current techniques and helpful as a resource in poorer countries for the rapid diagnosis of the virus.

2. Materials and Methods

2.1. Primer Design for PCR and LAMP Assays

The published sequence from Genbank (Accession number NC_045512.2) was used as the reference sequence for the primer design. In addition, full genome sequences of SARS-CoV-2 were collected from GISAID (Appendix A Table A1), and the N, M and E gene regions were identified. Multiple sequence alignment was conducted on each gene to identify the conserved regions of each gene. These regions were used as inputs in PrimerExplorer version5 http://primerexplorer.jp/lampv5e/index.html (accessed on 3 January 2021) to obtain the F1, B1, F2, B2, F3 and B3 sites for the LAMP primer design. Once desired regions were selected, loop primers were then generated using the same software.

For each gene, a set of primers was designed, consisting of two inner primers (FIP and BIP), two outer primers (F3 and B3) and two loop primers (LF and LB). The forward inner primer (FIP) was designed by a combination of the complementary sequence of F1 (F1c) and F2, linked by a poly-T linker. Additionally, and similarly, the backward inner primer (BIP) was a combination of B1c and B2 with a poly-T linker as well (Table 1).

Table 1. The four sets of primers were designed for different gene targets consisting of the inner, outer and loop primers, along with the size of the target amplicon. The "automatic judgement" feature and default parameters from Primer Explorer Version 5 (http://primerexplorer.jp/lampv5e/index.html, accessed on 3 January 2021) were used to design each set of primers.

Primer Set	Target Gene	Primer Name	Sequence ($5' \rightarrow 3'$)	Target Size (bp)
N1	Nucleocapsid	F3_N1	CCAGAATGGAGAACGCAGTG	202
		B3_N1	CCGTCACCACCACGAATT	
		FIP_N1	Biotin-AGCGGTGAACCAAGACGCAGTTTT GGCGCGATCAAAACAACG	
		BIP_N1	DIG-AATTCCCTCGAGGACAAGGCGTTTT AGCTCTTCGGTAGTAGCCAA	
		LF_N1	TTATTGGGTAAACCTTGGGGC	
		LB_N1	TTCCAATTAACACCAATAGCAGTCC	
N2	Nucleocapsid	F3_N2	AGATCACATTGGCACCCG	213
		B3_N2	CCATTGCCAGCCATTCTAGC	
		FIP_N2	Biotin-TGCTCCCTTCTGCGTAGAAGCTTTT CAATGCTGCAATCGTGCTAC	
		BIP_N2	FAM-GGCGGCAGTCAAGCCTCTTCTTTT CCTACTGCTGCCTGGAGTT	
		LF_N2	AGATCACATTGGCACCCG	
		LB_N2	CCATTGCCAGCCATTCTAGC	
M	Membrane	F3_M	TCTTCTCAACGTGCCACT	220
		B3_M	CTGAGTCACCTGCTACAC	
		FIP_M	Biotin-TACGAAGATGTCCACGAAGGATTTT TCAGACCGCTTCTAGAAAGT	
		BIP_M	FAM-GGACACCATCTAGGACGCTGTTTTT AATAAGAAAGCGTTCGTGATG	
		LF_M	CACAGCTCCGATTACGAGTTC	
		LB_M	TGACATCAAGGACCTGCCT	
E	Envelope	F3_E	TCATTCGTTTCGGAAGAGA	205
		B3_E	GAACTCTAGAAGAATTCAGA	
		FIP_E	Biotin-CGCAGTAAGGATGGCTAGTGTATTTT CAGGTACGTTAATAGTTAATAGCG	
		BIP_E	DIG-TCGATTGTGTGCGTACTGCTGTT TTTTTTAACACGAGAGTAAACGT	
		LF_E	CTAGCAAGAATACCACGAAAGC	
		LB_E	CAATATTGTTAACGTGAGTCTTGTA	

2.2. Preparation of DNA Template

The target N, M and E genic regions were amplified via PCR using the outer primers of each primer set. The PCR reactions were as described [30] with modifications. The reaction of 25 µL consisted of a 1X PCR buffer, 1.5 mM $MgCl_2$, 0.2 mM dNTP mix (Promega, Madison, WI, USA), 20 pmol of each outer primer and 0.2 U *Taq* DNA polymerase (Promega, Madison, WI, USA). The thermocycler protocol included an initial denaturation of 95 °C for 5 min, followed by 40 amplification cycles of 95 °C for 30 s, 55 °C for 30 s and 72 °C for 30 s, and a final extension at 72 °C for 10 min. PCR products were viewed on agarose gel, which were then excised and purified using a QIAquick Gel Extraction Kit (Qiagen, Germantown, MD, USA).

2.3. Recombinant Plasmid Construction for Positive Control for PCR and LAMP Analysis

The purified PCR products were each ligated into a pJET.2 Blunt Cloning Vector in accordance with the CloneJET Blunt End PCR Cloning Kit (Thermo Fisher Scientific, Waltham, MA, USA) instructions. The recombinant plasmids were transformed into 50 µL of a chemically competent *E. coli* strain TOP10, which were then cultured on an ampicillin-Luria Bertani (LB) agar plate at 37 °C for 16 h. Single colonies were selected and grown in an LB broth at 37 °C for 16 h, followed by plasmid extraction using a GeneJET Plasmid Miniprep Kit (Thermo Fisher Scientific, USA) according to the manufacturer's protocol. The plasmid was then verified through DNA sequencing.

2.4. RNA Synthesis for RT-LAMP Protocol Verification

Synthetic RNA for each gene was synthesized from their respective plasmid using a HiScribe ® T7 Quick High Yield RNA Synthesis Kit (New England Biolab, Hitchin, UK). The protocol was conducted as per the manufacturer's instructions. Synthetic RNA was purified using the lithium chloride protocol [31].

2.5. Optimization of LAMP and RT-LAMP Reaction Condition with UV Analysis

Initially, the LAMP assay was conducted modified based on a previously described protocol with a 30 µL reaction mixture containing a 1X Isothermal Amplification Buffer II (New England Biolab, UK), 0.4 M of Betaine (Sigma-Aldrich, St. Louis, MO, USA), 8 mM of $MgSO_4$, a 1.4 mM dNTP mix (Promega, USA), 10 U *Bst* of 3.0 DNA polymerase (New England Biolab, UK), 32 pmol of each inner primer, 8 pmol of each outer primer, 32 pmol of each loop primer, followed by 2 µL of template DNA or RNA [32]. Optimizations were performed by testing different ratios of outer, inner and loop primers, as well as the working concentrations of $MgSO_4$.

2.6. Sensitivity Test for the Detection of SARS-CoV-2 Nucleocapsid, Envelope and Membrane Genes Using End Point-PCR and Quantitative PCR

The recombinant plasmid DNA with each gene was serially diluted 10-fold to achieve 10^8 copies to one copy number [30]. Both the endpoint and quantitative PCR were conducted on each dilution of recombinant plasmid using the outer primers stated in Table 1. For the endpoint PCR, the protocol was as mentioned in Section 2.2. Subsequently, the amplification products were visualized on 2.0% agarose gel electrophoresis, stained with ethidium bromide, and observed under UV light. In addition, quantitative PCR (qPCR) was conducted with the addition of an SYBR Green stain using a real-time PCR machine (Biorad CFX96, Hercules, CA, USA).

2.7. Sensitivity Test for the Detection of SARS-CoV-2 Nucleocapsid, Envelope and Membrane Genes Using LAMP-UV, LAMP-SYBR Green and LAMP-LFD Analyses

The optimized LAMP protocol was conducted on the same set of 10-fold serial dilution positive control recombinant plasmids for 30 min at 65 °C UV, and SYBR Green and LFD analyses were used to visualize the amplification products of the LAMP assays. The 1.5% agarose gel electrophoresis was conducted on amplification products, stained in ethidium bromide and observed under UV conditions. Colorimetry using SYBR Green was conducted by the addition of 2 µL of 1:10, which was diluted SYBR Green I nucleic acid gel stain to all tubes containing LAMP products, and observations on the color changes were immediate [30]. As for LFD, the LAMP protocol was conducted using the primers stated in Table 1, where inner primers were labeled with specific antigens for LFD detection. The PCRD Flex Nucleic Acid-Based Immunoassay (Abingdon Health, York, UK) was used according to the manufacturer's instructions.

2.8. Specificity Test of LAMP Assay

The specificity of each set of LAMP primers was evaluated in silico and in vitro. The sequences of closely related coronavirus, regardless of the hosts, were downloaded from

Genbank (AY613950.1, KY352407.1, KY417144.1, NC_001451.1, NC_002645.1, NC_004718.3, NC_005831.2, NC_006213.1, NC_006577.2, NC_038294.1, NC_048213.1) and aligned on MEGA X [33]. Mismatches were observed on sequences of the primer binding regions on viral sequences that did not belong to SARS-CoV-2. For comparison, the synthetic DNA of SARS-CoV-1 and MERS-CoV genes, as well as the cDNA of Infectious Bronchitis Virus (IBV), were used for in vitro specificity tests as these were the ones available. The LAMP assays were performed at 65 °C for 30 min with 20 ng of each control.

3. Results

3.1. LAMP and RT-LAMP Optimization Using UV Analyses

The optimization of the LAMP protocol was performed with the purpose of shortening the time required for amplification without compromising on sensitivity. Positive LAMP reactions were indicated by the formation of ladder-like bands after performing agarose gel electrophoresis, stained with ethidium bromide, and these were observed under UV light. Four ratios of Outer:Inner:Loop primers were used, 1:1:1, 2:1:2, 4:1:4 and 8:1:8, to perform the LAMP assay on the recombinant plasmid DNA. Thus, we observed that a 4:1:4 primer ratio produced DNA amplification with every set of primers (Figure 1A). Subsequent to identifying the right ratio of the primers in use, the same protocol was repeated against synthetic RNA to test for the single enzyme one-step RT-LAMP protocol. To further optimize the reaction time, 4 mM, 6 mM, 8 mM and 10 mM of $MgSO_4$ was used for the assay and tested on an incubation time of 5 min intervals (5, 10, 15, 20, 25 and 30), which showed that 8 mM had the fastest reaction speed; this allowed for amplification without having amplifications at the non-template control (NTC). The results showed that the addition of 8 mM of $MgSO_4$ was able to produce an amplification within 10 min (Figure 1B). This was also the fastest reaction speed in comparison to the other $MgSO_4$ concentrations.

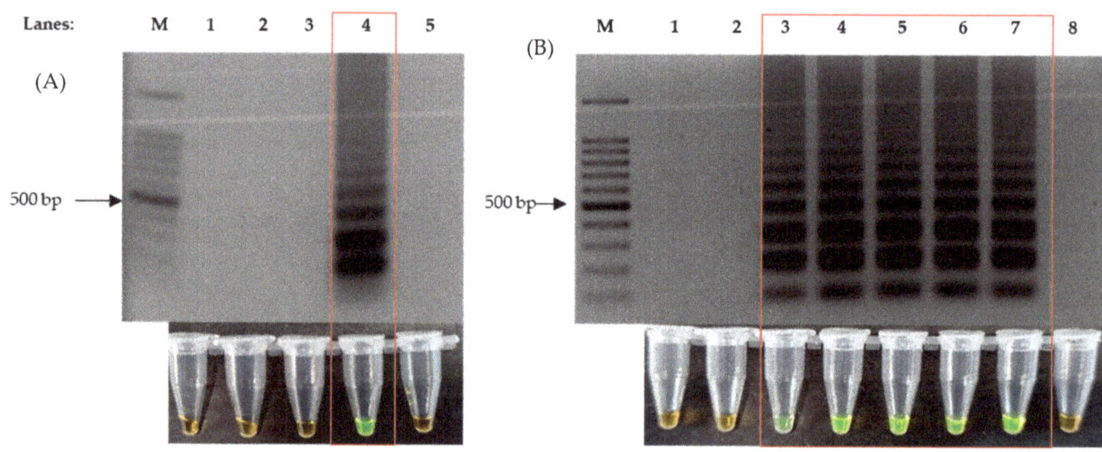

Figure 1. Formation of ladder-like bands on agarose gel electrophoresis produced by LAMP reaction and colorimetric changes of SYBR Green from orange to green, as highlighted in the red box. (**A**) The correct ratio of inner, outer and loop primers for successful LAMP reaction (LAMP-UV, top and LAMP-SYBR Green, bottom), lane 1: 100 bp DNA ladder, lane 2: primer ratio of 1:1:1, lane 3: primer ratio of 2:1:2, lane 4: primer ratio of 4:1:4 and lane 5: primer ratio of 8:1:8. (**B**) Agarose gel electrophoresis results of 8 mM of added $MgSO_4$ against time (LAMP-UV, top and LAMP-SYBR Green, bottom), lane M: 100 bp ladder, lane 1: 1 min, lane 2: 5 min, lane 3: 10 min, lane 4: 15 min, lane 5: 20 min, lane 6: 25 min, lane 7: 30 min and lane 8: non-template control at 30 min.

3.2. *Sensitivity Test of End Point-PCR, Quantitative PCR, LAMP-UV and LAMP-SYBR for the Detection of SARS-CoV-2 Genes*

The sensitivity test was successfully conducted using a set of SARS-CoV-2-positive control plasmids which were serially diluted 10-fold, from 10^8 copies to just one copy. The visualization methods of the PCR (end-point and quantitative) and LAMP (UV, SYBR Green and LFD) were observed to not have had any effect on the sensitivity. The sensitivity of PCR and the optimized LAMP protocol varied between the different target gene types. However, ultimately, the optimized LAMP protocol proved to be equally sensitive and, on certain genes, more sensitive when compared to the PCR.

For the detection of the nucleocapsid (N) gene, two sets of LAMP primers were designed to target two different regions, namely N1 and N2. However, following the sensitivity test, it was revealed that the sensitivity of both primers set across all assays was identical. As revealed in Figure 2A,B, the detection limit for both end-points and quantitative PCR was 10^2 copies of plasmids using an N1 primer set, with the corresponding LAMP-UV, LAMP-SYBR Green, and LAMP-LFD assays providing a detection limit of 10^2 copies as well (Figure 2D,F,H). Conversely, for the N2 primer set, the results showed that the detection limit across all assays (quantitative PCR, end-point PCR, LAMP-UV, LAMP-SYBR Green and LAMP-LFD) was 10^2 copies, respectively (Figure 2A,C,E,G,I).

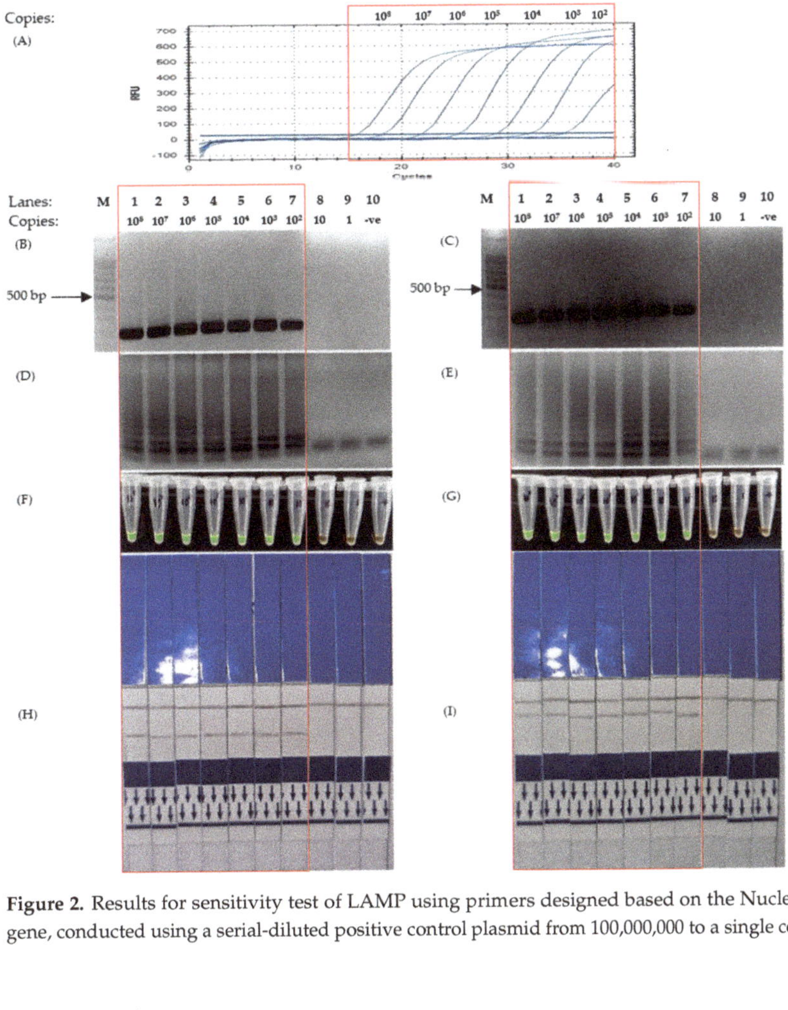

Figure 2. Results for sensitivity test of LAMP using primers designed based on the Nucleocapsid gene, conducted using a serial-diluted positive control plasmid from 100,000,000 to a single copy. The

red box highlights the positive detection of SARS-CoV-2 N genic regions. (**A**) Quantitative Polymerase Chain Reaction using the protocol mentioned in 2.6. (**B,D,F,H**) Corresponding results of End-point PCR, LAMP-UV, LAMP-SYBR Green and LFD, respectively, using primers targeting the N1 region of the Nucleocapsid gene. (**C,E,G,I**) Corresponding results of End-point PCR, LAMP-UV, LAMP-SYBR Green and LFD, respectively, using primers targeting the N2 region of the Nucleocapsid gene. Lane M: 100 bp ladder, lane 1: 10^8 copies, lane 2: 10^7 copies, lane 3: 10^6 copies, lane 4: 10^5 copies, lane 5: 10^4 copies, lane 6: 10^3 copies, lane 7: 10^2 copies, lane 8: 10 copies, lane 9: 1 copy, lane 10: non-template control.

The results for the sensitivity test of the M gene showed that the LAMP protocol was able to detect as little as 10^3 copies of plasmids, as visualized with both UV and colorimetry by SYBR Green staining and LFD (Figure 3C–E). This is similar to that of the PCR tests, which were able to detect 10^3 copies, as indicated by the graph and agarose gel photo in Figure 3A,B.

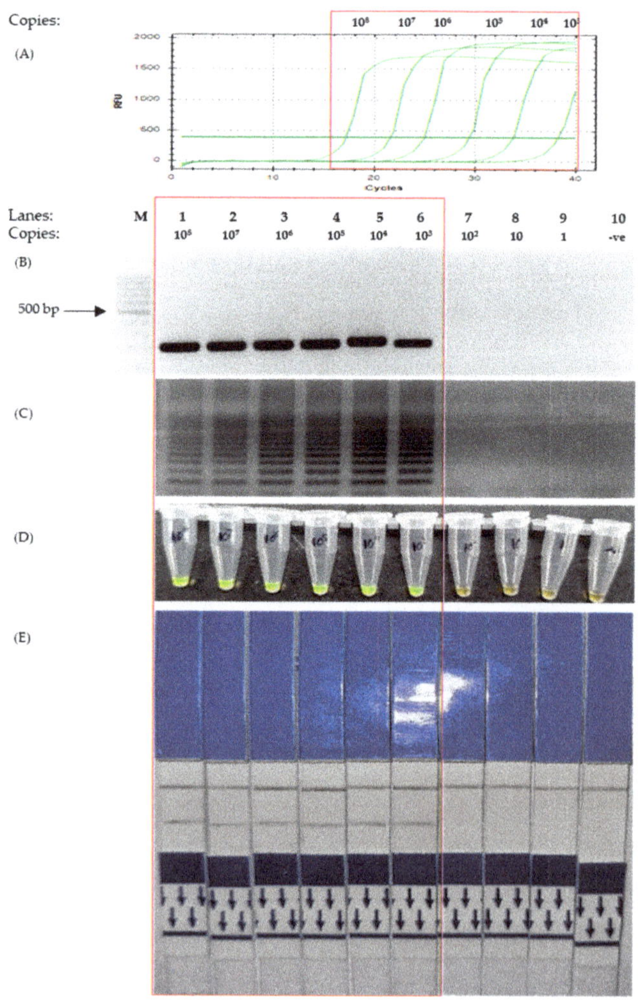

Figure 3. Sensitivity test to assess the detection limit of the SARS-CoV-2 membrane gene using different concentrations of positive control plasmids from 1 copy to 10^8 copies (as indicated on the top

of Figure 3B). The red box highlights the positive detection of SARS-CoV-2 M gene. (**A**) Quantitative PCR. (**B**) Agarose gel electrophoresis of PCR product under UV conditions and stained with EtBr. (**C**) Agarose gel electrophoresis of LAMP product under UV conditions with EtBr staining. (**D**) The corresponding LAMP products were stained with SYBR Green for colorimetric visualization. (**E**) The corresponding LAMP products are visualized on LFD. Lane M: 100 bp ladder, lane 1: 10^8 copies, lane 2: 10^7 copies, lane 3: 10^6 copies, lane 4: 10^5 copies, lane 5: 10^4 copies, lane 6: 10^3 copies, lane 7: 10^2 copies, lane 8: 10 copies, lane 9: 1 copy, lane 10: non-template control.

The sensitivity test showed that the detection limit of LAMP with regard to Envelope (E) gene primers was lower than that of the PCR. It can be observed that both qPCR and end-point PCR had a detection limit of 10^4 copies (Figure 4A,B). However, as depicted in Figure 4C–E, the corresponding photos showed that the detection limit of LAMP was 10-fold lower at 10^3 copies.

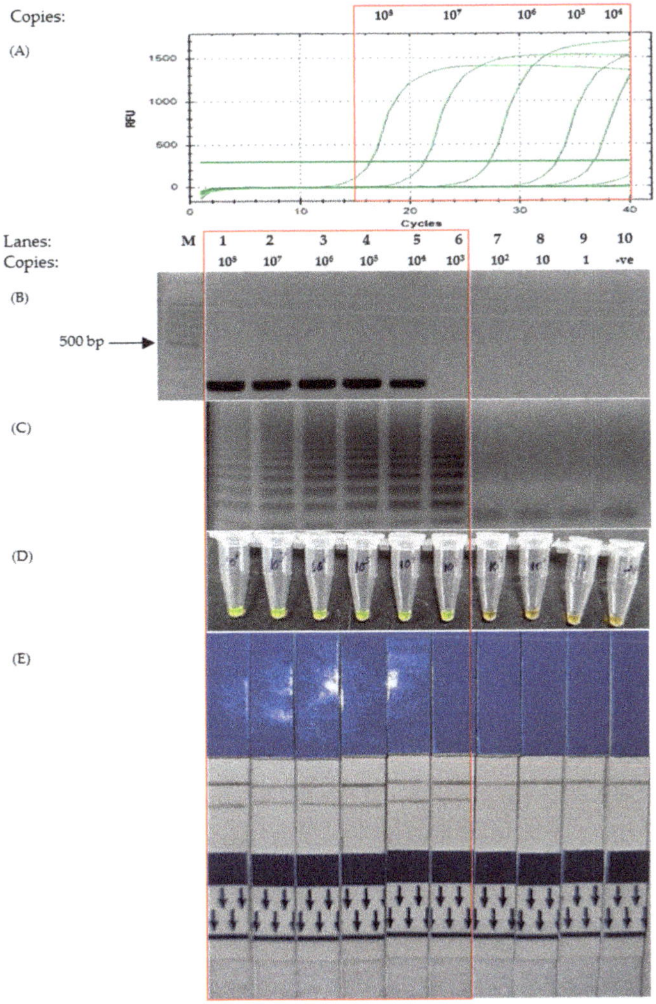

Figure 4. Comparison of the sensitivity test of the Envelope gene on various detection assays. The red box highlights the positive detection of SARS-CoV-2 E gene. Lane M: 100 bp ladder, lane 1: 10^8 copies,

lane 2: 10^7 copies, lane 3: 10^6 copies, lane 4: 10^5 copies, lane 5: 10^4 copies, lane 6: 10^3 copies, lane 7: 10^2 copies, lane 8: 10 copies, lane 9: 1 copy, lane 10: non-template control. (**A**) Quantitative PCR. (**B**) End-point PCR. (**C**) LAMP-UV. (**D**) LAMP-SYBR Green. (**E**) LAMP-LFD.

3.3. Specificity Test of LAMP-UV, LAMP-SYBR Green and LAMP-LFD

Prior to the development of LAMP, in silico screening was performed to design primers not only with conserved regions within SARS-CoV-2 variants but also with a low affinity toward the genes of closely related viral species. Specificity tests were successfully conducted on control plasmids with the gene inserts of various coronaviruses which had close genetic ties to SARS-CoV-2 and were available on hand, SARS-CoV-1, MERS-CoV and IBV.

Positive results indicated by ladder-like bands were observed only against the plasmids containing genes of SARS-CoV-2 (Figure 5). The DNA samples with genes of other coronaviruses produced negative results, as shown by the absence of ladder-like bands. Thus, the results from this specificity test for each set of LAMP primers implied that it was specific only to SARS-CoV-2.

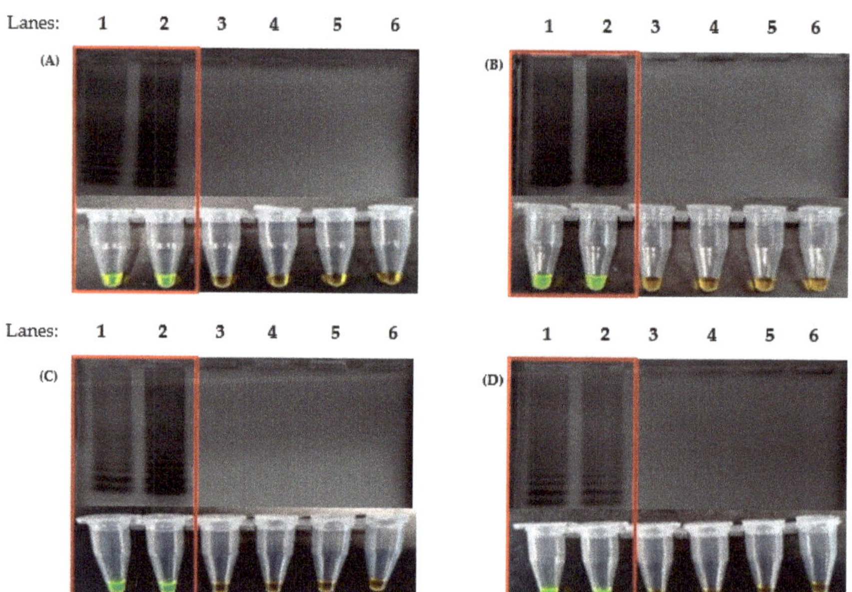

Figure 5. Specificity test to detect only the presence of SARS-CoV-2 genes without false positives, using the newly designed primers, N1 (**A**), N2 (**B**), M (**C**) and E (**D**). The red box highlights the positive detection. (Top) Agarose gel electrophoresis of the LAMP product under UV conditions and stained with ethidium bromide. (Bottom) The corresponding LAMP products were stained with SYBR Green for colorimetric visualization. Lane 1: 100 bp DNA ladder, lane 2: SARS-CoV-2 control plasmid, lane 3: SARS-CoV-2 synthetic RNA, lane 4: SARS-CoV-1 recombinant plasmid, lane 5: MERS-CoV recombinant plasmid, lane 6: cDNA of IBV, lane 7: non-template control.

3.4. Visualization of Multiplexed LAMP on a Single Strip of LFD

Four different combinations were made from the designed primers (Figure 6). This was because the differentiating labels were only digoxigenin (N1 and E) and fluorescein (N2 and M). Therefore, primers with the same labels could not be used in combination. As observed, in a multiplexed reaction, two test lines were observed on a single LFD.

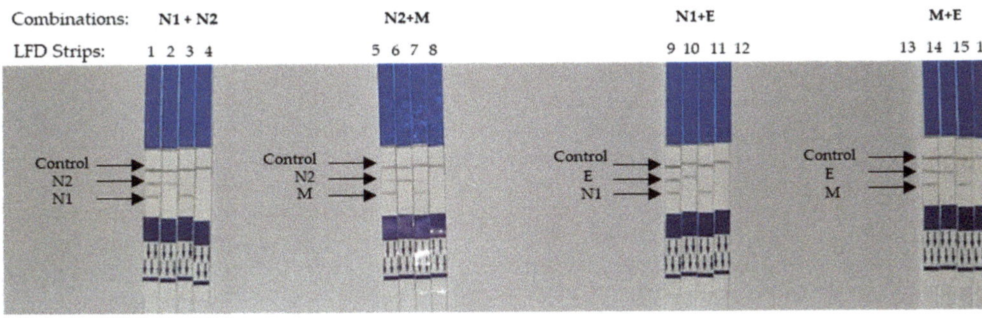

Figure 6. Visualization of multiplexed LAMP on LFD strips. LFD strips 1–4 are the results of multiplexed LAMP using primers N1 and N2. (Template used in strip 1: Control plasmid contained both N1 and N2 regions, strip 2: regions of N1 only, strip 3: regions of N2 only, strip 4: NTC). LFD strips 5–8 were multiplexed by LAMP using N1 and M primers. (Template used in strip 5: Control plasmid contained both N1 and M regions, strip 6: regions of N1 only, strip 7: regions of N2 only, strip 8: NTC). LFD strips 9–12 were multiplexed by LAMP using N2 and E primers. (Template used in strip 9: Control plasmid contained both N2 and E regions, strip 10: regions of E only, strip 11: regions of N2 only, strip 12: NTC). LFD strips 13–16 were multiplexed by LAMP using M and E primers. (Template used in strip 13: Control plasmid contained both M and E regions, strip 14: regions of E only, strip 15: regions of M only, strip 16: NTC).

4. Discussion

Due to the expensive, time-consuming, and tedious nature of RT-PCR tests, healthcare professionals are opting for the less reliable but fast antigen-based rapid detection tests. Therefore, a simple and fast yet accurate detection method is needed for the detection of SARS-CoV-2 at early stages of infection. Taking this into consideration, a multiplexed LAMP-based approach was optimized as fast but also specific and sensitive.

The genes selected for this study were the N, M and E genes (Table 1). This was due to the high read coverage among coronavirus genes when RNA was sequenced from cultured tissues infected with HCoV-299E coronavirus [34]. Others reported LAMP-based SARS-CoV-2 detection using N, RdRp, S, ORF1ab, ORF8 and E genes [35–41]. In addition to that, through GISAID, we obtained genomic sequences for screening with the conserved regions within these genes. This was to ensure that the primers designed were able to detect SARS-CoV-2 across all variants. Therefore, false negative diagnoses were avoided. It was well described that to avoid false negatives, amplicons should be selected from conserved regions or multiple regions at the same time [42]. This was especially difficult as the viral genome is constantly mutating. In addition, even the primers from the gold standard, RT-PCR, were found to produce false negatives [8]. This was due to the ever so rapid occurrence of mutations at the common commercially used primer binding site.

A wide range of genomic sequences of SARS-CoV-2 was downloaded from the GISAID database (Appendix A Table A1), along with the reference genome from GenBank (NC_045512.2), as mentioned in Section 2.1. Using multiple alignment tools, we managed to obtain two conserved regions in the N gene and a region from the M and E genes, respectively, ranging from 200 to 220 bp in length. Six primers were designed for each set of primers, a pair for the inner and outer primers, respectively, as well as a pair of loop primers. Loop primers were found to increase the amplification speed of LAMP [19,30,43] as well as increase its specificity [44].

The LAMP protocol described In this study was able to detect SARS-CoV-2 control plasmids and synthetic RNA in as little as 10 min (Figure 1). With the use of *Bst*, 3.0 DNA polymerase (NEB), the time required for a one-step RT-LAMP reaction was reduced. This is due to the high reverse transcription activity of *Bst* 3.0 DNA polymerase: a single enzyme reaction could be performed using a single temperature (65 °C) [45]. Therefore, simple apparatus could be used to conduct the test (for example, a water bath or heat block). For optimum sensitivity, 30 min of incubation is ideal. Other LAMP-based detection methods have been described using various temperatures in the range of 60–65 °C with an extensive incubation time of up to 60 min. Two-step RT-LAMP protocols, however, take away the rapidness, thus making it unsuitable for point-of-care tests.

Plasmid DNA and synthetic RNA were used for the in vitro testing in the development of this multiplexed RT-LAMP-based LFD. The approach to amplify the genes using the outer primers was subsequently cloned for use as a template for the test, as previously described by others [30,46]. This provided a more accurate quantitative approach compared to repeated RNA extraction from SARS-CoV-2 virions. Furthermore, repeated exposure to SARS-CoV-2 brought a risk of infection during the experimentations.

The colorimetry LAMP visualization method used in this study was the addition of SYBR Green. Color changes were observed to indicate positive (green) and negative (orange) results (Figure 1A Bottom). However, as seen in lane 6 (Figures 3D and 4D, respectively), the color changes we not as vivid. SYBR Green is a DNA intercalating dye with double-stranded DNA and showed color changes from orange to green [47,48]. This method is rapid in the sense that positive detection was observed with the naked eye without the need for agarose gel electrophoresis. However, at a low copy number of templates, the intermediate colors varied from individual observers as observed by others [49,50]. Therefore, colorimetric visualization is not the best option for LAMP. Other colorimetry dyes were documented for LAMP visualizations with varying success rates. This included phenol red [51,52], leuco crystal violet [53], calcein [54], and hydroxy-naphthol blue [55].

Another approach to minimize errors from calorimetry dyes is through the use of a lateral flow dipstick (LFD) in combination with the RT-LAMP, as described in this study. These clear distinct lines on LFD indicate the successful amplification of the N1, N2, M and E genes by LAMP. We used a carbon nanoparticle-based LFD with two distinct test lines targeting Biotin-fluorescein and Biotin–DIG complexes, respectively. The LFD presents tremendous prospects for point-of-care testing because it is straightforward, rapid and visual [56,57]. The genes amplified via LAMP with the primers listed in Table 1 exponentially increased the number of amplicons carrying Biotin-FAM or Biotin-DIG. Through capillary actions, the amplicons migrated through the LFD, where they bound to anti-biotin antibodies bonded with carbon nanoparticles, which were then mobilized [58]. Migrating amplicons carrying the carbon nanoparticle were immobilized at the test lines (coated with neutravidin or anti-DIG antibody). Thus, leaving the black lines observed on the test lines, a control line was formed due to the excess amplicons and/or biotin-labeled primers (no LAMP reaction) that were immobilized at the control line by unspecific antibodies. The use of RT-LAMP coupled with LFD reduced the need for potentially harmful carcinogens (for example, ethidium bromide stain in agarose gel electrophoresis), increased the accuracy (eliminates the use of colorimetric dyes), and even the use of expensive machinery (real-time PCR machine) [59–61].

The detection limit using N1, N2 and M primer sets was equal to that of PCR detection (Figures 2 and 3), with the exception of primers for the E gene (Figure 4). This was because the sensitivity test for the E gene indicated that LAMP was more sensitive than the PCR. Furthermore, our sensitivity test was conducted with just 30 min of incubation in an isothermal condition (65 °C). With regard to LAMP, several publications were reportedly able to detect lower concentrations of the template material but required more time or the conduction of the reverse transcription process separately [62–65]. We used the isothermal enzyme *Bst* 3.0 DNA polymerase (NEB) for both reverse transcription and LAMP. The *Bst* 3.0 DNA polymerase has a dual activity of reverse transcriptase and polymerase in a single temperature incubation [32]. The use of this single enzyme here was efficient and made the one-step RT-LAMP more economical.

It is also important to consider the fact that the detection of low copies of SARS-CoV-2 did not indicate that a person was currently infected with COVID-19 [66]. A higher viral load was required for the body to show symptoms and severity increase with viral load [67]. It has previously been reported that positive qPCR results were observed weeks after infection, the majority of which had a high C_t value, which was an indication of a low viral load [68–71].

For the specificity test, the designed primers were indeed specific only toward the intended targets. Only the SARS-CoV-2 positive control plasmids and synthetic RNA revealed ladder-like bands, and none of the negative controls (SARS-CoV-1, MERS-CoV and IBV) were amplified (Figure 5). However, the control plasmids used in this study were limited due to the difficulty in procuring the genetic material of infectious pathogens. A more comprehensive screening was needed to further validate the results. Therefore, the in silico screening conducted earlier during primer design played an important role in the specificity of these primers.

The detection of two genes with two test lines on a single LFD is extremely relevant in diagnosis as this reduces the chances of false negatives (Figure 6). LFD is useful in multiplex LAMP (mLAMP) reactions as the amplification products could not be differentiated when using gel electrophoresis or visualized with SYBR Green. Therefore, a binary (positive or negative) interpretation of results may be flawed when it comes to multiplexed reactions. The different labels modified on the inner primers play an important role in LFD detection. A single tube assay along with the LFD is convenient for diagnosing COVID-19. Typically, two genes are required to further increase the specificity by avoiding false negatives. This is especially useful as the risk of mutation at one of the primer binding sites is ever-present. Others have described multiplexed molecular amplification when diagnosing SARS-CoV-2, notably using RT-PCR. In RT-PCR, these amplicons could be differentiated using different labels on the probes used. A multiplexed RT-PCR-based protocol [72] was recommended by WHO, which utilized the RdRp and E genes. This highly accurate multiplex RT-PCR approach, which has been utilized in many commercial kits, is still time-consuming and expensive.

While we acknowledge the need for clinical testing, our data show that the RT-LAMP-LFD protocol described here was accurate and rapid. Efforts are currently ongoing to utilize clinical samples to provide a more comprehensive examination of the RT-LAMP-LFD protocol developed from this study. Ongoing and not yet presented results on limited positive samples on hand revealed great promises. Due to varying geographical outbreaks, it was not possible to collect clinical samples of all the variants. Therefore, we decided it would be of interest to other researchers to release LAMP primers and protocols. Therefore, they could begin to test the samples available to them.

To further validate the protocol, trials using samples collected from swabs or saliva could be encouraged. Trials using crude saliva could generate high usefulness in point-of-care settings, preferably with variability in the variants. The convenience of direct testing from crude samples (saliva [73,74] and nasopharyngeal swabs [75]) was documented as before. Additionally, *Bst* 3.0 DNA polymerase was robust and capable of sustaining its activities in the presence of inhibitors [76]. This was especially useful for saliva and other samples which are known to carry amplification inhibitors [77]. This would ensure that the RT-LAMP-LFD was useful in a clinical setting.

5. Conclusions

The current protocol that we have developed is indeed rapid, sensitive, and specific in regard to the study procedures. Therefore, the development of a multiplexed RT-LAMP-LFD reaction on crude VTM samples would be suitable for point-of-care diagnosis of COVID-19 in diagnostic laboratories as well as in private homes.

Author Contributions: Conceptualization, D.S.S. and V.S.K.; methodology, D.S.S. and V.S.K.; formal analysis, D.S.S. and V.S.K.; investigation, D.S.S.; resources, C.-W.Y. and V.S.K.; data curation, D.S.S.; writing—original draft preparation, D.S.S.; writing—review and editing, C.-W.Y. and V.S.K.; supervision, C.-W.Y. and V.S.K.; project administration, V.S.K.; funding acquisition, C.-W.Y. and V.S.K. All authors have read and agreed to the published version of the manuscript.

Funding: This research was funded by Dana Khas COVID-19, Universiti Malaysia Sabah (Grant code: DKC2005).

Institutional Review Board Statement: This article does not contain any studies with human participants or animal subjects.

Informed Consent Statement: Not applicable.

Data Availability Statement: Not applicable.

Acknowledgments: We thank Zarina Amin for providing assistance with the funding and sample material.

Conflicts of Interest: The authors declare no conflict of interest.

Appendix A

Table A1. Full list of genome sequences of SARS-CoV-2 collected from the GISAID database, together with the countries of origin of samples from which the sequences were derived.

No.	Accession Number	Country of Origin
1	EPI_ISL_416866	Malaysia
2	EPI_ISL_430441	Malaysia
3	EPI_ISL_455312	Malaysia
4	EPI_ISL_501182	Malaysia
5	EPI_ISL_719133	Malaysia
6	EPI_ISL_718138	Malaysia
7	EPI_ISL_718174	Malaysia
8	EPI_ISL_738086	Malaysia
9	EPI_ISL_807150	Malaysia
10	EPI_ISL_807153	Malaysia
11	EPI_ISL_936488	Malaysia
12	EPI_ISL_936495	Malaysia
13	EPI_ISL_615652	Denmark
14	EPI_ISL_616802	Denmark
15	EPI_ISL_641491	Denmark
16	EPI_ISL_581117	United Kingdom
17	EPI_ISL_601443	United Kingdom
18	EPI_ISL_678386	Australia
19	EPI_ISL_728189	Singapore
20	EPI_ISL_733573	Hong Kong
21	EPI_ISL_739662	Canada
22	EPI_ISL_745260	Canada
23	EPI_ISL_755593	USA
24	EPI_ISL_755594	USA
25	EPI_ISL_755595	USA
26	EPI_ISL_755627	New Zealand
27	EPI_ISL_763074	Brazil
28	EPI_ISL_794625	New Zealand
29	EPI_ISL_803963	Singapore
30	EPI_ISL_842652	Argentina
31	EPI_ISL_843071	United Kingdom
32	EPI_ISL_845923	United Kingdom
33	EPI_ISL_846595	United Kingdom
34	EPI_ISL_849760	Australia
35	EPI_ISL_852526	United Kingdom
36	EPI_ISL_678594	South Africa
37	EPI_ISL_978596	South Africa
38	EPI_ISL_678597	South Africa

Table A1. *Cont.*

No.	Accession Number	Country of Origin
39	EPI_ISL_745169	South Africa
40	EPI_ISL_762992	South Korea
41	EPI_ISL_770472	Botswana
42	EPI_ISL_825398	Japan
43	EPI_ISL_825489	South Africa
44	EPI_ISL_843196	New Zealand
45	EPI_ISL_852547	United Kingdom
46	EPI_ISL_855369	France
47	EPI_ISL_855514	Kenya
48	EPI_ISL_1562503	USA
49	EPI_ISL_2550714	Malaysia
50	EPI_ISL_2803686	Zambia
51	EPI_ISL_2815331	Malaysia
52	EPI_ISL_2839566	Australia
53	EPI_ISL_2854187	Malaysia
54	EPI_ISL_2868394	Botswana
55	EPI_ISL_2876397	South Africa
56	EPI_ISL_2924057	Malaysia
57	EPI_ISL_2931921	Malaysia
58	EPI_ISL_2984856	South Africa
59	EPI_ISL_3019329	India
60	EPI_ISL_3049843	Kenya
61	EPI_ISL_3050795	Australia
62	EPI_ISL_3060617	India
63	EPI_ISL_3066408	India
64	EPI_ISL_3066431	India
65	EPI_ISL_3066449	India
66	EPI_ISL_3067537	USA
67	EPI_ISL_3071976	USA
68	EPI_ISL_833366	Japan
69	EPI_ISL_1250700	New Zealand
70	EPI_ISL_1416322	Australia
71	EPI_ISL_1428640	Japan
72	EPI_ISL_1543939	Singapore
73	EPI_ISL_1931621	Japan
74	EPI_ISL_2349709	Singapore
75	EPI_ISL_2769807	Japan
76	EPI_ISL_2933406	France
77	EPI_ISL_2956430	Germany
78	EPI_ISL_2988020	Turkiye
79	EPI_ISL_3033191	USA
80	EPI_ISL_3043979	Germany
81	EPI_ISL_3050309	Brazil
82	EPI_ISL_3050508	Brazil
83	EPI_ISL_3050610	Brazil
84	EPI_ISL_3072221	Brazil
85	EPI_ISL_3072616	Brazil
86	EPI_ISL_3087264	Belgium
87	EPI_ISL_3089659	Canada
88	EPI_ISL_416036	Brazil
89	EPI_ISL_431180	Fujian
90	EPI_ISL_445380	Thailand
91	EPI_ISL_490026	Australia
92	EPI_ISL_508266	India
93	EPI_ISL_522491	South Korea
94	EPI_ISL_579320	New Zealand
95	EPI_ISL_591450	Japan

Table A1. *Cont.*

No.	Accession Number	Country of Origin
96	EPI_ISL_630998	United Kingdom
97	EPI_ISL_640129	South Africa
98	EPI_ISL_640130	South Africa
99	EPI_ISL_672711	Brazil
100	EPI_ISL_690818	Japan
101	EPI_ISL_728187	Singapore
102	EPI_ISL_732179	Portugal
103	EPI_ISL_733300	Russia
104	EPI_ISL_746686	Chile
105	EPI_ISL_779245	Japan
106	EPI_ISL_779617	Australia
107	EPI_ISL_801402	Brazil
108	EPI_ISL_850198	South Korea
109	EPI_ISL_875048	United Kingdom
110	EPI_ISL_877765	Italy
111	EPI_ISL_901605	Japan
112	EPI_ISL_920984	Northern Ireland
113	EPI_ISL_941896	Portugal
114	EPI_ISL_985178	Brazil
115	EPI_ISL_1004317	Switzerland
116	EPI_ISL_648527	USA
117	EPI_ISL_707800	New Zealand
118	EPI_ISL_717710	Australia
119	EPI_ISL_755638	New Zealand
120	EPI_ISL_768628	Singapore
121	EPI_ISL_779199	Japan
122	EPI_ISL_818613	Denmark
123	EPI_ISL_846181	United Kingdom
124	EPI_ISL_857314	Taiwan
125	EPI_ISL_860112	Japan
126	EPI_ISL_872584	Australia
127	EPI_ISL_873881	United Kingdom
128	EPI_ISL_904760	Aruba
129	EPI_ISL_905242	Aruba
130	EPI_ISL_956331	Taiwan
131	EPI_ISL_967766	USA
132	EPI_ISL_972791	Denmark
133	EPI_ISL_982043	USA
134	EPI_ISL_984780	USA
135	EPI_ISL_985140	USA
136	EPI_ISL_762449	United Kingdom
137	EPI_ISL_906277	Nigeria
138	EPI_ISL_944748	Australia
139	EPI_ISL_995301	Singapore
140	EPI_ISL_1168766	USA
141	EPI_ISL_1168768	USA
142	EPI_ISL_1173226	Nigeria
143	EPI_ISL_1583653	Brazil
144	EPI_ISL_1896666	Denmark
145	EPI_ISL_1914650	Singapore
146	EPI_ISL_2155777	Philippines
147	EPI_ISL_2242809	Nigeria
148	EPI_ISL_2385974	Australia
149	EPI_ISL_2535627	Malaysia
150	EPI_ISL_3031386	Kenya
151	EPI_ISL_3063476	Turkiye
152	EPI_ISL_3089260	USA
153	EPI_ISL_861280	USA

Table A1. *Cont.*

No.	Accession Number	Country of Origin
154	EPI_ISL_896394	USA
155	EPI_ISL_1158385	USA
156	EPI_ISL_1698346	United Kingdom
157	EPI_ISL_1699692	United Kingdom
158	EPI_ISL_1721838	Germany
159	EPI_ISL_1994447	USA
160	EPI_ISL_2254415	USA
161	EPI_ISL_2967806	Spain
162	EPI_ISL_3032634	Turkiye
163	EPI_ISL_1360328	India
164	EPI_ISL_1442952	Singapore
165	EPI_ISL_1547802	India
166	EPI_ISL_1623010	Rep. Ireland
167	EPI_ISL_1647348	South Korea
168	EPI_ISL_1663320	India
169	EPI_ISL_1847409	Germany
170	EPI_ISL_2710315	South Africa
171	EPI_ISL_2762283	Germany
172	EPI_ISL_2882750	USA
173	EPI_ISL_1111128	Peru
174	EPI_ISL_1111321	Peru
175	EPI_ISL_1111341	Peru
176	EPI_ISL_1445272	Brazil
177	EPI_ISL_1477056	Spain
178	EPI_ISL_1494722	Australia
179	EPI_ISL_2492441	Mexico
180	EPI_ISL_2508552	Chile
181	EPI_ISL_2837340	USA
182	EPI_ISL_2876943	South Africa

References

1. Zhu, N.; Zhang, D.; Wang, W.; Li, X.; Yang, B.; Song, J.; Zhao, X.; Huang, B.; Shi, W.; Lu, R.; et al. A novel coronavirus from patients with pneumonia in China, 2019. *N. Engl. J. Med.* **2020**, *382*, 727–733. [CrossRef] [PubMed]
2. Karami, H.; Sadeghi, K.; Zadheidar, S.; Saadatmand, F.; Mirsalehi, N.; Ardestani, N.H.; Mokhtari-Azad, T. Surveillance of endemic coronaviruses during the COVID-19 pandemic in Iran, 2021–2022. *Influ. Other Respir. Viruses* **2023**, *17*, e13128. [CrossRef]
3. Su, S.; Wong, G.; Shi, W.; Liu, J.; Lai, A.C.; Zhou, J.; Gao, G.F. Epidemiology, genetic recombination, and pathogenesis of coronaviruses. *Trends Microbiol.* **2016**, *24*, 490–502. [CrossRef]
4. Moguerza, J.M.; Oliver, S.P.; de Diego, I.M.; Aceña, V.; Lancho, C.; Cuesta, M.; Fernández, C.G. Health Sufficiency Indicators for Pandemic Monitoring. *Int. J. Environ. Res. Public Health* **2021**, *18*, 5358. [CrossRef] [PubMed]
5. Da Silva, S.J.R.; Pena, L. Collapse of the public health system and the emergence of new variants during the second wave of the COVID-19 pandemic in Brazil. *One Health* **2021**, *13*, 100287. [CrossRef] [PubMed]
6. Williams, G.H.; Llewelyn, A.; Brandao, R.; Chowdhary, K.; Hardisty, K.M.; Loddo, M. SARS-CoV-2 testing and sequencing for international arrivals reveals significant cross border transmission of high risk variants into the United Kingdom. *EClinicalMedicine* **2021**, *38*, 101021. [CrossRef]
7. Douglas, J.; Winter, D.; McNeill, A.; Carr, S.; Bunce, M.; French, N.; Geoghegan, J.L. Tracing the international arrivals of SARS-CoV-2 Omicron variants after Aotearoa New Zealand reopened its border. *Nat. Commun.* **2022**, *13*, 6484. [CrossRef]
8. Laine, P.; Nihtilä, H.; Mustanoja, E.; Lyyski, A.; Ylinen, A.; Hurme, J.; Meri, T. SARS-CoV-2 variant with mutations in N gene affecting detection by widely used PCR primers. *J. Med. Virol.* **2022**, *94*, 1227–1231. [CrossRef]
9. Rotondo, J.C.; Martini, F.; Maritati, M.; Caselli, E.; Gallenga, C.E.; Guarino, M.; Giorgio, R.D.; Mazziotta, C.; Tramarin, M.L.; Badiale, G.; et al. Advanced molecular and immunological diagnostic methods to detect SARS-CoV-2 infection. *Microorganisms* **2022**, *10*, 1193. [CrossRef]
10. Vasudevan, H.N.; Xu, P.; Servellita, V.; Miller, S.; Liu, L.; Gopez, A.; Chiu, C.Y.; Abate, A.R. Digital droplet PCR accurately quantifies SARS-CoV-2 viral load from crude lysate without nucleic acid purification. *Sci. Rep.* **2021**, *11*, 780. [CrossRef]
11. Jayamohan, H.; Lambert, C.J.; Sant, H.J.; Jafek, A.; Patel, D.; Feng, H.; Beeman, M.; Mahmood, T.; Nze, U.; Gale, B.K. SARS-CoV-2 pandemic: A review of molecular diagnostic tools including sample collection and commercial response with associated advantages and limitations. *Anal. Bioanal. Chem.* **2020**, *413*, 49–71. [CrossRef] [PubMed]

12. Kevadiya, B.D.; Machhi, J.; Herskovitz, J. Diagnostics for SARS-CoV-2 infections. *Nat. Mater.* **2021**, *20*, 593–605. [CrossRef] [PubMed]
13. Wu, S.Y.; Yau, H.S.; Yu, M.Y.; Tsang, H.F.; Chan, L.W.C.; Cho, W.C.S.; Yu, A.C.S.; Yim, A.K.Y.; Li, M.J.W.; Wong, Y.K.E.; et al. The diagnostic methods in the COVID-19 pandemic, today and in the future. *Expert Rev. Mol. Diagn.* **2020**, *20*, 985–993. [CrossRef] [PubMed]
14. Peaper, D.R.; Landry, M.L. Laboratory Diagnosis of Viral Infection. In *Handbook of Clinical Neurology*; Tselis, A.C., Booss, J., Eds.; Elsevier: Amsterdam, The Netherlands, 2014; Volume 123, pp. 123–147. [CrossRef]
15. Tang, Y.W.; Schmitz, J.E.; Persing, D.H.; Stratton, C.W. Laboratory diagnosis of COVID-19: Current issues and challenges. *J. Clin. Microbiol.* **2020**, *58*, e00512-20. [CrossRef]
16. Chaouch, M. Loop-Mediated Isothermal Amplification (LAMP): An Effective Molecular Point-of-Care Technique for the Rapid Diagnosis of Coronavirus SARS-CoV-2. *Rev. Med. Virol.* **2021**, *31*, e2215. [CrossRef] [PubMed]
17. Moore, K.J.; Cahill, J.; Aidelberg, G.; Aronoff, R.; Bektaş, A.; Bezdan, D.; Butler, D.J.; Chittur, S.V.; Codyre, M.; Federici, F.; et al. Loop-mediated isothermal amplification detection of SARS-CoV-2 and myriad other applications. *J. Biomol. Tech.* **2021**, *32*, 228. [CrossRef] [PubMed]
18. Notomi, T.; Okayama, H.; Masubuchi, H.; Yonekawa, T.; Watanabe, K.; Amino, N.; Hase, T. Loop-Mediated Isothermal Amplification of DNA. *Nucleic Acids Res.* **2000**, *28*, e63. [CrossRef]
19. Mori, Y.; Notomi, T. Loop-Mediated Isothermal Amplification (LAMP): A rapid, accurate, and cost-effective diagnostic method for infectious diseases. *J. Infect. Chemother.* **2009**, *15*, 62–69. [CrossRef]
20. Curtis, K.A.; Morrison, D.; Rudolph, D.L.; Shankar, A.; Bloomfield, L.S.; Switzer, W.M.; Owen, S.M. A multiplexed RT-LAMP assay for detection of group M HIV-1 in plasma or whole blood. *J. Virol. Methods* **2018**, *255*, 91–97. [CrossRef]
21. Kim, J.H.; Kang, M.; Park, E.; Chung, D.R.; Kim, J.; Hwang, E.S. A simple and multiplex Loop-Mediated Isothermal Amplification (LAMP) assay for rapid detection of SARS-CoV. *BioChip J.* **2019**, *13*, 341–351. [CrossRef]
22. Shirato, K.; Yano, T.; Senba, S.; Akachi, S.; Kobayashi, T.; Nishinaka, T.; Notomi, T.; Matsuyama, S. Detection of Middle East respiratory syndrome coronavirus using reverse transcription loop-mediated isothermal amplification (RT-LAMP). *Virol. J.* **2014**, *11*, 139. [CrossRef] [PubMed]
23. Geojith, G.; Dhanasekaran, S.; Chandran, S.P.; Kenneth, J. Efficacy of loop mediated isothermal amplification (LAMP) assay for the laboratory identification of *Mycobacterium tuberculosis* isolates in a resource limited setting. *J. Microbiol. Methods* **2011**, *84*, 71–73. [CrossRef] [PubMed]
24. Kokkinos, P.A.; Ziros, P.G.; Bellou, M.; Vantarakis, A. Loop-Mediated Isothermal Amplification (LAMP) for the Detection of Salmonella in Food. *Food Anal. Methods* **2014**, *7*, 512–526. [CrossRef]
25. Comini, S.; Bianco, G.; Boattini, M.; Iannaccone, M.; Casale, R.; Banche, G.; Cavallo, R.; Costa, C. Evaluation of the Amplex eazyplex SuperBug Acineto test for direct detection of multi-drug-resistant *Acinetobacter baumannii* bloodstream infections in high endemicity settings. *J. Hosp. Infect.* **2021**, *117*, 179–181. [CrossRef] [PubMed]
26. Scharmann, U.; Kirchhoff, L.; Schmidt, D.; Buer, J.; Steinmann, J.; Rath, P.M. Evaluation of a commercial Loop-mediated Isothermal Amplification (LAMP) assay for rapid detection of *Pneumocystis jirovecii*. *Mycoses* **2020**, *63*, 1107–1114. [CrossRef] [PubMed]
27. Zhao, J.; Xu, W.; Tu, G.; Zhou, Y.; Wu, X. Sensitive and Rapid Detection of *Ortleppascaris sinensis* (Nematoda: Ascaridoidea) by Loop-Mediated Isothermal Amplification. *PeerJ* **2019**, *7*, e7607. [CrossRef] [PubMed]
28. Tomlinson, J.A.; Barker, I.; Boonham, N. Faster, simpler, more-specific methods for improved molecular detection of *Phytophthora ramorum* in the field. *Appl. Environ. Microbiol.* **2007**, *73*, 4040–4047. [CrossRef] [PubMed]
29. Comini, S.; Bianco, G.; Boattini, M.; Banche, G.; Ricciardelli, G.; Allizond, V.; Cavallo, R.; Costa, C. Evaluation of a diagnostic algorithm for rapid identification of Gram-negative species and detection of extended-spectrum β-lactamase and carbapenemase directly from blood cultures. *J. Antimicrob. Chemother.* **2022**, *77*, 2632–2641. [CrossRef]
30. Rahman, A.M.A.; Ransangan, J.; Subbiah, V.K. Improvements to the rapid detection of the marine pathogenic bacterium, *Vibrio harveyi*, using Loop-Mediated Isothermal Amplification (LAMP) in combination with SYBR Green. *Microorganisms* **2022**, *10*, 2346. [CrossRef]
31. Walker, S.E.; Lorsch, J. RNA Purification–Precipitation Methods. In *Methods in Enzymology*; Lorch, J., Ed.; Academic Press: Cambridge, MA, USA, 2013; Volume 530, pp. 337–343. [CrossRef]
32. Silva, S.J.R.D.; Paiva, M.H.S.; Guedes, D.R.D.; Krokovsky, L.; Melo, F.L.D.; Silva, M.A.L.D.; Barreto-Vieira, D.F.; Santos, C.N.D.; Pena, L.J. Development and validation of Reverse Transcription Loop-Mediated Isothermal Amplification (RT-LAMP) for rapid detection of ZIKV in mosquito samples from Brazil. *Sci. Rep.* **2019**, *9*, 4494. [CrossRef]
33. Kumar, S.; Stecher, G.; Li, M.; Knyaz, C.; Tamura, K. MEGA X: Molecular evolutionary genetics analysis across computing platforms. *Mol. Biol. Evol.* **2018**, *35*, 1547–1549. [CrossRef] [PubMed]
34. Viehweger, A.; Krautwurst, S.; Lamkiewicz, K.; Madhugiri, R.; Ziebuhr, J.; Hölzer, M.; Marz, M. Direct RNA nanopore sequencing of full-length coronavirus genomes provides novel insights into structural variants and enables modification analysis. *Genome Res.* **2019**, *29*, 1545–1554. [CrossRef]
35. Yang, W.; Dang, X.; Wang, Q.; Xu, M.; Zhao, Q.; Zhou, Y.; Zhao, H.; Wang, L.; Xu, Y.; Wang, J.; et al. Rapid detection of SARS-CoV-2 using Reverse Transcription RT-LAMP method. *medRxiv* **2020**. [CrossRef]
36. Lu, R.; Wu, X.; Wan, Z.; Li, Y.; Jin, X.; Zhang, C. A novel Reverse Transcription Loop-Mediated Isothermal Amplification method for rapid detection of SARS-CoV-2. *Int. J. Mol. Sci.* **2020**, *21*, 2826. [CrossRef] [PubMed]

37. Mautner, L.; Baillie, C.K.; Herold, H.M.; Volkwein, W.; Guertler, P.; Eberle, U.; Ackermann, N.; Sing, A.; Pavlovic, M.; Goerlich, O.; et al. Rapid Point-of-Care detection of SARS-CoV-2 using Reverse Transcription Loop-Mediated Isothermal Amplification (RT-LAMP). *Virol. J.* **2020**, *17*, 160. [CrossRef] [PubMed]
38. Baek, Y.H.; Um, J.; Antigua, K.J.C.; Park, J.H.; Kim, Y.; Oh, S.; Kim, Y.; Choi, W.S.; Kim, S.G.; Jeong, J.H.; et al. Development of a Reverse Transcription-Loop-Mediated Isothermal Amplification as a Rapid Early-Detection method for novel SARS-CoV-2. *Emerg. Microbes Infect.* **2020**, *9*, 998–1007. [CrossRef]
39. Iijima, T.; Ando, S.; Kanamori, D.; Kuroda, K.; Nomura, T.; Tisi, L.; Kilgore, P.E.; Pecry, N.; Kohase, H.; Hayakawa, S.; et al. Detection of SARS-CoV-2 and the L452R Spike Mutation Using Reverse Transcription Loop-Mediated Isothermal Amplification Plus Bioluminescent Assay in Real-Time (RT-LAMP-BART). *PLoS ONE* **2022**, *17*, e0265748. [CrossRef]
40. Huang, W.E.; Lim, B.; Hsu, C.C.; Xiong, D.; Wu, W.; Yu, Y.; Jia, H.; Wang, Y.; Zeng, Y.; Ji, M.; et al. RT-LAMP for Rapid Diagnosis of Coronavirus SARS-CoV-2. *Microb. Biotechnol.* **2020**, *13*, 950–961. [CrossRef]
41. Mohon, A.N.; Oberding, L.; Hundt, J.; van Marle, G.; Pabbaraju, K.; Berenger, B.M.; Lisboa, L.; Griener, T.; Czub, M.; Doolan, C.; et al. Optimization and clinical validation of dual-target RT-LAMP for SARS-CoV-2. *J. Virol. Methods* **2020**, *286*, 113972. [CrossRef]
42. Zimmermann, F.; Urban, M.; Krüger, C.; Walter, M.; Wölfel, R.; Zwirglmaier, K. In vitro evaluation of the effect of mutations in primer binding sites on detection of SARS-CoV-2 by RT-qPCR. *J. Virol. Methods* **2022**, *299*, 114352. [CrossRef]
43. Cao, Y.T.; Wu, Z.H.; Jian, J.C.; Lu, Y.S. Evaluation of a loop-mediated isothermal amplification method for the rapid detection of Vibrio harveyi in cultured marine shellfish. *Lett. Appl. Microbiol.* **2010**, *51*, 24–29. [CrossRef] [PubMed]
44. Nagamine, K.; Hase, T.; Notomi, T. Accelerated reaction by loop-mediated isothermal amplification using loop primers. *Mol. Cell. Probes* **2002**, *16*, 223–229. [CrossRef] [PubMed]
45. Ghaith, D.M.; Ghazaleh, R.A. Carboxamide and N-alkylcarboxamide additives can greatly reduce non specific amplification in Loop-Mediated Isothermal Amplification for Foot-and-Mouth disease Virus (FMDV) using *Bst* 3.0 polymerase. *J. Virol. Methods* **2021**, *298*, 114284. [CrossRef] [PubMed]
46. Arunrut, N.; Prombun, P.; Saksmerprome, V.; Flegel, T.W.; Kiatpathomchai, W. Rapid and sensitive detection of infectious hypodermal and hematopoietic necrosis virus by loop-mediated isothermal amplification combined with a lateral flow dipstick. *J. Virol. Methods* **2011**, *171*, 21–25. [CrossRef]
47. Parida, M.; Sannarangaiah, S.; Dash, P.K.; Rao, P.V.L.; Morita, K. Loop mediated isothermal amplification (LAMP): A new generation of innovative gene amplification technique; perspectives in clinical diagnosis of infectious diseases. *Rev. Med. Virol.* **2008**, *18*, 407–421. [CrossRef]
48. Chen, D.; Liang, Z.; Ren, S.; Alali, W.; Chen, L. Rapid and visualized detection of virulence-related genes of Vibrio cholerae in water and aquatic products by loop-mediated isothermal amplification. *J. Food Prot.* **2022**, *85*, 44–53. [CrossRef]
49. Loo, K.Y.; Law, J.W.F.; Tan, L.T.H.; Pusparajah, P.; Letchumanan, V.; Lee, L.H. Diagnostic techniques for rapid detection of Vibrio species. *Aquaculture* **2022**, *561*, 738628. [CrossRef]
50. Garg, N.; Ahmad, F.J.; Kar, S. Recent advances in loop-mediated isothermal amplification (LAMP) for rapid and efficient detection of pathogens. *Curr. Res. Microb. Sci.* **2022**, *3*, 100120. [CrossRef]
51. Thi, V.L.D.; Herbst, K.; Boerner, K.; Meurer, M.; Kremer, L.P.; Kirrmaier, D.; Freistaedter, A.; Papagiannidis, D.; Galmozzi, C.; Stanifer, M.L.; et al. A colorimetric RT-LAMP assay and LAMP-Sequencing for detecting SARS-CoV-2 RNA in clinical samples. *Sci. Transl. Med.* **2020**, *12*, eabc7075. [CrossRef]
52. Zen, L.P.Y.; Lai, M.Y.; binti Rozlan, S.I.; Hamid, M.H.A.; Jelip, J.; Mudin, R.N.; Lau, Y.L. End-point detection of loop-mediated isothermal amplification (LAMP) on malaria by direct observation with colorimetric dyes. *Exp. Parasitol.* **2022**, *239*, 108310. [CrossRef]
53. Park, G.S.; Ku, K.; Baek, S.-H.; Kim, S.-J.; Kim, S.I.; Kim, B.T.; Maeng, J.-S. Development of Reverse Transcription Loop-Mediated Isothermal Amplification assays targeting Severe Acute Respiratory Syndrome Coronavirus 2 (SARS-CoV-2). *J. Mol. Diagn.* **2020**, *22*, 729–735. [CrossRef] [PubMed]
54. Luo, Z.; Ye, C.; Xiao, H.; Yin, J.; Liang, Y.; Ruan, Z.; Yang, Y.; Lin, Y.; Shen, Y.; Luo, D.; et al. Optimization of Loop-Mediated Isothermal Amplification (LAMP) assay for robust visualization in SARS-CoV-2 and emerging variants diagnosis. *Chem. Eng. Sci.* **2022**, *251*, 117430. [CrossRef] [PubMed]
55. Hu, X.; Deng, Q.; Li, J.; Chen, J.; Wang, Z.; Zhang, X.; Ding, X.; Yang, M.; Lv, S.; Xu, Y.; et al. Development and clinical application of a rapid and sensitive Loop-Mediated Isothermal Amplification test for SARS-CoV-2 infection. *mSphere* **2020**, *5*, e00808-20. [CrossRef] [PubMed]
56. Liu, J.; Mazumdar, D.; Lu, Y. A simple and sensitive "dipstick" test in serum based on lateral flow separation of aptamer-linked nanostructures. *Angew. Chem. Int. Ed.* **2006**, *45*, 7955–7959. [CrossRef] [PubMed]
57. Jung, J.H.; Oh, S.J.; Kim, Y.T.; Kim, S.Y.; Kim, W.J.; Jung, J.; Seo, T.S. Combination of multiplex reverse-transcription loop-mediated isothermal amplification with an immunochromatographic strip for subtyping influenza A virus. *Anal. Chim. Acta* **2015**, *853*, 541–547. [CrossRef]
58. Jiang, Y.; Chen, S.; Zhao, Y.; Yang, X.; Fu, S.; McKillip, J.L.; Fox, E.M.; Man, C. Multiplex loop-mediated isothermal amplification-based lateral flow dipstick for simultaneous detection of 3 food-borne pathogens in powdered infant formula. *J. Dairy Sci.* **2020**, *103*, 4002–4012. [CrossRef]
59. Lamb, L.E.; Bartolone, S.N.; Ward, E.; Chancellor, M.B. Rapid detection of novel coronavirus (COVID-19) by reverse transcription-loop-mediated isothermal amplification. *MedRxiv* **2020**. [CrossRef]

60. Zhu, X.; Wang, X.; Han, L.; Chen, T.; Wang, L.; Li, H.; Li, S.; He, L.; Fu, X.; Chen, S.; et al. Multiplex reverse transcription loop-mediated isothermal amplification combined with nanoparticle-based lateral flow biosensor for the diagnosis of COVID-19. *Biosens. Bioelectron.* **2020**, *166*, 112437. [CrossRef]
61. Quesada-González, D.; Merkoçi, A. Nanoparticle-based lateral flow biosensors. *Biosens. Bioelectron.* **2015**, *73*, 47–63. [CrossRef]
62. Nam, D.; Kim, S.; Kim, J.H.; Lee, S.; Kim, D.; Son, J.; Kim, D.; Cha, B.S.; Lee, E.S.; Park, K.S. Low-Temperature Loop-Mediated Isothermal Amplification Operating at Physiological Temperature. *Biosensors* **2023**, *13*, 367. [CrossRef]
63. Zhang, C.; Lv, J.; Cao, Y.; Yao, X.; Yin, M.; Li, S.; Zheng, J.; Liu, H. A triple-target reverse transcription loop-mediated isothermal amplification (RT-LAMP) for rapid and accurate detection of SARS-CoV-2 virus. *Anal. Chim. Acta* **2023**, *1255*, 341146. [CrossRef] [PubMed]
64. Bhadra, S.; Riedel, T.E.; Lakhotia, S.; Tran, N.D.; Ellington, A.D. High-surety isothermal amplification and detection of SARS-CoV-2. *MSphere* **2021**, *6*, e00911-20. [CrossRef] [PubMed]
65. El-Kafrawy, S.A.; El-Daly, M.M.; Hassan, A.M.; Harakeh, S.M.; Alandijany, T.A.; Azhar, E.I. Rapid and reliable detection of SARS-CoV-2 using direct RT-LAMP. *Diagnostics* **2022**, *12*, 828. [CrossRef] [PubMed]
66. Tom, M.R.; Mina, M.J. To interpret the SARS-CoV-2 test, consider the cycle threshold value. *Clin. Infect. Dis.* **2020**, *71*, 2252–2254. [CrossRef] [PubMed]
67. He, Y.; Xu, X.; Lu, Q.; Hu, Z.; Jiang, Y.; Song, C.; Chen, W.; Li, P.; Wang, W.; Xu, C.; et al. High viral load suggests increased COVID-19 severity in a longitudinal cohort. *Front. Med.* **2020**, *7*, 566888. [CrossRef]
68. Zou, L.; Ruan, F.; Huang, M.; Liang, L.; Huang, H.; Hong, Z.; Yu, J.; Kang, M.; Song, Y.; Xia, J.; et al. SARS-CoV-2 viral load in upper respiratory specimens of infected patients. *N. Engl. J. Med.* **2020**, *382*, 1177–1179. [CrossRef] [PubMed]
69. Young, B.E.; Ong, S.W.X.; Kalimuddin, S.; Low, J.G.; Tan, S.Y.; Loh, J.; Ng, O.-T.; Marimuthu, K.; Ang, L.E.; Mak, T.M.; et al. Epidemiologic features and clinical course of patients infected with SARS-CoV-2 in Singapore. *JAMA* **2020**, *323*, 1488–1494. [CrossRef]
70. COVID-19 Investigation Team; Kujawski, S.A.; Wong, K.K.; Collins, J.P.; Epstein, L.; Killerby, M.E.; Midgley, C.M.; Abedi, G.R.; Ahmed, N.S.; Almendares, O.; et al. First 12 patients with coronavirus disease 2019 (COVID-19) in the United States. *medRxiv* **2020**. [CrossRef]
71. He, X.; Lau, E.H.; Wu, P.; Deng, X.; Wang, J.; Hao, X.; Lau, Y.C.; Wong, J.Y.; Guan, Y.; Tan, X.; et al. Temporal dynamics in viral shedding and transmissibility of COVID-19. *Nat. Med.* **2020**, *26*, 672–675. [CrossRef]
72. Corman, V.M.; Landt, O.; Kaiser, M.; Molenkamp, R.; Meijer, A.; Chu, D.K.; Bleicker, T.; Brunink, S.; Schneider, J.; Schmidt, M.L.; et al. Detection of 2019 novel coronavirus (2019-nCoV) by real-time RT-PCR. *Eurosurveillance* **2020**, *25*, 2000045. [CrossRef]
73. Ranoa, D.R.E.; Holland, R.L.; Alnaji, F.G.; Green, K.J.; Wang, L.; Brooke, C.B.; Burke, M.D.; Fan, T.M.; Hergenrother, P.J. Saliva-Based molecular testing for SARS-CoV-2 that bypasses RNA extraction. *bioRxiv* **2020**. [CrossRef]
74. Ott, I.M.; Strine, M.S.; Watkins, A.E.; Boot, M.; Kalinich, C.C.; Harden, C.A.; Vogels, C.B.F.; Casanovas-Massana, A.; Moore, A.J.; Muenker, M.C.; et al. Simply saliva: Stability of SARS-CoV-2 detection negates the need for expensive collection devices. *medRxiv* **2020**. [CrossRef]
75. Alekseenko, A.; Barrett, D.; Pareja-Sanchez, Y.; Howard, R.J.; Strandback, E.; Ampah-Korsah, H.; Rovšnik, U.; Zuniga-Veliz, S.; Klenov, A.; Malloo, J.; et al. Direct detection of SARS-CoV-2 using non-commercial RT-LAMP reagents on heat-inactivated samples. *Sci. Rep.* **2021**, *11*, 1820. [CrossRef] [PubMed]
76. Lee, D.; Shin, Y.; Chung, S.; Hwang, K.S.; Yoon, D.S.; Lee, J.H. Simple and highly sensitive molecular diagnosis of Zika virus by lateral flow assays. *Anal. Chem.* **2016**, *88*, 12272–12278. [CrossRef]
77. Beckmann, J.F.; Fallon, A.M. Decapitation improves detection of *Wolbachia pipientis* (Rickettsiales: Anaplasmataceae) in *Culex pipiens* (Diptera: Culicidae) mosquitoes by the polymerase chain reaction. *J. Med. Entomol.* **2014**, *49*, 1103–1108. [CrossRef]

microorganisms

Article

The Role of Aspartate Aminotransferase-to-Lymphocyte Ratio Index (ALRI) in Predicting Mortality in SARS-CoV-2 Infection

José Manuel Reyes-Ruiz [1,*], Omar García-Hernández [2], Gustavo Martínez-Mier [1], Juan Fidel Osuna-Ramos [3], Luis Adrián De Jesús-González [4], Carlos Noe Farfan-Morales [5], Selvin Noé Palacios-Rápalo [6], Carlos Daniel Cordero-Rivera [6], Tatiana Ordoñez-Rodríguez [2] and Rosa María del Ángel [6]

1 Department of Research, Unidad Médica de Alta Especialidad, Hospital de Especialidades No. 14, Centro Médico Nacional "Adolfo Ruiz Cortines", Instituto Mexicano del Seguro Social (IMSS), Veracruz 91897, Mexico; gustavo.martinezmi@imss.gob.mx
2 Department of Internal Medicine, Unidad Médica de Alta Especialidad, Hospital de Especialidades No. 14, Centro Médico Nacional "Adolfo Ruiz Cortines", Instituto Mexicano del Seguro Social (IMSS), Veracruz 91897, Mexico; haro-green@hotmail.com (O.G.-H.); tatiana.ordonez@imss.gob.mx (T.O.-R.)
3 Facultad de Medicina, Universidad Autónoma de Sinaloa, Culiacán 80019, Mexico; osunajuanfidel.fm@uas.edu.mx
4 Unidad de Investigación Biomédica de Zacatecas, Instituto Mexicano del Seguro Social, Zacatecas 98000, Mexico; luis.dejesus@cinvestav.mx
5 Departamento de Ciencias Naturales, Universidad Autónoma Metropolitana (UAM), Unidad Cuajimalpa, Mexico City 05348, Mexico; carlos.farfan@cinvestav.mx
6 Department of Infectomics and Molecular Pathogenesis, Center for Research and Advanced Studies (CINVESTAV-IPN), Mexico City 07360, Mexico; selvin.palacios@cinvestav.mx (S.N.P.-R.); carlos.cordero@cinvestav.mx (C.D.C.-R.); rmangel@cinvestav.mx (R.M.d.Á.)
* Correspondence: jose.reyesr@imss.gob.mx

Citation: Reyes-Ruiz, J.M.; García-Hernández, O.; Martínez-Mier, G.; Osuna-Ramos, J.F.; De Jesús-González, L.A.; Farfan-Morales, C.N.; Palacios-Rápalo, S.N.; Cordero-Rivera, C.D.; Ordoñez-Rodríguez, T.; Ángel, R.M.d. The Role of Aspartate Aminotransferase-to-Lymphocyte Ratio Index (ALRI) in Predicting Mortality in SARS-CoV-2 Infection. *Microorganisms* **2023**, *11*, 2894. https://doi.org/10.3390/microorganisms11122894

Academic Editors: Shengxi Chen, Fabio Zicker and Josep M. Llibre

Received: 6 September 2023
Revised: 28 September 2023
Accepted: 10 October 2023
Published: 30 November 2023

Abstract: COVID-19 has a mortality rate exceeding 5.4 million worldwide. The early identification of patients at a high risk of mortality is essential to save their lives. The AST-to-lymphocyte ratio index (ALRI) is a novel biomarker of survival in patients with hepatocellular carcinoma, an organ susceptible to SARS-CoV-2 infection. For this study, the prognostic value of ALRI as a marker of COVID-19 mortality was evaluated. For this purpose, ALRI was compared with the main biomarkers for COVID-19 mortality (neutrophil-to-lymphocyte ratio [NLR], systemic immune-inflammation index [SII], platelet-to-lymphocyte ratio [PLR], lactate dehydrogenase (LDH)/lymphocyte ratio [LDH/LR]). A retrospective cohort of 225 patients with SARS-CoV-2 infection and without chronic liver disease was evaluated. In the non-survival group, the ALRI, NLR, SII, and LDH/LR were significantly higher than in the survival group ($p_{corrected} < 0.05$). ALRI had an area under the curve (AUC) of 0.81, a sensitivity of 70.37%, and a specificity of 75%, with a best cut-off value >42.42. COVID-19 patients with high ALRI levels had a mean survival time of 7.8 days. Multivariate Cox regression revealed that ALRI > 42.42 (HR = 2.32, 95% CI: 1.35–3.97; $p_{corrected} = 0.01$) was a prognostic factor of COVID-19 mortality. These findings prove that ALRI is an independent predictor of COVID-19 mortality and that it may help identify high-risk subjects with SARS-CoV-2 infection upon admission.

Keywords: COVID-19; SARS-CoV-2; aspartate aminotransferase-to-lymphocyte ratio index (ALRI); biomarker; mortality

1. Introduction

In December 2019, China (Wuhan, Hubei Province) reported cases of pneumonia caused by an unknown virus [1]. Later, on 11 February 2020, the new virus was named Severe Acute Respiratory Syndrome Coronavirus 2 (SARS-CoV-2) by the International Committee on Taxonomy of Viruses (ICTV), and the resulting disease was named COVID-19 by Word Health Organization (WHO) [2]. COVID-19 has characteristics of viral pneumonia that can lead to respiratory failure, respiratory distress syndrome, and, ultimately, death [1].

During SARS-CoV-2 infection, there is an immune response imbalance due to hyperinflammation, leading to a poor prognosis of COVID-19. Thus, due to the high COVID-19 mortality rate [3], it is important to screen all patients with hyperinflammation using hematologic parameters or ratios and to predict disease progression early on to decrease COVID-19 mortality. In this context, identifying biomarkers that can reflect inflammation and immune status are potential predictors for the prognosis of COVID-19 [4]. Inflammation parameters from routine bloodwork, such as leukocytes [5], neutrophiles [6], lymphocytes [6], neutrophil-to-lymphocyte ratio (NLR) [7], platelet-to-lymphocyte ratio (PLR) [7], systemic immune-inflammation index (SII) [8], lactate dehydrogenase (LDH)/lymphocyte ratio [9], aspartate aminotransferase (AST)-to-neutrophil ratio index (ANRI) [10], and AST-to-platelet ratio index (APRI) [11] have been demonstrated to be efficient prognostic biomarkers for COVID-19.

SARS-CoV-2 infection induces alterations in liver function tests and hepatic impairment, according to pathological findings concerning patients with COVID-19 [12]. An increase in AST levels at admission has been identified as an independent predictor of COVID-19 mortality [12]. The AST-to-lymphocyte ratio index (ALRI) is a novel inflammatory index for hepatocellular carcinoma, and it is related to the survival of patients with this condition [13]. A study conducted in a hospital in Asia (Ankara, Turkey) found no relationship between ALRI and COVID-19 severity [11]. However, its role in COVID-19 mortality has yet to be explored. Thus, this study aims to assess the prognostic value of ALRI regarding the mortality of SARS-CoV-2-infected patients upon hospital admission. For this purpose, a comparison of the main biomarkers for COVID-19 mortality (lymphocytes, platelets, NLR, SII, PLR, LDH/lymphocyte ratio) was also performed.

2. Materials and Methods

2.1. Study Design and Participants

A retrospective single-center observational and cross-sectional study was conducted at a tertiary care hospital from the Mexican Social Security Institute (IMSS): Unidad Médica de Alta Especialidad, Hospital de Especialidades No. 14, Centro Médico Nacional "Adolfo Ruiz Cortines". This study was approved by our local research ethics committee (R-2023-3001-111) and conducted following the Declaration of Helsinki. Patients without chronic liver disease and with a positive nucleic acid test for SARS-CoV-2 from May to September 2020 were included in the study. This study only included patients with in-hospital mortality due to COVID-19. Patients with other causes of mortality and with additional SARS-CoV-2 infection were excluded from the study. Also, patients <18 years of age, patients missing a baseline value of aspartate aminotransferase (AST) or lymphocytes, and patients with acquired immune deficiency syndrome, a malignant tumor, liver cirrhosis, or pregnancy were excluded.

2.2. Data Collection

The demographic data, laboratory results, clinical features, and medical history of the patients included in the study were obtained from electronic medical records. Data collected from laboratory results were defined using the first-time examination at admission (within 24 h after admission). Hematological data were tested in the same laboratory using the same standard. To analyze mortality risk, patients were followed up from admission to discharge (1 to 22 days). The follow-up data were collected through reviewing medical records using a double-blind method.

2.3. Variables

2.3.1. Outcome Variable: Mortality

The in-hospital mortality of patients hospitalized with a confirmed diagnosis of COVID-19 was evaluated as an outcome variable. Mortality after hospital discharge was not considered. These data were obtained by reviewing the virtual medical records of patients hospitalized from 14 May to 25 September 2020.

2.3.2. Exposure Variable: ALRI

The ALRI was calculated using the following: ALRI = aspartate aminotransferase (U/L)/lymphocyte [14].

Laboratory Biomarkers

APRI was calculated as AST (U/L)/upper limit of normal value (U/L)/platelet count × 100 [11]. AST-to-neutrophil ratio index (ANRI) was calculated by dividing the AST (U/L) by the neutrophil count [10]. Neutrophil–lymphocyte ratio (NLR) was obtained by dividing the total absolute neutrophil counts over the total lymphocyte counts [15]. Platelets–lymphocyte ratio (PLR) was determined using the following formula: PLR = absolute platelet count/absolute lymphocyte count [16]. SII = platelet count × neutrophil count/lymphocyte count ($\times 10^9$/L) [8]. The leukocyte glucose index (LGI) was defined as the product between blood leukocytes counts and glucose levels divided by 1000 [17]. Lactate dehydrogenase (LDH)/lymphocyte ratio was obtained as the ratio between LDH levels (U/L) and lymphocyte counts (cells/μL) [9]. The first measure of laboratory markers during the first 24 h of hospital admission were considered for this study.

2.3.3. Other Variables

Demographic characteristics such as age, sex, and comorbidities (diabetes, hypertension, obesity, chronic kidney disease (CKD), and chronic obstructive pulmonary disease (COPD)) were collected. The laboratory parameters measured in the first 24 h of hospital admission included the following: hemoglobin (g/dL), hematocrit (%), leukocytes ($\times 10^9$/L), platelets ($\times 10^9$/L), neutrophils ($\times 10^9$/L), lymphocytes ($\times 10^9$/L), glucose (mg/dL), urea (mg/dL), creatinine (mg/dL), LDH (U/L), bilirubin (mg/dL), liver enzymes (AST and alanine aminotransferase (ALT)), and electrolytes.

2.4. *Statistical Analysis*

Qualitative variables are described as numbers or percentages, and quantitative data are expressed as mean (±standard deviation, SD) for normally distributed and median (interquartile range, IQR) for non-normally distributed. The distribution of quantitative variables was assessed using the Shapiro–Wilk test. The Mann–Whitney test or Student's-t test was used to compare the quantitative variables between the different groups. The Chi-square test and Fisher's exact test were used to evaluate the differences between the qualitative variables between the groups. The Receiver Operator Characteristic (ROC) curve [18] was calculated to find the accuracy of the variables in predicting COVID-19 mortality. An area under the curve (AUC) values were considered adequate at 0.7 to 0.8. Youden's J index finds the cut-off value that maximizes the sum of sensitivity and specificity (or equally minimizes the sum of false positive and false negative errors) and is calculated as J = (sensitivity + specificity-1) [19]. The lymphocytes, PTT, AST, ALRI, NLR, SII, and LDH/lymphocyte ratio levels were dichotomized by defining the best cut-off obtained for Youden's-J statistic on the ROC curve. Survival curves were estimated using the Kaplan–Meier method [20], and the differences between groups were assessed using the log-rank test [21]. Univariate and multivariate Cox proportional hazard regression models were adopted to calculate the hazard ratio (HR) with a corresponding 95% confidence interval (CI) [22] for mortality risk. The covariates that were significant in the univariate Cox regression analysis were selected to be tested in the multivariate Cox regression analysis. The variance inflation factors (VIF) were calculated to evaluate multicollinearity. The statistical assumptions for the regression analysis were met when there was no multicollinearity. VIF > 5 was used to identify highly correlated variables. None of the variables included in the model had a VIF > 1.5, thus indicating that there were no issues with multicollinearity, i.e., no multicollinearity between the independent variables was found. All analyses were 2-tailed, and a p-value < 0.05 was considered statistically significant. p-values were adjusted using Bonferroni correction ($p_{corrected}$) to compensate for the effect of multiple hypothesis testing. The variables were filtered using <0.05 as a significance cut-off. Data analysis was performed using SPSS Statistics v.25 (SPSS Inc., Chicago, IL, USA), R v4.03 Statistical Software (R Foundation, Vienna, Austria), and MedCalc Statistical Software (MedCalc Software Ltd., Ostend, Belgium).

3. Results

3.1. Clinical Characteristics of the Study Subjects

The current study included 225 patients admitted to hospital with COVID-19, including 144 surviving patients and 81 non-surviving patients. The mean age of the non-survival group was 65 ± 12.705, and that of the survival group was 60.2 ± 13.515 (p = 0.004).The results of the laboratory tests at the time of hospitalization are summarized in Table 1. In the non-survival group, the urea, AST, ALRI, APRI, NLR, PLR, SII, and LDH/lymphocyte ratio were significantly higher than in the survival group ($p < 0.05$), while the lymphocyte count and partial thromboplastin time (PTT) values were significantly lower than that of the survival group ($p < 0.0001$). However, the age, urea, APRI, and PLR did not remain significant when multiple hypothesis correction was performed ($p_{corrected} \geq 0.05$).

Table 1. Characteristics of the study population according to their COVID-19 survival.

Variable	Total (n = 225)	Survival Group (n = 144)	Non-Survival Group (n = 81)	*p*-Value	Adjusted *p*-Value
Sex, female	99 (44%)	68 (47.2%)	31 (38.3%)	0.194	1
Age (years old)	61.97 ± 13.39	60.2 ± 13.515	65 ± 12.705	0.004	0.168
Obesity	50 (22.2%)	30 (20.8%)	20 (24.7%)	0.504	1
Diabetes	116 (51.6%)	80 (55.6%)	36 (44.4%)	0.109	1
Hypertension	143 (63.6%)	87 (60.4%)	56 (69.1%)	0.192	1
CKD	45 (25%)	27 (18.8%)	18 (22.2%)	0.532	1
COPD	6 (2.7%)	3 (2.1%)	3 (3.7%)	0.670	1
Cardiovascular disease	11 (4.9%)	9 (6.3%)	2 (2.5%)	0.335	1
Days of hospitalization	4 (4)	4 (4)	3 (6)	0.206	1
Hemoglobin (g/dL)	12.8 (2.7)	12.95 (2.3)	12.3 (3.5)	0.865	1
Hematocrit (%)	39 (6.2)	39 (6)	39 (8.9)	0.738	1
Leukocytes ($\times 10^9$/L)	10.2 (5.31)	10.29 (5.673)	10 (4.61)	0.615	1
Platelets ($\times 10^9$/L)	300 (176)	304.623 ± 126.045	298.253 ± 117.407	0.601	1
Neutrophils ($\times 10^9$/L)	7.23 (4.8)	7.145 (4.65)	8 (4.79)	0.140	1
Lymphocytes ($\times 10^9$/L)	1 (0.720)	1.065 (0.893)	0.8 (0.5)	**<0.0001**	**0.0009**
PT (s)	13 (3)	13 (3.2)	13 (2.9)	0.696	1
PTT (s)	29 (5)	30 (6)	28.3 (4.8)	**<0.0005**	**0.014**
INR	1.2 (0.2)	1.2 (0.2)	1.2 (0.2)	0.554	1
Glucose (mg/dL)	130 (86)	126 (83)	130 (82)	0.248	1
Urea (mg/dL)	40 (54)	40 (36.3)	50 (94)	0.014	0.588
Creatinine (mg/dL)	0.9 (1.1)	0.9 (0.8)	1 (1.7)	0.576	1
LDH (U/L)	360 (213)	363 (198)	351 (254)	0.438	1
TB (mg/dL)	0.9 (0.5)	0.9 (0.4)	0.8 (0.6)	0.753	1
DB (mg/dL)	0.4 (0.3)	0.4 (0.3)	0.4 (0.3)	0.908	1
IB (mg/dL)	0.5 (0.3)	0.5 (0.3)	0.5 (0.3)	0.564	1
AST (IU/L)	40 (28)	34.5 (26)	43 (30)	**<0.0005**	**0.009**
ALT (IU/L)	33 (26)	30.5 (31)	37 (22)	0.136	1
CPK (IU/L)	157 (240)	143 (184)	160 (308)	0.091	1
CK-MB (IU/L)	24 (28)	23.5 (26.8)	24 (32.2)	0.263	1
BNP (pg/mL)	56 (138)	56 (160.8)	50 (138)	0.870	1
MYO (ng/mL)	200 (176)	200 (179)	259 (162)	0.056	1
Na (mmol/L)	137 (5)	137 (5)	137 (5)	0.377	1
K (mmol/L)	4.5 (0.7)	4.5 (0.7)	4.5 (0.7)	0.595	1
Cl (mmol/L)	99 (3)	99 (3)	99 (2)	0.678	1
ALRI	37.31 (25.274)	31.428 (25.773)	51.136 (36.6)	**<0.0001**	**<0.0001**
APRI	0.323 (0.29)	0.284 (0.26)	0.403 (0.41)	0.001	0.050

Table 1. *Cont.*

Variable	Total (n = 225)	Survival Group (n = 144)	Non-Survival Group (n = 81)	*p*-Value	Adjusted *p*-Value
ANRI	4.777 (4.06)	4.6 (3.88)	4.78 (5.59)	0.127	1
NLR	8.315 (7.669)	7.481 (6.731)	10.302 (9.884)	**<0.0001**	**0.0007**
PLR	302.727 (286.31)	278.4 (256.31)	361.428 (281)	0.003	0.126
SII	2133 (2482)	1838 (2367.131)	2890 (3568)	**0.001**	**0.042**
LGI	73.75 (63.959)	72.894 (65.619)	76.9 (57.494)	0.721	1
LDH/LR	0.361 (0.304)	0.331 (0.271)	0.5 (0.333)	**<0.0001**	**<0.0001**

Data are presented as mean ± standard deviation, median (interquartile range), n (%). *p* values were calculated using Student's-*t* test, Mann–Whitney U test, Chi-squared test, or Fisher's exact test as appropriate. The bold values pertain to statistical significance ($p_{corrected} < 0.05$). Abbreviations: CKD, chronic kidney disease; COPD, chronic obstructive pulmonary disease; PT, prothrombin time; PTT, partial thromboplastin time; INR, international normalized ratio; LDH, lactate dehydrogenase; TB, total bilirubin; DB, direct bilirubin; IB, Indirect bilirubin; AST, aspartate aminotransferase; ALT, alanine aminotransferase; CPK, creatinine phosphokinase; CK-MB, creatine kinase myocardial band; BNP, B-type natriuretic peptide; MYO, myoglobin; Na, sodium; K, potassium; Cl, chloride; ALRI, aspartate aminotransferase-to-lymphocyte ratio index; APRI, aspartate aminotransferase-to-platelet ratio index; ANRI, aspartate aminotransferase-to-neutrophil ratio index; NLR, neutrophil-to-lymphocyte ratio; PLR, platelet-to-lymphocyte ratio; SII, systemic immune-inflammation index; LGI, leukocyte glucose index; LDH/LR, lactate dehydrogenase/lymphocyte ratio.

3.2. Use of the Best Cut-Off Values of Hematologic Parameters to Discern COVID-19 Mortality

The best cut-off values of the lymphocytes, PTT, AST, ALRI, NLR, SII, and LDH/lymphocyte ratio were calculated using ROC analysis to identify the non-surviving patients. The area under the curve (AUC) values regarding lymphocytes, PTT, and AST were 0.67 (SE = 0.03; 95% CI: 0.6–0.73), 0.64 (SE = 0.03; 95% CI: 0.57–0.7), and 0.64 (SE = 0.03; 95% CI: 0.58–0.71), respectively (Figure 1A). Regarding ratios, the AUC of ALRI, NLR, SII, and LDH/lymphocyte ratio were 0.81 (SE = 0.02; 95% CI: 0.76–0.86), 0.67 (SE = 0.03; 95% CI: 0.6–0.73), 0.63 (SE = 0.03; 95% CI: 0.57–0.69), and 0.68 (SE = 0.03; 95% CI: 0.61–0.74), respectively (Figure 1B).

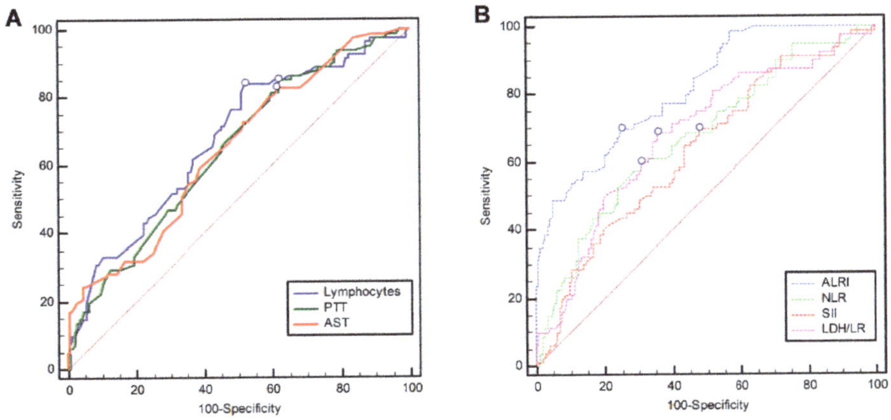

Figure 1. Receiver operating characteristic (ROC) curves of (**A**) lymphocytes (blue line), PPT (green line), AST (red line), and the ratios (**B**) ALRI (blue dotted line), NLR (green dotted line), SII (red dotted line), and (purple dotted line) LDH/LR in predicting mortality in patients with COVID-19. The blue circles indicate the best cut-off values. Abbreviations: PTT, partial thromboplastin time; AST, aspartate aminotransferase; ALRI, aspartate aminotransferase-to-lymphocyte ratio index; NLR, neutrophil-to-lymphocyte ratio; SII, systemic immune-inflammation index; LDH/LR, lactate dehydrogenase/lymphocyte ratio.

Lymphocytes, PTT, and AST could not be used in this study as potential diagnostic tools for subsequent analysis because their sensitivity or specificity was <50% (i.e., no better than chance) (Table 2). The best cut-off values for lymphocytes (≤1.21), PTT (≤31.9), AST (>26), ALRI (>42.42), NLR (>9.25), SII (>1857), and LDH/lymphocyte ratio (>0.369) were determined using Youden's J index.

Table 2. Receiver operating characteristic (ROC) curves, best cut-off points, and prognostic accuracy of hematologic and enzymatic parameters in COVID-19 mortality.

Variable	AUC	95% CI	*p*-Value	Best Cut-Off Point	Sensitivity (%)	Specificity (%)
Lymphocytes	0.67	0.6–0.73	**<0.0001**	≤1.21	83.95	47.22
PTT	0.64	0.57–0.7	**0.0001**	≤31.9	85.19	37.76
AST	0.64	0.58–0.71	**0.0001**	>26	82.72	38.89
ALRI	0.81	0.76–0.86	**<0.0001**	>42.42	70.37	75
NLR	0.67	0.6–0.73	**<0.0001**	>9.25	60.49	68.75
SII	0.63	0.57–0.69	**0.0004**	>1857	70.37	52.08
LDH/LR	0.68	0.61–0.74	**<0.0001**	>0.369	68.75	64

The bold values pertain to statistical significance ($p < 0.05$). Abbreviations: PTT, partial thromboplastin time; AST, aspartate aminotransferase; ALRI, aspartate aminotransferase-to-lymphocyte ratio index; NLR, neutrophil-to-lymphocyte ratio; SII, systemic immune-inflammation index; LDH/LR, lactate dehydrogenase/lymphocyte ratio.

3.3. Analysis of the Association of Hematologic and Enzymatic Parameters with COVID-19 Mortality

To identify the risk factors related to COVID-19 mortality, the seven statistically significant variables ($p_{corrected} < 0.05$) in Table 1 were included in our analysis by using Kaplan–Meier curves and the univariate Cox regression model. The Kaplan–Meier curves were created using the best cut-off points obtained by the Youden's J index for each variable. The mean survival time for the COVID-19 patients above the stated lymphocyte, PTT, AST, ALRI, NLR, SII, and LDH/lymphocyte ratio best cut-off values were 11.89 (SE = 0.71), 12.82 (SE = 0.81), 9.94 (SE = 0.83), 7.8 (SE = 0.86), 8.31 (SE = 0.78), 8.92 (SE = 0.76), and 7.9 (SE = 0.65) days, respectively. In comparison, the mean survival time for COVID-19 patients with below the stated lymphocyte, PTT, AST, ALRI, NLR, SII, and LDH/lymphocyte ratio best cut-off values were 8.05 (SE = 0.5), 9.58 (SE = 0.87), 11.57 (SE = 0.87), 11.69 (SE = 0.59), 13.66 (SE = 1.02), 13.62 (SE = 1.18), and 13.53 (SE = 1.21) days, respectively (Figure 2A–G). The differences in survival for patients with laboratory and enzymatic parameters above the stated best cut-off point compared to those below the cut-off point were statistically significant ($p < 0.05$).

Regarding the univariate Cox regression model elaborated with the seven variables that had statistically significant differences when multiple hypotheses correction was performed ($p_{corrected} < 0.05$), lower levels of lymphocyte (HR = 2.69, 95% CI: 1.48–4.88; $p_{corrected} = 0.007$) and PTT (HR = 2.68, 95% CI: 1.45–4.94; $p_{corrected} = 0.007$) and higher levels of ALRI (HR = 3.2, 95% CI: 1.98–5.17; $p_{corrected} < 0.0001$), NLR (HR = 2.01, 95% CI: 1.28–3.16; $p_{corrected} = 0.014$), and LDH/lymphocyte ratio (HR = 2.3, 95% CI: 1.43–3.69; $p_{corrected} = 0.004$) were predictors of in-hospital mortality (Table 3). Nevertheless, multivariate Cox regression reveled that ALRI [>42.42] (HR = 2.32, 95% CI: 1.35–3.97; $p_{corrected} = 0.01$) was a prognostic factor of COVID-19 mortality (Table 3).

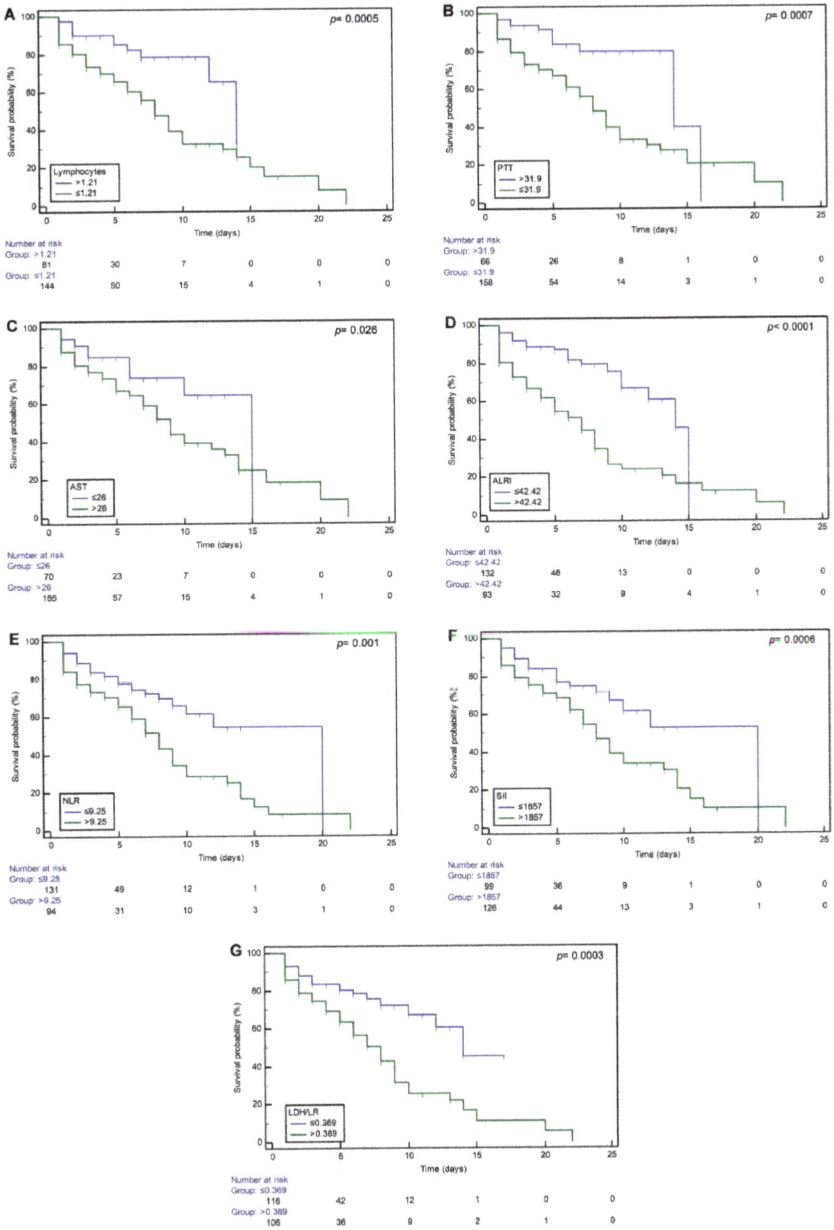

Figure 2. Kaplan–Meier survival curves of hospitalized COVID-19 patients according to the established best cut-off values of (**A**) lymphocytes, (**B**) PTT, (**C**) AST, (**D**) ALRI, (**E**) NLR, (**F**) SII, and (**G**) LDH/LR. A *p*-value <0.05 was considered statistically significant. The overall survival curves were estimated using the Kaplan–Meier method and *p* values (obtained via log-rank test). Abbreviations: PTT, partial thromboplastin time; AST, aspartate aminotransferase; ALRI, aspartate aminotransferase-to-lymphocyte ratio index; NLR, neutrophil-to-lymphocyte ratio; SII, systemic immune-inflammation index; LDH/LR, lactate dehydrogenase/lymphocyte ratio.

Table 3. Univariate and multivariate Cox regression analysis of the variables associated with COVID-19 mortality.

	Univariate				Multivariate			
Variable	**HR**	**95% CI**	***p*-Value**	**Adjusted *p*-Value**	**HR**	**95% CI**	***p*-Value**	**Adjusted *p*-Value**
Lymphocytes	2.69	1.48–4.88	**0.001**	**0.007**	1.2	0.56–2.54	0.627	1
PTT	2.68	1.45–4.94	**0.001**	**0.007**	2.08	1.08–4.01	0.028	0.14
AST	1.86	1.04–3.32	0.035	0.245	-	-	-	-
ALRI	3.2	1.98–5.17	**<0.0001**	**<0.0001**	2.32	1.35–3.97	**0.002**	**0.01**
NLR	2.01	1.28–3.16	**0.002**	**0.014**	1.31	0.8–2.15	0.275	1
SII	1.88	1.16–3.03	0.009	0.063	-	-	-	-
LDH/LR	2.3	1.43–3.69	**0.0006**	**0.004**	1.2	0.66–2.2	0.538	1

Univariate Cox regression analysis was performed using the seven variables (lymphocyte, PTT, AST, ALRI, NLR, SII, and LDH/lymphocyte ratio) that had statistically significant differences when multiple hypotheses correction was performed ($p_{corrected} < 0.05$). Candidate predictors with statistically significant differences ($p_{corrected} < 0.05$) in the univariate Cox regression analysis were included in our multivariate Cox regression analysis. Hazard ratios (HRs) and 95% Confidence Intervals (CI 95%) are reported. Variance inflation factors were computed for the final model to evaluate multicollinearity. The bold values pertain to statistical significance ($p_{corrected} < 0.05$). Abbreviations: PTT, partial thromboplastin time; AST, aspartate aminotransferase; ALRI, aspartate aminotransferase-to-lymphocyte ratio index; NLR, neutrophil-to-lymphocyte ratio; SII, systemic immune-inflammation index; LDH/LR, lactate dehydrogenase/lymphocyte ratio.

3.4. Demographic, Clinical, and Hematological Characteristics of COVID-19 Patients Stratified According to ALRI Values

Patients with lower ALRI levels were more frequently diabetic than those patients with high ALRI levels (58.3% vs. 41.9%, $p = 0.015$). However, diabetes did not remain significant when multiple hypothesis correction was performed ($p_{corrected} = 0.615$). Patients with higher ALRI values had increased mortality (61.3% vs. 18.2%, $p_{corrected} < 0.0001$) (Table 4). Moreover, patients with ALRI >42.42 had less lymphocytes ($p_{corrected} < 0.0001$) and lower PTT ($p_{corrected} = 0.004$) when compared with patients with ALRI $\leq$ 42.42, as well as increased values of AST ($p_{corrected} < 0.0001$), ALT ($p_{corrected} = 0.0003$), chloride ($p_{corrected} = 0.041$), APRI ($p_{corrected} = 0.0001$), ANRI ($p_{corrected} = 0.012$), NLR ($p_{corrected} < 0.0001$), PLR ($p_{corrected} = 0.0004$), SII ($p_{corrected} = 0.0001$), and LDH/lymphocyte ratio ($p_{corrected} = 0.0001$) (Table 4).

Table 4. Demographic, clinical, and laboratory characteristics of COVID-19 patients stratified by the ALRI.

Variable	ALRI ≤ 42.42 (n = 132)	ALRI > 42.42 (n = 93)	*p*-Value	Adjusted *p*-Value
Sex, female	59 (44.7%)	40 (43%)	0.802	1
Age (years old)	61.11 ± 13.39	63.14 ± 13.38	0.096	1
Obesity	25 (18.9%)	25 (26.9%)	0.158	1
Diabetes	77 (58.3%)	39 (41.9%)	0.015	0.615
Hypertension	88 (66.7%)	55 (59.1%)	0.248	1
CKD	26 (19.7%)	19 (20.4)	0.892	1
COPD	3 (2.3%)	3 (3.2%)	0.693	1
Cardiovascular disease	7 (5.3%)	4 (4.3%)	1	1
Non-survival	24 (18.2%)	57 (61.3%)	**<0.0001**	**<0.0001**
Hemoglobin (g/dL)	12.7 (3.1)	12.85 (2.7)	0.794	1
Hematocrit (%)	39 (6.6)	39 (5.8)	0.802	1
Leukocytes ($\times 10^9$/L)	10.36 (5.63)	9.6 (5.18)	0.797	1
Platelets ($\times 10^9$/L)	296 (158.5)	300 (177.75)	0.990	1
Neutrophils ($\times 10^9$/L)	7.15 (4.82)	7.58 (4.79)	0.093	1
Lymphocytes ($\times 10^9$/L)	1.21 (1.04)	0.79 (0.42)	**<0.0001**	**<0.0001**

Table 4. *Cont.*

Variable	ALRI ≤ 42.42 (n = 132)	ALRI > 42.42 (n = 93)	*p*-Value	Adjusted *p*-Value
PT (s)	13 (3)	13 (3)	0.960	1
PTT (s)	30 (6)	28.15 (5)	**0.0001**	**0.004**
INR	1.2 (0.2)	1.2 (0.2)	0.901	1
Glucose (mg/dL)	130 (100)	125.5 (68)	0.588	1
Urea (mg/dL)	40 (44)	42 (94)	0.040	1
Creatinine (mg/dL)	0.9 (0.6)	1.05 (1.8)	0.114	1
LDH (U/L)	347 (180)	400 (292)	0.003	0.123
TB (mg/dL)	0.8 (0.5)	0.95 (0.6)	0.154	1
DB (mg/dL)	0.4 (0.2)	0.4 (0.3)	0.347	1
IB (mg/dL)	0.5 (0.3)	0.5 (0.3)	0.293	1
AST (IU/L)	28 (23)	45 (25)	**<0.0001**	**<0.0001**
ALT (IU/L)	24 (26)	40 (21)	**<0.0001**	**0.0003**
CPK (IU/L)	127 (160)	178.5 (284)	0.009	0.369
CK-MB (IU/L)	20 (23)	29 (32.3)	0.007	0.287
BNP (pg/mL)	50 (92.5)	67 (150)	0.212	1
MYO (ng/mL)	200 (162)	236 (200)	0.200	1
Na (mmol/L)	138 (6)	137 (5)	0.072	1
K (mmol/L)	4.5 (0.7)	4.5 (0.51)	0.878	1
Cl (mmol/L)	99 (2)	100 (3)	**0.001**	**0.041**
APRI	0.25 (0.26)	0.41 (0.36)	**<0.0001**	**0.0001**
ANRI	4.09 (4.2)	5.53 (4.17)	**0.0003**	**0.012**
NLR	6 (6.51)	10.38 (8.7)	**<0.0001**	**<0.0001**
PLR	252.63 (255.02)	343.57 (270.8)	**<0.0001**	**0.0004**
SII	1730.11 (2489)	3030.62 (3333.93)	**<0.0001**	**<0.0001**
LGI	76.05 (84.11)	72.35 (44.17)	0.720	1
LDH/LR	0.3 (0.297)	0.517 (0.356)	**<0.0001**	**<0.0001**

Data are presented as mean ± standard deviation, median (interquartile range), n (%). *p*-values were calculated using Student's-*t* test, the Mann–Whitney U test, Chi-squared test, or Fisher's exact test as appropriate. The bold values pertain to statistical significance ($p_{corrected}$ < 0.05). Abbreviations: CKD, chronic kidney disease; COPD, chronic obstructive pulmonary disease; PT, prothrombin time; PTT, partial thromboplastin time; INR, international normalized ratio; LDH, lactate dehydrogenase; TB, total bilirubin; DB, direct bilirubin; IB, indirect bilirubin; AST, aspartate aminotransferase; ALT, alanine aminotransferase; CPK, creatinine phosphokinase; CK-MB, creatine kinase myocardial band; BNP, B-type natriuretic peptide; MYO, myoglobin; Na, sodium; K, potassium; Cl, chloride; ALRI, aspartate aminotransferase-to-lymphocyte ratio index; APRI, aspartate aminotransferase-to-platelet ratio index; ANRI, aspartate aminotransferase-to-neutrophil ratio index; NLR, neutrophil-to-lymphocyte ratio; PLR, platelet-to-lymphocyte ratio; SII, systemic immune-inflammation index; LGI, leukocyte glucose index; LDH/LR, lactate dehydrogenase/lymphocyte ratio.

3.5. Correlation between Hematological Parameters and ALRI Levels at Admission in COVID-19 Patients

Spearman's rank correlation coefficient was used to determine the correlation between ALRI and the hematologic parameters that had a significant difference in the comparison of patients with low and high ALRI levels. Most of the variables showed a weak correlation (value of r = −0.3 to −0.1 or 0.1 to 0.3). A statistically significant negative correlation was observed between lymphocytes (r = −0.514 [95% IC −0.604 to −0.410]; $p < 0.0001$), PTT (r = −0.265 [95% IC −0.383 to −0.139]; $p = 0.0001$), and chloride (r = −0.206 [95% IC −0.328 to −0.077]; $p = 0.001$). ALRI had a moderate correlation (value of r = 0.3 to 0.5) with ALT, APRI, NLR, PLR, and SII. Also, ALRI had a strong correlation with LDH/lymphocyte ratio (r = 0.547 [95% IC 0.448 to 0.633]; $p < 0.0001$) (Figure 3).

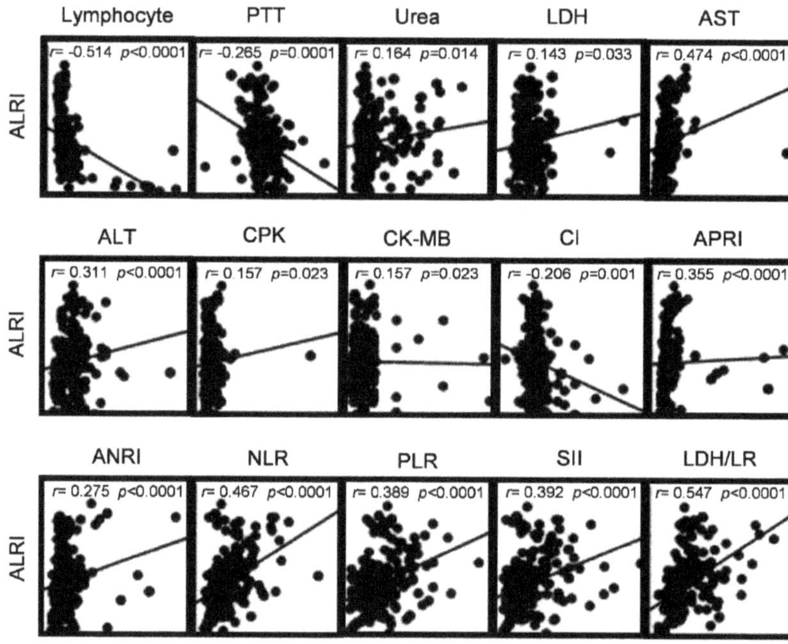

Figure 3. Spearman correlations between ALRI and hematologic parameters in COVID-19 patients. Scatter plots showing the correlation between ALRI and laboratory variables (data on admission). Weak correlation (value of r = −0.3 to −0.1 or 0.1 to 0.3); moderate correlation (value of r = −0.5 to −0.3 or 0.3 to 0.5); strong correlation (value of r = −1.0 to −0.5 or 0.5 to 1). Spearman's rank correlation coefficient was used to evaluate the correlations. A *p*-value < 0.05 was considered statistically significant. Abbreviations: PTT, partial thromboplastin time; LDH, lactate dehydrogenase; AST, aspartate aminotransferase; ALT, alanine aminotransferase; CPK, creatinine phosphokinase; CK-MB, creatine kinase myocardial band; Cl, chloride; ALRI, aspartate aminotransferase-to-lymphocyte ratio index; APRI, aspartate aminotransferase-to-platelet ratio index; ANRI, aspartate aminotransferase-to-neutrophil ratio index; NLR, neutrophil-to-lymphocyte ratio; PLR, platelet-to-lymphocyte ratio; SII, systemic immune-inflammation index; LDH/LR, lactate dehydrogenase/lymphocyte ratio.

4. Discussion

This study reported data from 225 patients consecutively admitted with COVID-19 at one tertiary hospital in Mexico. Of these, 81 (36%) patients did not survive during hospitalization. Dessie and Zewotir suggested that the mortality rate of COVID-19 ranges from 3.14 to 61.51% [3]. Demographic data show that non-surviving COVID-19 patients are significantly older than COVID-19 survivors; in a previous report, COVID-19 mortality was significantly related to age, with approximate rates of 40% at the age 70+ years [23]. We found a mean age of 65 ± 12.705 in the non-surviving patients. However, no statistically significant differences were observed. Although comorbidities such as diabetes, hypertension, chronic kidney disease (CKD), chronic obstructive pulmonary disease (COPD), and cardiovascular disease each increased the risk of mortality [24], in our study, these variables were not associated with COVID-19 mortality.

On the other hand, a decrease in blood platelet count and an increase in leukocytes and neutrophils have been associated with mortality in patients with COVID-19 [5,6]. Our results showed no significant differences in the mean or median regarding the platelet, leukocyte, and neutrophil concentrations of the non-surviving and surviving patients. In contrast, the median value of lymphocytes was significantly lower in non-surviving than in surviving patients, as previously reported [6]. PTT and AST had statistical significance between groups ($p_{corrected} < 0.05$), as reported by Mahdavi et al., Hodges et al., and Ding et al. [12,25,26]. However, this study could not use lymphocytes, PTT, and AST as potential diagnostic tools because their sensitivity or specificity was <50% (i.e., no better than chance) [27]. Other reports have identified NLR (>9.1) [7], PLR (356.6) [28], SII (1835) [8], LDH/lymphocyte ratio (0.21) [9] as independent risk factors for COVID-19 mortality. Our study's best cut-off points for NLR, SII, and LDH/lymphocyte ratio were >9.25, >362.5, >1857, and >0.369, respectively. NLR, SII, and LDH/lymphocyte ratio were not predictors of mortality despite a statistically significant AUC. Interestingly, PLR was also not a predictor of mortality in this study. These results on PLR and SII in COVID-19 mortality agree with that of previous reports [7,29]. Some studies that have reported an association between PLR and COVID-19 mortality included patients with thrombocytopenia, generating a confusion variable [4,30]. Furthermore, the results of a meta-analysis performed by Sarker et al. suggest that the prognostic value of PLR is heterogeneous and low-quality evidence [31].

Regarding liver enzyme-derived ratios, ANRI and APRI had no statistically significant difference between the study groups ($p_{corrected} \geq 0.05$). ANRI and APRI were also excluded from the univariate Cox regression analysis. Therefore, the liver enzyme-derived ratios analyzed in this study show little relevance (ANRI and APRI), except ALRI. Interestingly, APRI > 0.64 on admission was associated with severe COVID-19, whereas ALRI > 24.1 was not accepted as an independent risk factor for severe cases of COVID-19 [11]. In contrast, we found that ALRI > 42.42 could predict COVID-19 mortality. Hence, the role of ALRI as a predictor may depend on disease progression.

NLR, LDH/lymphocyte ratio, and ALRI showed a superior prognostic possibility for mortality with specificity and sensitivity > 60%. Kaplan–Meier survival curves obtained with best cut-off values obtained from ROC curves demonstrated that mortality was significantly associated with lymphocytes, PTT, AST, ALRI, NLR, SII, and LDH/lymphocyte ratio. Nevertheless, only the ALRI remained associated with mortality after multivariate Cox regression analysis.

ALRI includes one liver enzyme (AST) and one peripheral blood parameter (lymphocytes), comprehensively summarizing the balance of liver inflammation (the organ most susceptible to SARS-CoV-2 infection [32]) and host immune response. ALRI (>18.734) is related to a poor prognosis in hepatocellular carcinoma patients after hepatectomy [13]. In our study, the ALRI values in non-surviving patients were significantly higher than that in the surviving patients. Additionally, ALRI > 42.42 was identified as an independent risk factor (HR = 2.32, 95% CI: 1.35–3.97; $p_{corrected} = 0.01$) for COVID-19 mortality. Because there are no studies on the relationship between ALRI and COVID-19 mortality, we hypothesized that liver abnormalities in COVID-19 (association between early AST elevation and COVID-19 mortality), secondary to inflammation during SARS-CoV-2 infection [12] and lymphopenia (linked to the ability of SARS-CoV-2 to infect T cells [33]), could be an indicator of poor prognosis in COVID-19 patients [6] and suggest the relevance of ALRI increase in predicting patient mortality. Data from the COVID-19 Lombardy ICU Network in Italy show that age is the ultimate risk factor for COVID-19 mortality [34]. However, our study suggests that ALRI (>42.42) could be another risk factor for COVID-19 mortality compared to age (>61 years) upon admission in the Mexican population.

In our study, the patients with higher ALRI values were associated with increased values of ALT, chloride, APRI, ANRI, NLR, PLR, SII, and LDH/lymphocyte ratio. The association between these hematologic parameters and higher ALRI values requires further investigation. Interestingly, we found that ALRI levels showed significant positive moderate/strong correlations with ALT, APRI, NLR, PLR, SII, and LDH/lymphocyte ratio, indicating its potential value in evaluating COVID-19 mortality. Therefore, a combination of these inflammatory biomarkers may be beneficial for assessing and managing patients with COVID-19.

The present study has some limitations that need to be addressed. First, this study was a retrospective, single-center study with a relatively small sample size, which may affect the generalization of the results due to the enrollment limitations of enrolled patients.

Nevertheless, these results could be compared to other studies with a similar number of patients to determine clinical predictors for COVID-19 outcomes [35–37]. The results of this study should be validated via comparison with other studies or future efforts focused on prospective analyses to strengthen our understanding of the utility of ALRI. The second limitation of this study is that most of the studies on NLR, PLR, LDH/lymphocyte ratio, and SII are retrospective and from Asian countries; thus, the comparisons made in this study may not have been strong enough. On the other hand, our study has some strengths, namely (a) it demonstrates the predictive prognostic value of ALRI independently of comorbidities and (b) to the best of our knowledge, this study is the first to determine the prognostic role of ALRI compared with other biomarkers in a COVID-19 cohort free of confounders affecting the count of hematologic parameters and liver enzymes. Furthermore, we applied Bonferroni correction, a stringent procedure, to discard the slightly significant variables. Although comparing the strength of different ratios with the same divisor (lymphocyte count) makes little sense, we decided to evaluate which of all these ratios is a better predictor of COVID-19 mortality.

5. Conclusions

In summary, our study suggests that ALRI at admission is a biomarker of mortality in COVID-19 patients. Although the cut-off value of ALRI is yet to be standardized, the prognostic role of ALRI in COVID-19 mortality is promising. Assessing this enzymatic/hematologic parameter may help to identify patients with a high risk of mortality. Early medical intervention for high-risk patients may reduce the fatality rate associated with this disease.

Author Contributions: Conceptualization: J.M.R.-R., G.M.-M., J.F.O.-R., L.A.D.J.-G., S.N.P.-R., C.N.F.-M., C.D.C.-R., T.O.-R., and R.M.d.Á.; data curation: J.M.R.-R., G.M.-M., O.G.-H., J.F.O.-R., L.A.D.J.-G., and R.M.d.Á.; formal analysis: J.M.R.-R., G.M.-M., J.F.O.-R., L.A.D.J.-G., S.N.P.-R., and R.M.d.Á.; funding acquisition: J.M.R.-R. and R.M.d.Á.; investigation: J.M.R.-R., G.M.-M., S.N.P.-R., J.F.O.-R., C.N.F.-M., C.D.C.-R., L.A.D.J.-G., T.O.-R., and R.M.d.Á.; methodology: J.M.R.-R., G.M.-M., L.A.D.J.-G., J.F.O.-R., C.N.F.-M., C.D.C.-R., S.N.P.-R., T.O.-R., and R.M.d.Á.; project administration: J.M.R.-R. and R.M.d.Á.; resources: J.M.R.-R. and R.M.d.Á.; supervision: J.M.R.-R., G.M.-M., and R.M.d.Á.; validation: J.M.R.-R., G.M.-M., L.A.D.J.-G., S.N.P.-R., J.F.O.-R., C.N.F.-M., C.D.C.-R., T.O.-R., and R.M.d.Á.; visualization: J.M.R.-R., G.M.-M., S.N.P.-R., J.F.O.-R., C.N.F.-M., C.D.C.-R., and L.A.D.J.-G.; writing—original draft preparation: J.M.R.-R., G.M.-M., S.N.P.-R., J.F.O.-R., C.N.F.-M., C.D.C.-R., L.A.D.J.-G., T.O.-R., and R.M.d.Á.; writing—review and editing: J.M.R.-R., G.M.-M., S.N.P.-R., L.A.D.J.-G., J.F.O.-R., C.N.F.-M., C.D.C.-R., T.O.-R., O.G.-H., and R.M.d.Á. All authors have read and agreed to the published version of the manuscript.

Funding: This research was funded by FUNDACIÓN IMSS, A.C.

Institutional Review Board Statement: This study was conducted according to the guidelines of the Declaration of Helsinki and approved by the Research Ethical Committee of the UMAE H. E. No. 14 (R-2023-3001-111).

Informed Consent Statement: Given the retrospective nature of this study, the Research Ethical Committee of the UMAE, Hospital de Especialidades No. 14, CMN "Adolfo Ruiz Cortines" from Mexican Social Security Institute (Registry code: CONBIOÉTICA-30-CEI-003–20,180,412; COFEPRIS 17CI30193067) approved our request to waive the need for informed consent.

Data Availability Statement: Data supporting the reported results can be provided upon reasonable request by the corresponding author.

Acknowledgments: The authors would like to thank all of the health workers, doctors, nurses, and supportive staff working in the UMAE, H. E. No. 14, Centro Médico Nacional "Adolfo Ruiz Cortines", IMSS Veracruz, who were fighting against COVID-19 together.

Conflicts of Interest: The authors declare no conflict of interest.

References

1. Huang, C.; Wang, Y.; Li, X.; Ren, L.; Zhao, J.; Wu, W.; Guo, L.; Jin, Q.; Cao, B.; Wang, J.; et al. Clinical features of patients infected with 2019 novel coronavirus in Wuhan, China. *Lancet* **2020**, *395*, 497–506. [CrossRef] [PubMed]
2. Xu, X.; Chen, P.; Wang, J.; Feng, J.; Zhou, H.; Li, X.; Zhong, W.; Jao, P. Evolution of the novel coronavirus from the ongoing Wuhan outbreak and modeling of its spike protein for risk of human transmission. *Sci. China Life Sci.* **2020**, *63*, 457–460. [CrossRef] [PubMed]
3. Dessie, Z.G.; Zewotir, T. Mortality-related risk factors of COVID-19: A systematic review and meta-analysis of 42 studies and 423,117 patients. *BMC Infect. Dis.* **2021**, *21*, 855. [CrossRef] [PubMed]
4. Mousavi, S.A.; Rad, S.; Rostami, T.; Rostami, M.; Mousavi, S.A.; Mirhoseini, S.A.; Kiumarsi, A. Hematologic Predictors of Mortality in Hospitalized Patients with COVID-19: A Comparative Study. Hematology. Published online 30 October 2020. Available online: https://www.tandfonline.com/doi/abs/10.1080/16078454.2020.1833435 (accessed on 2 September 2023).
5. Zhao, X.; Wang, K.; Zuo, P.; Liu, Y.; Zhang, M.; Xie, S.; Zhang, H.; Chen, X. Early decrease in blood platelet count is associated with poor prognosis in COVID-19 patients—Indications for predictive, preventive, and personalized medical approach. *EPMA J.* **2020**, *11*, 139–145. [CrossRef]
6. Ghobadi, H.; Mohammadshahi, J.; Javaheri, N.; Fouladi, N.; Mirzazadeh, Y.; Aslani, M.R. Role of leukocytes and systemic inflammation indexes (NLR, PLR, MLP, dNLR, NLPR, AISI, SIR-I, and SII) on admission predicts in-hospital mortality in non-elderly and elderly COVID-19 patients. *Front. Med.* **2022**, *9*, 916453. Available online: https://www.frontiersin.org/articles/10.3389/fmed.2022.916453 (accessed on 2 September 2023). [CrossRef]
7. Citu, C.; Gorun, F.; Motoc, A.; Sas, I.; Gorun, O.M.; Burlea, B.; Tuta-Sas, I.; Tomescu, L.; Neamtu, R.; Malita, D.; et al. The Predictive Role of NLR, d-NLR, MLR, and SIRI in COVID-19 Mortality. *Diagnostics* **2022**, *12*, 122. [CrossRef]
8. Fois, A.G.; Paliogiannis, P.; Scano, V.; Cau, S.; Babudieri, S.; Perra, R.; Ruzzittu, G.; Zinellu, E.; Pirina, P.; Carru, C.; et al. The Systemic Inflammation Index on Admission Predicts In-Hospital Mortality in COVID-19 Patients. *Molecules* **2020**, *25*, 5725. [CrossRef]
9. Serin, I.; Sari, N.D.; Dogu, M.H.; Acikel, S.D.; Babur, G.; Ulusoy, A.; Onar, M.I.; Gokce, E.C.; Altunok, O.; Mert, F.Y.; et al. A new parameter in COVID-19 pandemic: Initial lactate dehydrogenase (LDH)/Lymphocyte ratio for diagnosis and mortality. *J. Infect. Public Health* **2020**, *13*, 1664–1670. [CrossRef]
10. Huang, D.; Yang, H.; Yu, H.; Wang, T.; Chen, Z.; Yao, R.; Liang, Z. Diagnostic Value of Hematological and Biochemical Parameters Combinations for Predicting Coronavirus Disease 2019 (COVID-19) in Suspected Patients. *Am. J. Med. Sci.* **2021**, *362*, 387–395. [CrossRef]
11. Gemcioglu, E.; Davutoglu, M.; Catalbas, R.; Karabuga, B.; Kaptan, E.; Aypak, A.; Kalem, A.K.; Özdemir, M.; Yeşilova, N.Y.; Kalkan, E.A.; et al. Predictive values of biochemical markers as early indicators for severe COVID-19 cases in admission. *Future Virol.* **2021**, *16*, 353–367. [CrossRef]
12. Ding, Z.-Y.; Li, G.-X.; Chen, L.; Shu, C.; Song, J.; Wang, W.; Wang, Y.-W.; Chen, Q.; Jin, G.-N.; Liu, T.-T.; et al. Association of liver abnormalities with in-hospital mortality in patients with COVID-19. *J. Hepatol.* **2021**, *74*, 1295–1302. [CrossRef] [PubMed]
13. Liu, K.; Lv, Y.; Niu, Y.; Bu, Y. Prognostic value of γ-glutamyl transpeptidase to albumin ratio combined with aspartate aminotransferase to lymphocyte ratio in patients with hepatocellular carcinoma after hepatectomy. *Medicine* **2020**, *99*, e23339. [CrossRef] [PubMed]
14. Jin, J.; Zhu, P.; Liao, Y.; Li, J.; Liao, W.; He, S. Elevated preoperative aspartate aminotransferase to lymphocyte ratio index as an independent prognostic factor for patients with hepatocellular carcinoma after hepatic resection. *Oncotarget* **2015**, *6*, 19217–19227. [CrossRef] [PubMed]
15. Fu, J.; Kong, J.; Wang, W.; Wu, M.; Yao, L.; Wang, Z.; Jin, J.; Wu, D.; Yu, X. The clinical implication of dynamic neutrophil to lymphocyte ratio and D-dimer in COVID-19: A retrospective study in Suzhou China. *Thromb. Res.* **2020**, *192*, 3–8. [CrossRef] [PubMed]
16. Yang, A.P.; Liu, J.; Tao, W.; Li, H. The diagnostic and predictive role of NLR, d-NLR and PLR in COVID-19 patients. *Int. Immunopharmacol.* **2020**, *84*, 106504. [CrossRef]
17. Ramos-Hernández, W.M.; Soto, L.F.; Del Rosario-Trinidad, M.; Farfan-Morales, C.N.; De Jesús-González, L.A.; Martínez-Mier, G.; Osuna-Ramos, J.F.; Bastida-González, F.; Bernal-Dolores, V.; Del Ángel, R.M.; et al. Leukocyte glucose index as a novel biomarker for COVID-19 severity. *Sci. Rep.* **2022**, *12*, 14956. [CrossRef]
18. Zou, K.H.; O'Malley, A.J.; Mauri, L. Receiver-Operating Characteristic Analysis for Evaluating Diagnostic Tests and Predictive Models. *Circulation* **2007**, *115*, 654–657. [CrossRef]
19. Youden, W.J. Index for rating diagnostic tests. *Cancer* **1950**, *3*, 32–35. [CrossRef]
20. Kaplan, E.L.; Meier, P. Nonparametric Estimation from Incomplete Observations. *J. Am. Stat. Assoc.* **1958**, *53*, 457–481. [CrossRef]
21. Peto, R.; Pike, M.C.; Armitage, P.; Breslow, N.E.; Cox, D.R.; Howard, S.V.; Mantel, N.; McPherson, K.; Peto, J.; Smith, P.G. Design analysis of randomized clinical trials requiring prolonged observation of each patient, I.I. Analysis and examples. *Br. J. Cancer* **1977**, *35*, 1–39. [CrossRef]
22. Cox, D.R. Regression Models and Life-Tables. *J. R. Stat. Soc. Ser. B* **1972**, *34*, 187–202. [CrossRef]
23. Polidori, M.C.; Sies, H.; Ferrucci, L.; Benzing, T. COVID-19 mortality as a fingerprint of biological age. *Ageing Res. Rev.* **2021**, *67*, 101308. [CrossRef] [PubMed]

24. Singh, A.K.; Gillies, C.L.; Singh, R.; Singh, A.; Chudasama, Y.; Coles, B.; Seidu, S.; Zaccardi, F.; Davies, M.J.; Khunti, K. Prevalence of co-morbidities and their association with mortality in patients with COVID-19: A systematic review and meta-analysis. *Diabetes Obes. Metab.* **2020**, *22*, 1915–1924. [CrossRef]
25. Mahdavi, M.; Choubdar, H.; Zabeh, E.; Rieder, M.; Safavi-Naeini, S.; Jobbagy, Z.; Ghorbani, A.; Abedini, A.; Kiani, A.; Khanlarzadeh, V.; et al. A machine learning based exploration of COVID-19 mortality risk. *PLoS ONE* **2021**, *16*, e0252384. [CrossRef] [PubMed]
26. Hodges, G.; Pallisgaard, J.; Olsen, A.M.S.; McGettigan, P.; Andersen, M.; Krogager, M.; Kragholm, K.; Køber, L.; Gislason, G.H.; Torp-Pedersen, C.; et al. Association between biomarkers and COVID-19 severity and mortality: A nationwide Danish cohort study. *BMJ Open* **2020**, *10*, e041295. [CrossRef] [PubMed]
27. Smith, C.J. Diagnostic tests (1)—sensitivity and specificity. *Phlebology* **2012**. Published online. [CrossRef] [PubMed]
28. Simon, P.; Le Borgne, P.; Lefevbre, F.; Cipolat, L.; Remillon, A.; Dib, C.; Hoffmann, M.; Gardeur, I.; Sabah, J.; Kepka, S.; et al. Platelet-to-Lymphocyte Ratio (PLR) Is Not a Predicting Marker of Severity but of Mortality in COVID-19 Patients Admitted to the Emergency Department: A Retrospective Multicenter Study. *J. Clin. Med.* **2022**, *11*, 4903. [CrossRef]
29. Regolo, M.; Vaccaro, M.; Sorce, A.; Stancanelli, B.; Colaci, M.; Natoli, G.; Russo, M.; Alessandria, I.; Motta, M.; Santangelo, N.; et al. Neutrophil-to-Lymphocyte Ratio (NLR) Is a Promising Predictor of Mortality and Admission to Intensive Care Unit of COVID-19 Patients. *J. Clin. Med.* **2022**, *11*, 2235. [CrossRef]
30. Eslamijouybari, M.; Heydari, K.; Maleki, I.; Moosazadeh, M.; Hedayatizadeh-Omran, A.; Vahedi, L.; Ghasemian, R.; Sharifpour, A.; Alizadeh-Navaei, R. Neutrophil-to-Lymphocyte and Platelet-to-Lymphocyte Ratios in COVID-19 Patients and Control Group and Relationship with Disease Prognosis. *Caspian J. Intern. Med.* **2020**, *11* (Suppl. 1), 531–535. [CrossRef]
31. Sarkar, S.; Kannan, S.; Khanna, P.; Singh, A.K. Role of platelet-to-lymphocyte count ratio (PLR), as a prognostic indicator in COVID-19: A systematic review and meta-analysis. *J. Med. Virol.* **2022**, *94*, 211–221. [CrossRef]
32. Wanner, N.; Andrieux, G.; Badia-i-Mompel, P.; Edler, C.; Pfefferle, S.; Lindenmeyer, M.T.; Schmidt-Lauber, C.; Czogalla, J.; Wong, M.N.; Okabayashi, Y.; et al. Molecular consequences of SARS-CoV-2 liver tropism. *Nat. Metab.* **2022**, *4*, 310–319. [CrossRef] [PubMed]
33. Chan, A.S.; Rout, A. Use of Neutrophil-to-Lymphocyte and Platelet-to-Lymphocyte Ratios in COVID-19. *J. Clin. Med. Res.* **2020**, *12*, 448–453. [CrossRef] [PubMed]
34. Grasselli, G.; Zangrillo, A.; Zanella, A.; Antonelli, M.; Cabrini, L.; Castelli, A.; Cereda, D.; Coluccello, A.; Foti, G.; Fumagalli, R.; et al. Baseline Characteristics and Outcomes of 1591 Patients Infected With SARS-CoV-2 Admitted to ICUs of the Lombardy Region, Italy. *JAMA* **2020**, *323*, 1574–1581. [CrossRef] [PubMed]
35. Gu, Y.; Wang, D.; Chen, C.; Lu, W.; Liu, H.; Lv, T.; Song, Y.; Zhang, F. PaO2/FiO2 and IL-6 are risk factors of mortality for intensive care COVID-19 patients. *Sci. Rep.* **2021**, *11*, 7334. [CrossRef] [PubMed]
36. Song, C.Y.; Xu, J.; He, J.Q.; Lu, Y.Q. Immune dysfunction following COVID-19, especially in severe patients. *Sci. Rep.* **2020**, *10*, 15838. [CrossRef] [PubMed]
37. Xu, J.B.; Xu, C.; Zhang, R.B.; Wu, M.; Pan, C.K.; Li, X.J.; Wang, Q.; Zeng, F.F.; Zhu, S. Associations of procalcitonin, C-reaction protein and neutrophil-to-lymphocyte ratio with mortality in hospitalized COVID-19 patients in China. *Sci. Rep.* **2020**, *10*, 15058. [CrossRef]

microorganisms

Article

Analysis of the Frequency of Mutations at Diagnostic Oligonucleotide Sites and Their Impact on the Efficiency of PCR for HIV-1

Elena V. Bogoslovskaya *, Galina M. Tsyganova, Anastasiia O. Nosova and German A. Shipulin

Federal State Budgetary Institution "Centre for Strategic Planning and Management of Biomedical Health Risks" of the Federal Medical Biological Agency, 119121 Moscow, Russia; tsyganova@cspmz.ru (G.M.T.); anosova@cspfmba.ru (A.O.N.); shipulin@cspmz.ru (G.A.S.)
* Correspondence: lenabo2@mail.ru

Abstract: The development of effective diagnostic kits for HIV-1 remains a pressing concern. We designed diagnostic oligonucleotides for HIV-1 real-time PCR to target the most conserved region of the HIV-1 genome and assessed the mutation frequency at annealing sites. Two databases of nucleotide sequences, Los Alamos and NCBI, were analyzed, revealing that more than 99% of the sequences either lack mutations or contain 1–2 mutations at the binding site of the forward and reverse primers. Additionally, 98.5% of the sequences either lack mutations or contain 1–2 mutations at the binding site of the TaqMan probe. To evaluate the efficiency of primers and the probe in real-time PCR in the case of mutations at their binding sites, we constructed several plasmids containing the most common mutations and, in a model experiment, showed how different mutations affect the efficiency of PCR. Our analysis demonstrated that about 98.5% of HIV-1 strains can be efficiently detected using a single pair of selected primers. For the remaining 1.5% of strains, a more careful selection of the second target is needed.

Keywords: HIV-1; real-time PCR; mutations

Citation: Bogoslovskaya, E.V.; Tsyganova, G.M.; Nosova, A.O.; Shipulin, G.A. Analysis of the Frequency of Mutations at Diagnostic Oligonucleotide Sites and Their Impact on the Efficiency of PCR for HIV-1. *Microorganisms* **2023**, *11*, 2838. https://doi.org/10.3390/microorganisms11122838

Academic Editors: Shengxi Chen and Fabio Zicker

Received: 23 October 2023
Revised: 13 November 2023
Accepted: 21 November 2023
Published: 22 November 2023

1. Introduction

HIV-1 viral load (VL) is a critical diagnostic marker for prognosis and evaluating the effectiveness of ARV therapy in HIV-infected patients. Commercial kits for the quantitative detection of HIV-1 RNA (VL) have been developed and used throughout the world for many years. The most common method for determining HIV-1 VL is real-time PCR.

The high variability of HIV-1 leads to the regular emergence of new variants of the virus [1], some of which may be ineffectively detected using long-used commercial kits. In these cases, VL is determined incorrectly (or HIV-1 RNA is not detected at all), which leads to inadequate diagnosis and treatment. To ensure accurate diagnosis and treatment, it is extremely important to regularly analyze existing kits for their effectiveness in the context of an ever-changing virus population, as well as to develop new, more effective kits.

When developing new real-time PCR kits, it is important not only to select the most conserved region of the HIV-1 genome, but also to correctly position the primers and probes to minimize the number of mismatches at their binding sites. Due to the extreme variability of the virus, it is not possible to completely avoid mismatches, but you can reduce their impact on PCR efficiency through varying the position of the primer relative to the mismatch.

The objectives of this study included selecting new diagnostic oligonucleotides for the quantitative detection of HIV-1 RNA using real-time PCR. An important consideration when choosing primers for PCR is the sampling of nucleotide sequences used to analyze the conservation of the selected regions of the HIV-1 genome. The presence of international databases of nucleotide sequences greatly simplifies the task of selecting PCR primers, with the Los Alamos database being one of the best freely available resources. However, the completeness of nucleotide sequence databases varies depending on the genotype and often depends on the volume of sequenced samples.

The prevalence of different genotypes varies significantly in different regions of the world. While genotype B is dominant in the USA and Western European countries, subtype C is the prevalent genotype in Africa. In Russia and Eastern European countries, genotype A is more common [2], and in China, predominantly recombinant forms circulate [3]. As a result, the completeness of databases heavily depends on how actively each country sequences samples and uploads sequences to international publicly available databases. That is why there is a certain bias towards HIV-1 subtypes B and C in international databases, with these subtypes being underrepresented in some countries, particularly in Russia. Therefore, it was also important for us to assess how conserved the selected regions of the HIV-1 genome are among Russian isolates. In addition to theoretical analysis, our task was to evaluate the efficiency of PCR using selected oligonucleotides in model experiments.

2. Materials and Methods

2.1. Nucleotide Sequence Databases and Alignments

Analysis of the selected oligonucleotides was carried out using two databases: NCBI [4] and Los Alamos [5]. The first database was used to generate data on the diversity of genetic variants of HIV-1 isolated on the territory of the Russian Federation; the second database made it possible to obtain information on HIV-1 sequences from all over the world.

All records of HIV-1 nucleotide sequences containing the keyword "Russia" in the "country" field were downloaded from the NCBI database [4]. From the resulting sample of sequences (9031 records), records that did not contain regions of interest in the sequence (primer and probe binding sites) were excluded. Multiple alignment of nucleotide sequences was performed using MAFFT [6]. A separate alignment for each oligonucleotide was created from the resulting sample. The final alignment for the forward and reverse primer regions was 1206 and 1171, respectively, and for the probe region was 1202.

The second sample was formed in a similar way using ready-made sequence alignment ("web alignment") of the HIV-1 pol gene published in the Los Alamos HIV database [5]. Sequences containing deletions in the studied regions were excluded. The final sample consisted of 6329 sequences. From the resulting sample separate alignments were created for each oligonucleotide, limiting the available sequences to the region of a particular primer or probe.

For convenient presentation of data in each alignment, sequences were clustered using the CD-HIT program [7] (identity threshold—100%) and the resulting clusters were sorted according to the number of representatives included in them in descending order.

2.2. DNA Constructs and Site-Directed Mutagenesis

To obtain DNA PCR calibrators, we used the pBH10 plasmid (obtained through the NIH HIV Reagent Program, Division of AIDS, NIAID, NIH: Human Immunodeficiency Virus 1 (HIV-1) BH10 Non-Infectious Molecular Clone (pBH10), ARP-90, contributed by Dr. Beatrice Hahn and Dr. George M. Shaw), which was diluted to the required concentrations using a series of 10-fold dilutions.

To obtain a control DNA construct, a target fragment of the HIV genome region, including the HIV integrase gene 412 bp long, was generated via PCR using specific primers: Int_for4 CCCTACAATCCCCAAAGTCARGGAGT and Int_rev5 CATCACCTGCCATCTGTTTTTCCATARTC and plasmid pBH10 as a DNA template.

The resulting amplicon was ligated into the pGEM vector (Promega, Madison, WI, USA). The ligase mixture was introduced through transformation into *E. coli* cells, strain DH10b. Screening of the obtained clones was carried out using the PCR method. The structure of the target fragment was confirmed using Sanger sequencing.

To obtain DNA constructs with mutations in the binding sites of the primers and probe, the control DNA construct was used, into which mutations were introduced using the QuikChange II Site-Directed Mutagenesis kit (Agilent, Santa Clara, CA, USA) in accordance with the manufacturer's recommendations. The resulting DNA constructs were verified via Sanger sequencing.

2.3. Determination of Concentrations of DNA Constructs

The concentration of DNA constructs was measured using primers and a probe for the region of the ampicillin resistance gene (the bla gene, which is part of all plasmid constructs and encodes the beta-lactamase protein):

Bla1 tacgggagggcttaccatctg

Bla2 ggctggtttattgctgataaatctg

Bla3 R6G-accgcgagacccacgctcacc-BHQ1

Concentration measurements were conducted using the Digital PCR system, which includes the automatic loading of samples into a QX200™ AutoDG™ droplet generator (BioRad, Hercules, CA, USA) and a C1000 Touch Thermal Cycler (Bio-Rad, Hercules, CA, USA). To perform the analysis, we used the ddPCR™ reagent kit for sample preparation with TaqMan QX100/QX200™ probes (Bio-Rad, Hercules, CA, USA). Next, all samples were diluted to the same concentration value, taking into account the previously obtained values. The concentration of the prepared samples was measured in 5 replicates, and the median of the obtained concentrations in copies per milliliter was used for further studies. The concentration in DNA calibrators was determined using the same method.

2.4. Real-Time PCR

For real-time PCR, selected primers and a probe in the region of the HIV integrase gene, in-house reagents, and a RotorGene Q device (Qiagen, Hilden, Germany) were used. To evaluate quantitative results, DNA calibrators with DNA concentrations of 10^5, 10^4, and 10^3 copies per ml were used.

2.5. Statistics

For all quantitative values, the mean value and standard deviation were calculated. The significance of differences between data groups was assessed using the Mann–Whitney U-test. Differences were considered statistically significant if the p-value was <0.001.

3. Results

3.1. Oligonucleotide Selection for Real-Time PCR

The target for real-time PCR of HIV-1 was chosen in the integrase gene since it is the most conserved region of HIV-1 [8]. When choosing, we took into account the conservation of the oligos binding site, GC composition, melting temperature, and other factors that may affect the efficiency of PCR.

When selecting primers and probes, we used alignments of HIV-1 nucleotide sequences presented in the Los Alamos database, in particular, the HIV Sequence Compendium 2018 [9]. As a result, a pair of primers was selected to amplify a 237 bp fragment of HIV-1 and TaqMan probe for the detection of amplification products (supplementary materials).

The length and, accordingly, the melting temperature of the primers were selected with higher-than-optimal values so that during PCR, 1–2 mutations at the binding sites would not affect the amplification efficiency.

3.2. HIV-1 Nucleotide Sequence Analysis

The selected primers and probe were analyzed using two databases: Los Alamos and NCBI. The alignment of all HIV-1 nucleotide sequences from the Los Alamos database resulted in 6328 nucleotide sequences in the forward primer region, 6324 sequences in the reverse primer region, and 6326 sequences bounded by the probe region. The most prevalent genotype was B, representing over 33% of the sequences. Following closely was genotype C, accounting for approximately 17%. The recombinant form 01_AE was the next most common, making up 9.8% of the sequences. These data further confirm the predominance in international databases of HIV sequences distributed in the United States and African countries, where the highest HIV prevalence is observed.

Additionally, the primers and probe were analyzed using the NCBI database limited to the code word "Russia". The final alignment in the forward and reverse primer regions contained 1206 and 1171 sequences, respectively. Probe region analysis was performed for 1202 sequences. In this sample, the predominant genotype was A (65.8%), and the second most common was genotype B (7.2%). More than 10% of the sample was occupied by circulating recombinant forms (CRFs) with a predominance of CRF63_02A.

Estimation of Mutation Frequency at Primer and Probe Binding Sites

We analyzed the frequency of mutations at the binding sites of the primers and probe and also assessed the position of each mutation in the oligonucleotide. The results are presented in tables and the supplementary materials.

The analysis of nucleotide sequences from the Los Alamos database at the TaqMan probe binding site showed that more than 70% of the sequences are completely complementary to the probe sequence (Figure 1). The remaining 23% of sequences have single mutations at the probe binding site, with clusters 1 to 6 (the most numerous clusters in terms of the number of sequences) accounting for about 21% of all analyzed sequences (supplementary materials, Figure S1). At the same time, two mutations at the probe binding site occur with a frequency of less than 4%, with the most frequent corresponding to clusters 7, 8 and 10 and a total of 1.64%. At the same time, three mutations occur with a frequency of 1.5%, the most frequent ones correspond to clusters 12 and 14, with a total of 0.55%.

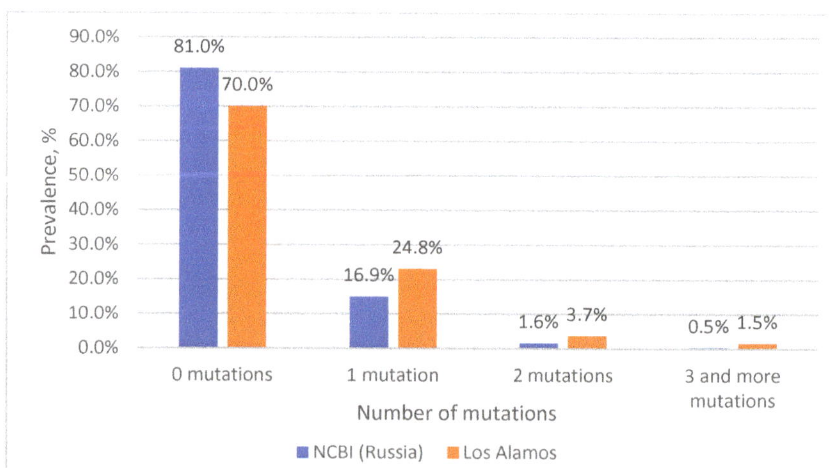

Figure 1. Prevalence of HIV-1 strains with mutations in the probe binding site.

When analyzing the nucleotide sequences of Russian isolates from the NCBI database at the binding site of the TaqMan probe, it turned out that more than 81% of the sequences were completely complementary to the probe sequence (Figure 1). Sequences included in clusters 1 to 10 contain single mutations and in total account for 15% (supplementary materials, Figure S2). The two mutations simultaneously occur most frequently in sequences corresponding to cluster 11 (0.42%), which corresponds to cluster 8 from the Los Alamos database (0.63%). With a low frequency (0.17%), two mutations occur simultaneously in the sequences of cluster 17, which corresponds to cluster 10 from the Los Alamos database, with a frequency of occurrence of 0.33%. Three mutations among Russian isolates are extremely rare, accounting for a total of 0.3%.

From the presented data, it is clear that 93–96% of sequences have no substitutions or have one substitution at the probe binding site and theoretically should be effectively detected by the selected probe; however, its effectiveness in the case of single mismatches in the probe binding area requires experimental validation. For this purpose, we subsequently constructed plasmids containing an insertion of a 412 bp fragment of the integrase gene, into which the necessary mutations were introduced. At the same time, two mutations at the probe binding site are also of interest, since they account for about 4% of all sequences. It is worth noting that the most common combination of two mutations among the sequences in the Los Alamos database belongs to cluster 7, which includes sequences of viruses, most of which belong to group O (86%), which are poorly represented in the world and are not found in Russia. At the same time, among Russian isolates from the NCBI database with two mutations, representatives of genotype B (cluster 11) are the most common. Three mutations at the same time are quite rare, no more than 1.5%, so the analysis of such sequences is of rather theoretical interest. However, we also further investigated such variants experimentally.

We designed the forward primer with one degenerate position C/T (fifth nucleotide from the 3′ end) because nucleotide C occurs with a frequency of 6.7% among Russian isolates and with a frequency of 34.8% in the Los Alamos database, and nucleotide T occurs with a frequency of 91.6% among Russian isolates and with a frequency of 63.5% in the Los Alamos database. In further analysis, we did not consider the degenerate position as a mismatch.

The analysis of nucleotide sequences from the Los Alamos database showed that about 73% of the sequences are completely complementary to the forward primer sequence (Figure 2). The remaining 23% consist of sequences containing single mutations, most of which are located closer to the 5′ end, which is known to have little effect on the efficiency of primers. At the same time, two mutations at the site of the forward primer occur with a frequency of less than 4%. Three or more mutations occur with a frequency of less than 0.5%.

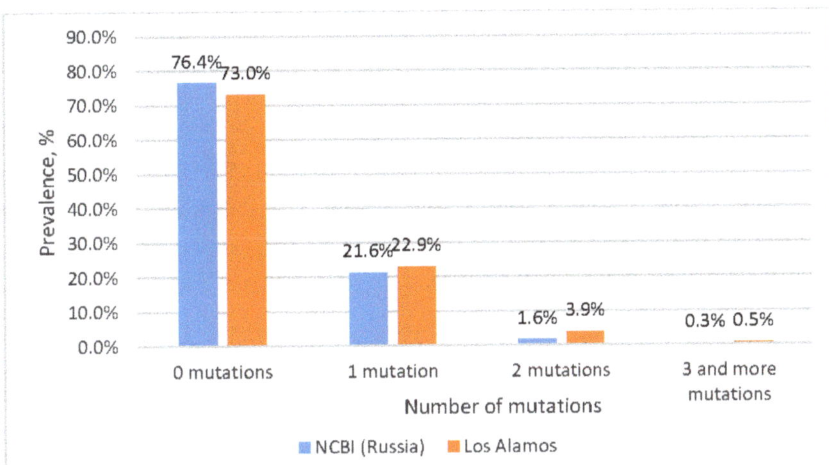

Figure 2. Prevalence of HIV-1 strains with mutations in the forward primer binding site.

When analyzing the nucleotide sequences of Russian isolates from the NCBI database, it turned out that more than 76% of the sequences are fully complementary to the sequence of the forward primer (Figure 2). About 21% of the sequences included in the analysis have one mutation. At the same time, two and three mutations in the forward primer region among Russian isolates occur with a frequency of 1.6% and 0.3%, respectively.

From the data presented in Figure 2, it is clear that more than 99% of the sequences either do not contain mutations or contain 1–2 mutations at the binding site of the forward primer. According to our calculations, the presence of two mutations at the binding site of the selected primers should not affect the efficiency of amplification. For subsequent experimental evaluation we constructed plasmids containing a fragment of the integrase gene into which 3 or more mutations were introduced.

When choosing a reverse primer, we faced the fact that in two positions there was a strong discrepancy in the frequency of occurrence of nucleotides between isolates from the international database and Russian isolates (supplementary materials). Thus, in position 2 from the 5′ end, nucleotide C occurs with a frequency of 84% among Russian isolates and with a frequency of 16% in the international database. A similar situation is observed at position 11 from the 5′ end, in which the nucleotide C occurs with a frequency of 82.8% among Russian isolates and with a frequency of 4.4% in the international database. It was decided to leave nucleotide C at position 2 at the 5′ end, since mutations in this position have virtually no effect on the efficiency of amplification and since this nucleotide is more common among Russian isolates. At position 11 (i.e., almost the middle of the primer) we made a degenerate position C/T. In further analysis, the degenerate position was not considered a mismatch.

When analyzing nucleotide sequences from the Los Alamos database, it is clear that complete complementarity to the reverse primer is observed in 13% of cases (Figure 3); this is due to the fact that we left the mutation in position 2 from the 5′ end of the primer. Additionally, 68% of the analyzed sequences had a single mutation at the primer binding site, with position 2 from the 5′ end of the primer accounting for 62%.

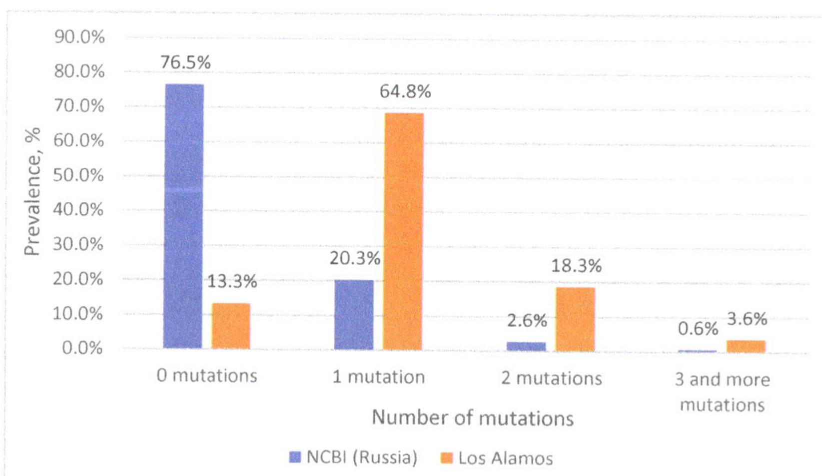

Figure 3. Prevalence of HIV-1 strains with mutations in the reverse primer binding site.

Simultaneously, two mutations at the site of the reverse primer occur with a frequency of 18%. At the same time, three or more mutations occur with a frequency of 3.6%.

Although a total of three or more mutations occurred quite often (3.6%), each combination occurred no more often than 0.2%. The exception was sequences from cluster 8 (0.65%), in which four mutations occur simultaneously; most of the sequences belong to the group O virus (85%). Therefore, when constructing plasmids containing mutations for model experiments, preference was given to those mutations that could maximally reduce the efficiency of the primer, i.e., were located closer to the 3′ end or there were four and five mutations.

When analyzing the nucleotide sequences of Russian isolates from the NCBI database, it turned out that more than 76% of the sequences are completely complementary to the reverse primer (Figure 3). Among sequences with mutations, the most common are variants with single substitutions—20%. The highest percentage of occurrence is for sequences with adenine in position 2 from the 5′ end of the primer (12.6%). Simultaneously, two mutations are found in 2.6% of sequences, of which the most common variant unites about 1% of sequences (cluster 4). Three or more mutations are rare, accounting for a total of 0.6% of sequences, and, moreover, each specific variant is represented in the NCBI database by a single sequence.

3.3. Efficiency of Primers and Probe in Real-Time PCR in Case of Mismatches at Binding Sites

In order to evaluate the efficiency of primers and probes in real-time PCR in the case of mutations at their binding sites, we constructed several plasmids containing the amplification region of the integrase gene fragment, into which the most common mutations were introduced. The resulting DNA plasmids with equal concentration were tested together with the mutation-free plasmid in real-time PCR using the selected primers and probe. In parallel, DNA calibrators were tested in each setup to obtain quantitative values for each sample.

Table 1 presents the constructed plasmids, along with details of the introduced mutations and the frequency of nucleotide sequences with these mutations in two databases.

Table 1. DNA-constructs with mutations in the oligonucleotides binding sites.

Construct Name	Binding Sites	Russian Strains (NCBI)				International Strains (Los Alamos)			
		%	Claster	Predominant Genotype		%	Claster	Predominant Genotype	
DNA constructs with mutations in the TaqMan probe binding site									
Mut 5	gggattggggg**a**tacagtgcaggg	2.41%	2	B	48%	9.8%	1	B	47%
Mut 3-2	gggattgggggtacagtgcagg**a**	2.16%	3	B	50%	4.41%	2	B	37.9%
Mut 3-1	gggattgggggtaca**c**tgcaggg	1.25%	4	A6/B	40%/40%	0.87%	6	B	43.4%
Mut 6	gggattgggggtac**t**gtgcaggg	0.58%	6	B	100%	0.22%	19	B	75%
Mut 9	gggattggggg**a**tacagtgcagg**a**	0.42%	11	B	80%	0.63%	8	B	60.5%
Mut 3	gggattgggggtaca**c**tgcagg**a**	0	ND	ND	ND	0.68%	7	O	90%
Mut 1a	gggattggggact**ac**agtgcaggg	0.17%	17	A6/ CRF63_02A1	50%/50%	0.33%	10	B	47.4%
Mut 10	gggattggggg**a**ta**t**agtgcaggg	0	ND	ND	ND	0.24%	16	B	46%
Mut 8	gggattgggg**aa**tacagtgcagg**a**	0.25%	14	B	100%	0.03%	49	B	100%
Mut 1	gggattgggg**ac**tac**t**gtgcaggg	0	ND	ND	ND	0.28%	12	B/01_AE	37.5%/37.5%
Mut 2	gggattgggg**act**a**t**agtgcaggg	0.08% *	40	A1	100%	0.27%	14	B	33%
Mut 4	gggattgggg**acc**acagtgcaggg	0	ND	ND	ND	0.02% *	133	75_BF1	100%
DNA constructs with mutations in the primers binding sites									
Primer Int for	cagcagtacaaatggcagtattcatycaca								
Mut 1	cagcagtacaaatggcagtatt**tgtg**caca	0	ND	ND	ND	0.02% *	209	O	100%
Mut 2	cagcagtacaaatggcagt**c**tt**tg**tccaca	0	ND	ND	ND	0.06%	43	O	100%
Mut 3	cagcagtac**t**aatggcagtat**a**catcca**t**a	0	ND	ND	ND	0.02% *	104	B	100%
Mut 4	ca**a**cagt**c**caaat**aa**ca**a**tattcatccaca	0	ND	ND	ND	0	ND	ND	ND
Mut 5	cagcagt**g**caaatggc**gg**t**t**ttcatccaca	0	ND	ND	ND	0.08%	33	N	100%
Primer Int rev	cctgtattacyactgccccttcacctttcca								
Mut 6	c**tt**g**g**attaccactgccccttc**t**cctttcca	0	ND	ND	ND	0.16%	20	N	100%
Mut 7	c**tt**gtat**g**actactgc**t**cc**c**tcacctttcca	0	ND	ND	ND	0.65%	8	O	89.7%
Mut 8	c**tt**gtat**g**actactgcccctt**t**acctt**t**tca	0	ND	ND	ND	0	ND	ND	ND

* one sequence in the data base, ND—no data, introduced mutations are indicated in bold.

In the case of the probe, we made the largest number of plasmids, dividing them into three groups. The first group included constructs with the most common single mutations at the probe binding site, the second group included constructs with two mutations, and the third group, three mutations. The third group was of only theoretical interest, since it was obvious that the probe would not work effectively in the case of three mutations.

Since we deliberately chose primers with a higher melting temperature, taking into account the possible presence of 1–2 mutations at the binding site, plasmids were constructed containing 3 or more mutations at the binding sites of the primers (Table 1). The frequency of such sequences in both analyzed databases did not exceed 3.6%.

The results for each construct obtained during PCR were quantified using DNA calibrators and a calibration curve constructed on their basis. Thus, the result was obtained in copies of DNA per PCR sample. The obtained quantitative results were then normalized to the values obtained in ddPCR for the bla gene. To assess the reliability of the differences for most DNA constructs, the experiment was repeated 3–4 times. The exceptions were those constructs for which a significant decrease in amplification efficiency was observed (constructs with three mutations in the probe binding region).

The quantitative results for DNA constructs with mutations are presented in Figures 4 and 5 and in the supplementary materials (Table S1).

Plasmids with single mutations showed quantitative results similar to the control, except for mut3-1, where a two-fold decrease in concentration was observed. For constructs with two mutations in the probe binding region, a slight decrease in quantitative values was observed, by a factor of 1.5–2.7, depending on the construct. As expected, constructs with three mutations at the probe binding site are poorly detected; results are underestimated by a factor of four or more. However, it is worth noting that in the Los Alamos database, three mutations occur simultaneously with a frequency of 1.5%, and among Russian isolates, 0.3%.

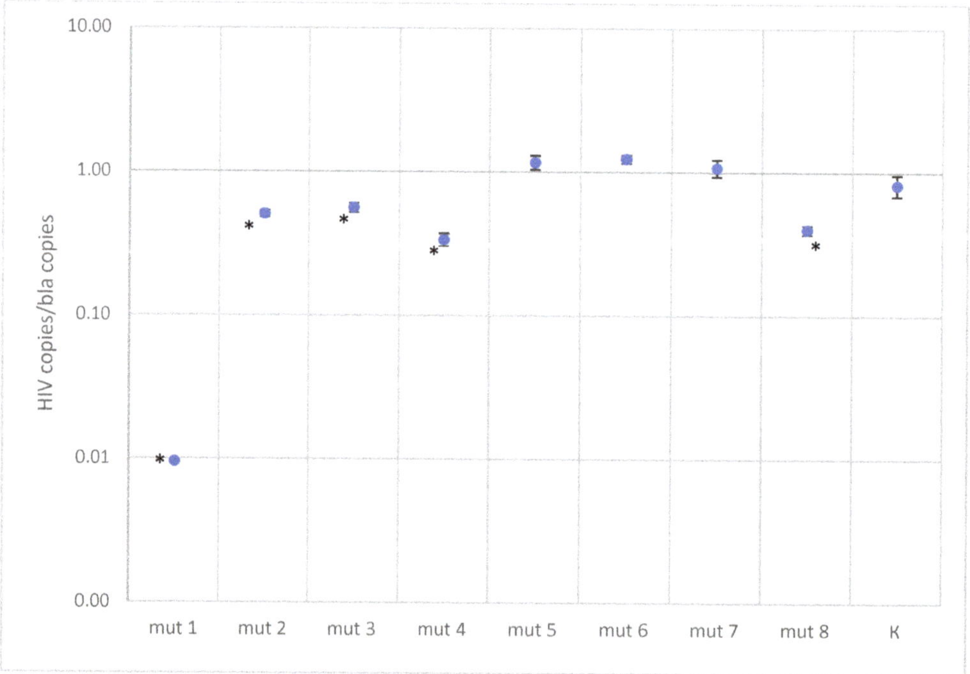

Figure 4. Results of quantitative measurements of constructs with mutations in the primer binding region, normalized to the bla gene (K—control plasmid without mutations. A statistically significant of the difference (p-value < 0.001 according to the Mann–Whitney U-test) relative to the control plasmid is shown with asterisk *).

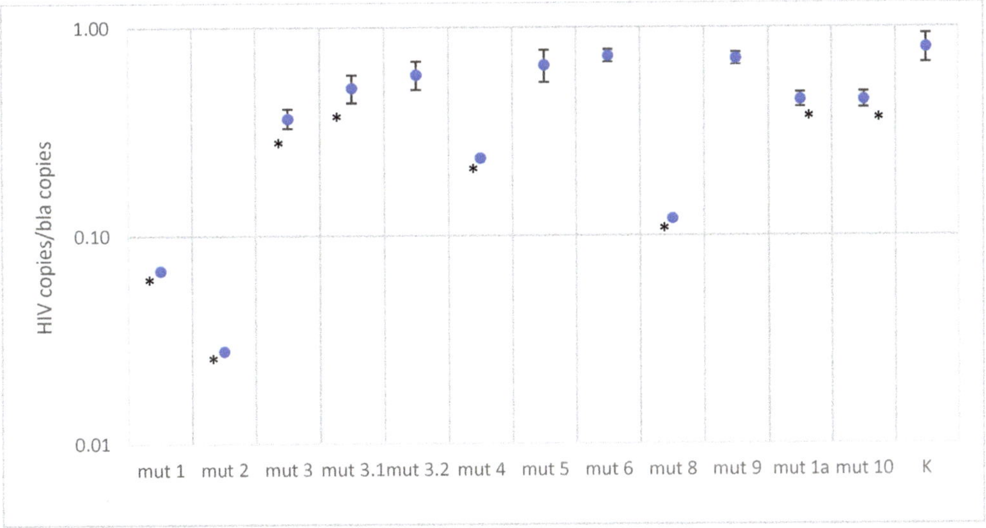

Figure 5. Results of quantitative measurements of constructs with mutations in the probe binding region, normalized to the bla gene (K—control plasmid without mutations. A statistically significant of the difference (p-value < 0.001 according to the Mann–Whitney U-test) relative to the control plasmid is shown with asterisk *).

In the case of testing plasmids containing mutations in the primer binding sites, the situation turned out to be more optimistic. Most constructs were effectively detected using primers. The only exception was one construct, mut 1, which had the closest location of substitutions to the 3′ end of the primer and to each other, forming two consecutive TG mismatches and another one across the nucleotide—the quantitative result was underestimated by a factor of 100. This sequence variant in the Los Alamos database is represented by the only isolate belonging to group O. A reduced amplification efficiency (2.5–3 times) was also recorded when testing the mut4 construct, which has five substitutions, two of which are located one after another in the middle of the forward primer binding site, and mut8 construct, which has four substitutions in the reverse primer region. Sequences similar to these two constructs were not found in the analyzed databases.

Despite the statistically significant difference between the control plasmid and the mut2 and mut3 constructs, the quantitative result was underestimated by less than two times.

3.4. Selection of a Conserved Target for Discordant Sequences

To effectively identify discordant samples, we attempted to select an additional target for real-time PCR. For this purpose, we aligned 86 nucleotide sequences with three or more mutations in the probe binding region and conducted a search for conserved regions to select the primers and probe. The most conserved region was found to be the same region of integrase in the immediate vicinity of the first target. Within this region, we selected the primers int_for_41 and int_rev_40 and the probe int_Pr_rev10 (Figure 6). The analysis of the frequency of mutations in the region of the primer and probe showed that in the probe region, only one sequence has three mutations, which belongs to HIV group N, and in the primers region, there is one sequence each with four mutations.

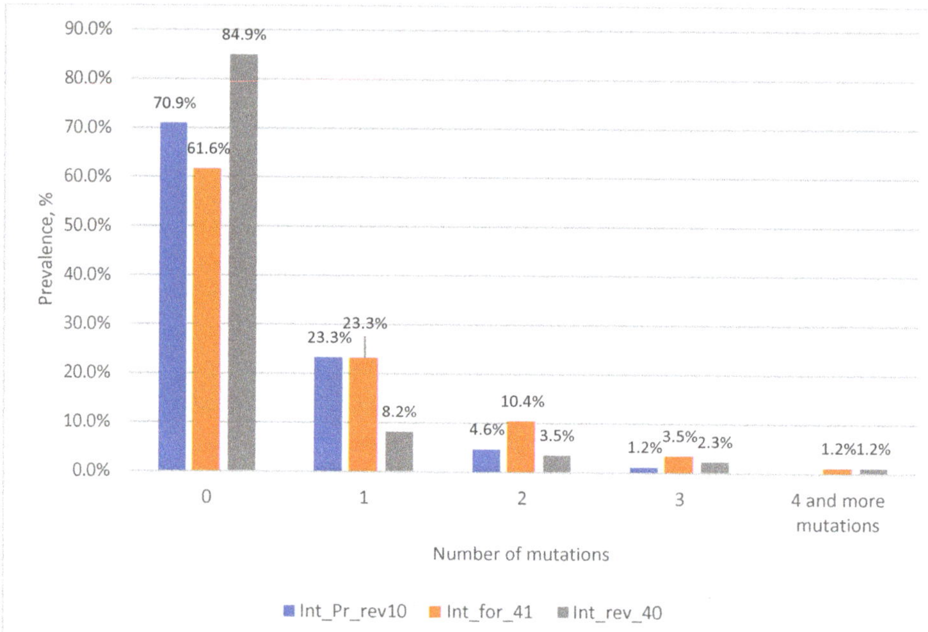

Figure 6. Prevalence of HIV-1 strains with mutations in primers and probe binding sites.

4. Discussion

Due to the high variability of HIV-1, finding a single highly conserved target sequence for the effective detection of HIV-1 RNA via Real-Time The peal-Time PCR is challenging. In recent years, a number of experts have expressed the idea of the need to select at least two targets for HIV-1 amplification, thereby ensuring oneself in the event of mutations in one of the targets.

In our work, we have shown that the task of choosing primers for the effective amplification of most HIV-1 variants is quite feasible. The amplification conditions were selected so that 2–3 mismatches in the primer annealing region did not lead to a decrease in amplification efficiency. A more complex situation is the choice of a TaqMan probe, since its selection requires more stringent requirements, in particular, short length, high melting temperature, and a minimum of mutations. Our findings demonstrate that, in most cases, 1–2 mutations at the probe binding site do not significantly impact detection efficiency; therefore, according to our data, such a probe will detect approximately 98.5% of all viruses from the analyzed databases with similar efficiency. The problem remains with the remaining 1.5% of viruses, in the sequence of which there are three or more mutations at the probe binding site and for which a significant decrease in PCR efficiency has been shown. In this case, the most obvious tactic is to select a second target. On the one hand, this is an insignificant percentage of strains that will be under-detected throughout the world, while in one specific region, for example, in Russia, this percentage is much lower (0.3%). On the other hand, achieving 100% detection using one target is unlikely, since it is impossible to predict the occurrence of certain mutations.

Developers of HIV-1 viral load kits predominantly target the polymerase gene, particularly its highly conservative integrase gene (examples include Abbott m2000 sp/rt RealTime HIV-1, Alinity m HIV-1 assay, Siemens Versant® kPCR HIV-1, and Hologic Panther® Aptima HIV-1). Only the Roche cobas® AmpliPrep/cobas TaqMan (CAP/CTM) HIV-1 v2.0 kits, following an older tradition, utilize a target in the gag region. As this region is not highly conservative, Roche Molecular Systems was among the first to introduce a second target into their kits, specifically in the LTR region. Subsequently, other manufacturers also started incorporating a second target, predominantly in the LTR region (for example, Alinity m HIV-1 assay), considering it to be sufficiently conservative. This approach has notably impacted Russian kit developers. However, in our opinion, this approach has a significant drawback: the number of LTR nucleotide sequences in international databases is much smaller than, for example, in the polymerase region. Thus, in the Los Alamos database, there are 6802 "web complete sequences" in the pol region and 1305 in LTR. If you search the NCBI database using the keywords "HIV-1 and pol gene", more than 580,000 sequences are downloaded; a similar search for the LTR gene yields about 10,000 sequences. Evaluating the conservation level of this target proves to be quite challenging. Moreover, it is now becoming clear that Roche Molecular Systems was forced to take this step, since a number of publications have accumulated showing that kits with a single target in the gag gene under-detect a number of strains [10]. The addition of a second target improved the analytical performance of the kit and increased the efficiency of identifying discordant samples [11]. However, it did not completely resolve the issue. Thus, when comparing two kits, COBAS TaqMan HIV-1 Test v2.0 (uses two targets—gag and LTR) and Abbott m2000 Real-Time HIV-1 assay (uses one target—integrase), one sample related to HIV subtype C was identified. This sample gave a lower viral load value of more than 1 lg in the COBAS TaqMan HIV-1 Test v2.0, as it had several mismatches in both targets [12]. In contrast, the Abbott m2000 Real-Time HIV-1 assay kit, containing only one target, showed higher stability in identifying different subtypes of HIV-1.

Due to the high variability of HIV-1, the need to select a second target for PCR diagnostics is obvious, but this requires a more thorough and meaningful approach aimed at identifying those viruses that are not detected with the first target. We used one of the approaches in our work: we made an alignment for discordant sequences, found conserved regions only for them, and then estimated how broadly the selected target would cover the remaining sequences. We aligned sequences (86 pieces) that were discordant at the probe binding site (three or more mutations). It turned out that, in general, it is possible to find conserved regions for most sequences (with the exception of one sequence from group N with three mutations in the probe binding region and one sequence from group O with four mutations in the forward primer binding region). The most conserved regions are located approximately in the same region of the integrase gene as the first target, for which we were able to select primers and a probe and demonstrate their effectiveness on plasmids with mutations at the binding site of the first probe. In this case, it is even difficult to talk about the second target. However, in our opinion, this is a more meaningful approach than choosing a target in the LTR region. In particular, of the 86 sequences, only 58 had full-genome sequences and only 20 had full-length LTR fragments. However, we were able to make an alignment for 54 sequences in the LTR region, but no conserved regions for primer and probe were found.

More generally, the problem of identifying all possible HIV-1 strains could be solved through choosing two maximally conservative targets independently of each other. In this case, it is desirable that two targets be comparable in terms of sample size. Next, it will be necessary to compare the discordant sequences from the two regions to ensure that one primer pair effectively detects the discordant sequences from the second target. The task is not trivial, but with modern bioinformatics capabilities, it is feasible.

5. Conclusions

Thus, in this work, we were able to select a conservative target, select primers and a probe for real-time PCR in it, and prove their high efficiency for identifying the majority of HIV-1 strains (98.5% of the analyzed sequences). And for the remaining discordant samples, we were able to select an additional pair of primers and probe. Obviously, other approaches to selecting a second target can be used.

Supplementary Materials: The following supporting information can be downloaded at: https://www.mdpi.com/article/10.3390/microorganisms11122838/s1, Figure S1: Los Alamos Database. Region of the HIV-1 genome corresponding to the TaqMan probe; Figure S2: NCBI Database (Russian isolates). Region of the HIV-1 genome corresponding to the TaqMan probe; Figure S3: Los Alamos Database. Region of the HIV-1 genome corresponding to the forward primer; Figure S4: NCBI Database (Russian isolates). Region of the HIV-1 genome corresponding to the forward primer; Figure S5: Los Alamos Database. Region of the HIV-1 genome corresponding to the reverse primer; Figure S6: NCBI Database (Russian isolates). Region of the HIV-1 genome corresponding to the reverse primer; Table S1: Quantitative results obtained in Real-Time PCR and digital PCR for DNA constructs.

Author Contributions: E.V.B. designed the experiments, analyzed the data, and wrote the paper. G.M.T. designed experiments and conducted the experiments. A.O.N. conducted the experiments. G.A.S. designed experiments and reviewed the paper. All authors have read and agreed to the published version of the manuscript.

Funding: This research received no external funding.

Data Availability Statement: Data supporting reported results can be requested from the authors.

Conflicts of Interest: The authors declare no conflict of interest.

References

1. Maksimenko, L.V.; Totmenin, A.V.; Gashnikova, M.P.; Astakhova, E.M.; Skudarnov, S.E.; Ostapova, T.S.; Yaschenko, S.V.; Meshkov, I.O.; Bocharov, E.F.; Maksyutov, R.; et al. Genetic Diversity of HIV-1 in Krasnoyarsk Krai: Area with High Levels of HIV-1 Recombination in Russia. *BioMed Res. Int.* **2020**, *2020*, 9057541. [CrossRef] [PubMed]
2. Beloukas, A.; Psarris, A.; Giannelou, P.; Kostaki, E.; Hatzakis, A.; Paraskevis, D. Molecular epidemiology of HIV-1 infection in Europe: An overview. *Infect. Genet. Evol.* **2016**, *46*, 180–189. [CrossRef] [PubMed]
3. He, X.; Xing, H.; Ruan, Y.; Hong, K.; Cheng, C.; Hu, Y.; Xin, R.; Wei, J.; Feng, Y.; Hsi, J.H.; et al. A comprehensive mapping of HIV-1 genotypes in various risk groups and regions across China based on a nationwide molecular epidemiologic survey. *PLoS ONE* **2012**, *7*, e47289. [CrossRef] [PubMed]
4. GenBank. Available online: https://www.ncbi.nlm.nih.gov/genbank (accessed on 22 December 2021).
5. HIV Sequence Database. Available online: https://www.hiv.lanl.gov/content/sequence/NEWALIGN/align.html (accessed on 9 March 2022).
6. Katoh, K.; Misawa, K.; Kuma, K.I.; Miyata, T. MAFFT: A novel method for rapid multiple sequence alignment based on fast Fourier transform. *Nucleic Acids Res.* **2002**, *30*, 3059–3066. [CrossRef] [PubMed]
7. Fu, L.; Niu, B.; Zhu, Z.; Wu, S.; Li, W. CD-HIT: Accelerated for clustering the next-generation sequencing data. *Bioinformatics* **2012**, *28*, 3150–3152. [CrossRef] [PubMed]
8. Li, G.; Piampongsant, S.; Faria, N.R.; Voet, A.; Pineda-Peña, A.-C.; Khouri, R.; Lemey, P.; Vandamme, A.-M.; Theys, K. An integrated map of HIV genome-wide variation from a population perspective. *Retrovirology* **2015**, *12*, 18. [CrossRef] [PubMed]
9. HIV Sequence Compendium 2021. Available online: https://www.hiv.lanl.gov/content/sequence/HIV/COMPENDIUM/2021 compendium.html (accessed on 21 March 2018).
10. Korn, K.; Weissbrich, B.; Henke-Gendo, C.; Heim, A.; Jauer, C.M.; Taylor, N.; Eberle, J. Single-point mutations causing more than 100-fold underestimation of human immunodeficiency virus type 1 (HIV-1) load with the Cobas TaqMan HIV-1 real-time PCR assay. *J. Clin. Microbiol.* **2009**, *47*, 1238–1240. [CrossRef] [PubMed]
11. Damond, F.; Avettand-Fenoel, V.; Collin, G.; Roquebert, B.; Plantier, J.C.; Ganon, A.; Sizmann, D.; Babiel, R.; Glaubitz, J.; Chaix, M.L.; et al. Evaluation of an upgraded version of the Roche Cobas AmpliPrep/Cobas TaqMan HIV-1 test for HIV-1 load quantification. *J. Clin. Microbiol.* **2010**, *48*, 1413–1416. [CrossRef] [PubMed]
12. Karasi, J.; Dziezuk, F.; Quennery, L.; Förster, S.; Reischl, U.; Colucci, G.; Schoener, D.; Seguin-Devaux, C.; Schmit, J. High correlation between the Roche COBAS®AmpliPrep/COBAS®TaqMan®HIV-1, v2.0 and the Abbott m2000 RealTime HIV-1 assays for quantification of viral load in HIV-1 B and non-B subtypes. *J. Clin. Virol.* **2011**, *52*, 181–186. [CrossRef] [PubMed]

microorganisms

Article

Molecular Characterization of Dengue Virus Strains from the 2019–2020 Epidemic in Hanoi, Vietnam

Juthamas Phadungsombat [1], Huong Thi Thu Vu [2], Quynh Thi Nguyen [3], Ha Thi Van Nguyen [2], Ha Thi Nhu Nguyen [2], Bich Thi Dang [2], Emi E. Nakayama [1], Azumi Ishizaki [3], Hiroshi Ichimura [3], Tatsuo Shioda [1,*] and Thach Ngoc Pham [2]

[1] Department of Viral Infections, Research Institute for Microbial Diseases, Osaka University, Osaka 565-0871, Japan; juthamas@biken.osaka-u.ac.jp (J.P.); emien@biken.osaka-u.ac.jp (E.E.N.)

[2] National Hospital for Tropical Disease, Hanoi 100000, Vietnam; huongvu13@gmail.com (H.T.T.V.); vumanhhungha@gmail.com (H.T.V.N.); nhuha.niitd@yahoo.com (H.T.N.N.); bich.vaac@gmail.com (B.T.D.); phamngocthachnhtd@gmail.com (T.N.P.)

[3] Department of Viral infection and International Health, Graduate School of Medical Science, Kanazawa University, Kanazawa 920-8640, Japan; rubynguyen.hmu@gmail.com (Q.T.N.); aishizak@staff.kanazawa-u.ac.jp (A.I.); ichimura@med.kanazawa-u.ac.jp (H.I.)

* Correspondence: shioda@biken.osaka-u.ac.jp

Abstract: Dengue virus (DENV), which has circulated in Vietnam for several decades, has multiple serotypes and genotypes. A 2019 dengue outbreak resulted in a larger number of cases than any other outbreak. We conducted a molecular characterization using samples collected in 2019–2020 from dengue patients in Hanoi and nearby cities located in northern Vietnam. The circulating serotypes were DENV-1 (25%, n = 22) and DENV-2 (73%, n = 64). Phylogenetic analyses revealed that all DENV-1 (n = 13) were genotype I and clustered to local strains circulating during the previous outbreak in the 2017, whereas DENV-2 consisted of two genotypes: Asian-I (n = 5), related to local strains from 2006–2022, and cosmopolitan (n = 18), the predominant genotype in this epidemic. The current cosmopolitan virus was identified as having an Asian-Pacific lineage. The virus was closely related to strains in other recent outbreaks in Southeast Asian countries and China. Multiple introductions occurred in 2016–2017, which were possibly from maritime Southeast Asia (Indonesia, Singapore, and Malaysia), mainland Southeast Asia (Cambodia and Thailand), or China, rather than from an expansion of localized Vietnamese cosmopolitan strains that were previously detected in the 2000s. We also analyzed the genetic relationship between Vietnam's cosmopolitan strain and recent global strains reported from Asia, Oceania, Africa, and South America. This analysis revealed that viruses of Asian-Pacific lineage are not restricted to Asia but have spread to Peru and Brazil in South America.

Keywords: DENV-1; DENV-2; DENV genotype; phylogenetic tree; Vietnam; Hanoi

Citation: Phadungsombat, J.; Vu, H.T.T.; Nguyen, Q.T.; Nguyen, H.T.V.; Nguyen, H.T.N.; Dang, B.T.; Nakayama, E.E.; Ishizaki, A.; Ichimura, H.; Shioda, T.; et al. Molecular Characterization of Dengue Virus Strains from the 2019–2020 Epidemic in Hanoi, Vietnam. *Microorganisms* **2023**, *11*, 1267. https://doi.org/10.3390/microorganisms11051267

Academic Editors: Shengxi Chen and Fabio Zicker

Received: 19 April 2023
Revised: 9 May 2023
Accepted: 10 May 2023
Published: 11 May 2023

1. Introduction

Dengue is a mosquito-borne disease transmitted to humans via the bite of infected mosquitos and commonly occurs in over 100 tropical and subtropical countries. Clinically, dengue cases vary in severity from mild to fatal. The World Health Organization (WHO) categorizes the disease as dengue (with or without warning signs) and severe dengue [1]. Symptoms, such as high fever, severe headache, muscle and joint pain, nausea, and rash, manifest for 2–7 days, and warning signs of severe disease include severe abdominal pain, persistent vomiting, rapid breathing, bleeding gums, and nosebleed [2]. The incidence of dengue cases reported to the WHO increased from approximately 500,000 cases in 2009 to 5 million cases in 2019 [2], resulting in significant global human and economic impacts.

The causative agent of dengue fever is dengue virus (DENV) (genus *Flavivirus*, family *Flaviviridae*). DENVs can be classified into four serotypes (DENV-1, DENV-2, DENV-3, and

DENV-4). In primary DENV infection, long-term host immunity is specifically generated against the actual serotype, with short-term protection against other serotypes. Subsequent secondary infection with a different serotype can lead to antibody-dependent enhancement, increasing the risk of severe dengue [3]. Within the dengue serotypes, 4–6 genotypes have been genetically defined [4]. Serotype/genotype replacement has occurred concurrently with the increase in the number of patients [5].

Vietnam, located in Southeast Asia and bordered by Laos, Cambodia, and China, is a hyperendemic country for dengue. Periodic and cyclic DENV circulation patterns have occurred in Vietnam, but DENV-1 and -2 are the most common serotypes [5,6]. In 2019, the highest recorded number of dengue cases was reported worldwide, particularly affecting Asia, with 420,000, 320,000, 131,000, and 101,000 cases reported in the Philippines, Vietnam, Malaysia, and Bangladesh, respectively [2]. The previous molecular epidemiological studies in Vietnam described an increasing in the dengue incidence associated with the changes in DENV-2 genotypes [5]. Recently, the 2017 outbreak was shown to be caused by local DENV-1 strains [7]. To date, the DENV characteristics responsible for the more recent outbreak in the 2019–2020 season have not been well elucidated. Therefore, we conducted a molecular characterization of DENV isolates from patient samples collected between 2019 and 2020 in Hanoi, the capital city of Vietnam located in the northern part of the country, and nearby provincial cities. We also present clinical data regarding these cases.

2. Materials and Methods

2.1. Patients, Sample Processing, and Laboratory Testing

Clinical samples were collected from patients with suspected dengue infection who were aged 18 years or older and had a fever (within 5 days of the study). Samples were collected according to the usual diagnostic procedure of the National Hospital for Tropical Disease (NHTD), Hanoi, Vietnam. Ethical Approval was obtained from the Ethics Committee of the NHTD, and all participants gave informed consent. Serum from participating patients was locally analyzed by Dengue Duo (SD Bioline, Seoul, Republic of Korea) to detect the NS1 antigen and dengue-specific antibodies (IgM and IgG). RNA extraction and dengue serotyping were conducted for all NS1-positive serum samples. Viral RNA was extracted from 140 μL of each serum sample using a QIAamp viral mini kit (Qiagen, Hilden, Germany), and 5 μL was used to determine the dengue serotype by multiplex real-time RT-PCR using a Genesig kit (Genesig, Chandler's Ford, UK).

2.2. Virus Isolation

Thirty microliters of serotyping-positive patient serum was inoculated in C6/36, an *Aedes albopictus* cell line, as described previously [8]. A total of 140 μL of culture supernatant was harvested on days 7 and 14 post-infection, and DENV RNA was extracted and analyzed using real-time RT-PCR [9].

2.3. Envelope Nested RT-PCR and Whole-Genome RT-PCR

To amplify the DENV envelope region, 5 μL of extracted RNA from patient serum was primarily reverse transcribed and amplified (1°PCR), targeting the first and second halves of the envelope region using a One-step RT-PCR kit (Qiagen) with primers covering the envelope region (Table S1). The PCR master mix was prepared according to the manufacturer's protocol at a final primer concentration of 0.6 μM. To increase the DNA concentration, the primary PCR product was subjected to secondary PCR (2°PCR) using GXL polymerase (Takara, Shiga, Japan) and primers (Table S1) that amplified the first and second halves of the envelope region product.

To amplify the whole DENV genome, 5.5 μL of extracted RNA of each isolate was used to synthesize cDNA with two primers specific to the first (5′half) and second (3′half) halves of the genome using SuperScript IV (Invitrogen, Vilnius, Lithuania). The RT reaction mixture was prepared according to the manufacturer's protocol. The 5′half and 3′half cDNAs were amplified using corresponding primers [8] and GXL polymerase (Takara, Shiga, Japan).

The amplified products were examined by gel red staining (Biotium, Fremont, CA, USA) and agarose gel electrophoresis at 100 volts in TBE buffer. The target band was purified using Nucleospin kit (MACHEREY-NAGEL, Düren, Germany). DNA concentration was quantified using either a Nanodrop (Thermo Fisher Scientific, Waltham, MA, USA) or Qubit system (Invitrogen, Eugene, OR, USA).

2.4. Envelope and Whole-Genome Sequencing

The nucleotide sequence of the genome region encoding the envelope was determined by Sanger sequencing using an ABI sequencer. The sequencing reactions were prepared using an ABI Prism BigDye Terminator Cycle Sequencing Ready Reaction kit, version 3.1 (Applied Biosystems, Vilnius, Lithuania). Briefly, 2 μL of DNA template was mixed with 2 μL of reaction mix, 1 μL of 5× sequencing buffer, 1.6 μL of specific primer (Table S1), and 3.4 μL of nuclease-free water. The amplification cycle conditions were programmed according to the manufacturer's protocol.

The whole genome sequences of DENV isolates obtained in the present study were determined by next-generation sequencing (NGS) on a Miseq platform (Illumina, San Diego, CA, USA). Preparation of the NGS library using NexteraXT (Illumina, San Diego, CA, USA) was described previously [8]. The FASTQ results were examined using CLC Genomics Workbench software, version 21 (CLC Bio, QIAGEN, Valencia, CA, USA). Whole-genome assembly was conducted using Map read reference command (DENV-1 Mochizuki AB074760.1 and DENV-2 16681 NC_001474.2 were used as reference strains).

All newly obtained sequences of the envelope and the whole genome were deposited in GenBank with accession numbers OQ832560-OQ832594 and OQ832609-OQ832627, respectively.

2.5. Phylogenetic and Genetic Diversity Analyses

DENV sequences analyzed in the present study were retrieved by February 2023 from GenBank and The Bacterial and Viral Bioinformatics Resource Center (BV-BRC) (https://www.bv-brc.org/ (accessed on 15 February 2023)) [10] and combined with the new Vietnam sequences (Table S2). Alignment and translation were conducted in AliView v1.26 [11]. The maximum-likelihood (ML) tree and model selection were estimated in IQ-TREE for DENV-1 and DENV-2 using the envelope region [12]. The molecular clock phylogeny of DENV-2 cosmopolitan based on the envelope region was estimated using BEAST, v1.10 [13]. The temporal signal dataset was examined by root-to-tip using Tempest and showed R^2 of 0.87 (Figure S1) [14]. Triplicate BEAST runs using settings for the SRD06 and Skygrid coalescent model with either a strict clock or uncorrelated clock model were employed for >50 million Markov Chain Monte Carlo cycles with 5 million cycle burn-ins, and marginal-likelihood estimation was performed using path sampling and stepping stone sampling analyses, which suggested that the strict clock was the best fit for this analysis (Table S3) [15,16]. The run convergence was accessed in Tracer, v1.7.1 [17]. Triplicate runs obtaining an effective sample size of >200 were combined in Logcombiner, and the maximum clade credibility (MCC) phylogeny was generated using TreeAnnotator and visualized in Figtree, v1.4.4.

To analyze amino acid polymorphism in Hanoi DENVs, the coding sequences were compared to previous Vietnam sequences available in GenBank. The amino acid position was annotated corresponding to the reference viruses: NC_001477.1, NC_001274.2, and GQ398263 for DENV-1 genotype I, DENV-2 genotype Asian-I, and DENV-2 genotype cosmopolitan, respectively.

2.6. Statistical Tests

Differences in categorical and continuous variables were statistically evaluated by Fisher's Exact test and Mann–Whitney U test, respectively. All the statistical calculations were performed by GraphPad Prism (GraphPad Prism Software, Boston, MA, USA).

3. Results

3.1. DENVs in Hanoi, Vietnam, in 2019–2020

Between October and December 2019 and September and December 2020, a total of 266 DENV clinical suspected samples were collected at the NHTD (2019, n = 30; 2020, n = 103), with 201 patient serum samples from Hanoi, the capital city, and 24 from nearby provinces (Hoa Bihn, Phu Tho, Vinh Phuc, Bac Ninh, Hung Yen, Hai Dung, Ha Nam, Thai Binh, Bac Giang, Bac Kan, Son La, and Thanh Hoa) in northern Vietnam (Figure 1); no data were available for 41 serum samples. Of these samples, 133 were DENV NS1–positive (49.61%), 7 were DENV IgM–positive (2.71%), and 11 were DENV IgG-positive samples (4.26%). The DENV-positivity rate was 52.47% and 20.83% for Hanoi city and the provincial cities (Hai Duong, Hoa Bihn, Phu Tho, and Thanh Hoa), respectively.

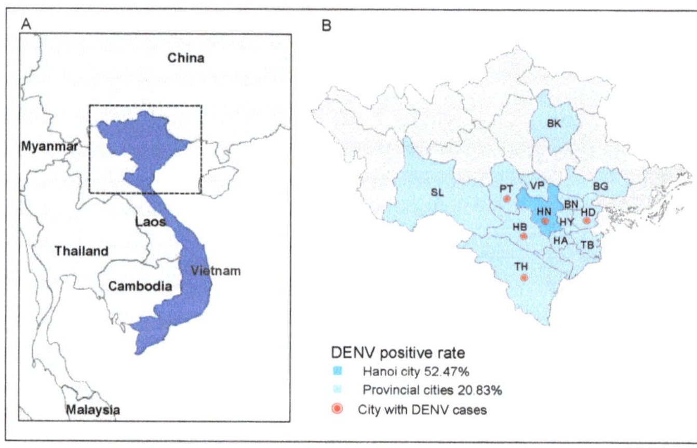

Figure 1. Map of Vietnam showing sample sites and DENV-positivity rates. (**A**) Vietnam is highlighted in blue. (**B**) Map of northern Vietnam; Hanoi (HN) is highlighted in blue. The provincial cities are highlighted in light blue and shown as follows: Hoa Bihn (HB), Phu Tho (PT), Vinh Phuc (VP), Bac Ninh (BN), Hung Yen (HY), Hai Dung (HD), Ha Nam (HA), Thai Binh (TB), Bac Giang (BG), Bac Kan (BK), Son La (SL), and Thanh Hoa (TH). Cities with DENV cases are marked on the map with a red dot.

Among the 133 DENV-positive patients, case records were available for some, as described in Table 1.

Table 1. Demographic and clinical characteristics of DENV patients.

	Total DENV (n = 133)		DENV-1 (n = 22)		DENV-2 (n = 64)		p-Value
	Frequency	Percentage	Frequency	Percentage	Frequency	Percentage	
Sex							
Male	55	41.35	11	50.00	28	43.75	0.6290 [b]
Female	51	38.35	8	36.36	22	34.38	>0.9999 [b]
No data	27	20.30	3	13.64	14	21.88	0.5413 [b]

Table 1. *Cont.*

	Total DENV (n = 133)		DENV-1 (n = 22)		DENV-2 (n = 64)		p-Value
	Frequency	Percentage	Frequency	Percentage	Frequency	Percentage	
Age							
18–29	73	54.89	14	63.64	31	48.44	0.3225 [b]
30–49	41	30.83	8	36.36	23	35.94	>0.9999 [b]
>50	13	9.77	0	0.00	8	12.50	0.1071 [b]
N/A	6	4.51	0	0.00	2	3.13	>0.9999 [b]
Symptoms [a]							
Fever	99	98.02	17	94.44	48	97.96	0.4681 [b]
Fatigue	91	88.35	17	94.44	43	87.76	0.6645 [b]
Muscle pain	55	53.40	9	50.00	21	42.86	0.7823 [b]
Joint pain	26	25.24	5	27.78	11	22.45	0.7488 [b]
Skin rash	7	6.80	2	11.11	3	6.12	0.6051 [b]
Headache	6	5.83	2	11.11	1	2.04	0.1735 [b]
Tourniquet	5	4.85	1	5.56	3	6.12	>0.9999 [b]
Abdomen pain	4	3.88	2	11.11	2	4.08	0.2909 [b]
Joint swelling	3	2.91	1	5.56	1	2.04	0.4681 [b]
Vomit	2	1.94	0	0.00	2	4.08	>0.9999 [b]
Other (gingival bleeding, ascites, and reduction in urination)	3	2.91	0	0.00	3	6.12	0.5581 [b]
Chemical [a]	Median	IQR	Median	IQR	Median	IQR	
Platelets ($10^3/\mu L$)	158	126–185	131.00	115.50–194.75	162	129–183	0.2470 [c]
White blood cells ($10^3/\mu L$)	4.6	3–5.9	3.8	2.98–5.98	4.9	3.15–6.3	0.3337 [c]
Hemoglobin (g/L)							
Male	148	141–155.5	146	140–155	147	141–153	0.9431 [c]
Female	131	125–141.25	124	111–145	131	126–134.25	0.1245 [c]
Hematocrit (%)							
Male	43.3	41.4–45.3	43	41.6–44.9	42.5	40.8–45.3	0.9557 [c]
Female	38.35	36.75–41.35	36.6	34–43.8	38.45	36.95–39.9	0.2858 [c]

[a] Data available for only 18 DENV-1 and 49 DENV-2 cases. [b] Fisher's exact test. [c] Mann–Whitney test.

Among the 88 DENV serotyped cases, 22 (25%), 64 (72.73%), and 2 (2.27%) were identified as DENV-1, DENV-2, and DENV-4, respectively. DENV-2 was the predominant serotype in 2019 and 2020. The demographic and clinical characteristics of DENV-1 and DENV-2 cases are summarized in Table 1, and there was no significant difference in demographic and clinical laboratory data between DENV-1 and DENV-2 cases. Molecular characterization was performed for 36 cases, consisting of 13 DENV-1 and 23 DENV-2 strains, as presented in Table 2. Unfortunately, the DENV-4 sequence was unobtainable due to the low viral load.

Table 2. DENV strains characterized in the present study.

Strain	Resident Area	Country	Collection Date	NS1 Ag	Anti-IgM	Anti-IgG	Serotype	Genotype	Accession Number	
									Envelope	Full Genome
DENVN19_075	ND	Vietnam	November 2019	POS	NEG	NEG	DENV-1	I	OQ832560	OQ832609
DENVN19_080	Hanoi	Vietnam	December 2019	POS	NEG	NEG	DENV-1	I	OQ832561	OQ832610
DENVN19_129	ND	Vietnam	October 2019	POS	NEG	NEG	DENV-1	I	ud	OQ832611
DENVN19_137	ND	Vietnam	October 2019	POS	NEG	NEG	DENV-1	I	OQ832562	OQ832612
DENVN20_020	Phu Tho	Vietnam	September 2020	POS	POS	NEG	DENV-1	I	OQ832563	ud

Table 2. *Cont.*

Strain	Resident Area	Country	Collection Date	NS1 Ag	Anti-IgM	Anti-IgG	Serotype	Genotype	Accession Number	
									Envelope	Full Genome
DENVN20_032	Hanoi	Vietnam	September 2020	POS	NEG	NEG	DENV-1	I	OQ832564	ud
DENVN20_059	Thanh Hoa	Vietnam	September 2020	POS	NEG	NEG	DENV-1	I	OQ832565	ud
DENVN20_066	Hanoi	Vietnam	October 2020	POS	NEG	NEG	DENV-1	I	OQ832566	ud
DENVN20_111	Hanoi	Vietnam	September 2020	POS	NEG	NEG	DENV-1	I	OQ832567	OQ832613
DENVN20_112	Hanoi	Vietnam	September 2020	POS	NEG	NEG	DENV-1	I	OQ832568	OQ832614
DENVN20_120	Hanoi	Vietnam	September 2020	POS	NEG	NEG	DENV-1	I	OQ832569	OQ832615
DENVN20_124	Hanoi	Vietnam	September 2020	POS	NEG	NEG	DENV-1	I	OQ832570	ud
DENVN20_226	Hanoi	Vietnam	October 2020	POS	NEG	NEG	DENV-1	I	OQ832571	ud
DENVN19_004	ND	Vietnam	October 2019	POS	NEG	NEG	DENV-2	Cosmopolitan	OQ832572	OQ832616
DENVN19_006	ND	Vietnam	October 2019	POS	NEG	NEG	DENV-2	Cosmopolitan	OQ832573	ud
DENVN19_010	ND	Vietnam	October 2019	POS	NEG	NEG	DENV-2	Asian-I	OQ832574	OQ832617
DENVN19_011	ND	Vietnam	October 2019	POS	NEG	NEG	DENV-2	Cosmopolitan	OQ832575	OQ832618
DENVN19_013	ND	Vietnam	October 2019	POS	NEG	NEG	DENV-2	Cosmopolitan	OQ832576	OQ832619
DENVN19_015	ND	Vietnam	October 2019	POS	NEG	NEG	DENV-2	Cosmopolitan	OQ832577	OQ832620
DENVN19_078	ND	Vietnam	December 2019	POS	NEG	NEG	DENV-2	Cosmopolitan	OQ832578	OQ832621
DENVN19_089	Hanoi	Vietnam	December 2019	POS	NEG	NEG	DENV-2	Cosmopolitan	OQ832579	OQ832622
DENVN19_140	ND	Vietnam	October 2019	POS	NEG	NEG	DENV-2	Asian-I	OQ832580	ud
DENVN19_142	ND	Vietnam	October 2019	POS	NEG	NEG	DENV-2	Cosmopolitan	OQ832581	ud
DENVN19_143	ND	Vietnam	October 2019	POS	NEG	NEG	DENV-2	Asian-I	OQ832582	OQ832623
DENVN19_144	ND	Vietnam	October 2019	POS	NEG	NEG	DENV-2	Cosmopolitan	OQ832583	ud
DENVN20_019	Hanoi	Vietnam	September 2020	POS	NEG	NEG	DENV-2	Cosmopolitan	OQ832584	ud
DENVN20_021	Hanoi	Vietnam	September 2020	POS	NEG	NEG	DENV-2	Cosmopolitan	OQ832585	ud
DENVN20_049	Hanoi	Vietnam	September 2020	POS	NEG	NEG	DENV-2	Cosmopolitan	OQ832586	ud
DENVN20_074	Hanoi	Vietnam	October 2020	POS	NEG	NEG	DENV-2	Cosmopolitan	OQ832587	ud
DENVN20_106	Hanoi	Vietnam	September 2020	POS	NEG	NEG	DENV-2	Cosmopolitan	OQ832588	ud
DENVN20_107	Hanoi	Vietnam	September 2020	POS	NEG	NEG	DENV-2	Cosmopolitan	OQ832589	OQ832624
DENVN20_113	Hanoi	Vietnam	September 2020	POS	NEG	NEG	DENV-2	Cosmopolitan	OQ832590	OQ832625
DENVN20_118	Hanoi	Vietnam	September 2020	POS	NEG	NEG	DENV-2	Asian-I	OQ832591	OQ832626
DENVN20_127	Hanoi	Vietnam	September 2020	POS	NEG	NEG	DENV-2	Asian-I	OQ832592	OQ832627
DENVN20_210	Hanoi	Vietnam	October 2020	POS	NEG	NEG	DENV-2	Cosmopolitan	OQ832593	ud
DENVN20_220	Hanoi	Vietnam	October 2020	POS	NEG	NEG	DENV-2	Cosmopolitan	OQ832594	ud

ND: no data; POS: positive; NEG: negative; ud: undetermined.

3.2. *Persistence of DENV-1 Genotype I in Vietnam*

The ML tree of the envelope gene region (1485 bp) of the newly obtained sequences was reconstructed together with sequences from GenBank, including DENV-1 genotype references (I–V) and the related sequences from BLASTN searches (Figure 2). All DENV-1 isolates belonged to genotype I and were separated into two clades. First, two DENV-1 isolates (DENVN20_120 and DENVN20_20) were the most closely related to the Vietnam strains collected in 2022 and clustered with Vietnam strains circulating in 2015–2017. Second, the remaining DENV-1 isolates formed the majority (DENVN19_129, DENVN19_137, DENVN19_075, DENVN19_080, DENVN20_124, DENVN20_032, DENVN20_059, DENVN 20_066, and DENVN20_226) of those related to the 2017 strains collected in China and Vietnam.

3.3. *Co-Circulation of the Asian-I and Cosmopolitan Genotype of DENV-2 in Vietnam*

The ML tree of the DENV-2 envelope sequences (Figure 3) consisted of the Hanoi 2019–2020 strains reported in the present study, their related sequences obtained by BLASTN searches, other Vietnam strains available in the database, and the genotype representative strains. Five distinct genotypes were separated with branch support (bootstrap 100%). The Hanoi DENV-2 isolates were classified into two genotypes. First, five isolates, including DENVN19_010, DENVN19_140, DENVN19_143, DENVN20_118, and DENVN20_127, belonged to the Asian-I genotype and clustered closely with recently reported viruses isolated from Vietnam in 2017–2022, China in 2017–2019, and Cambodia in 2019. Second, the vast majority (18 strains shown in Table 1) were characterized as the cosmopolitan genotype. Of these cosmopolitan isolates, DENVN19_015 and DENVN19_144, which were collected in 2019, clustered to a small clade with the most-related strain from Thailand collected in 2016 and 2018, whereas the remaining 16 strains (2019–2020) formed a clade together with other Vietnam strains circulating in 2018–2021 and Australia strains collected in 2019.

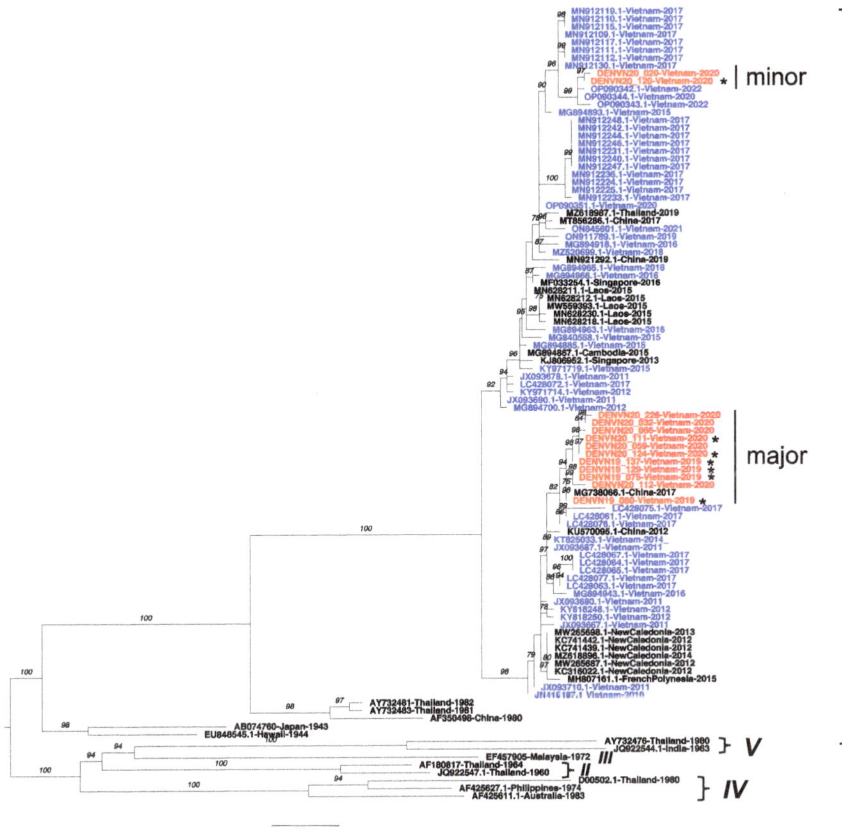

Figure 2. Maximum-likelihood tree of DENV-1 isolates from Vietnam based on the envelope gene. The best nucleotide substitution model was TIM2 + F + G4. The Vietnam strains sequenced in the present study are labeled in red. Of them, the sequences obtained from the whole-genome analysis are marked with an asterisk. Public Vietnam sequences from GenBank are labeled in blue. Genotypes of DENV-1 are indicated to the right. Numbers on branches indicate bootstrap support values (>75%).

Since we observed that the DENV-2 cosmopolitan genotype predominated in the Hanoi epidemic of 2019–2020, we collected public sequences to determine the historical distribution of DENV-2 genotypes in Vietnam from 1988–2022 (Figure 4). The results indicated that there was no circulation of the Asian-American genotype after the middle of the 2000s, but Asian-I was the major genotype over most of the 2005–2013 and 2016–2019 period. Regarding the phylogeny of DENV-2 (Figure 3), Asian-I envelope sequences sampled between 2003 and 2022 formed clusters and descended from earlier Vietnam strains, which is consistent with the larger dataset analysis summarized in Figure S2. This genetic relationship suggested that the Vietnam Asian-I lineage was localized and maintained until the present. On the other hand, the cosmopolitan genotype had multiple introductions. It was detected sporadically in 2006–2007, 2009, and 2011 (Figure 4). Notably, the cosmopolitan genotype dominated for a short time in 2014 and 2015, at 100% and 84%, respectively. The cosmopolitan genotype recently re-emerged between 2018 and 2022 and exhibited a trend of increasing proportion during this period, at 5, 17, 79, 50, and 89%. Based on our results, we observed the cosmopolitan genotype at 75% and 83% of DENV-2 samples in 2019 and 2020, respectively.

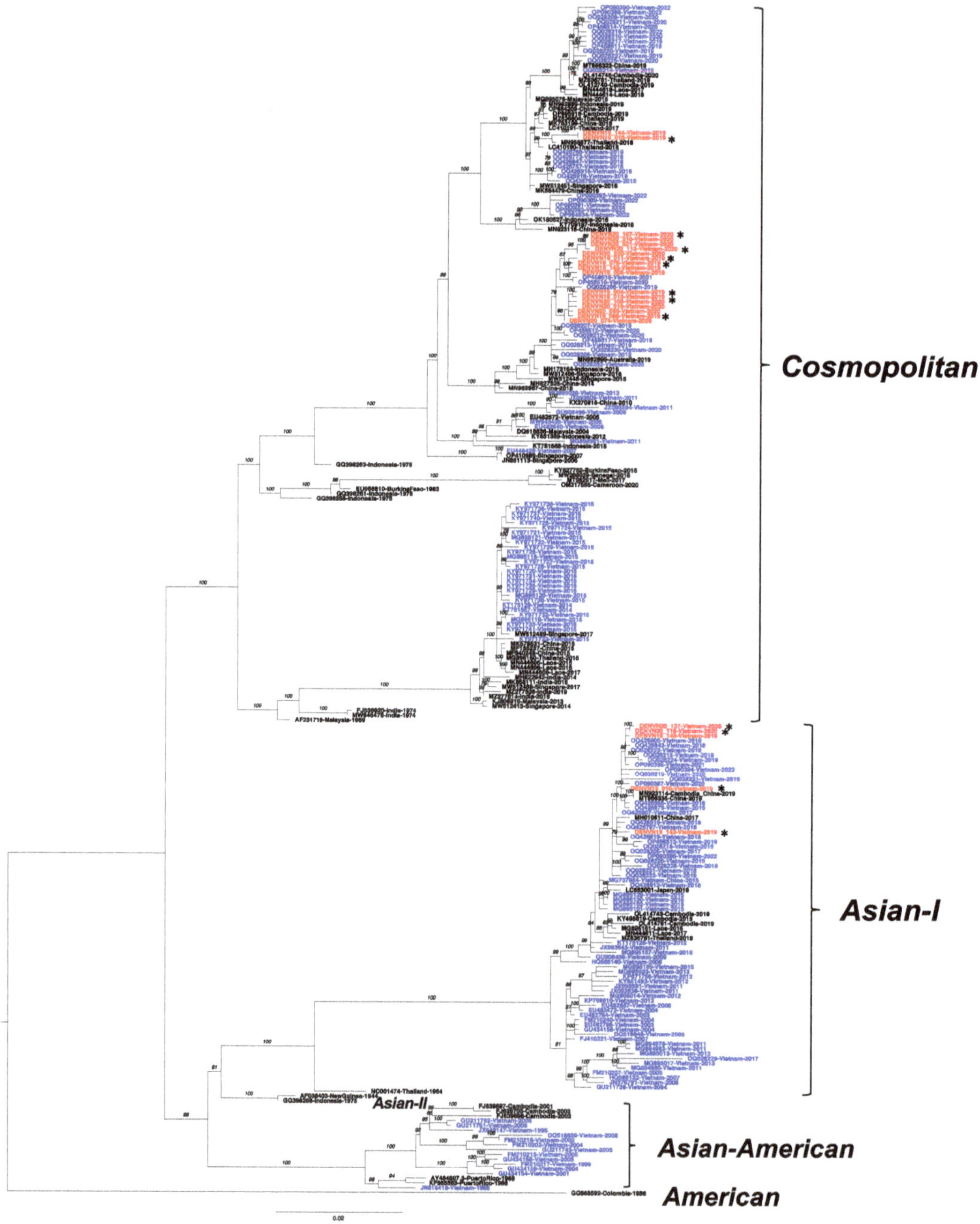

Figure 3. Maximum-likelihood tree of DENV-2 isolates from Vietnam based on the envelope gene. The best nucleotide substitution model was TIM2 + F + I + G4. The Vietnam strains sequenced in the present study are labeled in red. Of them, the sequences obtained from whole genome analysis are marked with an asterisk. Public Vietnam sequences from GenBank are labeled in blue. Genotypes of DENV-2 are indicated to the right. Numbers on branches indicate bootstrap support values (>75%).

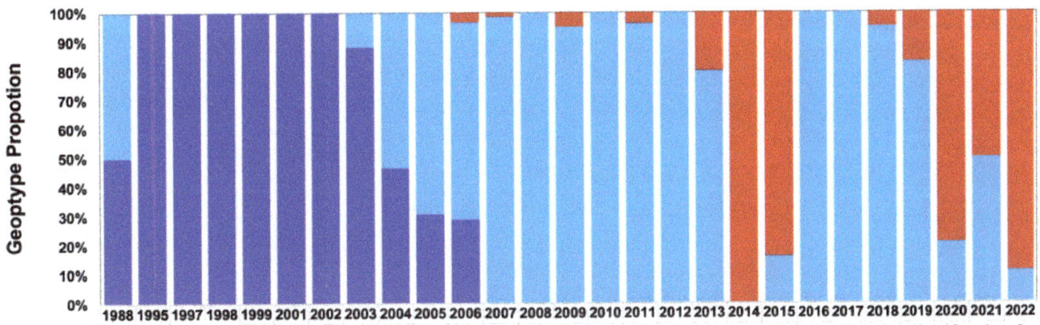

	1988	1995	1997	1998	1999	2001	2002	2003	2004	2005	2006	2007	2008	2009	2010	2011	2012	2013	2014	2015	2016	2017	2018	2019	2020	2021	2022
Cosmopolitan	0	0	0	0	0	0	0	0	0	0	3	1	0	1	0	3	0	1	2	26	0	0	7	19	19	1	8
Asian-I	1	0	0	0	0	0	0	4	16	9	61	68	29	19	14	77	24	4	0	5	1	3	133	92	5	1	1
Asian-American	1	2	9	7	1	4	7	30	14	4	26	0	0	0	0	0	0	0	0	0	0	0	0	0	0	0	0

Figure 4. Proportion of the DENV-2 genotype in Vietnam from 1988 to 2022 based on sequences available in GenBank.

3.4. Emergence of DENV-2 Cosmopolitan Lineage C (Asian-Pacific) in Vietnam

The time to the most recent common ancestor (tMRCA) of the DENV-2 genotype cosmopolitan was estimated using BEAST, and the result is summarized in an MCC tree in Figure 5. The dataset consisted of the reference strains representing the DENV-2 cosmopolitan lineages previously described [8,18], all available Vietnam and related strains from the database, and strains reported during 2015–2022. Three lineages separated with a posterior of 1, including lineage A (African), lineage B (Indian), and lineage C (Asian-Pacific). Within lineage C (Asian-Pacific), six phylogenetic clades (Clades i–vi) were observed during the 2014–2022 period. The basal strains of each clade were cosmopolitan viruses that circulated prior to mid-2010 in countries in maritime Southeast Asia, including Singapore, Indonesia, and Malaysia, which were cosmopolitan virus hotspots.

In Figure 5, the Vietnam cosmopolitan viruses formed seven clusters. Cluster 1 strains from 2006–2011 fell into lineage C from China in 2010, and the tMRCA was estimated at 2004.2 (2003.3–2005.1). Cluster 2 strains from 2014–2015 were in lineage B and related to the Indian strain from 2014. The tMRCA was 2012.3 (2011.5–2013.2). Notably, recent Vietnam strains collected during 2018–2022 classified in lineage C separated in clusters 3–7. Cluster 3 was related to the Indonesia strains from 2016, and the tMRCA was estimated at 2016.0 (2015.2–2016.8). The vast majority of our Hanoi strains from 2019–2020 were in this cluster together with the Ho Chi Minh City strain collected in 2018–2020. Furthermore, a Vietnam strain collected in 2020 (OQ028232) in Cluster 3 was most closely related to Australia 2019 strains (MN982899-MN982901). Viruses of Clusters 4, 5, and 6 were most closely related to viruses from China, Thailand, and Cambodia that circulated in 2018–2020, and the tMRCA of these viruses was 2016.7 (2015.8–2017.7), 2017.0 (2016.4–2017.6), and 2018.8 (2018.6–2019.1), respectively. Interestingly, a minor group of Hanoi isolates (DENVN19_015 and DENVN19_144) belonged to Cluster 6 and was distinctly separated from the majority of Hanoi viruses (cluster 3). The ancestral strain of these isolates was closely related to Thailand strains (MN955677 and LC410190). Five Vietnam strains collected in 2022 were in Cluster 7 in which the tMRCA was dated at 2019.8 (2018.7–2020.9). These strains were related to the Indonesia strains from 2016.

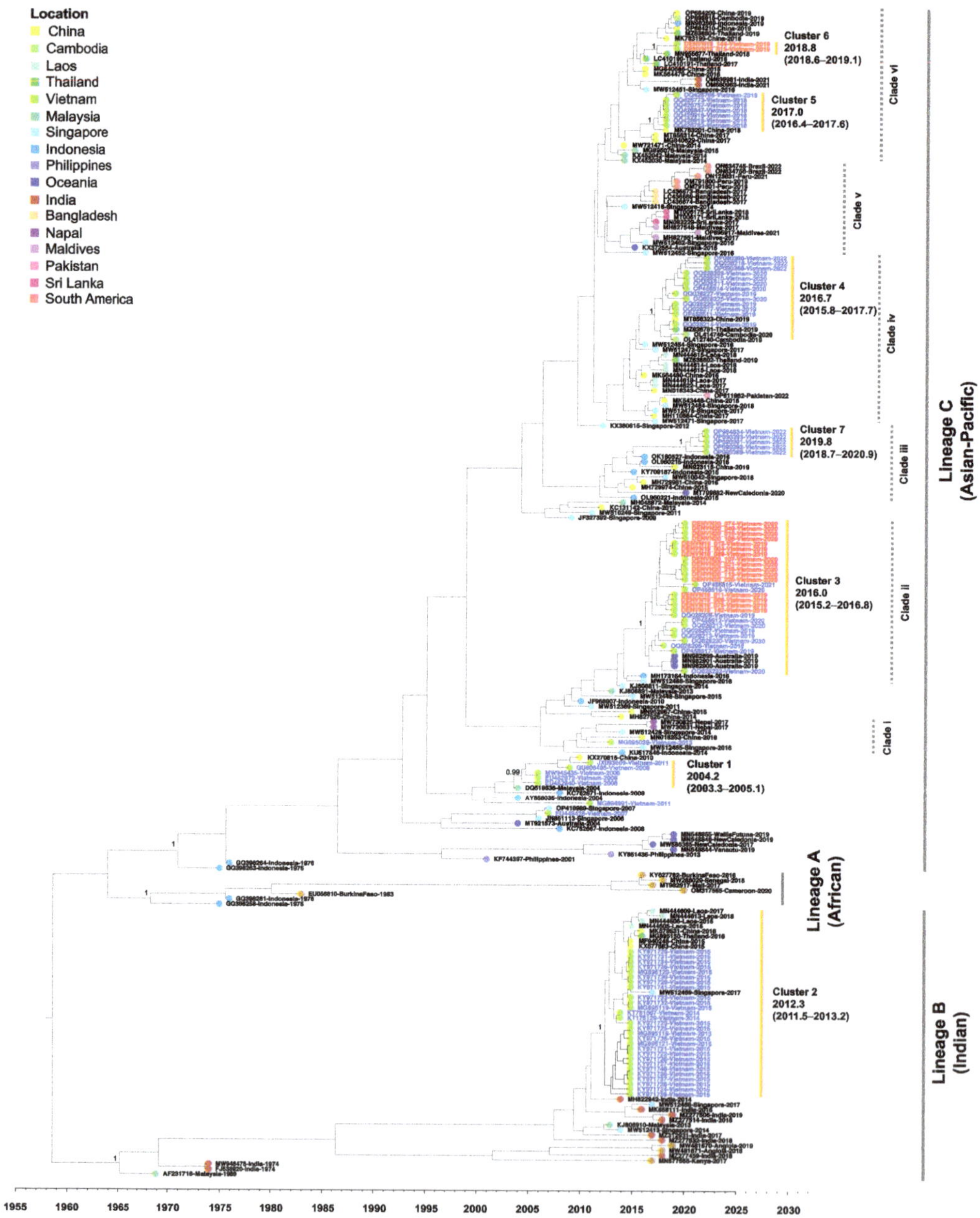

Figure 5. Maximum clade credibility tree of DENV-2 genotype cosmopolitan based on the envelope gene. The Vietnam sequences obtained in the present study and from GenBank are labeled in red and blue, respectively. The grey and dotted grey brackets indicate the lineage and clade, respectively. The yellow bracket indicates Vietnam clusters with the most recent common ancestor (tMRCA) and a 95% highest posterior density interval (HPD). The numbers of posterior probability (PP) support are shown adjacent to the key nodes. The branch tip color corresponds to the location indicated. The timescale in years is shown on the *x*-axis at the bottom.

3.5. Amino Acid Polymorphisms within Genotypes of Hanoi DENV Strains

The whole genomes of DENV isolates obtained in the present study were analyzed for amino acid polymorphisms and compared to the previous Vietnam strains (Figure 6). The envelope sequences analyzed above showed complete consistency with the corresponding region in the whole genome sequences for each sample. Among seven strains of DENV-1 genotype I, a total of 22 amino acid differences were observed. DENVN20_111 had the most newly observed mutations, with 4 amino acid differences compared to other strains. For DENV-2 genotype Asian-I strains, 22 different amino acids were observed along the entire coding sequence. The major clade strains had mutations similar to those of the Vietnam strains reported in 2017–2018. Among DENV-2 genotype cosmopolitan viruses, 35 different amino acid substitutions located in both structural and nonstructural genes were observed. The DENVN19_015 isolate, classified in Cluster 6, showed six newly observed mutations, whereas the remaining isolates (DENVN19_004, DENVN19_011, DENVN19_013, DENVN19_078, DENVN19_089, DENVN20_107, and DENVN20_113) that were grouped in Cluster 3 shared the seven specific mutations of NS1-146I, NS1-178L, NS2A-137I, NS2A-171I, NS3-31F, NS3-519V, and NS5-648E.

A. DENV-1 genotype I

Strain	Collection Year	Clade	preM		ENV				NS1			NS2A				NS3	NS4A		NS5					
			127	136	52	320	409	485	173	213	246	38	118	143	171	323	76	97	65	199	200	642	647	833
NC_001477.1	1974	-	A	I	N	V	A	V	D	A	I	A	N	H	I	V	R	M	F	K	H	V	K	E
Vietnam (n=66)	2017	-	A	I/M	N/D	V	A	V	D	A/T	I/V	A/V	N	H/Y	V/I	V/M	K/R	M/V	F/L	K/R	H/Y	V/A	E/G	G
DENVN19_129	2019	major	A	I	N	V	A	V	D	T	V	A	N	H	I	M	R	M	L	R	H	A	G	G
DENVN19_075	2019	major	A	I	N	V	A	V	D	T	V	A	N	H	I	M	R	M	L	R	H	A	G	G
DENVN19_080	2019	major	A	I	N	V	A	V	G	T	V	A	N	H	I	M	R	M	L	R	H	A	G	G
DENVN20_112	2020	major	A	I	N	V	A	V	G	T	V	A	S	H	I	M	R	M	L	R	H	A	G	G
DENVN19_137	2019	major	V	I	N	V	V	V	D	T	V	A	N	H	I	M	K	M	L	R	H	A	G	E
DENVN20_111	2020	major	V	I	N	I	V	V	D	T	V	A	N	Y	I	M	R	M	L	R	H	A	G	E
DENVN20_120	2020	minor	A	M	D	V	A	I	D	A	I	V	N	H	V	V	K	V	F	K	Y	V	E	G

B. DENV-2 genotype Asian-I

Strain	Collection Year	Clade	Capsid	ENV		NS1					NS2A				NS3		NS5							
			72	160	347	129	177	202	246	351	36	104	109	167	577	590	5	76	176	286	400	628	631	898
NC_001474.2	1964	-	I	K	V	H	V	I	N	T	V	T	V	V	V	K	I	K	T	Q	T	V	S	V
Vietnam (130)	2006–2008	-	I	K/Q/M	V	H/Y	A	I/M	N	T	V/A	T	V	V	V/M	K	I/M	K	T	Q	T	V	N/S	V
Vietnam (216)	2017–2020	-	V/I	M	A/V	Y	V/A/T	I	N/S	S/T	V	T	V	V	V	K	I	K	T	E	T	I/V	N	V
DENVN19_143	2019	minor	V	M	A	Y	T	I	N	S	V	A	A	A	V	R	I	K	I	E	I	V	N	V
DENVN19_010	2019	major	V	M	A	Y	A	I	S	S	A	T	V	V	V	K	I	R	T	E	T	I	N	V
DENVN20_118	2020	major	V	M	A	Y	A	M	S	S	V	T	V	V	M	K	I	K	T	E	T	I	S	A
DENVN20_127	2020	major	V	M	A	Y	A	M	S	S	V	T	V	V	M	K	I	K	T	E	T	I	S	V

C. DENV-2 genotype Cosmopolitan

Strain	Collection Year	Cluster	Capsid			PreM	Env			NS1						NS2A				
			14	63	104	15	173	345	484	80	96	109	146	178	251	29	119	137	171	189
GQ398263.1	1975	-	N	A	M	R	A	R	V	S	I	T	T	L	V	T	T	V	I	A
Vietnam (n=3)	2006	1	N	A	I	R	A	R	V	S	I	T	T	F	V	T	T	V	T	A
Vietnam (n=6)	2018	5	N	A	I	R	A	R	V	T	I	T	T	S	V	T	T	V	A	T
Vietnam (n=4)	2019	4, 5	N	A	I	R	A	R	V	T	I	T	T	S	V	T	T	V	A	T
Vietnam (n=3)	2020	3, 4	N	A	I	R	A	R	V/I	S/T	I	T	T/I	S/L	V	T	T	V/I	A/I	T/A
DENVN19_004	2019	3	N	A	I	R	A	R	I	S	I	T	I	L	V	T	T	I	I	A
DENVN19_011	2019	3	S	A	I	K	A	K	I	S	I	T	I	L	V	T	T	I	I	A
DENVN19_013	2019	3	N	A	I	R	A	R	I	S	I	T	I	L	V	T	N	I	I	A
DENVN19_078	2019	3	N	T	I	R	A	K	I	S	I	T	I	L	V	T	T	I	I	A
DENVN19_089	2019	3	N	A	I	R	A	R	I	S	V	T	I	L	V	M	T	I	I	A
DENVN20_107	2020	3	N	A	I	R	A	R	V	S	I	T	I	L	V	T	T	I	I	A
DENVN20_113	2020	3	N	A	I	R	A	R	V	S	I	T	I	L	A	T	T	I	I	A
DENVN19_015	2019	6	N	T	V	R	V	R	V	T	I	I	T	S	V	T	T	V	A	T

Strain	NS2B			NS3					NS4A	NS5							
	21	59	99	15	29	31	395	519	89	271	388	637	648	676	698	865	898
GQ398263.1	L	V	I	K	G	L	I	I	I	I	K	V	A	S	R	T	V
Vietnam (n=3)	L	V	I	K	G	L	I	I	I	T	K	V	A	S	R	T	V
Vietnam (n=6)	F	I	I	R	G	L	T	I	V	A	K	A	V	N	K	T	V
Vietnam (n=4)	F	I	I	R	G	L	I/T	I	V	A/V	K/E	A	V	N	R/K	T	V
Vietnam (n=3)	F/L	V/I	I	K/R	G	L/F	I	I/V	I/V	T/V	K	V/A	V/E	S/N	R	T	V
DENVN19_004	L	V	I	K	G	F	I	V	I	T	E	V	E	S	R	T	V
DENVN19_011	L	V	I	K	G	F	I	V	I	T	K	V	E	S	R	T	V
DENVN19_013	L	V	I	K	G	F	I	V	I	T	K	V	E	S	R	T	V
DENVN19_078	L	V	I	K	G	F	I	V	I	T	K	V	E	S	R	I	V
DENVN19_089	L	V	I	K	G	F	T	V	I	T	K	V	E	S	R	T	V
DENVN20_107	L	V	I	K	G	F	I	V	I	T	K	V	E	S	R	T	V
DENVN20_113	L	V	V	K	G	F	I	V	I	T	K	V	E	S	R	T	V
DENVN19_015	F	I	I	R	R	L	I	I	V	A	K	A	V	N	K	T	A

Figure 6. Amino acid polymorphisms within genotypes of Hanoi DENV strains compared to the previous Vietnam strains. The codon numbering with respect to the reference viruses of NC_001477.1, NC_001274.2, and GQ398263.1 for DENV-1 genotype I, DENV-2 genotype Asian-I, and DENV-2 genotype cosmopolitan, respectively. The variations newly observed in the present study and the amino acids specific to certain viral clusters are labeled in blue and red, respectively. The number of previous Vietnam strains is indicated in the parenthesis.

4. Discussion

In 2019, more than 300,000 dengue cases were reported in Vietnam, an approximately three-fold increase from the previous year [19]. In the present study, DENV patients in Hanoi city and nearby provinces in the northern part of Vietnam during October-December 2019 and September-October 2020 were investigated. The DENV NS1-positivity rate was higher in urban areas, with most DENV cases detected in Hanoi involving no travel history, and four cases occurred in rural areas. These positivity rates indicate that the transmission significantly occurred in the more densely populated areas [20]. Most of the DENV patients in the present study had primary infection with mild illness showing fever, fatigue, and muscle pain as the typical symptoms. A cohort study of hospitalized adult patients in northern Vietnam during 2016–2019 reported that dengue with warning signs in primary infection were found in 33% of patients in the 2017–2018 season and in 17% in the 2018–2019 season. Dengue with warning signs was found more often in secondary infection, at 40% in the 2017–2018 season and at 21% in the 2018–2019 season, while there were few cases of severe dengue [21]. DENV-1 was reported as the primary serotype responsible for the 2017 epidemic [7,22], whereas DENV-2 was the predominant serotype in the 2019–2020 season. These results suggested that the change in serotype influenced the clinical profile seen in the present study.

DENV-1 and DENV-2 were the major serotypes, and very few DENV-4 could be detected in the present study. DENV-2 was the predominant serotype both in 2019 and 2020. In the 2017 epidemic, DENV-1 was the dominant serotype in locations in northern Vietnam such as Hanoi, Ha Nam, and Hai Duong. Subsequently, DENV-1 decreased gradually from 2017 to 2019 perhaps due to the presence of serotype-specific immunity in the human population [7,21–23]. Notably, the serotype shift from DENV-1 to DENV-2 noted in our study occurred primarily in northern Vietnam in 2019–2020. However, other regions, such as central Vietnam, had a different predominant type during the period from December 2018 to February 2019, with mostly DENV-4 and a small proportion of DENV-2 [24]. Serotype replacement occurred in southern Vietnam as DENV-2 observed between 2003 and 2006 [5] switched to DENV-1 between 2006 and 2008 [20], whereas in northern Vietnam, and particularly Hanoi, the proportions of DENV-1 and DENV-2 were reportedly equal in 2008 [25]. Dengue transmission is high in Ho Chi Minh City and southern Vietnam, with the annual wave of cases typically peaking in the dry season [26]. Unlike in southern Vietnam, Hanoi has a subtropical climate with four seasons. Dengue transmission in northern Vietnam driven by DENVs from the southern region is usually interrupted by a seasonal bottleneck [27]. However, recent changes in climate have altered the situation. This has led to a greater frequency of spread in the northern region and a higher incidence of dengue, especially between June and November, when the highest temperatures of the year occur [28,29]. Thus, DENV surveillance is required routinely to monitor the new serotype such as DENV-2 in Hanoi through the winter since the serotype shift would link to an increase in the patient number.

We detected DENV-1 as a minor serotype with a single genotype as all 13 DENV-1 isolates were phylogenetically classified as genotype I. Although these viruses clustered in two separate clades, they were related and clustered with local strains that were also collected in 2017, 2019–2020, and 2022. A total of 59,063 dengue fever cases were reported in the 2017 epidemic in northern Vietnam, which was eight times the number of cases reported in 2016 that were associated with DENV-1 genotype I [7,30]. Our results showed that the 2019–2020 DENV-1 genotype I isolates descended from viruses that caused the 2017 epidemic that persisted in this region. Indeed, DENV-1 genotype I has existed in Vietnam since it was first introduced from Thailand during the late 1980s or early 1990s, and it has been imported multiple times from Cambodia in the 2000s [20,27]. DENV-1 genotype I subsequently circulated as the dominant genotype and has now descended into several local clades associated with the latter outbreaks [7].

Among DENV-2 isolates detected in the present study, the cosmopolitan genotype was predominant over the Asian-I genotype during 2019–2020. Genotype replacement has occurred in Vietnam over the last decade. Asian-I, which initially arose as the new type in 2003, subsequently replaced the Asian-American type, the previous local type, between 2003 and 2007 [5]. Since then, the Asian-I type has circulated sustainably until the present [31,32]. However, the co-circulation of Asian-I and cosmopolitan viruses was observed not only in the present study but also in a previous study involving a DENV-infected visitor who had a travel background in Vietnam; that study described the trend in genotype distribution during 2003–2016 [33]. Co-circulation was detected in the single years 2007 and 2011 and continuously between 2013 and 2015. Regarding our DENV-2 phylogenetic tree, Vietnam cosmopolitan viruses were observed in several distinct clades from 2006–2022. The Vietnam 2006–2011 virus (Cluster 1, which was part of a small cluster within lineage C (Asian-Pacific)) and the 2014–2015 virus (Cluster 2, which was within lineage B (Indian)) are both no longer detected. The Vietnam 2018–2022 cosmopolitan viruses were associated with lineage C (Asian-Pacific). In particular, the viruses of clusters 4–6 are closely related to virus strains from bordering countries, including Cambodia, Thailand, and China, that were isolated in 2018–2019 and are probable origins. In contrast, viruses of Clusters 3 and 7 had Indonesia and Singapore strains isolated in 2016 as ancestral strains, indicating that multiple, independent introductions occurred in Vietnam. In the case of Hanoi, most of the Hanoi viruses were in Cluster 3 and were very close to the Ho Chi Minh City strain, the earliest of which was detected in 2018. Moreover, Cluster 3 viruses shared specific amino acid mutations (NS1-146I, NS1-178L, NS2A-137I, NS2A-171I, NS3-31F, NS3-519V, and NS5-648E). This suggests that the virus that spread across southern and northern Vietnam and Ho Chi Minh City was likely the origin of the Hanoi cosmopolitan virus.

Global cosmopolitan isolates sampled from public databases and isolated in different geographic regions, including Asia, Oceania, Africa, and South America, also fall into lineage C (Asian-Pacific), indicating a global spread with increased DENV cases or outbreaks. Their emergence in southeast Asia, including Thailand 2016–2017, Cambodia 2019–2020, and Vietnam 2018–2022, led to co-circulation with Asian-I, the local genotype [8,34]. In areas of southern Asia, including Nepal 2017, Sri Lanka 2017–2018, Bangladesh 2017–2018, the Maldives 2017–2019, India 2021, and Pakistan 2022 [35–37], the cosmopolitan lineage C virus emerged, and this is where the lineage B (Indian) virus had previously circulated. Furthermore, the cosmopolitan lineage C virus spread for the first time to South America, with local transmission reported in Peru in 2019 [38] and Brazil in 2022 [39], both of which were mostly related to the Bangladesh 2017 virus [37]. It is possible that the cosmopolitan lineage C virus successfully adapted to humans, particularly in transmission fitness, resulting in its spread to a wider area and an increase in cases of infection. In the case of Vietnam, more than 300,000 individuals were affected in 2019. Monitoring DENV lineages could provide a better assessment of the epidemic risk as well as advice on resource allocation and guided control actions to minimize transmission intensity.

Multiple DENV clades are often observed in hyperendemic areas. Here, we reported the prevalent clades circulating in Hanoi and the amino acid polymorphisms. Regarding the diversity within DENV genotypes, differences at the nucleotide and amino acid levels were less than 6% and 3%, respectively [40]. DENV has acquired viral fitness or vector competence, and both these mechanisms are known to be associated with viral turnover events such as the persistence or replacement of clades/lineages [4]. For instance, the K160Q/M mutation in the DENV-2 genotype Asian-I viruses that emerged in Vietnam in 2008–2011 caused higher viremia in patients but increased neutralization sensitivity in DENV-2 genotype Asian-American [41]. Interestingly, the K160M mutation was still maintained and detected in the Vietnam Asian-I lineage in the present study. In addition, we have identified new mutations that were suspected to play certain roles in fitness. Their phenotypic effects are still unknown and should be investigated further. Several studies have explained that there was greater replication of the major/dominant clade virus in native mosquitos, thereby enhancing the local transmission. This has been seen in the DENV-1 genotype I clade/lineage shift in Thailand and Cambodia and the DENV-2 Asian-American dominant clade replacement in Nicaragua [42–44]. As mentioned above, the newly introduced cosmopolitan lineage C (Asian-pacific) is invading into new areas and subsequently co-circulating with the local genotype, such as Asian I, as shown in the present study. However, the mechanisms of viral evolution are still unclear. Further studies are required to explore the mechanisms or factors involving viral turnovers.

The limitations of this study include the lack of travel history of patients that could illustrate the virus transmission more clearly. Detailed demographic and clinical laboratory data were not available for some patients. The available Vietnam DENV sequences in the database are also limited in some periods. Therefore, more sequence data are critical to monitor the spread of the virus. In conclusion, our study characterized DENV strains associated with a large dengue outbreak in 2019–2020 in northern Vietnam and enhanced our understanding of the recent dynamics of DENV transmission. DENV-1 genotype I and DENV-2 genotype Asian-I are still maintained, but DENV-2 genotype cosmopolitan has re-emerged. We demonstrated that the Hanoi cosmopolitan strain was associated with lineage C (Asian-Pacific) and related to viruses from neighboring countries. In addition, identifying the dengue genotype and estimating the time the outbreak strain emerged are important for determining the origin, routes of transmission, and circulation of DENVs as well as for evaluating vaccine performance and virus control efforts.

Supplementary Materials: The following supporting information can be downloaded at https://www.mdpi.com/article/10.3390/microorganisms11051267/s1. Table S1: PCR and sequencing primers targeting the envelope gene region; Table S2: Sequences analyzed in the present study; Table S3: The model-fit comparison of the log marginal likelihood estimation (MLE) using path sampling (PS) and stepping-stone sampling (SS); Figure S1: Temporal signal analysis of regression of root-to-tip divergence against date; Figure S2: Phylogenetic tree of Vietnam strains from 1988 to 2022. The envelope sequences of 760 Vietnam DENV-2 strains from 1988 to 2022 and DENV-2 genotype reference strains were constructed for the maximum-likelihood tree under TIM2 + F + I + G4. The DENV-2 genotypes and the Bootstrap (>80%) are indicated at the adjacent branch. The tree branch is colored corresponding to genotypes. Vietnam taxa from a public database and obtained in the present study are labeled in blue and red, respectively.

Author Contributions: Conceptualization, H.T.N.N., E.E.N., and A.I.; methodology, J.P.; validation, H.T.T.V., Q.T.N., A.I., and T.S.; formal analysis, J.P.; investigation, J.P.; resources, H.T.T.V., H.T.V.N., H.T.N.N., and B.T.D.; data curation, H.T.T.V., Q.T.N., and E.E.N.; writing—original draft preparation, J.P.; writing—review and editing, H.T.T.V., A.I., H.I., T.S., and T.N.P.; visualization, J.P.; supervision, H.T.T.V., E.E.N., A.I., H.I., T.S., and T.N.P.; project administration, H.T.T.V., A.I., H.I., T.S., and T.N.P.; funding acquisition, E.E.N. and T.S. All authors have read and agreed to the published version of the manuscript.

Funding: This research was funded by the Japan Agency for Medical Research and Development under grant numbers 20wm0225010h0101 and 21wm0225010h0102. The APC was funded by the Research Institute for Microbial Diseases, Osaka University.

Institutional Review Board Statement: Ethical approval was obtained from the Ethics Committee of the National Hospital for Tropical Disease, Hanoi, Vietnam on 12 September 2019 (approval number: 09/HDDD-NDTU), the Medical Ethics Committee of Kanazawa University (approval number: 3123-1), and the institutional review board of the Research Institute for Microbial Diseases, Osaka University on 22 March 2022 (approval number: 2020-6-3).

Informed Consent Statement: Informed consent was obtained from all subjects involved in the study.

Data Availability Statement: The newly obtained sequences were deposited in GenBank with accession numbers OQ832560-OQ832594 and OQ832609-OQ832627. The viral sequences analyzed in the present study were retrieved using GenBank and BV-BRC and are listed in Table S2.

Acknowledgments: We thank all of the patients who participated in this study. We also thank the staff members of the National Hospital for Tropical Disease, Hanoi, Vietnam for their collaborative support in this study. We thank Silmi Rahmani and Koichiro Seko for their help with sample organization and sequencing. We also thank Kumi Yamamoto for her assistance and generous support.

Conflicts of Interest: The authors declare no conflict of interest. The funders had no role in the design of the study; in the collection, analyses, or interpretation of data; in the writing of the manuscript; or in the decision to publish the results.

References

1. World Health Organization. *Dengue Guidelines for Diagnosis, Treatment, Prevention and Control: New Edition;* World Health Organization: Geneva, Switzerland, 2009; p. 147.
2. World Health Organization. Dengue and Severe Dengue. Available online: https://www.who.int/news-room/fact-sheets/detail/dengue-and-severe-dengue (accessed on 3 April 2023).
3. Guzman, M.G.; Alvarez, M.; Halstead, S.B. Secondary infection as a risk factor for dengue hemorrhagic fever/dengue shock syndrome: An historical perspective and role of antibody-dependent enhancement of infection. *Arch. Virol.* **2013**, *158*, 1445–1459. [CrossRef]
4. Chen, R.; Vasilakis, N. Dengue—Quo tu et quo vadis? *Viruses* **2011**, *3*, 1562–1608. [CrossRef]
5. Vu, T.T.; Holmes, E.C.; Duong, V.; Nguyen, T.Q.; Tran, T.H.; Quail, M.; Churcher, C.; Parkhill, J.; Cardosa, J.; Farrar, J.; et al. Emergence of the Asian 1 genotype of dengue virus serotype 2 in viet nam: In vivo fitness advantage and lineage replacement in South-East Asia. *PLoS Negl. Trop. Dis.* **2010**, *4*, e757. [CrossRef]
6. Aguas, R.; Dorigatti, I.; Coudeville, L.; Luxemburger, C.; Ferguson, N.M. Cross-serotype interactions and disease outcome prediction of dengue infections in Vietnam. *Sci. Rep.* **2019**, *9*, 9395. [CrossRef] [PubMed]
7. Takemura, T.; Nguyen, C.T.; Pham, H.C.; Nguyen, T.T.; Hoang, V.M.P.; Nguyen, L.K.H.; Nabeshima, T.; Nguyen, T.T.T.; Le, T.Q.M.; Moi, M.L.; et al. The 2017 Dengue virus 1 outbreak in northern Vietnam was caused by a locally circulating virus group. *Trop. Med. Health* **2022**, *50*, 3. [CrossRef] [PubMed]
8. Phadungsombat, J.; Lin, M.Y.; Srimark, N.; Yamanaka, A.; Nakayama, E.E.; Moolasart, V.; Suttha, P.; Shioda, T.; Uttayamakul, S. Emergence of genotype Cosmopolitan of dengue virus type 2 and genotype III of dengue virus type 3 in Thailand. *PLoS ONE* **2018**, *13*, e0207220. [CrossRef]
9. Shu, P.Y.; Chang, S.F.; Kuo, Y.C.; Yueh, Y.Y.; Chien, L.J.; Sue, C.L.; Lin, T.H.; Huang, J.H. Development of group- and serotype-specific one-step SYBR green I-based real-time reverse transcription-PCR assay for dengue virus. *J. Clin. Microbiol.* **2003**, *41*, 2408–2416. [CrossRef]
10. Olson, R.D.; Assaf, R.; Brettin, T.; Conrad, N.; Cucinell, C.; Davis, J.J.; Dempsey, D.M.; Dickerman, A.; Dietrich, E.M.; Kenyon, R.W.; et al. Introducing the Bacterial and Viral Bioinformatics Resource Center (BV-BRC): A resource combining PATRIC, IRD and ViPR. *Nucleic Acids Res.* **2023**, *51*, D678–D689. [CrossRef]
11. Larsson, A. AliView: A fast and lightweight alignment viewer and editor for large datasets. *Bioinformatics* **2014**, *30*, 3276–3278. [CrossRef]
12. Trifinopoulos, J.; Nguyen, L.T.; von Haeseler, A.; Minh, B.Q. W-IQ-TREE: A fast online phylogenetic tool for maximum likelihood analysis. *Nucleic Acids Res.* **2016**, *44*, W232–W235. [CrossRef]
13. Suchard, M.A.; Lemey, P.; Baele, G.; Ayres, D.L.; Drummond, A.J.; Rambaut, A. Bayesian phylogenetic and phylodynamic data integration using BEAST 1.10. *Virus Evol.* **2018**, *4*, vey016. [CrossRef] [PubMed]
14. Rambaut, A.; Lam, T.T.; Max Carvalho, L.; Pybus, O.G. Exploring the temporal structure of heterochronous sequences using TempEst (formerly Path-O-Gen). *Virus Evol.* **2016**, *2*, vew007. [CrossRef]
15. Baele, G.; Lemey, P.; Suchard, M.A. Genealogical Working Distributions for Bayesian Model Testing with Phylogenetic Uncertainty. *Syst. Biol.* **2016**, *65*, 250–264. [CrossRef] [PubMed]
16. Gill, M.S.; Lemey, P.; Faria, N.R.; Rambaut, A.; Shapiro, B.; Suchard, M.A. Improving Bayesian population dynamics inference: A coalescent-based model for multiple loci. *Mol. Biol. Evol.* **2013**, *30*, 713–724. [CrossRef]
17. Rambaut, A.; Drummond, A.J.; Xie, D.; Baele, G.; Suchard, M.A. Posterior summarisation in Bayesian phylogenetics using Tracer 1.7. *Syst. Biol.* **2018**, *67*, 901–904. [CrossRef]
18. Fourie, T.; El Bara, A.; Dubot-Peres, A.; Grard, G.; Briolant, S.; Basco, L.K.; Ouldabdallahi Moukah, M.; Leparc-Goffart, I. Emergence of dengue virus serotype 2 in Mauritania and molecular characterization of its circulation in West Africa. *PLoS Negl. Trop. Dis.* **2021**, *15*, e0009829. [CrossRef] [PubMed]
19. World Health Organization. Regional Office for the Western, P. Dengue Situation Updates 2020. Available online: https://iris.wpro.who.int/handle/10665.10661/14461 (accessed on 3 April 2023).
20. Raghwani, J.; Rambaut, A.; Holmes, E.C.; Hang, V.T.; Hien, T.T.; Farrar, J.; Wills, B.; Lennon, N.J.; Birren, B.W.; Henn, M.R.; et al. Endemic dengue associated with the co-circulation of multiple viral lineages and localized density-dependent transmission. *PLoS Pathog.* **2011**, *7*, e1002064. [CrossRef] [PubMed]
21. Lytton, S.D.; Nematollahi, G.; van Tong, H.; Xuan Anh, C.; Hung, H.V.; Hoan, N.X.; Diez, G.; Schumacher, T.; Landt, O.; Melchior, W.; et al. Predominant secondary dengue infection among Vietnamese adults mostly without warning signs and severe disease. *Int. J. Infect. Dis.* **2020**, *100*, 316–323. [CrossRef]
22. Dang, T.T.; Pham, M.H.; Bui, H.V.; Van Le, D. Whole genome sequencing and genetic variations in several dengue virus type 1 strains from unusual dengue epidemic of 2017 in Vietnam. *Virol. J.* **2020**, *17*, 7. [CrossRef]
23. Gibbons, R.V.; Kalanarooj, S.; Jarman, R.G.; Nisalak, A.; Vaughn, D.W.; Endy, T.P.; Mammen, M.P., Jr.; Srikiatkhachorn, A. Analysis of repeat hospital admissions for dengue to estimate the frequency of third or fourth dengue infections resulting in admissions and dengue hemorrhagic fever, and serotype sequences. *Am. J. Trop. Med. Hyg.* **2007**, *77*, 910–913. [CrossRef]
24. Phan, D.Q.; Nguyen, L.D.N.; Pham, S.T.; Nguyen, T.; Pham, P.T.T.; Nguyen, S.T.H.; Pham, D.T.; Pham, H.T.; Tran, D.K.; Le, S.H.; et al. The Distribution of Dengue Virus Serotype in Quang Nam Province (Vietnam) during the Outbreak in 2018. *Int. J. Environ. Res. Public Health* **2022**, *19*, 1285. [CrossRef] [PubMed]

25. Fox, A.; Le, N.M.; Simmons, C.P.; Wolbers, M.; Wertheim, H.F.; Pham, T.K.; Tran, T.H.; Trinh, T.M.; Nguyen, T.L.; Nguyen, V.T.; et al. Immunological and viral determinants of dengue severity in hospitalized adults in Ha Noi, Viet Nam. *PLoS Negl. Trop. Dis.* **2011**, *5*, e967. [CrossRef] [PubMed]
26. Cuong, H.Q.; Vu, N.T.; Cazelles, B.; Boni, M.F.; Thai, K.T.; Rabaa, M.A.; Quang, L.C.; Simmons, C.P.; Huu, T.N.; Anders, K.L. Spatiotemporal dynamics of dengue epidemics, southern Vietnam. *Emerg. Infect. Dis.* **2013**, *19*, 945–953. [CrossRef] [PubMed]
27. Rabaa, M.A.; Simmons, C.P.; Fox, A.; Le, M.Q.; Nguyen, T.T.; Le, H.Y.; Gibbons, R.V.; Nguyen, X.T.; Holmes, E.C.; Aaskov, J.G. Dengue virus in sub-tropical northern and central Viet Nam: Population immunity and climate shape patterns of viral invasion and maintenance. *PLoS Negl. Trop. Dis.* **2013**, *7*, e2581. [CrossRef]
28. Cuong, H.Q.; Hien, N.T.; Duong, T.N.; Phong, T.V.; Cam, N.N.; Farrar, J.; Nam, V.S.; Thai, K.T.; Horby, P. Quantifying the emergence of dengue in Hanoi, Vietnam: 1998–2009. *PLoS Negl. Trop. Dis.* **2011**, *5*, e1322. [CrossRef]
29. Thanh Toan, D.T.; Hu, W.; Quang Thai, P.; Hoat, L.N.; Wright, P.; Martens, P. Hot spot detection and spatio-temporal dispersion of dengue fever in Hanoi, Vietnam. *Glob. Health Action* **2013**, *6*, 18632. [CrossRef]
30. Nguyen, H.V.; Than, P.Q.T.; Nguyen, T.H.; Vu, G.T.; Hoang, C.L.; Tran, T.T.; Truong, N.T.; Nguyen, S.H.; Do, H.P.; Ha, G.H.; et al. Knowledge, Attitude and Practice about Dengue Fever among Patients Experiencing the 2017 Outbreak in Vietnam. *Int. J. Environ. Res. Public Health* **2019**, *16*, 976. [CrossRef]
31. Quyen, D.L.; Thanh Le, N.; Van Anh, C.T.; Nguyen, N.B.; Hoang, D.V.; Montgomery, J.L.; Kutcher, S.C.; Hoang Le, N.; Hien, N.T.; Hue Kien, D.T.; et al. Epidemiological, Serological, and Virological Features of Dengue in Nha Trang City, Vietnam. *Am. J. Trop. Med. Hyg.* **2018**, *98*, 402–409. [CrossRef]
32. Van Tuan, L.; Van, N.T.T.; Quan, N.H.; Duoc, P.T. Phylogeny of Dengue virus type 2 isolated in the Central Highlands, Vietnam. *Rev. Biol. Trop.* **2017**, *65*, 819–826.
33. Yang, C.F.; Chang, S.F.; Hsu, T.C.; Su, C.L.; Wang, T.C.; Lin, S.H.; Yang, S.L.; Lin, C.C.; Shu, P.Y. Molecular characterization and phylogenetic analysis of dengue viruses imported into Taiwan during 2011–2016. *PLoS Negl. Trop. Dis.* **2018**, *12*, e0006773. [CrossRef]
34. Brook, C.E.; Li, Y.; Yek, C.; Northrup, G.R.; Lay, S.; Chea, S.; Ahyong, V.; Parker, D.M.; Man, S.; Pacheco, A.R.; et al. The Perfect Storm of 2019: An immunological and phylodynamic analysis of Cambodia's unprecedented dengue outbreak. *medRxiv* **2022**. [CrossRef]
35. Ngwe Tun, M.M.; Muthugala, R.; Nabeshima, T.; Rajamanthri, L.; Jayawardana, D.; Attanayake, S.; Soe, A.M.; Dumre, S.P.; Ando, T.; Hayasaka, D.; et al. Unusual, neurological and severe dengue manifestations during the outbreak in Sri Lanka, 2017. *J. Clin. Virol.* **2020**, *125*, 104304. [CrossRef] [PubMed]
36. Ngwe Tun, M.M.; Pandey, K.; Nabeshima, T.; Kyaw, A.K.; Adhikari, M.; Raini, S.K.; Inoue, S.; Dumre, S.P.; Pandey, B.D.; Morita, K. An Outbreak of Dengue Virus Serotype 2 Cosmopolitan Genotype in Nepal, 2017. *Viruses* **2021**, *13*, 1444. [CrossRef] [PubMed]
37. Suzuki, K.; Phadungsombat, J.; Nakayama, E.E.; Saito, A.; Egawa, A.; Sato, T.; Rahim, R.; Hasan, A.; Lin, M.Y.; Takasaki, T.; et al. Genotype replacement of dengue virus type 3 and clade replacement of dengue virus type 2 genotype Cosmopolitan in Dhaka, Bangladesh in 2017. *Infect. Genet. Evol.* **2019**, *75*, 103977. [CrossRef] [PubMed]
38. Garcia, M.P.; Padilla, C.; Figueroa, D.; Manrique, C.; Cabezas, C. Emergence of the Cosmopolitan genotype of dengue virus serotype 2 (DENV2) in Madre de Dios, Peru, 2019. *Rev. Peru. Med. Exp. Salud Publica.* **2022**, *39*, 126–128. [CrossRef] [PubMed]
39. Giovanetti, M.; Pereira, L.A.; Santiago, G.A.; Fonseca, V.; Mendoza, M.P.G.; de Oliveira, C.; de Moraes, L.; Xavier, J.; Tosta, S.; Fristch, H.; et al. Emergence of Dengue Virus Serotype 2 Cosmopolitan Genotype, Brazil. *Emerg. Infect. Dis.* **2022**, *28*, 1725–1727. [CrossRef]
40. Pollett, S.; Melendrez, M.C.; Maljkovic Berry, I.; Duchene, S.; Salje, H.; Cummings, D.A.T.; Jarman, R.G. Understanding dengue virus evolution to support epidemic surveillance and counter-measure development. *Infect. Genet. Evol.* **2018**, *62*, 279–295. [CrossRef]
41. Wang, C.; Katzelnick, L.C.; Montoya, M.; Hue, K.D.; Simmons, C.P.; Harris, E. Evolutionarily Successful Asian 1 Dengue Virus 2 Lineages Contain One Substitution in Envelope That Increases Sensitivity to Polyclonal Antibody Neutralization. *J. Infect. Dis.* **2016**, *213*, 975–984. [CrossRef]
42. Lambrechts, L.; Fansiri, T.; Pongsiri, A.; Thaisomboonsuk, B.; Klungthong, C.; Richardson, J.H.; Ponlawat, A.; Jarman, R.G.; Scott, T.W. Dengue-1 virus clade replacement in Thailand associated with enhanced mosquito transmission. *J. Virol.* **2012**, *86*, 1853–1861. [CrossRef]
43. O'Connor, O.; Ou, T.P.; Aubry, F.; Dabo, S.; Russet, S.; Girault, D.; In, S.; Minier, M.; Lequime, S.; Hoem, T.; et al. Potential role of vector-mediated natural selection in dengue virus genotype/lineage replacements in two epidemiologically contrasted settings. *Emerg. Microbes. Infect.* **2021**, *10*, 1346–1357. [CrossRef]
44. Quiner, C.A.; Parameswaran, P.; Ciota, A.T.; Ehrbar, D.J.; Dodson, B.L.; Schlesinger, S.; Kramer, L.D.; Harris, E. Increased replicative fitness of a dengue virus 2 clade in native mosquitoes: Potential contribution to a clade replacement event in Nicaragua. *J. Virol.* **2014**, *88*, 13125–13134. [CrossRef] [PubMed]

microorganisms

MDPI

Article

Comparative Evaluation of the Activity of Various Lentiviral Vectors Containing Three Anti-HIV Genes

Olga Vladimirovna Orlova *, Dina Viktorovna Glazkova, Ramil Rafailovich Mintaev, Galina Mikhailovna Tsyganova, Felix Anatolevich Urusov, German Alexandrovich Shipulin and Elena Vladimirovna Bogoslovskaya

Federal State Budgetary Institution "Centre for Strategic Planning and Management of Biomedical Health Risks" of the Federal Medical Biological Agency, 119121 Moscow, Russia

* Correspondence: orlovaov@gmail.com

Abstract: A promising direction in the treatment of HIV infection is a gene therapy approach based on the insertion of antiviral genes aimed at inhibiting HIV replication into the genome of host cells. We obtained six constructs of lentiviral vectors with different arrangements of three antiviral genes: microRNAs against the CCR5 gene, the gene encoding the C-peptide, and the gene encoding the modified human TRIM5a protein. We found that despite containing the same genes, these vectors were produced at different titers and had different effects on cell viability, transduction efficiency, and expression stability. Comparative evaluation of the antiviral activity of three of the six developed vectors that showed stable expression was carried out using the continuous SupT1 lymphocytic cell line. All of the vectors protected cells from HIV infection: the viral load was several orders of magnitude lower than in control cells, and with one vector, complete cessation of virus growth in modified cells was achieved.

Keywords: lentiviral vectors; TRIM5a; human immunodeficiency virus type 1 (HIV-1); C-peptide; miRNA

Citation: Orlova, O.V.; Glazkova, D.V.; Mintaev, R.R.; Tsyganova, G.M.; Urusov, F.A.; Shipulin, G.A.; Bogoslovskaya, E.V. Comparative Evaluation of the Activity of Various Lentiviral Vectors Containing Three Anti-HIV Genes. *Microorganisms* **2023**, *11*, 1053. https://doi.org/10.3390/microorganisms11041053

Academic Editors: Shengxi Chen and Fabio Zicker

Received: 16 March 2023
Revised: 14 April 2023
Accepted: 16 April 2023
Published: 18 April 2023

1. Introduction

One of the promising strategies in the development of drugs for the treatment of HIV infection is a gene therapy approach based on the insertion of genes that protect against viral infection into the genome of host cells [1,2]. The target cells are usually CD4+ T lymphocytes, the main population of cells affected by HIV, or their precursors, i.e., hematopoietic stem cells (HSCs). Currently, lentiviral vectors (LVs) are considered the most effective means of gene delivery, ensuring the integration of genes into the cellular genome. Integration is important when working with proliferating immune cells because it guarantees the transfer of genes to daughter cells and their protection from the virus [3–5].

To date, many genes with anti-HIV activity have already been proposed, some of which have shown high efficiency in preclinical studies and have even been evaluated in clinical studies [6]. Of particular interest are genes with products that act at the early stages of the HIV life cycle and prevent the integration of the virus into the cell genome; such genes are considered potentially highly effective [7]. Among them, genes aimed at turning off the CCR5 chemokine receptor (which is necessary for penetration of the R5-tropic virus), the C-peptide (which inhibits the fusion of HIV with a cell), and the modified human TRIM5a protein are notable.

Three cases of curing HIV infection associated with bone marrow transplantation from donors carrying a non-functional CCR5Δ32 gene [8] evidence that transplantation of immune cells with the disabled CCR5 gene is safe and can be used for HIV treatment. Many options have been proposed to knockdown the CCR5 receptor gene in patient cells [8], including the use of RNA interference. For example, we previously described a tandem miRNA that can effectively enhance expression of the CCR5 gene [9] and inhibit HIV replication in T lymphocytes containing such a tandem [10].

C-peptides are 36–46 aa peptides with a structure similar to that of the C-terminal helical domain of the HIV-1 gp41 protein. C-peptide interacts with the HIV Env protein, disrupting its structure and preventing the virus from merging with a cell. One of the promising C-peptides is the V2O peptide with reduced immunogenicity and high activity against a wide range of HIV strains [11]. A membrane-anchored form of C-peptide ensures its expression on the cell surface and is shown to be the most effective for the gene therapy of HIV infection [7,11].

The TRIM5a protein is an intracellular factor involved in protection against retroviruses. The spectrum of retroviruses neutralized by TRIM5a is specific for different mammalian species. Rhesus monkey TRIM5a completely protects cells from HIV infection, in contrast to human TRIM5a [12]. It has been shown that the TRIM5a-HRH chimeric protein, in which 11 amino acids of human TRIM5a are replaced by 13 amino acids from TRIM5a of rhesus monkey, is able to effectively inhibit HIV replication [12–14]. Importantly, this protein acts before the retrovirus enters the nucleus and prevents the insertion of proviral DNA into the genome.

It is advisable to use several antiviral genes simultaneously, as suppression of the virus at different stages of its life cycle increases effectiveness and prevents formation of gene-resistant variants of the virus. In the present work, six variants of a lentiviral vector that carried three genes, including a tandem of artificial microRNAs against the CCR5 gene, the TRIM5a-HRH gene, and the gene encoding the C-peptide, were generated and characterized, and their antiviral activity was studied in vitro.

2. Materials & Methods

2.1. Cell Lines Used

Human embryonic kidney cells HEK293FT (ThermoFisher, Waltham, MA, USA) and human SupT1 T lymphoblasts (ATCC CRL-1942) were used in this work.

HEK293FT human embryonic kidney cells were cultured in standard DMEM containing 10% fetal bovine serum (FBS) (HyClone, Logan, UT, USA); 4 mM L-glutamine (Gibco, Franklin Lakes, NJ, USA); 1 mM sodium pyruvate (Gibco, USA); 0.1 mM essential amino acids MEM NEAA 100× (Gibco, USA); and streptomycin and penicillin at concentrations of 100 μg/mL and 100 units/mL, respectively (Gibco, USA). Suspension cells of the SupT1 lymphoblastoid line were cultured in Advanced RPMI medium (Gibco, USA) with the addition of 2% FBS. All cultures were incubated at a constant temperature of 37 °C and a CO_2 content of 5% in air.

2.2. Plasmid Constructs

Sequences encoding the TRIM5a-HRH gene, miRNAs, pPGK and Ef1a promoters, and the EGFP reporter gene were amplified, using previously obtained plasmids as templates. For the TRIM5a-HRH, GFP, pPGK, and Ef1a genes, we used the pT plasmid [14]; for the mic13lg + mic1002 gene, we used the plasmid mic13lg + mic1002-Puro [10]. To generate a sequence of the V2O gene coding membrane-anchored peptide, a transmembrane domain and a secretion signal sequence were added to the codon-optimized sequence of the V2O-peptide described in the article by L. Egerer et al. [11]. For codon optimization, the OPTIMIZER web service (http://genomes.urv.es/OPTIMIZER, accessed on 1 March 2021) and a codon frequency table were used (https://www.kazusa.or.jp/codon/cgi-bin/showcodon.cgi?species=9606, accessed on 1 March 2021). The V2O gene was assembled by overlap extension PCR, resulting in the pV2O plasmid.

The primers and fragments used to generate LV N1–LV N6 plasmids are shown in Table S1 (Supplementary).

All genetic engineering manipulations were performed according to standard protocols [15]. Isolation of plasmid DNA for transfection of HEK293FT cells was performed using the Plasmid Maxi Kit (QIAGEN) under sterile conditions.

2.3. LV Particles Production

The second-generation packaging system [14] consisting of the two packaging plasmids pCMV-dR8.91 (encoding the HIV proteins GAG, POL, TAT, and REV) and pCMV-VSV-G (encoding the envelope glycoprotein of the vesicular stomatitis virus VSV-G) was used to produce LV particles [14].

The day before transfection, HEK293FT cells were seeded at a density of 0.7×10^5 cells/cm^2 into 2 Hyperflask (Corning, New York, NY, USA). The ratio between the plasmid DNA of the vector and the DNA of the packaging plasmids pCMV-dR8.91 and pCMV-VSV-G was 6:6:1. A solution of linear polyethyleneimine PEI MAX 40000 (Polysciences, Niles, IL, USA) was used for transfection. DNA samples (238.3 μg) in 17,200 μL of 150 mM NaCl were mixed with PEI working solution, which was prepared by diluting 953 μL of PEI stock solution (1 g/L) to 17,200 μL with 150 mM NaCl solution. The resulting mixture was transfected into each Hyperflask. The day after transfection, the medium was replaced with serum-free OptiMEM (Gibco, USA). The virus suspension was collected at 72 h after transfection and centrifuged at 3500 rpm for 5 min using a CM-6MT centrifuge (Elmi, Riga, Latvia). The supernatant was filtered through a 0.45 μm PES filter (Corning, USA) and incubated with 25 U/mL benzonase (Merc, Hackensack, NJ, USA) at 37 °C for 1 h.

2.4. Tangential Flow Filtration and Ultracentrifugation

The concentration in tangential flow was performed with a KrosFlo KR2i device (Repligen, Waltham, MA, USA) and a column with a PES filter and pore diameter of 500 kDa preliminarily equilibrated with phosphate-buffered saline (PBS) and 1% fetal bovine serum (FBS). Then, it was dialyzed against phosphate-buffered saline (PBS) with 1% FBS.

The concentrated viral suspension was ultracentrifuged (UC) through a cushion of 2 mL of 20% sucrose solution for 2 h at a speed of 25,000 rpm at a temperature of +4 °C using an Optima XPN-100 ultracentrifuge ("Beckman Coulter", Brea, CA, USA). After UC, the virus pellets were resuspended in 600 μL of RPMI growth medium.

2.5. Determination of Infectious Titer

The infectious titer of the LV particles (transduction units/mL, or TU/mL) was determined using the SupT1 cell line. The cells were transduced with a series of dilutions of freshly thawed lentiviral particles in growth medium supplemented with 2 μg/mL of polybrene. The next day, the growth medium was replaced with fresh medium. The titer was determined 48 h after transduction by the percentage of the fluorescent EGFP+ cells in the population, using a Novo-Cyte Quanteon flow cytometer (Agilent Technologies, Santa Clara, CA, USA).

2.6. Transduction of Cells by LV Particles

SupT1 cells were plated in a 24-well plate at a concentration of 2×10^5 cells/mL in 0.5 mL per well. Then, viral particles were added in the amount necessary to obtain the desired multiplicity of infection (MOI). The cells were incubated overnight in the presence of polybrene (2 μg/mL) at 37 °C in a CO_2 incubator. The next day, the growth medium was replaced with fresh medium.

2.7. Determination of Cell Viability

The number of viable and dead cells in suspensions was determined by staining with trypan blue (Sigma, St. Louis, MO, USA). An equal volume of the cell suspension was added to 25 μL of PBS with 0.4% trypan blue and mixed. After 2 min, the number of blue cells (dead cells) and transparent cells was counted using a Luna II cell counter (Logos Biosystems, Anyang, Republic of Korea). The percentage of live and dead cells was also calculated by a cell counter.

2.8. Measurement of the Percentage of Transduced Cells by Flow Cytometry

Before flow cytometry, cells were stained with a LIVE/DEAD™ Fixable Violet Dead Cell Stain Kit (Invitrogen, Waltham, MA, USA), precipitated, and diluted in PBS to a density of 1×10^6 cells/mL at a rate of 1 µL of staining solution per 1 mL of cell suspension. Staining was carried out for 30 min at +4 °C. The percentage of EGFP+ fluorescent cells in the living cell population and the mean fluorescence intensity (FI) were determined using a NovoCyte Quanteon flow cytometer (Agilent Technologies, USA) in the FITC channel. The measurement results were processed using NovoExpress 1.4.1 software (Agilent Technologies, USA).

2.9. Measurement of Vector Copy Number and Quantification of V2O and TRIM5a-HRH Mrna Abundance in Supt1 Cells

SupT1 cells transduced with various lentiviral vectors were cultured for 3 weeks to remove residual plasmid DNA introduced into the culture along with the lentiviral particles. Then, 10^5 cells of each sample were collected and treated with Benzonase (Merck, Darmstadt, Germany). Nucleic acid extraction was carried out by precipitation with isopropanol, using the AmpliTest RIBO-prep reagent kit (Centre for Strategic Planning of FMBA, Moscow, Russia) according to the instructions. The integrated vector copy was determined by TaqMan quantitative PCR (qPCR) with the primers ccttgtataaatcctggttgctgtct and ggaaaggagctgacaggtggt and the probe R6G-tcaggcaacgtggcgtggtgtg-BHQ2 that target the wpre element. The human β-globin gene copy number was detected using the primers gtcagggcagagcctctattgct and ccacatgcccagtttctattggtct and the probe Fam-tgcccaggggcctcaccacca-BHQ1. The vector copy number was normalized to the human β-globin gene copy number, obtaining the average number of vector copies per cell.

To quantify the V2O and TRIM5a-HRH mRNA abundance, one-step reverse transcription qPCR (RT-qPCR) was performed using in-house reagents and the following primers and probes: for TRIM—primers gacaagtgagctctccgaaacca and gagagcccaggatgccagtaca, probe R6G—accctgttcaccttccccagcctgac-BHQ1; for V2O—primers gctgattgcgctggtgacca and cggttccagttccaggcgt, probe R6G—tgctggcggtgctgggcattaccg-BHQ1. The primers specified for TRIM detected only the transgenic TRIM5a-HRH gene. The human β-glucuronidase (GUS) gene was used as a control gene for the normalization of gene-expression levels (primers catttggaattttgccgatttcat and gtttcattggcaatcttccagtatct and probe Fam-cggcagagacaaccaaaaagtgcag-BHQ1). Before RT-qPCR, the samples were treated with DNase (DNase-1, Invitrogen, USA).

2.10. Determination of the HIV p24 Antigen

The amount of p24 antigen in samples was measured using the HIV-1 p24-antigen-ELISA-BEST kit (Vector BEST, Novosibirsk, Russia). To quantify the concentration of p24 antigen in the samples, standard samples (calibrators) with known p24 antigen concentrations were analyzed in each run. Calibrator samples were obtained by diluting recombinant HIV-1-p24 capsid protein (HIV-1-p24 Antigen AG 6054, Aalto Bio Reagent, Ireland).

2.11. HIV-1 Production and Cell Infection

Plasmid pNL4-3, containing the full-length genome of the NL4-3 HIV strain, was obtained from the United States under the NIH AIDS Research and Reference Reagent Program. To obtain the virus, HEK 293FT cells were transfected with the corresponding plasmid. After 2 days, the culture liquid containing infectious viral particles was collected, and the concentration of the p24 antigen was determined.

First, 25 µL (0.35 ng) of the NL4-3 virus was added to 1 mL of SupT1 cell suspension in Advanced RPMI + 2% FBS medium at a density of 10^6 cells/mL. Then, the cells were incubated at 37 °C for 2 h, with stirring every 30 min; after 2 h, the cells were washed twice with fresh Advanced RPMI medium without serum, and Advanced RPMI + 2% FBS was added to a density of 10^6 cells/mL. On the 2nd, 4th, 6th, 10th, 13th, 17th, 20th, and

23rd days, the cells were counted and subcultured to a density of 0.5×10^6 cells/mL. On the same days, samples of culture media were taken for p24 antigen concentration analysis.

2.12. Statistical Data Processing

Mean values and the standard deviation (SD) were calculated using Microsoft Excel 2016 software (Microsoft Corporation, Redmond, DC, USA).

3. Results

3.1. Design of Lentiviral Vectors Containing Three Anti-HIV Genes

The following anti-HIV genes were used: the mic131g-mic1002 gene encoding a tandem of two artificial microRNAs against CCR5; the TRIM5α-HRH gene encoding the modified human TRIM5α protein; and the V2O gene encoding a C-peptide, a fusion inhibitor. As gene expression can be affected by the arrangement of genes in a vector, as well as the elements regulating their expression, six different LVs were produced, each containing genes and regulatory elements in a different order (Figure 1). The genes were divided into two reading frames, each of which was under the control of its own constitutive promoter. The first frame was under the control of the elongation factor-1 alpha (EF1α) gene promoter; the second was under the control of the phosphoglycerate kinase (pPGK) gene promoter. One of the promoters regulated two proteins separated by a short 2a peptide (T2A derived from *Thosea asigna* virus), and this configuration allowed for a reduction in the size of the vector. In addition to antiviral genes, each construct included a marker gene encoding the green fluorescent protein GFP. In all constructs, the mic13lg + mic1002 miRNA tandem was located immediately after the EF1α promoter, before the open reading frame encoding the V2O or TRIM5α-HRH gene (hereinafter referred to as TRIM).

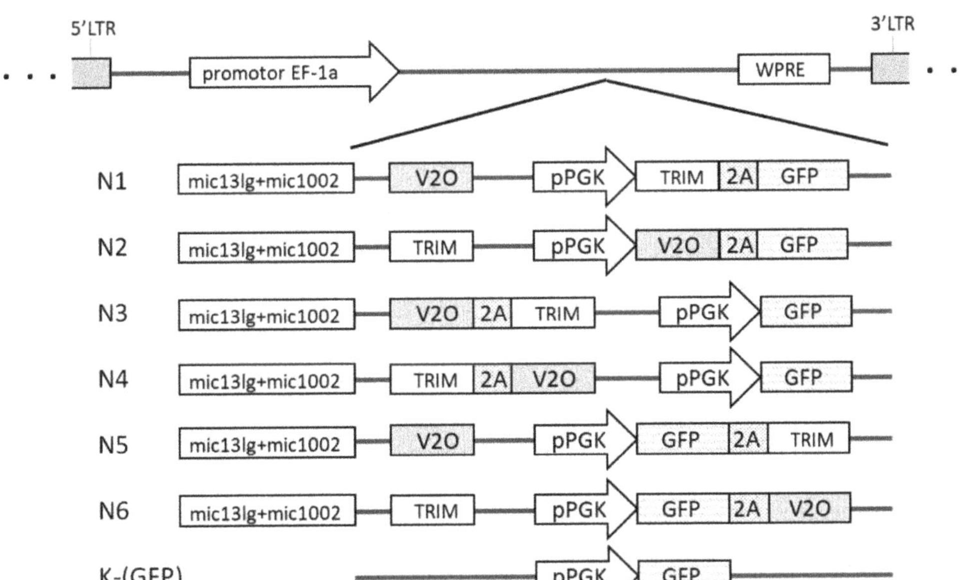

Figure 1. Scheme of the lentiviral vectors. Designations: 5′LTR and 3′LTR—5′ and 3′ long terminal repeats of HIV (long terminal repeats); WPRE—woodchuck hepatitis virus posttranscriptional regulatory element, EF-1a—elongation factor-1 alpha gene promoter, pPGK—phosphoglycerate kinase gene promoter, 2A—protein chain self-separation peptide.

3.2. Efficiency of Lentiviral Vector Production

For all described vectors, plasmid constructs were obtained and used to produce LV particles, together with the control vector K-(GFP). The titer of the obtained particles was evaluated before and after their concentration and purification.

According to the results presented in Figure 2a, the LV titers differed significantly, ranging from 2.5×10^4 to 1.2×10^6 infectious particles per ml, and were 2.5–120 times lower than the titer of the control GFP vector. The titers of the N3 and N4 vectors were the highest; for the N1, N5, and N6 vectors, average titers were obtained. The lowest titer of 2.5×10^4 was obtained for N2. After purification and concentration, LV titers increased by approximately two orders of magnitude (Figure 2b), and their ratio remained similar.

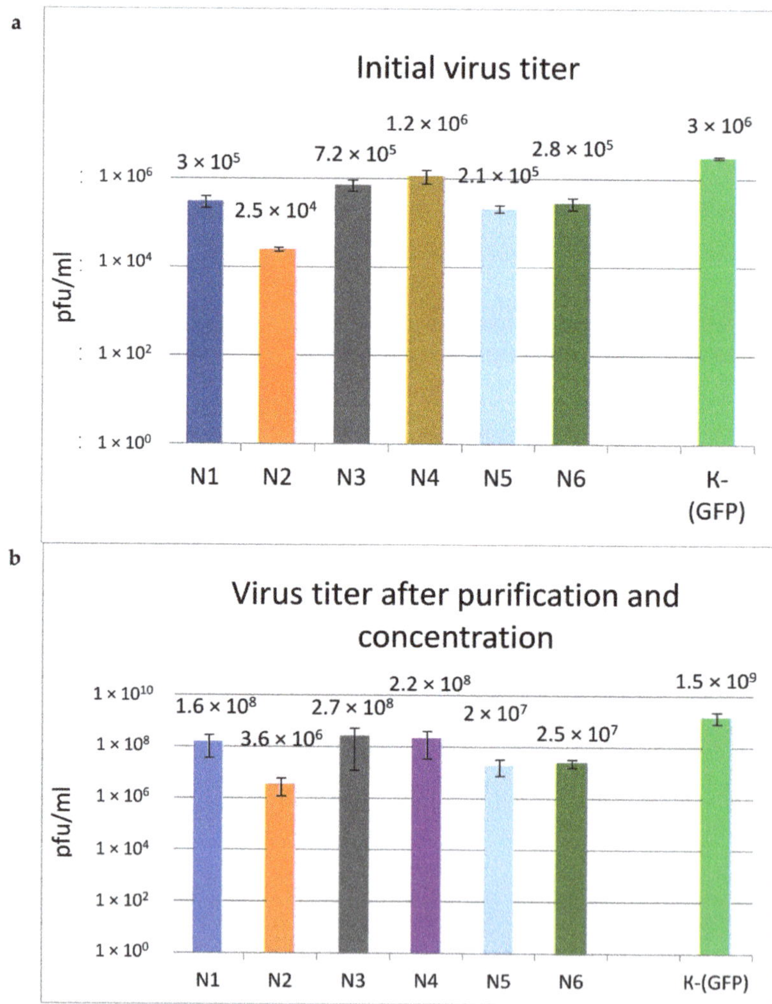

Figure 2. Comparison of titers of different LVs. (**a**)—Initial titer of viruses. (**b**)—Virus titer after purification and concentration. Columns reflect the average value of the +−SD (n = 4) titer of LV particles.

3.3. Transduction Efficiency, Transgene Stability in Cell Culture, and Influence of Different LV on the Viability of SupT1 Cells

The transduction efficiency, transgene stability, and viability of transduced cells were evaluated using SupT1 cells, which were transduced with the test vectors at an MOI of 1 and cultured for 5 weeks. The transduction efficiency was assessed at 4 days after transduction (Figure 3).

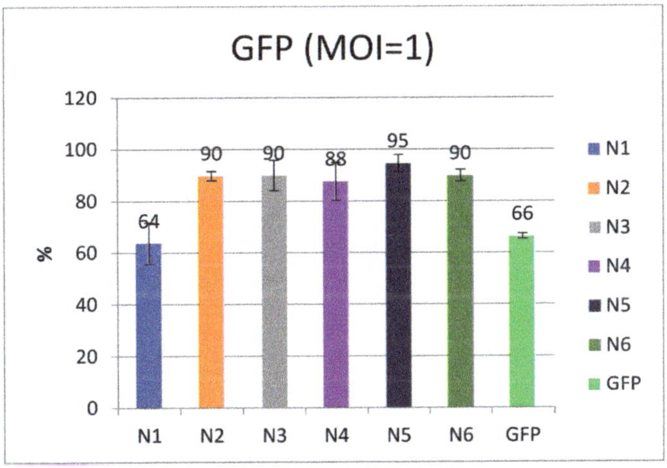

Figure 3. Comparison of the efficiency of transduction by different LVs at MOI = 1 (n = 3).

As shown in Figure 3, the transduction efficiency of most vectors was 88–95%. The lowest efficiency was observed for N1, but was comparable to the control.

The stability of the transgenes in culture was assessed by the dynamics of the changes in the proportion of cells containing the vector in culture over time (Figure 4a). For some of the LVs, a pronounced drop in the percentage of GFP+ cells was observed. For the N2 vector, the percentage of transduced cells decreased by 50% in 15 days; for N5 and N6, the decrease in the proportion of GFP+ cells was slower, but also reached 47 and 50%, respectively, by day 36. The decrease in the percentage of GFP+ cells was accompanied by low cell viability in culture (Figure 4b). This likely indicates the cytotoxicity of these vectors, which leads to the death of cells containing the LVs. For cells transduced with N1, N3, and N4, the drop in % of GFP+ cells was weakly pronounced and amounted to 8, 10, and 15%, respectively, with the main decrease occurring in the first week. This initial slight decrease may be due to the fact that the vector failed to integrate into some of the cells and was lost rapidly during cell division. No negative impact on viability was observed.

To obtain a higher percentage of cells carrying the vector, including after long-term cultivation, we performed transduction with the same LVs using MOI = 3 (Figure 5). We did not include the N2 vector in this experiment due to its low titer and high toxicity. At MOI = 3, the transduction efficiencies for all LVs were comparable and ranged from 83 to 96%.

As expected, an increase in MOI led to a greater drop in viability in cultured cells transduced with N5 and N6. Moreover, the decrease in the percentage of cells with the vector during cultivation was even more pronounced than with MOI = 1. In cultures transduced with vectors N1, N3, and N4, the dynamics of the decrease in the percentage of GFP+ cells repeated the dynamics of the decrease when using MOI = 1, decreasing by the end of the experiment by 7%, 11%, and 15%, respectively. Additionally, viability analysis revealed a negative effect on the cells by the N3 and N4 vectors, but not the N1 vector; thus, the use of MOI = 3 for transduction revealed the advantage of N1 as a vector with reduced toxicity.

a

b

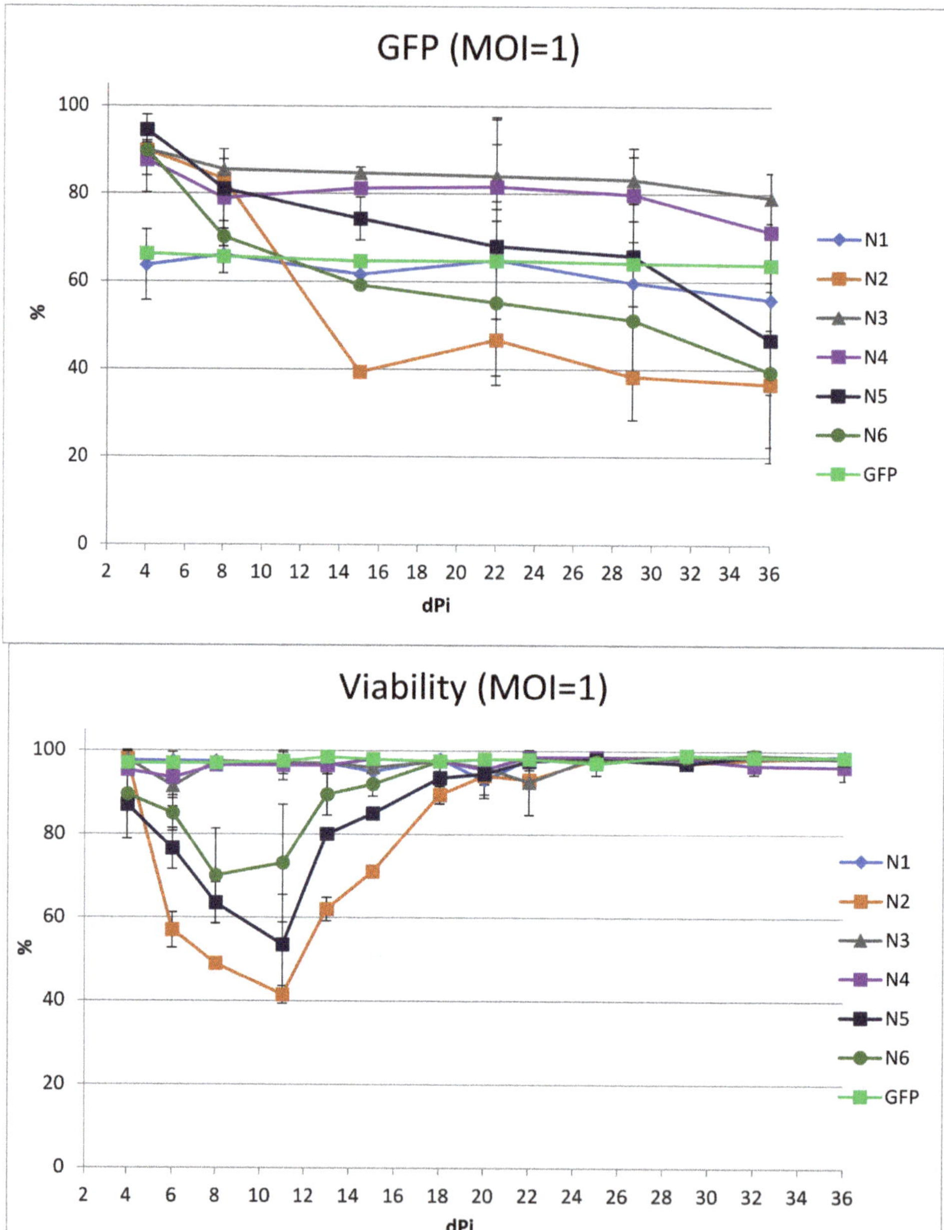

Figure 4. Analysis of SupT1 cells after transduction with different LVs. SupT1 cells were infected with LVs at MOI = 1. Then 2–3 times a week, the percentage of GFP-positive cells and the cell viability were measured. (**a**) Percentage of GFP-positive cells in culture within 5 weeks after transduction and (**b**) Cell viability within 5 weeks after transduction; $n = 3$.

a

b

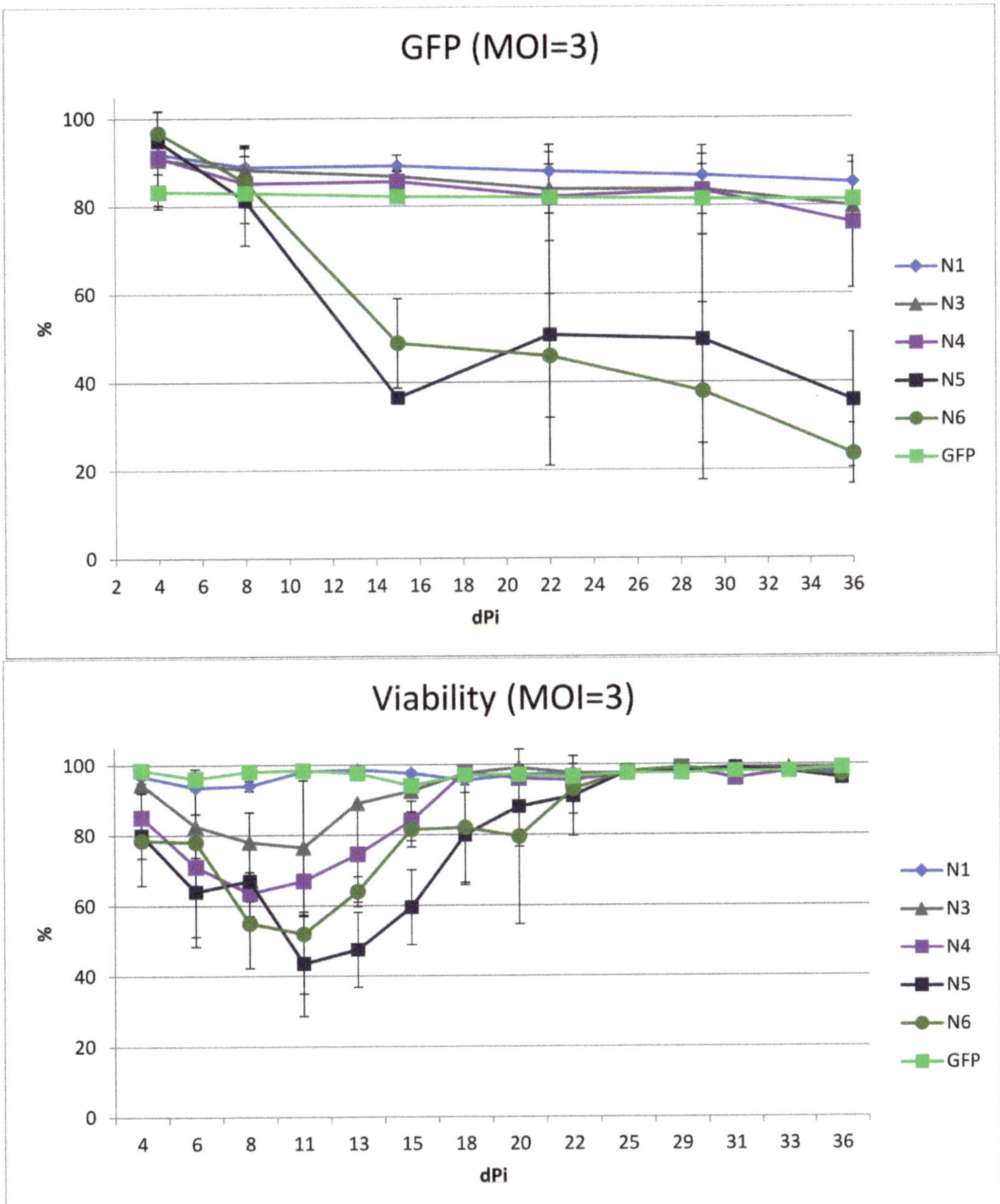

Figure 5. Analysis of SupT1 cells after transduction with different LVs. SupT1 cells were infected with LVs at MOI = 3. Then 2–3 times a week, the percentage of GFP-positive cells and the cell viability were measured. (**a**) Percentage of GFP-positive cells in culture within 5 weeks after transduction and (**b**) Cell viability within 5 weeks after transduction; $n = 3$.

Three weeks after transduction (MOI = 1), we measured the levels of V2O and TRIM5a-HRH mRNA, as well as the average vector copy numbers in the cell cultures. The amount of mRNA was normalized to the average vector copy number. The TRIM5a-HRH expression did not differ for cells transduced with different vectors. For V2O, a significant difference was found only between the N1 and N3 transduced cell lines (Figure S1). For N1, the expression level was lower than for N3.

3.4. Comparisons of Antiviral Activity of LVs in SupT1 Cell Culture

LVs N1, N3, and N4, which showed the lowest toxicity, were selected for evaluation of the antiviral activity. At 10 days after transduction with N1, N3, and N4 vectors at a dose of 1 MOI, SupT1 cells were infected with the HIV-1 NL4-3 strain. The cells were observed for three weeks after infection to determine the growth rate of the virus via the p24 antigen accumulation in the growth medium (Figure 6).

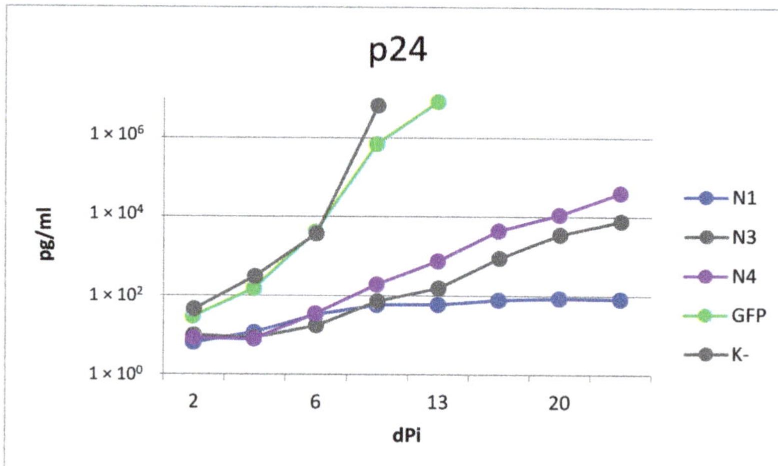

Figure 6. Dynamics of changes in the concentration of HIV p24 antigen in SupT1 cells transduced with N1, N3, N4, and GFP vectors and in the control non-transduced culture (K−).

The percentages of transduced cells for N1, N3, N4, and GFP were 71, 74, 78, and 95%, respectively, at 5 days before infection. In the cultures of non-transduced cells and cells transduced with the control GFP vector, an increase in the concentration of p24 antigen was observed, which indicated the active growth of the virus, leading to cell death on the 8th and 11th days after infection, respectively. In cells transduced with the vectors N1, N3, and N4, the virus growth rate was significantly reduced compared to control cells. Overall, the N1 vector provided maximum protection against the virus: after the 6th day, the amount of p24 antigen in the cell culture did not increase appreciably. The slight increase in the amount of p24 antigen on the 6th day after infection can be explained by virus replication in non-transduced cells; after the elimination of this factor, a tendency toward a reduced concentration of the antigen in the medium, caused by its slow degradation, occurred.

4. Discussion

The currently used antiretroviral therapy for HIV infection restrains development of the disease, but does not lead to a cure; hence, the search for new approaches to HIV therapy, one of which is gene therapy, is relevant. More than a dozen different antiviral genes have been proposed, the effectiveness of which has been demonstrated in vitro. However, one challenge of fighting HIV is its high variability, leading to the formation of variants that are resistant to individual genes [16]. As with the use of antiretroviral chemicals, this problem can be solved by combining several anti-HIV genes in a gene therapy drug. In addition, inclusion of genes that suppress virus replication at different stages of the life cycle in one vector increases antiviral activity. Therefore, we attempted to create a lentiviral vector for gene therapy that includes several genes. Three genes (anti-CCR5 microRNA gene; TRIM5a-HRH gene encoding a modified human TRIM5a protein; and V2O gene encoding C-peptide, a fusion inhibitor) that inhibit the virus at the initial stages of its life cycle and prevent insertion of proviral DNA were selected, as such genes can more effectively protect against infection [10–12].

Lentiviral vectors are currently the most efficient way to deliver genes to target cells. However, obtaining a vector that includes several genes can present certain problems [17]. It should be noted that to obtain a high percentage of transduced primary cells (lymphocytes and HSCs), it is necessary to use a high multiplicity of infection. As this is possible only at high titers of the vector, the production efficiency is an important parameter that further determines the feasibility of the practical use of LVs.

We constructed six lentiviral vectors containing the same three anti-HIV genes located in a different order and differentially regulated using promoters and the 2a peptide sequence. Analysis of the properties of these vectors showed that during production under the same conditions, the titers differed by more than 40 times, ranging from 2.5×10^4 to 1.2×10^6, and were 2.5–120 times lower than that of the control GFP vector.

The reduction in vector titer compared to the control vector was predictable and could be due to several factors. These include an increase in LV size [18]; the influence of the TRIM5a protein, which can lead to rapid degradation of Gag capsid proteins, suppressing virion assembly [19,20]; and promoter interference [13,21]. It has recently been shown that the Ef1a promoter in a similar vector adversely affected the titer, presumably by suppressing the promoter in the LTR, which regulates vector RNA production. Interference between Ef1a and the downstream PGK promoter is also possible [20]. For example, in constructs N3 and N4, the distance between Ef1a and pPGK is greater than in the other constructs, and the vectors obtained using these constructs showed a maximum titer.

In addition to the difference in the titer, the vectors differed in their effect on cell viability, which in turn led to a decrease in the proportion of cells containing the vector during cultivation. The toxicity of some vectors can be caused by various reasons. Antiviral proteins can affect cell viability, for example, high expression of the TRIM5a gene was shown to affect cell viability [22]. Therefore, the variation in viability among the cells transduced with different vectors could potentially be related to variations in gene expression. However, we were unable to see differences in the mRNA levels of the Trim5a-HRH and V2O genes in SupT1 cells transduced with different vectors. Unfortunately, we could not determine the amount of transgenic TRIM5a-HRH protein by a Western blot assay due to a low HRH-TRIM5a expression level in SupT1 cells, which cannot be distinguished from the background expression of the endogenous human Trim5a [20]. It can be assumed that some hybrid protein products resulting from incomplete cleavage of the 2a peptide may be toxic. It cannot be ruled out that the toxicity might be caused by a low titer of the vector, as it was necessary to add more vector, the preparation of which could contain impurities that negatively affect viability. It is notable that the vectors with the lowest titer after concentration and purification had the highest toxicity. In general, it was shown that at MOI = 1, vectors N1, N3, and N4 did not affect cell viability; at MOI = 3, only the N1 vector was nontoxic.

Analysis of the N1, N3, and N4 vectors' antiviral activity showed the N1 vector to be the most promising for further research, as it provided maximum cell protection with minimal toxicity. To understand the potential value of the resulting lentiviral vector for HIV therapy, further studies of the vector in vitro on primary human lymphocytes and in vivo in a humanized mouse model are needed.

Supplementary Materials: The following supporting information can be downloaded at: https://www.mdpi.com/article/10.3390/microorganisms11041053/s1, Figure S1: Estimation of the mRNA level of the V2O and TRIM5a-HRH genes. The cells were transduced with various lentivectors, after 3 weeks the cells were collected, treated with benzonase, then mRNA was isolated and OT-qPCR was performed with primers to the V2O(A) and TRIM(B) genes. Normalization was performed on the average number of vector per cell, which was determined using qPCR as WPRE normalized to Globin. * $p \leq 0.05$; Table S1: Plasmids and primers used to obtain LVs.

Author Contributions: Conceptualization, O.V.O., D.V.G. and E.V.B.; Methodology, O.V.O. and D.V.G.; Validation, O.V.O., D.V.G. and E.V.B.; Formal analysis, O.V.O. and D.V.G.; Investigation, O.V.O., R.R.M., G.M.T. and F.A.U.; Data curation, D.V.G. and E.V.B.; Writing—original draft, O.V.O.; Writing—review & editing, D.V.G. and E.V.B.; Supervision, D.V.G., G.A.S. and E.V.B.; Project administration, G.A.S. and E.V.B.; Funding acquisition, G.A.S. All authors have read and agreed to the published version of the manuscript.

Funding: This research received no external funding.

Data Availability Statement: The data will be available upon reasonable request to the corresponding authors.

Conflicts of Interest: The authors declare no conflict of interest.

References

1. Rossi, J.J.; June, C.H.; Kohn, D.B. Genetic therapies against HIV. *Nat. Biotechnol.* **2007**, *25*, 1444–1454. [CrossRef] [PubMed]
2. Stan, R.; Zaia, J.A. Practical Considerations in Gene Therapy for HIV Cure. *Curr. HIV/AIDS Rep.* **2014**, *11*, 11–19. [CrossRef] [PubMed]
3. Bogoslovskaya, E.V.; Glazkova, D.V.; Pokrovskii, V.V.; Shipulin, G.A. *Gene Therapy for HIV Infection*; Kolibri: Moscow, Russia, 2019.
4. Milone, M.C.; O'Doherty, U. Clinical use of lentiviral vectors. *Leukemia* **2018**, *32*, 1529–1541. [CrossRef] [PubMed]
5. Naldini, L. Lentiviruses as gene transfer agents for delivery to non-dividing cells. *Curr. Opin. Biotechnol.* **1998**, *9*, 457–463. [CrossRef] [PubMed]
6. Cornu, T.I.; Mussolino, C.; Müller, M.C.; Wehr, C.; Kern, W.V.; Cathomen, T. HIV Gene Therapy: An Update. *Hum. Gene Ther.* **2021**, *32*, 52–65. [CrossRef] [PubMed]
7. Lisa Egerer, L.; Kiem, H.-P.M.D.; von Laer, D. C Peptides as Entry Inhibitors for Gene Therapy. *Gene Ther. HIV Chronic Infect.* **2015**, *848*, 191–209.
8. Mohamed, H.; Gurrola, T.; Berman, R.; Collins, M.; Sariyer, I.K.; Nonnemacher, M.R.; Wigdahl, B. Targeting CCR5 as a Component of an HIV-1 Therapeutic Strategy. *Front. Immunol.* **2022**, *12*, 816515. [CrossRef] [PubMed]
9. Glazkova, D.V.; Vetchinova, A.S.; Bogoslovskaya, E.V.; IuA, Z.; Markelov, M.L.; Shipulin, G.A. Downregulation of human CCR5 receptor gene expression using artificial microRNAs. *Mol. Biol.* **2013**, *47*, 475–485. [CrossRef]
10. Felix Urusov, F.; Glazkova, D.; Omelchenko, D.; Bogoslovskaya, E.; Tsyganova, G.; Kersting, K.; Shipulin, G.; Pokrovsky, V. Optimization of Polycistronic Anti-CCR5 Artificial microRNA Leads to Improved Accuracy of Its Lentiviral Vector Transfer and More Potent Inhibition of HIV-1 in CD4+ T-Cells. *Cells* **2018**, *7*, 10. [CrossRef] [PubMed]
11. Brauer, F.; Schmidt, K.; Zahn, R.C.; Richter, C.; Radeke, H.H.; Schmitz, J.E.; von Laer, D.; Egerer, L. A Rationally Engineered Anti-HIV Peptide Fusion Inhibitor with Greatly Reduced Immunogenicity. *Antimicrob. Agents Chemother.* **2013**, *57*, 679–688. [CrossRef] [PubMed]
12. Stremlau, M.; Owens, C.M.; Perron, M.J.; Kiessling, M.; Autissier, P.; Sodroski, J. The cytoplasmic body component TRIM5α restricts HIV-1 infection in Old World monkeys. *Nature* **2004**, *427*, 848–853. [CrossRef] [PubMed]
13. Anderson, J.; Akkina, R. Human immunodeficiency virus type 1 restriction by human-rhesus chimeric tripartite motif 5alpha (TRIM 5alpha) in CD34(+) cell-derived macrophages in vitro and in T cells in vivo in severe combined immunodeficient (SCID-hu) mice transplanted with human fetal tissue. *Hum. Gene Ther.* **2008**, *19*, 217–228. [PubMed]
14. Omelchenko, D.O.; Glazkova, D.V.; Bogoslovskaya, E.V.; Urusov, F.A.; Zhogina, Y.A.; Tsyganova, G.M.; Shipulin, G.A. Protection of Lymphocytes Against HIV using Lentivirus Vector Carrying a Combination of TRIM5a-HRH Genes and microRNA Against CCR5. *Mol. Biol.* **2018**, *52*, 294–305. [CrossRef]
15. Maniatis, T.; Fritch, E.; Sambrook, J. Methods of Genetic Engineering. In *Molecular Cloning*; Mir: Moscow, Russia, 1984.
16. Rimsky, L.T.; Shugars DCMatthews, T.J. Determinants of Human Immunodeficiency Virus Type 1 Resistance to gp41-Derived Inhibitory Peptides. *J. Virol.* **1998**, *72*, 986–993. [CrossRef] [PubMed]
17. Escors, D.; Breckpot, K. Lentiviral vectors in gene therapy: Their current status and future potential. *Arch. Immunol. Ther. Exp.* **2010**, *58*, 107–119. [CrossRef] [PubMed]
18. Kirsten Canté-Barrett, K.; Mendes, R.D.; Smits, W.K.; van Helsdingen-van Wijk, Y.M.; Pieters, R.; Meijerink, J.P.P. Lentiviral gene transfer into human and murine hematopoietic stem cells: Size matters. *BMC Res. Notes* **2016**, *9*, 312. [CrossRef] [PubMed]
19. Sakuma, R.; Noser, J.A.; Ohmine, S.; Ikeda, Y. Rhesus monkey TRIM5a restricts HIV-1 production through rapid degradation of viral Gag polyproteins. *Nat. Med.* **2007**, *13*, 631–635. [CrossRef] [PubMed]

20. FA Urusov, F.A.; Glazkova, D.V.; Tsyganova, G.M.; Pozdyshev, D.V.; Bogoslovskaya, E.V.; Shipulin, G.A. The Titer of the Lentiviral Vector Encoding Chimeric TRIM5a-HRH Gene is Reduced Due to Expression of TRIM5a-HRH in Producer Cells and the Negative Effect of Efla Promoter. *Mol. Biol.* **2022**, *56*, 147–156. [CrossRef]
21. Palmer, A.C.; Egan, J.B.; Shearwin, K.E. Transcriptional interference by RNA polymerase pausing and dislodgement of transcription factors. *Transcription* **2010**, *2*, 9–14. [CrossRef] [PubMed]
22. Fletcher, A.J.; Vaysburd, M.; Maslen, S.; Zeng, J.; Skehel, J.M.; Towers, G.J.; James, L.C. Trivalent RING Assembly on Retroviral Capsids Activates TRIM5a Ubiquitination and Innate Immune Signaling. *Cell Host Microbe* **2018**, *24*, 761–775. [CrossRef] [PubMed]

microorganisms

MDPI

Article

Combined Analysis of the Whole Transcriptome of Piglets Infected with SADS−CoV Virulent and Avirulent Strains

Qianniu Li [1,2,†], Xiaoyu Tang [1,2,†], Ling Zhou [1,2], Xiaocheng Lv [1,2], Long Gao [1,2], Tian Lan [1,2], Yuan Sun [1,2,*] and Jingyun Ma [1,2,*]

1 Guangdong Provincial Key Laboratory of Agro−Animal Genomics and Molecular Breeding, College of Animal Science, South China Agricultural University, Guangzhou 510642, China
2 Guangdong Laboratory for Lingnan Modern Agriculture, Guangzhou 510642, China
* Correspondence: sunyuan@scau.com.cn (Y.S.); majy2400@scau.edu.cn (J.M.)
† These authors contributed equally to this work.

Abstract: When piglets are infected by virulent and avirulent strains of swine acute diarrhea syndrome coronavirus (SADS−CoV), there are obvious differences in their clinical symptoms; however, the specific mechanisms of pathogenicity and the immune regulation of highly pathogenic and low pathogenic strains are unknown. We collected intestinal tissues from SADS−CoV−infected piglets, performed a whole transcriptome sequencing analysis, including mRNA, miRNA, lncRNA, cicrRNA, and TUCP, and performed functional and correlation analyses of differentially expressed RNAs. Our results showed that the differentially expressed RNAs in group A versus group B (AvsB), group A versus group C (AvsC), and group B versus group C (BvsC) were relevant to immune and disease−related signaling pathways that participate in the organisms' viral infection and immune regulation. Furthermore, data obtained from the HAllA analysis suggested that there was a strong correlation between the differentially expressed RNAs. Specifically, LNC_011487 in the P set was significantly negatively correlated with ssc−miR−215, and LNC_011487 was positively correlated with PI3. Moreover, we also constructed a differentially expressed RNA association network map. This study provides a valuable resource for studying the SADS−CoV transcriptome and pathogenic mechanism from the perspective of RNA to understand the differences in and consistency of the interaction between virulent and attenuated SADS−CoV strains and hosts.

Keywords: SADS−CoV; RNA; avirulent strains; virulent strains; HAllA

Citation: Li, Q.; Tang, X.; Zhou, L.; Lv, X.; Gao, L.; Lan, T.; Sun, Y.; Ma, J. Combined Analysis of the Whole Transcriptome of Piglets Infected with SADS−CoV Virulent and Avirulent Strains. *Microorganisms* **2023**, *11*, 409. https://doi.org/10.3390/microorganisms11020409

Academic Editors: Fabio Zicker and Shengxi Chen

Received: 28 December 2022
Revised: 28 January 2023
Accepted: 31 January 2023
Published: 6 February 2023

1. Introduction

Swine acute diarrhea syndrome coronavirus (SADS−COV) is a newly discovered porcine coronavirus, that was first reported in China in 2017 [1]. Accordingly, SADS−CoV can cause swine acute diarrhea syndrome in piglets. Its clinical outcomes include acute diarrhea, vomiting, rapid weight loss in piglets less than one week old, and high mortality (more than 90%) [2,3]. SADS−CoV is a single−stranded, positive−stranded envelope virus, belonging to the genus *Alphacoronavirus* of the family *Coronaviridae* [1]. The diameter of SADS−CoV virus particles is about 100 nm. Moreover, filamentous protrusions similar to those of the coronavirus have been found on the surface of SADS−CoV's envelope. The SADS−CoV genome length is approximately 27 kb, and its genome encodes five non−structural proteins, ORF1a, ORF1b, NS3a, NS7a, and NS7b, and four structural proteins, S, E, M, and N [1,3].

There is a complex relationship between a virus and its host. Hence, understanding this relationship can help uncover the underlying mechanism of virus replication and infection in hosts and also is beneficial for developing disease prevention and control strategies. Viruses have evolved to manipulate multiple signal transduction pathways in favor of their own reproduction. It has been recorded that the SADS−CoV N protein could antagonize IFN−β production by inducing RIG−I ubiquitination [4]. Furthermore, our early study found that SADS−CoV could inhibit IFN−β production by targeting IPS−1 [5]. Mechanistically, the SADS−CoV N protein is an IFN−β antagonist, and it inhibits IFN−β production by targeting TBK1 to interfere with the interaction between TRAF3 and TBK1 [6]. Zhang et al. found that the inhibition of ZERK activity was associated with reduced apoptosis in SADS−CoV−infected cells, suggesting a critical role for ERK signaling in the SADS−CoV life cycle [7]. Edwards et al. found that rSADS−CoV did not use human coronavirus ACE−2, DPP4, or CD13 receptors for docking or entry [8]. At present, there are few studies about the mechanism of SADS−CoV infection, or hosts and their potential receptors. Therefore, studying the interaction mechanism between a virus and its host is particularly pivotal.

Biosafety precautions must be taken against the virus, as there is no vaccine for its treatment or defense. Only one study has stated that ZDHHC17 may be a potential drug target for SADS−CoV infection [9]. However, the pathogenic mechanism of SADS−CoV is still unclear. In this context, our laboratory created an attenuated SADS−CoV strain by serial passage. Moreover, virulent and avirulent strains were employed for animal challenge experiments. After attacks by avirulent strains, the tissue lesions, anal swab detoxification, and tissue tropism were relatively weakened [10]. To better understand the relationship between virulent strains and weak virulent strains, a whole transcriptome analysis, including mRNA, miRNA, lncRNA, cicrRNA, and TUCP, was performed, and the data were subjected to a basic bioinformatics analysis as well as expression/abundance detection. The results of differential expression ($p \leq 0.05$) were obtained after a differential expression analysis between the samples, from which the results of significant differences were filtered out, and the molecular events related to pathogenicity (P set) and immune response (I–V set) were detected. The results elucidate the difference and consistency of the interaction between the strong and weak strains of SADS−CoV and hosts from the perspective of RNA, allow us to study its pathogenic mechanism, and provide reference and support for the development of a vaccine.

2. Materials and Methods

2.1. Virus Strain

The passaged strains SADS−CoV−CN/GDWT/2017−P7 and SADS−CoV−CN/GDWT/2017−P83 of the SADS−CoV−CN/GDWT/2017 strain were isolated and preserved by our laboratory. Serum−free DMEM contained 10 μg/mL trypsin (Invitrogen, CA, USA) Infected cells were incubated at 37 °C in an incubator containing 5% CO_2 and monitored daily for cytopathic effects (CPEs). When 90% of cells showed CPEs, the cells were harvested, frozen and thawed three times, and centrifuged at 8000× *g* for ten minutes. The supernatant was collected and stored at −80 °C as a stock solution for the next passaging. Using the same approach, 82 subsequent passages were performed in Vero cells. They were purified every ten generations, starting from the 8th generation of cell culture. Viral titers were determined by TCID50 every five or ten generations.

2.2. Piglet Challenge Experiments

The piglets were purchased from a farm owned by Huanong Wen's Co., Ltd. In the experiment, we selected piglets weighing around two kilograms, which were responsive and energetic, had a good appetite, and were sensitive to external stimuli. They were divided into three groups: A (high pathogenicity group), B (low pathogenicity group), and C (control group) after ear tags were added. Piglets were housed in separate animal rooms to ensure no cross−contamination. Anal swabs were collected before the challenge. No specific virus infection was detected by RT−PCR, and SADS−CoV, porcine deltacoronavirus (PDCoV), transmissible gastroenteritis virus (TGEV), porcine epidemic diarrhea virus (PEDV), porcine rotavirus (RV), and other diarrhea pathogenic viruses were not detected. The piglets were sent to the animal room when they were 5 days old, and the challenge test was carried out after they had adapted to their environment. Group A was infected with 12 mL of virulent strain (P7) solution containing $1.0 \times 10^{6.3}$ TCID50/mL administered orally (9 piglets), group B was infected with 12 mL of avirulent strain (P83) solution containing $1.0 \times 10^{6.3}$ TCID 50/mL administered orally (9 piglets), and 8 control piglets were administered the same volume of DMEM.

The peak incidence period was from day 3 to day 5 post−challenge. Per group, three piglets were randomly selected to be bled and slaughtered, and the intestines were exposed after dissection of the piglets' abdominal cavity, followed by sampling of the intestines. The intestine was the ileum of the small intestine tissue, and 0.2 g was collected per sample. Intestinal tissues were lysed by grinding after we added TRIZOL reagent (Invitrogen, CA, USA). The tissue was fully ground, and then freeze−thawed three times for nucleic acid extraction.

2.3. Library Construction and Sequencing Process

Firstly, the quality of the sample was tested, the RNA integrity of the sample and the presence of DNA contamination were analyzed, and the RNA purity (OD260/280 and OD260/230 ratio) and the accurate detection of RNA integrity were examined [11]. After the sample was qualified, the Small RNA Sample Pre Kit (Illumina, San Diego, CA) was used to construct the library, which is designed to generate small RNA libraries directly from total RNA. Using the special structure of the 3′ and 5′ ends of the small RNA and with the total RNA as the starting sample, we directly added the adapters to both ends of the small RNA. After that, the cDNA synthesis was performed using kit. After PCR amplification, PAGE gel electrophoresis was conducted to separate the target DNA fragments, and then recovered cDNA library was obtained by cutting the gel [12]. The LncRNA/TUCP library uses strand−specific library construction [13]. We used the method of reverse transcription to synthesize the first strand of cDNA for the general library construction method of NEB with slight modulation. In contrast, when the second strand was synthesized, dTTP in dNTPs was replaced by dUTP, followed by cDNA end repair, A−tailing, ligation of sequencing adapters, and length screening [14,15]. After utilizing the USER enzyme to degrade the second strand of cDNA containing U, PCR amplification was conducted to obtain the library [16]. After the library was constructed, Qubit2.0 was used for preliminary quantification, then the library was diluted to 1 ng/L, and Agilent 2100 was used to detect the insert size of the library. After the insert size met our expectations, the effective concentration of the library was quantified by qPCR method (the effective concentration of the library was >2 nM) to ensure the quality of the library. After that, the different libraries were pooled according to the requirements of effective concentration and target data volume, and then HiSeq/MiSeq sequencing was performed.

2.4. Differentially Expressed RNA Screening

The overall distribution of differentially expressed RNAs was inferred by a volcano plot, and the differentially expressed RNAs were screened by evaluating them from the fold change and the corrected significance level (padj/qvalue). The default screening condition for differentially expressed RNA was padj < 0.05. The default differentially expressed RNA screening conditions were qvalue < 0.01 & | log2(fold change) | > 1.

2.5. GO and KEGG Enrichment Analysis

The GO enrichment analysis was conducted using GO seq, based on Wallenius' non−central hypergeometric distribution, compared with the ordinary hypergeometric distribution, the characteristic of which is that the probability of extracting an individual from a certain category is different from the probability of extracting an individual from outside a certain category. This difference in probability is obtained by estimating the preference of gene length, so that the probability of GO term being enriched by candidate target genes can be calculated more accurately [17].

In organisms, different genes coordinate with each other to perform their biological functions. Pathway significant enrichment can determine the most important biochemical metabolic pathways and signal transduction pathways involved in candidate target genes.

The KEGG (Kyoto Encyclopedia of Genes and Genomes) is the main public database on pathways [18]. Pathway significant enrichment analysis takes KEGG Pathway as its unit and applies hypergeometric test to find pathways that are significantly enriched in candidate target genes compared with the whole genome background.

2.6. Multi−Omics Association Analysis

Hierarchical all−against−all association (HAllA) is a computational method for finding multi−resolution associations in high−dimensional, heterogeneous datasets, and for discovering significant relationships between data features with high power [19]. HAllA has strong robustness of data types, operating on continuous and categorical values, and on homogeneous datasets (all measurements are of the same type, such as gene expression microarrays) and heterogeneous (contains measurements with different units or types, such as patient clinical data) datasets both work well.

We used the HAllA multi−omics association method (R package: halla) to perform multi−omics association analysis according to the following comparison schemes: circRNA vs mRNA, circRNA vs miRNA, lncRNA vs miRNA, miRNA vs mRNA, and lncRNA/TUCP vs mRNA [20]. A, B, and C represent the lethality group, the low−pathogenicity group, and the control group, respectively. The molecular events associated with pathogenicity in AvsC are consistent with those in AvsB. The immune response generated in AvsC contains a response to bacterial infection secondary to SADS−CoV, and the immune response generated in BvsC is purely a host immune response to SADS−CoV. Systematic errors were randomized. Data for each group were used to correlate molecular events related to pathogenicity (the intersection of AvsC and AvsB represents molecular events associated with viral potential pathogenicity, lethality, and inflammatory responses secondary to bacterial infection, P set) and immune response the intersection of BvsC and AvsC represents the host immune response to SADS−CoV, I–V set). The correlation of the correlation results, based on the pairwise Pearson correlation coefficients, were adjusted by false discovery rate (FDR < 0.05).

2.7. Construction of Differentially Expressed RNA Association Network

According to the association analysis results of HAllA of P set and I–V set, we used Cytoscape v3.2.1 to build an association network graph.

3. Results

3.1. Expression Profiles of Differentially Expressed mRNAs

After analyzing the differentially expressed mRNAs in different groups, it was found that there were 1955 differentially expressed mRNAs in the AvsB group, 890 differentially expressed mRNAs in the AvsC group, and 2074 mRNAs in the BvsC group with significant changes (Figure 1A–C). Among them, the expression of SPARCL1 (SPARC like 1) (https://www.genecards.org/cgi$-$bin/carddisp.pl?gene=SPARCL1&keywords=SPARCL1 accessed on 27 January 2023) was up−regulated 7.9 times in the AvsB group. However, in the other groups, it did not change significantly in expression. The expression of HNRNPH1 (Heterogeneous Nuclear Ribonucleoprotein H1) (https://www.genecards.org/cgi$-$bin/carddisp.pl?gene=HNRNPH1 accessed on 27 January 2023) was down−regulated 7.2 times. The expression of SLA−1 only in the AvsC group was up−regulated 8.08 times. In the BvsC group, the upregulation of SLC5A1 (Solute Carrier Family 5 Member 1) (https://www.genecards.org/cgi$-$bin/carddisp.pl?gene=SLC5A1 accessed on 27 January 2023) was the most pronounced. After the GO enrichment analysis, it was found that the mRNA of the AvsB group was mainly related to protein activity and metabolism−related processes. The mRNA of the AvsC group was mainly enriched in immune response and catalysis. In the BvsC group, mRNAs were enriched in protein metabolism−related processes (Figure 1D–F). The results of the KEGG analysis showed that the mRNA was significantly enriched in metabolic pathways in the AvsB group, and the mRNA in AvsC group was significantly enriched in disease−related pathways and antigen processing and presentation (Figure 1G,H). In the BvsC group, mRNAs were significantly enriched in immune−related signaling pathways, such as the Toll−like receptor signaling pathway, NF−kappa B signaling pathway, and T cell receptor signaling pathway (Figure 1I).

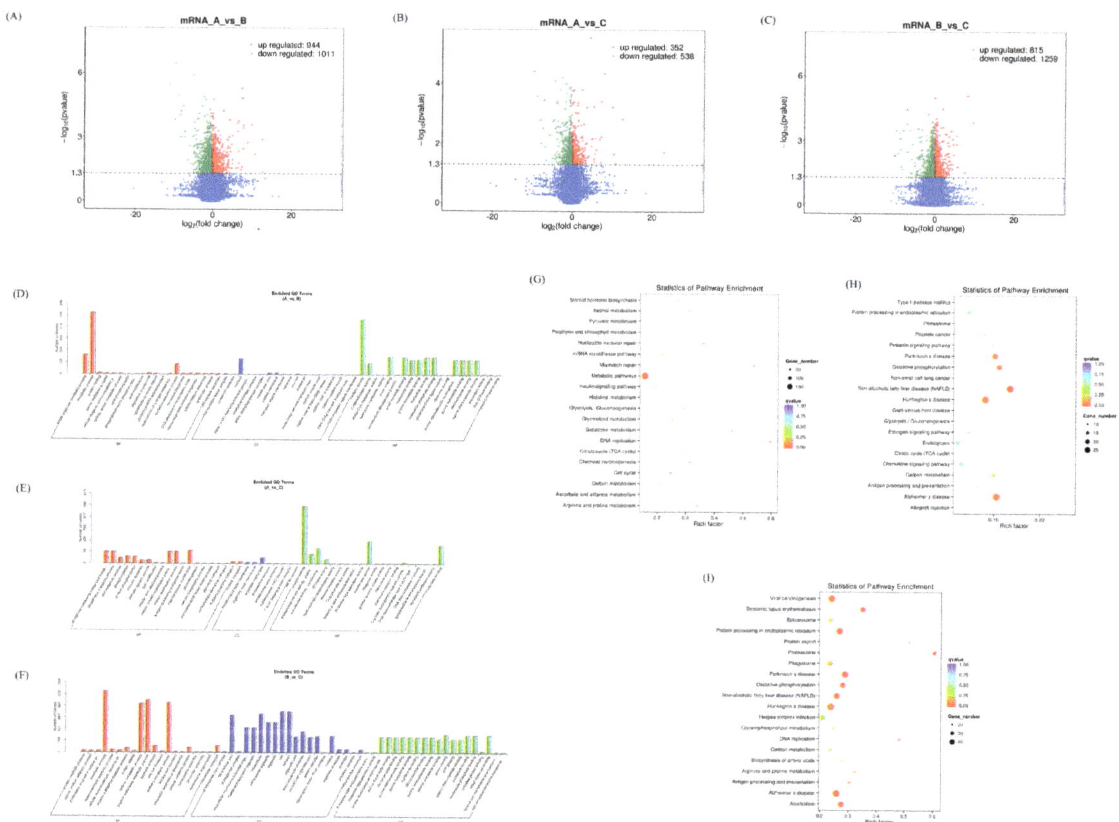

Figure 1. Expression profiles of differentially expressed mRNAs. (**A–C**) Volcano plot of differentially expressed mRNAs in AvsB group (**A**), AvsC group (**B**), and BvsC group (**C**). (**D,F**) GO analysis of differentially expressed mRNAs in AvsB group (**D**), AvsC group (**E**), and BvsC group (**F**). (**G,I**) KEGG pathway analysis of differentially expressed mRNAs in AvsB group (**G**), AvsC group (**H**), and BvsC group (**I**).

The P set had 70 intersection mRNAs, of which TGM3 (Transglutaminase 3) (https://www.genecards.org/Search/Keyword?queryString=TGM3) is related to a virus's entry into host cells and the virus cycle. FUT2 (Fucosyltransferase 2) (https://www.genecards.org/Search/Keyword?queryString=FUT2 accessed on 27 January 2023), CCL5 (C−C Motif Chemokine Ligand 5) (https://www.genecards.org/Search/Keyword?queryString=CCL5 accessed on 27 January 2023), SLA−DQA1, and CXCL9 (C−X−C Motif Chemokine Ligand 9) (https://www.genecards.org/Search/Keyword?queryString=CXCL9 accessed on 27 January 2023) in the intersection mRNA of the I–V set are all related to immune response, so these intersection mRNAs can be used as the research target of immune response (Supplementary Materials Table S1).

3.2. Expression Profiles of Differentially Expressed miRNAs

After analyzing the differentially expressed miRNAs in different groups, it was found that there were 66 differentially expressed miRNAs in the AvsB group, 51 differentially expressed miRNAs in the AvsC group, and 8 in the BvsC group. Among them, the expression of ssc−miR−10390 (https://www.mirbase.org/textsearch.shtml?q=ssc$-$miR$-$10390 accessed on 27 January 2023) in the AvsB group and AvsC group was down−regulated. In the AvsC group, the expression of ssc−miR−10390 was up−regulated. The expression of the newly discovered miRNA novel_536 was down−regulated in the AvsC group. In the BvsC group, the expression of ssc−miR−21−3p (https://www.mirbase.org/textsearch.shtml?q=ssc$-$miR$-$21$-$3p accessed on 27 January 2023) was significantly down−regulated. After the GO enrichment analysis, it was found that miRNAs in the AvsB group were mainly related to binding, autophagy, and junction processes (Figure 2D). The miRNAs in the AvsC group were mainly enriched in enzyme activator activity, the cytokinetic process, and other processes (Figure 2E). In the BvsC group, miRNAs were enriched in immune−related signaling pathways and phosphors metabolic processes. However, in the GO enrichment analysis of the BvsC group, the corrected p−value was >0.05. It is mentioned here for reference only. The KEGG analysis showed that the miRNAs in the AvsB group were significantly enriched in natural−killer−cell−mediated cytotoxicity, the B cell receptor signaling pathway, the NF−kappa B signaling pathway, and other pathways (Figure 2F). It is worth noting that the miRNAs in the AvsB group are significantly enriched in Fc gamma R−mediated phagocytosis and the Fc epsilon RI signaling pathway; all Fc γ receptors belong to the immunoglobulin superfamily; and Fc γ receptors are involved in multiple immune system functions [21]. Similar to the AvsB group, the significantly enriched pathways in the AvsC group were mostly related to immunity (Figure 2G). Like the AvsB group, the BvsC group was also enriched in the Fc gamma R−mediated phagocytosis and Fc epsilon RI signaling pathways (Figure 2H). In addition to this, the most significant pathway enriched in this group was the intestinal immune network for IgA production, which may be related to intestinal lesions after SADS−CoV infection.

The P set had 24 intersection miRNAs, and these miRNAs may be related to molecular events related to the potential pathogenicity, lethality, and inflammatory response of secondary bacterial infection. There were four miRNAs in the intersection of the I–V set, which may have an impact on the immune response of the host to SADS−CoV. Therefore, miRNAs are involved in regulation during SADS−CoV infection (Supplementary Materials Table S2).

We analyzed the association of miRNA and mRNA by the HAllA association method, and the results showed that in the pathogenic set (P set), ssc−miR−215 (https://www.mirbase.org/textsearch.shtml?q=ssc$-$miR$-$215 accessed on 27 January 2023) had the highest association with the mRNA ABHD3, which was cut into a positive correlation. The second highest correlation was that of ssc−miR−1285 (https://www.mirbase.org/textsearch.shtml?q=ssc$-$miR$-$1285 accessed on 27 January 2023) with S100G and SLC30A10. ssc−miR−133a−3p (https://www.mirbase.org/textsearch.shtml?q=ssc$-$miR$-$133a$-$3p accessed on 27 January 2023) was negatively correlated with GSTA1, indicating that these molecules have a high correlation in the P set (Figure 3A). In the immune response collections (IV set), ssc−miR−1388 (https://www.mirbase.org/textsearch.shtml?q=ssc$-$miR$-$1388 accessed on 27 January 2023) had the most significant association with LMNA, followed by ssc−miR−1285 and FUT2, both of which were positively correlated, indicating that these molecules were included in the immune response collection. It shows that these molecules have a high positive correlation in the immune response set (Figure 3B).

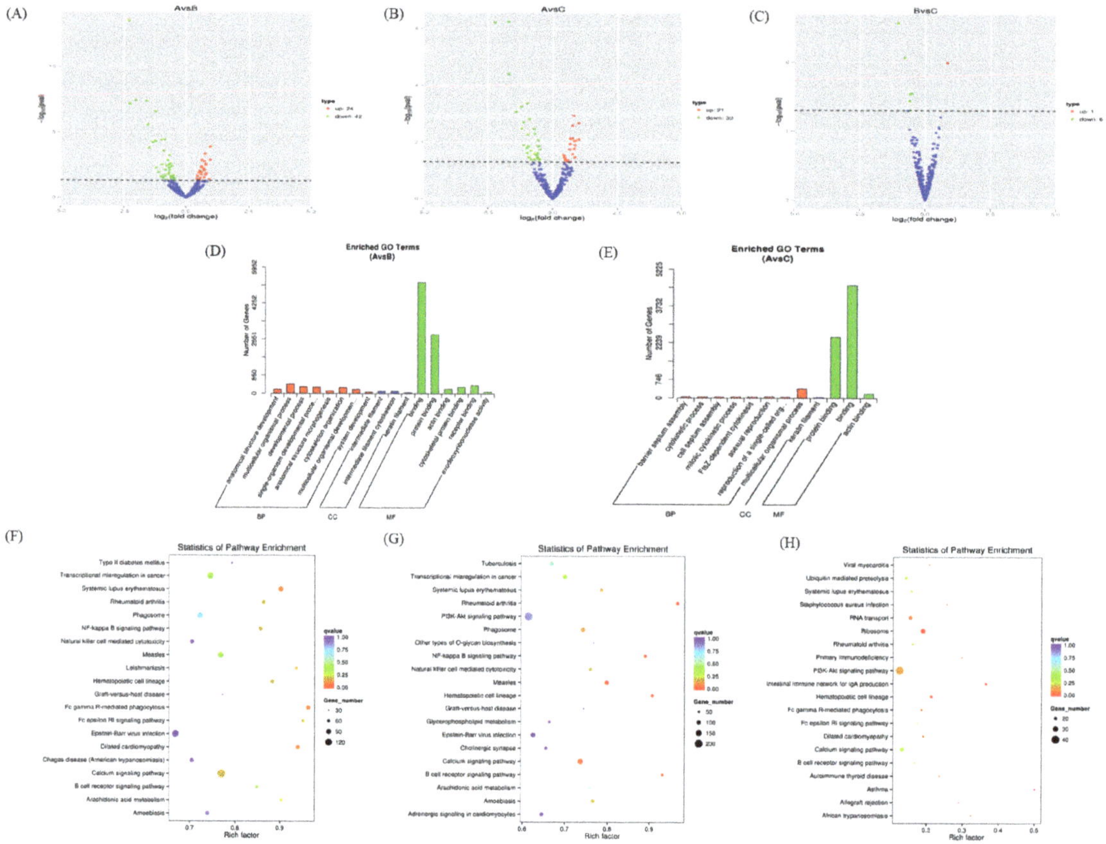

Figure 2. Expression profiles of differentially expressed miRNAs. (**A–C**) Volcano plot of differentially expressed miRNAs in AvsB group (**A**), AvsC group (**B**), and BvsC group (**C**). (**D**,**E**) GO analysis for neighbor gene functions of miRNA in AvsB group (**D**) and AvsC group (**E**). (**F–H**) KEGG annotation for neighbor gene functions of miRNA in AvsB group (**F**), AvsC group (**G**), and BvsC group (**H**).

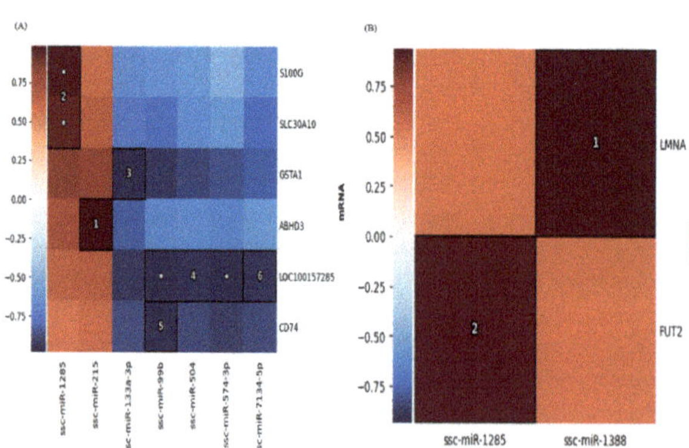

Figure 3. Association analysis of miRNAs and mRNAs. (**A**) Association results between miRNA and mRNA in the P set. (**B**) Association results between miRNA and mRNA in the I–V set. Color bars represent correlation coefficients: red for positive correlation and blue for negative correlation.

3.3. Expression Profiles of Differentially Expressed cicrRNAs

Meanwhile, we predicted new circRNAs. There were 80 new circRNAs differentially expressed in the AvsB group, 74 new circRNAs differentially expressed in the AvsC group, and 63 new circRNAs in the BvsC group that were significantly changed (Figure 4A–C). The results of the GO analysis showed that the target genes of the AvsB group were enriched in the negative regulation of cell adhesion, negative regulation of cell development, and description processes, which were mostly related to the negative regulation of biological processes (Figure 4D). The AvsC group was related to processes such as the endomembrane system, microtubule motor activity, and GMP biosynthetic process (Figure 4E). The target genes of the BvsC group were mainly involved in pathogenesis and enzymatic activity (Figure 4F). Most of the target genes in the AvsB group were related to material metabolism pathways (Figure 4G). The AvsC group was significantly enriched in protein export, the cGMP−PKG signaling pathway, and basal transcription factors pathways (Figure 4H). The target genes of the BvsC group were mainly enriched in the p53 signaling pathway, cell cycle, and mismatch repair pathway (Figure 4I).

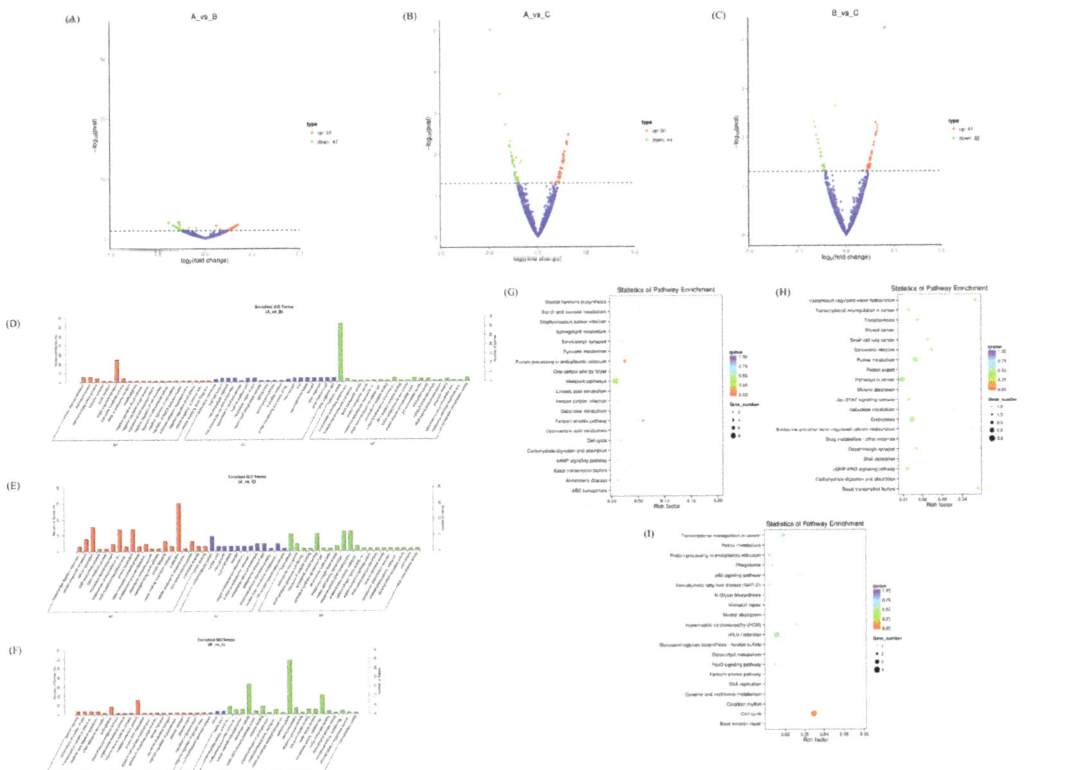

Figure 4. Expression profiles of differentially expressed cicrRNAs. (**A–C**) Volcano plot of differentially expressed cicrRNAs in AvsB group (**A**), AvsC group (**B**), and BvsC group (**C**). (**D–F**) GO analysis for neighbor gene functions of cicrRNAs in AvsB group (**D**), AvsC group (**E**), and BvsC group (**F**). (**G–I**) KEGG annotation for neighbor gene functions of cicrRNAs in AvsB group (**G**), AvsC group (**H**), and BvsC group (**I**).

The P set had 60 intersection circRNAs. There were 46 circRNAs in the intersection of the I–V set (Supplementary Materials Table S3). These predicted new circRNAs were involved in different molecular events during SADS−CoV infection, which can provide references for follow−up studies. The HAllA association analysis was performed on these circRNAs and mRNAs. In the P set, novel_circ_0006647, novel_circ_0009773, novel_circ_0007348 and novel_circ_0008394 all had the highest association with ROCK2 (Rho associated coiled−coil−containing protein kinase 2) (https://www.genecards.org/Search/Keyword?queryString=ROCK2 accessed on 27 January 2023) and were positively correlated (Figure 5A). In the I–V set, the highest correlation was between novel_circ_0002913 and PSMB9 (Proteasome 20S subunit beta 9) (https://www.genecards.org/Search/Keyword?queryString=PSMB9 accessed on 27 January 2023), CD3D (CD3 delta subunit of T cell receptor complex) (https://www.genecards.org/Search/Keyword?queryString=CD3D), CACYBP (calcyclin−binding protein) (https://www.genecards.org/Search/Keyword?queryString=CACYBP accessed on 27 January 2023), CD3G (CD3 gamma subunit of T cell receptor complex) (https://www.genecards.org/Search/Keyword?queryString=CD3G accessed on 27 January 2023), and there was a significant negative correlation between them (Figure 5B). A notable phenomenon is that most of the top circRNAs and mRNAs in the I–V set have negative correlations (Figure 5C). The results of the association analysis between circRNAs and miRNAs showed that in the I–V set, circRNA novel_circ_0009547 had the highest correlation with ssc−miR−1388, which were negatively correlated. In the P set, miRNA novel_424 (new prediction) had the highest correlation and a positive correlation with circRNA novel_circ_0000365, novel_circ_0005562, and novel_circ_0012577. Secondly, the miRNAs ssc−miR−504 (https://www.mirbase.org/textsearch.shtml?q=ssc$-$miR$-$504 accessed on 27 January 2023) and ssc−miR−574−3p (https://www.mirbase.org/textsearch.shtml?q=ssc$-$miR$-$574$-$3p accessed on 27 January 2023) were highly correlated with the circRNA novel_circ_0000931, implying a high positive correlation between these molecules in the P set (Figure 5D).

3.4. Expression Profiles of Differentially Expressed lncRNAs

There were 132 differentially expressed lncRNAs in the AvsB group, 51 differentially expressed lncRNAs in the AvsC group, and 146 lncRNAs in the BvsC group with significant changes (Figure 6A–C). More lncRNAs are involved in immune response regulation compared to the low pathogenic group. The GO results of the three groups were similar, and most of them were related to protein binding and metabolism (Figure 6D–F). The KEGG analysis showed that lncRNAs in the AvsB group were significantly enriched in Fc gamma R−mediated phagocytosis, mTOR signaling, and some metabolism−related pathways (Figure 6G). In the AvsC group, lncRNAs were significantly enriched in T cell receptor signaling, p53 signaling, and Jak−STAT signaling pathways (Figure 6H). In the BvsC group, lncRNAs were enriched in Fc gamma R−mediated phagocytosis, the T cell receptor signaling pathway, phagosomes, and other pathways (Figure 6I). Our results also suggest that differentially expressed lncRNAs can be involved in regulating multiple pathways, such as immune responses, cellular metabolism, and thus the body's response to viruses. In addition, we found that the differentially expressed lncRNAs of the three groups were significantly enriched in the inflammatory bowel disease pathway, which may be related to the ability of SADS−CoV to cause diarrhea and intestinal damage.

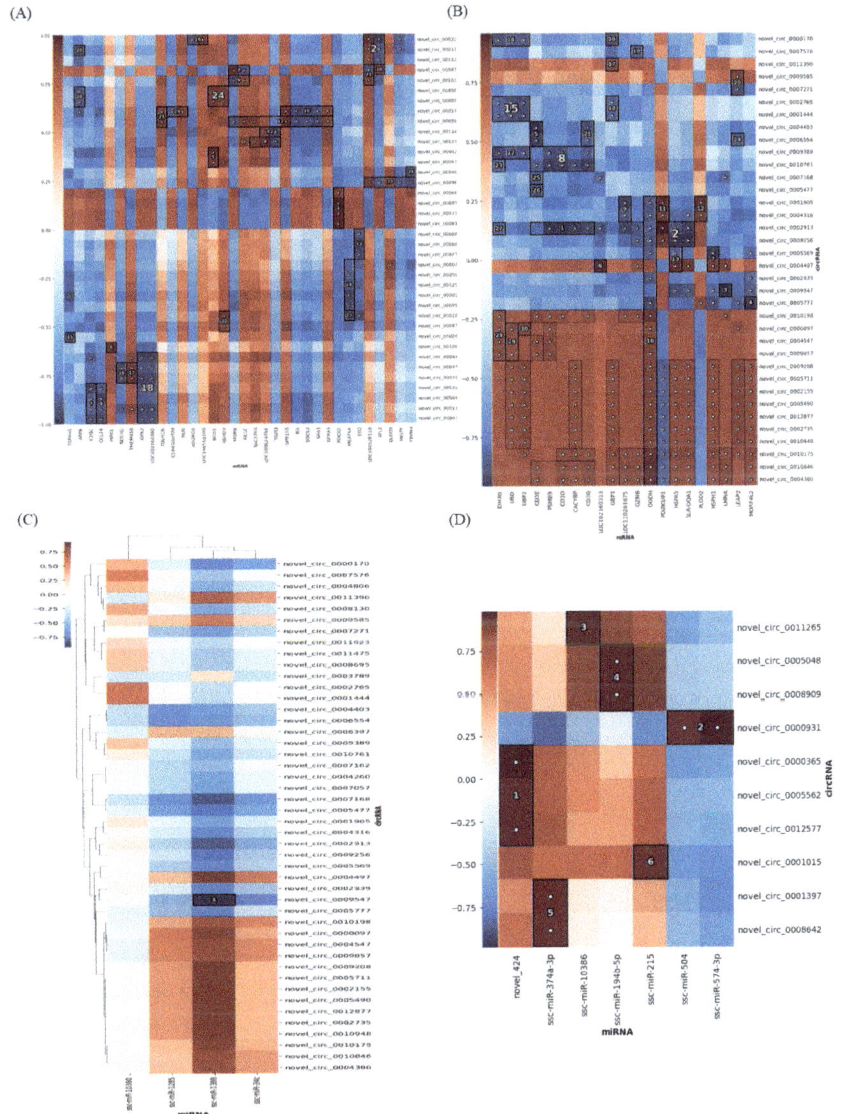

Figure 5. Association analysis of cicrRNAs−mRNAs and cicrRNAs−miRNAs. (**A**) Association results between cicrRNAs and mRNA in the P set. (**B**) Association results between cicrRNAs and mRNA in the I−V set. (**C**) Association results between cicrRNAs and miRNA in the I−V set. (**D**) Association results between cicrRNAs and miRNA in the P set. Color bars represent correlation coefficients: red for positive correlation and blue for negative correlation.

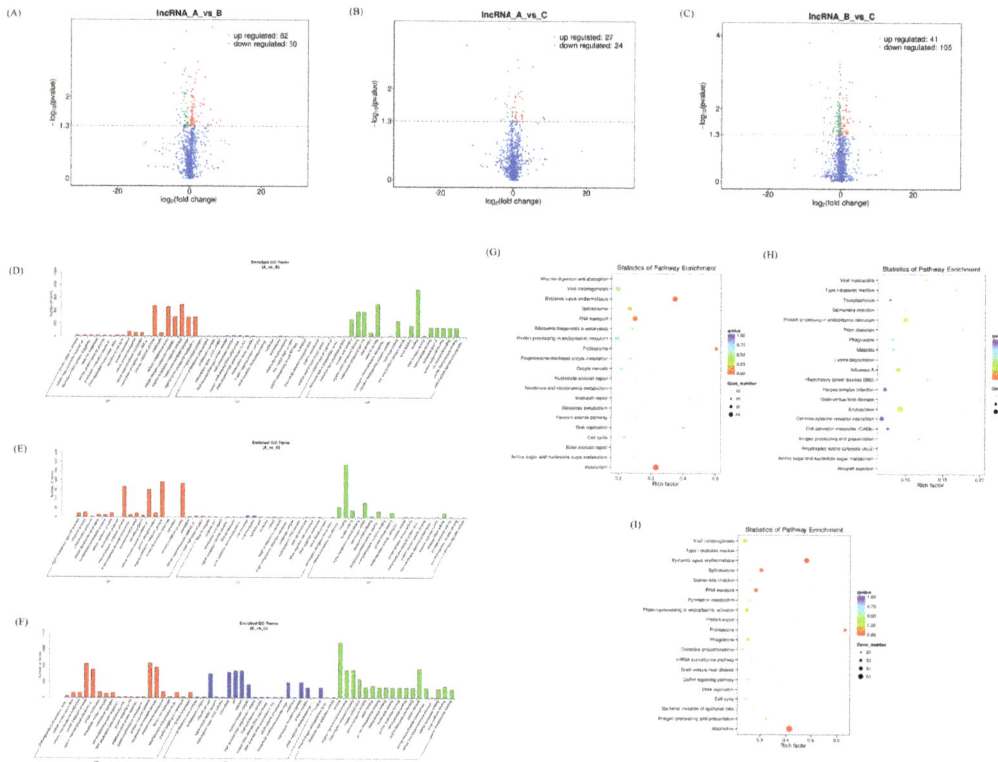

Figure 6. Expression profiles of differentially expressed lncRNAs. (**A–C**) Volcano plot of differentially expressed lncRNAs in AvsB group (**A**), AvsC group (**B**), and BvsC group (**C**). (**D–F**) GO analysis for neighbor gene functions of lncRNAs in AvsB group (**D**), AvsC group (**E**), and BvsC group (**F**). (**G–I**) KEGG annotation for neighbor gene functions of lncRNAs in AvsB group (**G**), AvsC group (**H**), and BvsC group (**I**).

There are seven intersection lncRNAs in the P set, and one intersection lncRNA in the I–V set, which is related to the host's immune response to SADS−CoV (Supplementary Materials Table S4). Then we analyzed the association between lncRNA and miRNA by the HAllA association method, and the results showed that in the P set, LNC_008160 was significantly positively correlated with ssc−miR−10391 and novel_424. LNC_011487 was significantly negatively correlated with ssc−miR−215, and further studies are needed on whether there is a competitive endogenous RNA relationship between the two genes (Figure 7A). The association analysis between lncRNA and mRNA showed that LNC_011487 was positively associated with PI3 and negatively associated with SLC37A3 in the P set (Figure 7B). The above results indicate that differentially expressed lncRNAs can play a role in the molecular events related to the potential pathogenicity, lethality, and inflammatory response of secondary bacterial infections in combination with miRNAs and mRNAs.

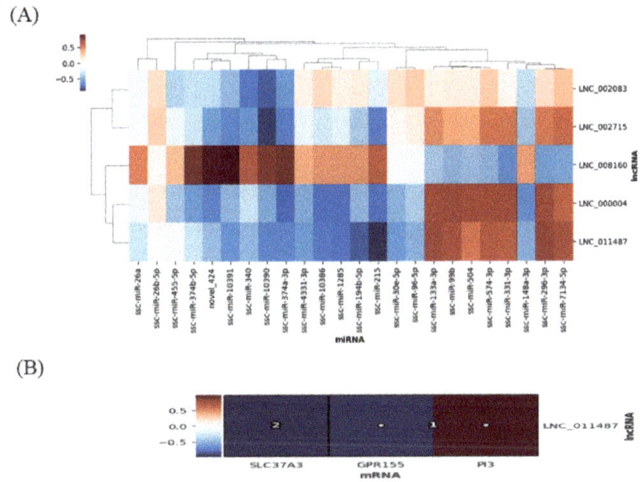

Figure 7. Association analysis of lncRNAs−miRNAs and lncRNAs−mRNAs. (**A**) Association results between lncRNAs and miRNA in the P set. (**B**) Association results between lncRNAs and mRNA in the P set. Color bars represent correlation coefficients: red for positive correlation and blue for negative correlation.

3.5. Expression Profiles of Differentially Expressed TUCP

Transcripts of uncertain coding potential (TUCP) are part of long noncoding RNAs, which include short open reading frames and can be translated into small peptides [22]. After analyzing the differentially expressed TUCPs, it was found that there were 66 differentially expressed TUCPs in the AvsB group, 25 differentially expressed TUCPs in the AvsC group, and 92 TUCPs in the BvsC group with significant changes (Figure 8A–C). The results showed that there were more differentially expressed TUCPs involved in the regulation of immune response. After the GO enrichment analysis of TUCP co−expressed genes, it was found that the TUCP in the AvsB group was mainly related to RNA binding and viral genome replication, indicating that the differential expression of TUPC in the AvsB group was related to pathogenicity (Figure 8D). The TUCP in the AvsC group was mainly related to inflammatory response, the viral envelope, viral membrane, and other processes, indicating that the differential expression of TUCP in this group indeed corresponds to the basis of the grouping and is related to pathogenicity and immune response (Figure 8E). In the BvsC group, the TUCP was associated with binding and the proteasome complex (Figure 8F). The KEGG enrichment analysis of TUCP co−expressed genes showed that TUCP was significantly enriched in p53 signaling, RNA transport, mTOR signaling, and other pathways in the AvsB group (Figure 8G). The AvsC group was significantly enriched for immune−related regulation, such as T cell receptor signaling, primary immunodeficiency, and NF−kappa B signaling (Figure 8H). Like the AvsB group, the BvsC group was also significantly enriched in immune−related pathways, indicating that the differential expression of TUCP between the AvsC and BvsC groups was more involved in the body's immune regulation (Figure 8I).

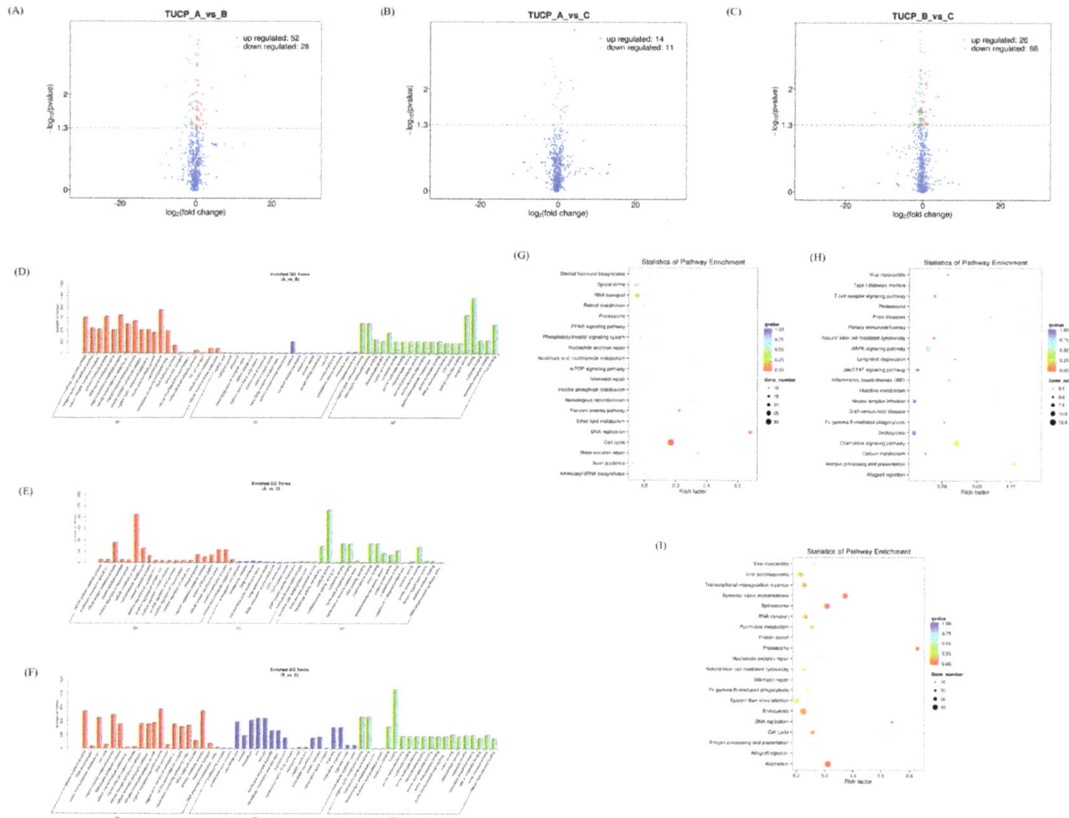

Figure 8. Expression profiles of differentially expressed TUCP. (**A–C**) Volcano plot of differentially expressed cicrRNAs in AvsB group (**A**), AvsC group (**B**), and BvsC group (**C**). (**D–F**) GO analysis for neighbor gene functions of TUCP in AvsB group (**D**), AvsC group (**E**), and BvsC group (**F**). (**G–I**) KEGG annotation for neighbor gene functions of TUCP in AvsB group (**G**), AvsC group (**H**), and BvsC group (**I**). AvsC and AvsB have three intersection TUCPs, and BvsC and AvsC have five intersection TUCPs (Supplementary Materials Table S5).

The results of the HAllA association analysis showed that in the P set, TUCP_002240 had the strongest association with LYZ and TSPAN1, and was positively correlated (Figure 9A). In the I–V set, TUCP_03109, TUCP_000239, TUCP_003235, TUCP_003225, and TUCP_003878 were all significantly and positively correlated with multiple mRNAs (Figure 9B). This shows that these TUCPs can play a role in association with mRNA.

3.6. Combined Analysis of mRNAs, miRNAs, lncRNAs, TUCP, cicrRNAs

We performed a joint analysis of the interaction between differentially expressed RNAs. In the I–V set, 67 interrelated differentially expressed RNAs were obtained to construct an association network. The network consists of 67 nodes (28 mRNAs, 32 circRNAs, 3 miRNAs, and 4 TUCPs) and 93 edges (Figure 10A). In the P collection, 134 interrelated differentially expressed RNAs were obtained to construct an association network. The network consists of 134 nodes (53 mRNAs, 57 circRNAs, 4 lncRNAs, 17 miRNAs, and 3 TUCPs) and 327 edges (Figure 10B).

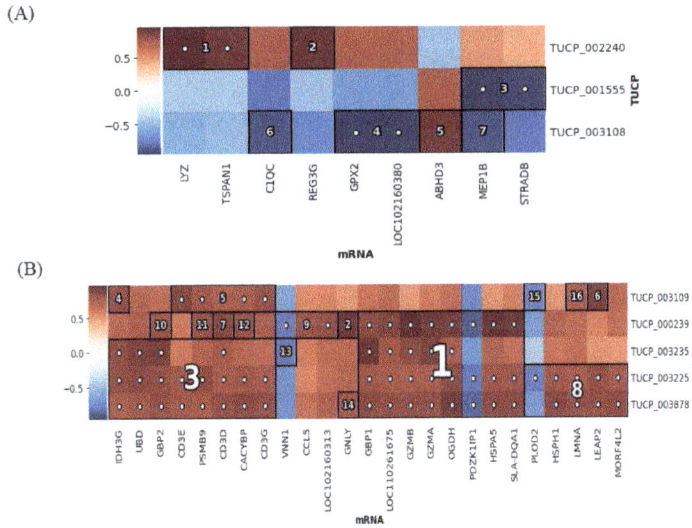

Figure 9. Association analysis of TUCP and mRNAs. (**A**) Association results between TUCP and mRNAs in the P set. (**B**) Association results between TUCP and mRNAs in the I–V set. Color bars represent correlation coefficients: red for positive correlation and blue for negative correlation.

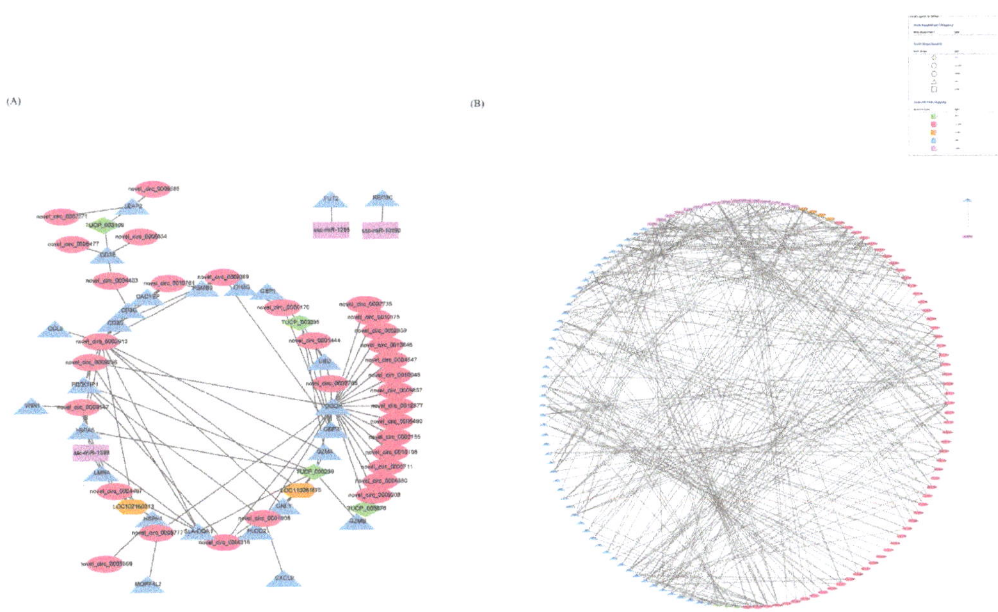

Figure 10. Combined analysis of mRNAs, miRNAs, lncRNAs, TUCP, and cicrRNAs. (**A**) The network of I–V set. (**B**) The network of P set.

4. Discussion

To our knowledge, this study is the first to comprehensively detect differentially expressed RNAs by RNA−Seq, including from three subgroups (virulent strain group vs control group, virulent strain group vs attenuated strain group, and attenuated strain group vs control group) of mRNA, lncRNA and miRNA, and cicrRNA and TUCP. Based on these differentially expressed RNAs, we performed an intersection analysis of molecular events related to pathogenicity and immune response and analyzed the associations between differentially expressed RNAs and mRNA, lncRNA, miRNA, cicrRNA, and TUCP by a HAllA association. The differences between the virulent strain and the control group, the virulent strain and the attenuated strain, and the attenuated strain and the control group were compared for each RNA. We also compared the intersection of AvsC and AvsB for the differentially expressed RNAs, denoted as set P, and the intersection of BvsC and AvsC, denoted as set I–V. We conducted an in−depth analysis of the associations between RNAs (circRNA−miRNA, circRNA−mRNA, mRNA−miRNA, TUCP−mRNA, lncRNA−miRNA, and lncRNA−mRNA) from two dimensions of set P and set I–V through a HALLA association.

From the analysis of differential mRNAs in the P set, we found that TGM3 was associated with virus entry into host cells and the virus cycle. In our setting, the P set represents the molecular events related to the potential pathogenicity, lethality, and inflammatory response of the secondary bacterial infection, so TGM3 can be used as the representative mRNA of the P set. Studies have shown that among the differential miRNAs, ssc−miR−1285 deserves attention, although the existing research results have not indicated its function [23]. However, ssc−miR−1285 was significantly positively correlated with S100G in the correlation analysis of miRNA and mRNA HALLA in the P set in this study. In the absence of IL−10, S100G expression was elevated, inflammatory changes were reduced after the induction of colitis, and S100G inhibited monocyte chemoattractant protein−1 (MCP−1) production by inhibiting NF−κB activation, which is an important anti−inflammatory mediator in fibroblasts after the induction of colitis [24]. Whether ssc−miR−1285 can interact with S100G in terms of anti−inflammatory response still needs to be explored in further research.

In the association analysis of differentially expressed cicrRNA, it was found that the mRNA with the highest correlation with cicrRNA was ROCK2. ROCK2 (Rho−related coiled−coil−containing protein kinase−2) is a protein with multiple functions. Its activity is up−regulated in acute inflammatory injury and chronic diseases, such as diabetes, metabolic syndrome and idiopathic pulmonary fibrosis. The disruption of ROCK2 functions to restore homeostasis by inhibiting pro−inflammatory TH17 cells and promoting regulatory T cells [25]. The miRNA with the highest association with cicrRNA is the newly predicted miRNA novel_424, and the association with multiple cicrRNAs is the highest. The differentially expressed lncRNA LNC_008160 is also significantly associated with novel_424. In the P collection, LNC_011487 was significantly negatively correlated with ssc−miR−215, and LNC_011487 was positively correlated with PI3. The specific role of LNC_011487−ssc−miR−215−PI3 needs further study. The mRNAs most associated with TUCP_002240 are LYZ and TSPAN1; LYZ is a component of innate immunity, and TSPAN1 has a regulatory role in cancer [26,27]. These results indicate that there are multiple molecular mechanisms that regulate the potential pathogenicity, lethality of the virus, and inflammatory response to secondary bacterial infection. This study has contributed to a better understanding of the inflammatory response to pathogenic, lethal, and secondary bacterial infections.

FUT2, CCL5, SLA−DQA1, and CXCL9 in the differential mRNA of the I–V intersection were all related to immune response. This could be explained by the fact that the intersection is associated with the host immune response to SADS−CoV. The differentially expressed miRNA ssc−miR−1285 is associated with FUT2, an enzyme known to be responsible for the addition of fucose to proteins or lipids through α−1,2−fucosylation on the intestinal mucosa, which can act as an attachment site and carbon source for gut bacteria [28]. Tong et al. analyzed the gut microbiota of healthy subjects and they found altered gut microbiota in Fut2 non−secretors [29]. Interestingly, previous studies have shown that in patients with primary sclerosing cholangitis, Fut2 non−secretors exhibit a different composition of bile microbiota compared to that of Fut2 secreters [30]. These findings suggest a potential regulatory role of Fut2 on the gut microbiota. So far, no one has investigated the regulatory effect of virulent and attenuated SADS−CoV strains on the intestinal flora, and it is worth considering whether FUT2 plays a role in this. The functions of most differentially expressed TUPCs are still unknown, but their functions can be mined through their associated RNAs. Among the differentially expressed TUPC−associated mRNAs in the I–V set, PSMB9 is an immunoproteasome component [31]. CCL5 induces the in vitro migration and recruitment of T cells, dendritic cells, eosinophils, NK cells, mast cells, and basophils [32]. Functional polymorphisms in the regulatory regions of the VNN1 gene are associated with susceptibility to inflammatory bowel disease [33]. The above results indicate that we can focus on some highly related RNAs in the I–V collection, and the regulation between them may be very helpful for the study of immune and pathogenic mechanisms.

The GO and KEGG analyses of differentially expressed RNAs in the AvsB group found that most RNAs were involved in protein binding, material metabolism, and autophagy [34]. Among them, autophagy is a key cellular response to pathogen infection. Studies have shown that the process of autophagy regulates intestinal microbiota, and an increase in microbiota is associated with diarrhea [35,36]. In the study of our group, it was found that the structure of gut microbiota changed significantly after SADS−CoV infection in piglets (data unpublished). At the same time, autophagy has also been identified to play a role in the immune system, participating in viral suppression or enhancement processes [37]. The differentially expressed RNAs of AvsB were mainly enriched in Fc gamma R−mediated phagocytosis, Fc epsilon RI signaling pathway, mTOR signaling, and some metabolism−related pathways. Phagocytosis is a key event in the immune system, allowing cells to engulf and eliminate pathogens, and Fc gamma receptors can mediate phagocytosis, suggesting that differentially expressed RNAs may be indirectly involved in the regulation of the immune system [38,39]. Zhang et al. showed by comprehensive functional analysis of SADS−CoV on host mRNA profiles that SADS−CoV infection induced strong immune responses, including innate immunity and cytokine–cytokine receptor interactions, and similar findings were indeed found in our study [40]. The analysis of differentially expressed RNAs of AvsC showed that differentially expressed RNAs were mainly related to immune response, catalysis, metabolism, and other processes, and were enriched in disease−related pathways, antigen processing and presentation, T cell receptor signaling, p53 signaling, Jak−STAT signaling, primary immunodeficiency, NF−kappa B signaling, and other pathways. The differential mRNAs in the AvsC group are involved in more immune and disease−related signaling pathways, which are basically consistent with the molecular events expected by our grouping. The differentially expressed RNA GO enrichment term in the BvsC group was a little different from the other groups, and most of them were related to metabolism and immunity. Toll−like receptor signaling, NF−kappa B signaling, T cell receptor signaling, and other pathways are found in KEGG enrichment. Like the AvsB group, the BvsC group was also significantly enriched in the Fc gamma R−mediated phagocytosis and Fc epsilon RI signaling pathways. In addition, the BvsC group had the most significant enrichment of differentially expressed miRNAs in the intestinal immune pathway network for IgA production, indicating that infection with attenuated strains stimulates intestinal immunity [41].

After the association analysis of the differentially expressed RNAs, a transcriptome-wide association network graph was constructed, and in the P set, the core differentially expressed RNA was OGDH, which was significantly associated with 20 differentially expressed RNAs. In the pathological state of sepsis, OGDH amplifies the inflammatory response through the MAPK pathway, releases pro−inflammatory factors, and induces acute lung injury [42]. In the I–V set, the core differentially expressed RNA is novel_circ_0001750, which is significantly associated with 13 differentially expressed RNAs, including 12 mRNAs and 1 miRNA, but it has not been studied in detail yet. The association network diagram facilitates our study of the role of association RNAs during SADS−CoV infection.

Of course, our research has certain limitations. First, due to the influence of African swine fever, experimental animals are rare and valuable. We only performed three biological replicates during sampling, and more replicates should be performed to ensure the reliability of the experiment. Secondly, we currently only predict the function of differentially expressed RNA, which we will further study in the future.

Supplementary Materials: The following supporting information can be downloaded at: https://www.mdpi.com/article/10.3390/microorganisms11020409/s1, Table S1: The mRNAs of P set and I−V set. Table S2: The miRNAs of P set and I−V set. Table S3: The cicrRNAs of P set and I−V set. Table S4: The lncRNAs of P set and I−V set. Table S5: The TUCPs of P set and I−V set.

Author Contributions: J.M. conceived and designed the study and critically revised the manuscript. X.T. wrote the manuscript. X.T., Q.L. and X.L. performed the experiments and conducted data analysis. Y.S., L.Z., L.G. and T.L. helped in experimental implementation. All authors have read and agreed to the published version of the manuscript.

Funding: The authors would like to acknowledge the Natural Science Foundation of Guangdong Province (2020A1515010220, 2022A1515012473).

Institutional Review Board Statement: This study was carried out in accordance with the recommendations of National Standards for Laboratory Animals of the People's Republic of China (GB149258−2010). The protocol was approved by Animal Research Committees of South China Agricultural University. Pigs used for the study were handled in accordance with good animal practices required by the Animal Ethics Procedures and Guidelines of the People's Republic of China.

Informed Consent Statement: Not applicable.

Data Availability Statement: The original contributions presented in the study are included in the article. Further inquiries can be directed to the corresponding authors.

Conflicts of Interest: The authors declare that they have no competing interest.

References

1. Zhou, P.; Fan, H.; Lan, T.; Yang, X.-L.; Shi, W.-F.; Zhang, W.; Zhu, Y.; Zhang, Y.-W.; Xie, Q.-M.; Mani, S.; et al. Fatal swine acute diarrhoea syndrome caused by an HKU2-related coronavirus of bat origin. *Nature* **2018**, *556*, 255–258. [CrossRef] [PubMed]
2. Chattha, K.S.; Roth, J.A.; Saif, L.J. Strategies for Design and Application of Enteric Viral Vaccines. *Annu. Rev. Anim. Biosci.* **2015**, *3*, 375–395. [CrossRef] [PubMed]
3. Li, K.; Li, H.; Bi, Z.; Gu, J.; Gong, W.; Luo, S.; Zhang, F.; Song, D.; Ye, Y.; Tang, Y. Complete Genome Sequence of a Novel Swine Acute Diarrhea Syndrome Coronavirus, CH/FJWT/2018, Isolated in Fujian, China, in 2018. *Microbiol. Resour. Announc.* **2018**, *7*, e01259-18. [CrossRef]
4. Liu, Y.; Liang, Q.-Z.; Lu, W.; Yang, Y.-L.; Chen, R.; Huang, Y.-W.; Wang, B. A Comparative Analysis of Coronavirus Nucleocapsid (N) Proteins Reveals the SADS-CoV N Protein Antagonizes IFN-β Production by Inducing Ubiquitination of RIG-I. *Front. Immunol.* **2021**, *12*, 688758. [CrossRef]
5. Zhou, Z.; Sun, Y.; Yan, X.; Tang, X.; Li, Q.; Tan, Y.; Lan, T.; Ma, J. Swine acute diarrhea syndrome coronavirus (SADS-CoV) antagonizes interferon-β production via blocking IPS-1 and RIG-I. *Virus Res.* **2019**, *278*, 197843. [CrossRef]
6. Zhou, Z.; Sun, Y.; Yan, X.; Tang, X.; Zhou, L.; Li, Q.; Lan, T.; Ma, J. Swine Acute Diarrhea Syndrome Coronavirus Nucleocapsid Protein Antagonizes Interferon-βProduction via Blocking the Interaction Between TRAF3 and TBK1. *Front. Immunol.* **2021**, *12*. [CrossRef]
7. Zhang, J.; Zhang, L.; Shi, H.; Feng, S.; Feng, T.; Chen, J.; Zhang, X.; Han, Y.; Liu, J.; Wang, Y.; et al. Swine acute diarrhea syndrome coronavirus replication is reduced by inhibition of the extracellular signal-regulated kinase (ERK) signaling pathway. *Virology* **2021**, *565*, 96–105. [CrossRef]
8. Edwards, C.E.; Yount, B.L.; Graham, R.L.; Leist, S.R.; Hou, Y.; Dinnon III, K.H.; Sims, A.C.; Swanstrom, J.; Gully, K.; Scobey, T.D.; et al. Swine acute diarrhea syndrome coronavirus replication in primary human cells reveals potential susceptibility to infection. *Proc. Natl. Acad. Sci. USA* **2020**, *117*, 26915–26925. [CrossRef] [PubMed]
9. Luo, Y.; Tan, C.W.; Xie, S.; Chen, Y.; Yao, Y.; Zhao, K.; Zhu, Y.; Wang, Q.; Liu, M.; Yang, X.; et al. Identification of ZDHHC17 as a Potential Drug Target for Swine Acute Diarrhea Syndrome Coronavirus Infection. *mBio* **2022**, *12*, e02342-21. [CrossRef]
10. Sun, Y.; Cheng, J.; Luo, Y.; Yan, X.L.; Wu, Z.X.; He, L.L.; Tan, Y.R.; Zhou, Z.H.; Li, Q.N.; Zhou, L.; et al. Attenuation of a virulent swine acute diarrhea syndrome coronavirus strain via cell culture passage. *Virology* **2019**, *538*, 61–70. [CrossRef]
11. Wei, H.Y.; Huang, S.; Wang, J.; Gao, F.; Jiang, J. Comparison of methods for library construction and short read annotation of shellfish viral metagenomes. *Genes Genom.* **2018**, *40*, 281–288. [CrossRef]
12. Linsen, S.E.; de Wit, E.; Janssens, G.; Heater, S.; Chapman, L.; Parkin, R.K.; Fritz, B.; Wyman, S.K.; de Bruijin, E.; Voest, E.E.; et al. Limitations and possibilities of small RNA digital gene expression profiling. *Nat. Methods* **2009**, *6*, 474–476. [CrossRef]

13. Zhao, X.; Zhang, Y.; Huang, D.; Meng, J.; Wei, Z. Detecting RNA modification using direct RNA sequencing: A systematic review. *Comput. Struct. Biotechnol. J.* **2022**, *20*, 5740–5749. [CrossRef]
14. Levin, J.Z.; Yassour, M.; Adiconis, X.; Nusbaum, C.; Thompson, D.A.; Friedman, N.; Gnirke, A.; Regev, A. Comprehensive comparative analysis of strand-specific RNA sequencing methods. *Nat. Methods* **2010**, *7*, 709–715. [CrossRef] [PubMed]
15. Parkhomchuk, D.; Borodina, T.; Amstislavskiy, V.; Banaru, M.; Hallen, L.; Krobitsch, S.; Lehrach, H.; Soldatov, A. Transcriptome analysis by strand-specific sequencing of complementary DNA. *Nucleic Acids Res.* **2009**, *37*, e123. [CrossRef]
16. Garalde, D.R.; Snell, E.A.; Jachimowicz, D.; Sipos, B.; Lloyd, J.H.; Bruce, M.; Pantic, N.; Admassu, T.; James, P.; Warland, A.; et al. Highly parallel direct RNA sequencing on an array of nanopores. *Nat. Methods* **2018**, *15*, 201–206. [CrossRef]
17. Young, M.D.; Davidson, N.; Wakefield, M.J.; Smyth, G.K.; Oshlack, A. Goseq: Gene Ontology testing for RNA-seq datasets. *R Bioconductor* **2012**, *8*, 1–25.
18. Kanehisa, M.; Araki, M.; Goto, S.; Hattori, M.; Hirakawa, M.; Itoh, M.; Katayama, T.; Kawashima, S.; Okuda, S.; Tokimatsu, T.; et al. KEGG for linking genomes to life and the environment. *Nucleic Acids Res.* **2008**, *36*, 480–484. [CrossRef] [PubMed]
19. Ghazi, A.R.; Sucipto, K.; Rahnavard, A.; Franzosa, E.A.; Mclver, L.J.; Lloyd-Price, J.; Schwager, E.; Weingart, G.; Moon, Y.S.; Morgan, X.C.; et al. High-sensitivity pattern discovery in large, paired multiomic datasets. *Bioinformatics* **2022**, *38* (Suppl. S1), i378–i385. [CrossRef] [PubMed]
20. Chen, S.; Zhang, Y.; Ding, X.; Li, W. Identification of lncRNA/circRNA-miRNA-mRNA ceRNA Network as Biomarkers for Hepatocellular Carcinoma. *Front. Genet.* **2022**, *13*, 838869. [CrossRef] [PubMed]
21. Greenberg, S.; Grinstein, S. Phagocytosis and innate immunity. *Curr. Opin. Immunol.* **2002**, *14*, 136–145. [CrossRef] [PubMed]
22. Xu, Z.; Hu, Q.; Zang, X.; Zhou, C.; Liu, D.; Liu, G.; Hong, L. Analysis of Transcripts of Uncertain Coding Potential Using RNA Sequencing During the Preattachment Phase in Goat Endometrium. *DNA Cell Biol.* **2021**, *40*, 998–1008. [CrossRef]
23. Zhu, X.H.; Han, L.; Zhang, R.; Zhang, P.; Chen, F.; Yu, J.; Luo, H.; Han, X. The functional activity of donor kidneys is negatively regulated by microribonucleic acid-451 in different perfusion methods to inhibit adenosine triphosphate metabolism and the proliferation of HK2 cells. *Bioengineered* **2022**, *13*, 12706–12717. [CrossRef] [PubMed]
24. Ishiguro, K.; Watanabe, O.; Nakamura, M.; Yamamura, T.; Ando, T.; Goto, H.; Hirooka, Y. S100G expression and function in fibroblasts on colitis induction. *Int. Immunopharmacol.* **2016**, *39*, 92–96. [CrossRef] [PubMed]
25. ROCK2 agent stops GVHD. *Nat. Biotechnol.* **2021**, *39*, 1323. [CrossRef]
26. Zhou, C.; Liang, Y.; Zhou, L.; Yan, Y.; Liu, N.; Zhang, R.; Huang, Y.; Wang, M.; Yang, Y.; Ali, D.W.; et al. TSPAN1 promotes autophagy flux and mediates cooperation between WNT-CTNNB1 signaling and autophagy via the MIR454-FAM83A-TSPAN1 axis in pancreatic cancer. *Autophagy* **2021**, *17*, 3175–3195. [CrossRef]
27. Moreno-Navarrete, J.M.; Latorre, J.; Lluch, A.; Ortega, F.J.; Comas, F.; Arnoriaga-Rodriguez, M.; Ricart, W.; Fernadez-Real, J.M. Lysozyme is a component of the innate immune system linked to obesity associated-chronic low-grade inflammation and altered glucose tolerance. *Clin. Nutr.* **2021**, *40*, 1420–1429. [CrossRef]
28. Ferrer-Admetlla, A.; Sikora, M.; Laayouri, H.; Esteve, A.; Roubinet, F.; Blancher, A.; Calafell, F.; Bertranpetit, J.; Casals, F. A Natural History of FUT2 Polymorphism in Humans. *Mol. Biol. Evol.* **2009**, *26*, 1993–2003. [CrossRef]
29. Tong, M.; McHardy, I.; Ruegger, P.; Goudarzi, M.; Kashyap, P.C.; Haritunians, T.; Li, X.; Graeber, T.G.; Schwager, E.; Huttenhower, C.; et al. Reprograming of gut microbiome energy metabolism by the FUT2 Crohn's disease risk polymorphism. *ISME J.* **2014**, *8*, 2193–2206. [CrossRef]
30. Maroni, L.; Hohenester, S.D.; van de Graaf, S.F.J.; Tolenaars, D.; van Lienden, K.; Verheij, J.; Marzioni, M.; Karisen, T.H.; Elferink, R.P.J.O.; Beuers, U. Knockout of the primary sclerosing cholangitis-risk gene Fut2 causes liver disease in mice. *Hepatology* **2017**, *66*, 542–554. [CrossRef]
31. Choi, J.-H.; Jo, H.S.; Lim, S.; Kim, H.-T.; Lee, K.W.; Moon, K.H.; Ha, T.; Kwak, S.S.; Kim, Y.; Lee, E.J.; et al. mTORC1 accelerates retinal development via the immunoproteasome. *Nat. Commun.* **2018**, *9*, 2502. [CrossRef]
32. Schall, T.J.; Jongstra, J.; Dryer, B.J.; Jorgensen, J.; Clayberger, C.; Davis, M.M.; Krensky, A.M. A human T cell-specific molecule is a member of a new gene family. *J. Immunol.* **1988**, *141*, 1018–1025. [CrossRef] [PubMed]
33. Gensollen, T.; Bourges, C.; Rihet, P.; Rostan, A.; Millet, V.; Noguchi, T.; Bourdon, V.; Sobol, H.; Dubuquoy, L.; Bertin, B.; et al. Functional Polymorphisms in the Regulatory Regions of the VNN1 Gene Are Associated with Susceptibility to Inflammatory Bowel Diseases. *Inflamm. Bowel Dis.* **2013**, *19*, 2315–2325. [CrossRef]
34. De Oliveira, A.P.; Lopes, A.L.F.; Pacheco, G.; de Sá Guimarães Nolêto, I.R.; Nicolau, L.A.D.; Medeiros, J.V.R. Premises among SARS-CoV-2, dysbiosis and diarrhea: Walking through the ACE2/mTOR/autophagy route. *Med. Hypotheses* **2020**, *144*, 110243. [CrossRef] [PubMed]
35. Cadwell, K.; Patel, K.K.; Komatsu, M.; Virgin, H.W.; Stappenbeck, T.S. A common role for Atg16L1, Atg5 and Atg7 in small intestinal Paneth cells and Crohn disease. *Autophagy* **2009**, *5*, 250–252. [CrossRef]
36. Cadwell, K.; Liu, J.Y.; Brown, S.L.; Miyoshi, H.; Loh, J.; Lennerz, J.K.; Kishi, C.; Wumesh, K.C.; Carrero, J.A.; Hunt, S.; et al. A key role for autophagy and the autophagy gene Atg16l1 in mouse and human intestinal Paneth cells. *Nature* **2009**, *456*, 259–263. [CrossRef] [PubMed]
37. Dreux, M.; Chisari, F.V. Viruses and the autophagy machinery. *Cell Cycle* **2010**, *9*, 1295–1307. [CrossRef]
38. Rivas-Fuentes, S.; García-García, E.; Nieto-Castañeda, G.; Rosales, C. Fcγ receptors exhibit different phagocytosis potential in human neutrophils. *Cell. Immunol.* **2010**, *263*, 114–121. [CrossRef]

39. Kang, J.; Park, K.-H.; Kim, J.-J.; Jo, E.-K.; Han, M.-K.; Kim, U.-H. The Role of CD38 in Fcγ Receptor (FcγR)-mediated Phagocytosis in Murine Macrophages. *J. Biol. Chem.* **2012**, *287*, 14502–14514. [CrossRef]
40. Zhang, F.; Yuan, W.; Li, Z.; Zhang, Y.; Ye, Y.; Li, K.; Ding, Z.; Chen, Y.; Cheng, T.; Wu, Q.; et al. RNA-Seq-Based Whole Transcriptome Analysis of IPEC-J2 Cells During Swine Acute Diarrhea Syndrome Coronavirus Infection. *Front. Vet. Sci.* **2020**, *7*, 492. [CrossRef]
41. Tezuka, H.; Ohteki, T. Regulation of IgA Production by Intestinal Dendritic Cells and Related Cells. *Front. Immunol.* **2019**, *10*, 1891. [CrossRef] [PubMed]
42. Hao, Y.; Wang, Z.; Wang, X.; Zhan, W.; Wu, D. OGDH is involved in sepsis induced acute lung injury through the MAPK pathway. *J. Thorac. Dis.* **2021**, *13*, 5042–5054. [CrossRef] [PubMed]

microorganisms

MDPI

Article

Capture ELISA for KPC Detection in Gram-Negative Bacilli: Development and Standardisation

André Valencio [1,†], Miriam Aparecida da Silva [2,†], Fernanda Fernandes Santos [1], Juliana Moutinho Polatto [2], Marcelo Marcondes Ferreira Machado [1], Roxane Maria Fontes Piazza [2] and Ana Cristina Gales [1,*]

1 Division of Infectious Diseases, Department of Internal Medicine, Escola Paulista de Medicina, Universidade Federal de São Paulo, São Paulo 04039-032, Brazil; andrebiomedicovalencio@gmail.com (A.V.); marcelo.marcondes@unifesp.br (M.M.F.M.)
2 Laboratório de Bacteriologia, Instituto Butantan, São Paulo 05503-900, Brazil
* Correspondence: ana.gales@unifesp.br
† André Valencio and Miriam Aparecida da Silva contribuited equally to this work.

Abstract: The detection of KPC-type carbapenemases is necessary for guiding appropriate antibiotic therapy and the implementation of antimicrobial stewardship and infection control measures. Currently, few tests are capable of differentiating carbapenemase types, restricting the lab reports to their presence or not. The aim of this work was to raise antibodies and develop an ELISA test to detect KPC-2 and its D179 mutants. The ELISA-KPC test was designed using rabbit and mouse polyclonal antibodies. Four different protocols were tested to select the bacterial inoculum with the highest sensitivity and specificity rates. The standardisation procedure was performed using 109 previously characterised clinical isolates, showing 100% of sensitivity and 89% of specificity. The ELISA-KPC detected all isolates producing carbapenemases, including KPC variants displaying the ESBL phenotype such as KPC-33 and -66.

Keywords: carbapenemases; *Klebsiella pneumoniae*; antibodies; immunodiagnostic

Citation: Valencio, A.; da Silva, M.A.; Santos, F.F.; Polatto, J.M.; Machado, M.M.F.; Piazza, R.M.F.; Gales, A.C. Capture ELISA for KPC Detection in Gram-Negative Bacilli: Development and Standardisation. *Microorganisms* **2023**, *11*, 1052. https://doi.org/10.3390/microorganisms11041052

Academic Editors: Shengxi Chen and Fabio Zicker

Received: 17 March 2023
Revised: 3 April 2023
Accepted: 5 April 2023
Published: 18 April 2023

1. Introduction

The World Health Organization (WHO) has declared antimicrobial resistance (AMR) as one of the top 10 public health threats [1]. Infections caused by antimicrobial-resistant pathogens occur in all geographic regions, increasing the length and cost of hospitalisation, and especially, impacting the morbidity and mortality rates [2,3]. Tacconelli et al. (2018), in collaboration with the WHO, published a list of the bacterial pathogens to be considered as priorities for research and development of new antimicrobials. *Acinetobacter baumannii*, *Pseudomonas aeruginosa*, and Enterobacterales resistant to carbapenems were the top three priority pathogens [4].

Among Enterobacterales, the production of carbapenemases is the main mechanism of carbapenem resistance. Carbapenemases present a great concern worldwide due to their wide variety and dissemination among different species through horizontal gene transfer (HGT). These enzymes belong to classes A, B, and D of the Ambler molecular classification and usually hydrolyse almost all β-lactams [5,6]. Although the frequency of New Delhi Metallo-beta-lactamase (NDM—class B) has increased during the COVID-19 pandemic, *Klebsiella pneumoniae* carbapenemase (KPC—class A) is still the most frequent carbapenemase in many geographic regions [7,8]. The rapid and successful dissemination of bla_{KPC}-like has been associated with the spread of endemic *K. pneumoniae* CC258 clones. In addition, bla_{KPC}-like has been acquired by distinct Inc plasmids facilitating its spread [9].

Early adequate antimicrobial therapy has been associated with lower mortality rates. Kumar et al. (2006) reported that for every hour of delay in the adequate prescription of antimicrobials, there was a 7.6% increase in mortality rates of patients with sepsis [10]. Unfortunately, the infections caused by KPC-2-producing *K. pneumoniae* isolates have been associated with high mortality rates [11,12]. New combinations of β-lactamase inhibitor–β-lactams have been developed and are already available for hospital use. However, these combinations only have activity against Gram-negative bacilli (GNB) producers of class A and OXA-48 carbapenemases (class D) [13]. The detection of β-lactamase-encoding genes via the PCR technique, followed by DNA sequencing or whole-genome sequencing, is considered the gold standard for the identification of β-lactamase-encoding genes. Different methodologies were proposed to investigate the production of carbapenemase using a clinical isolate in a short period of time and with a lower cost than PCR. However, tests such as CarbaNP or BlueCarba generally do not differentiate the carbapenemase classes [14,15]. In contrast, immunochromatographic tests have shown high sensitivity and specificity for distinguishing the class of most frequent carbapenemases with a rapid turnaround time for results without requiring technical expertise [16]. Unfortunately, these tests are expensive for low- and middle-income countries such as Brazil (about USD 10.00 per test). In addition, some KPC variants have not been accurately detected via immunochromatographic tests [17]. Although KPC variants not identified with such tests usually show resistance to ceftazidime–avibactam, they are susceptible to carbapenems. However, clinical failure has been observed when carbapenems were prescribed as monotherapy for the infections caused by such variants due to the occurrence of a heterogeneous population [18]. To date, no ELISA tests have been developed for the detection of carbapenemases in GNB. Herein, we aimed to develop an immunoenzymatic test to detect KPC production in GNB clinical isolates, given the extensive need for the detection of carbapenemase production in different Enterobacterales.

2. Materials and Methods

2.1. Expression and Purification of Recombinant KPC (rKPC-2)

For the development of anti-KPC antibodies, the animals (New Zealand rabbit and Balb/c mice) were immunised with purified recombinant KPC-2 enzyme (rKPC-2). The *K. pneumoniae* strain ATCC BAA-1705 (American Type Culture Collection, Manassas, VA, USA) producer of KPC-2 was commercially obtained and subcultured twice on blood agar (Oxoid, Basingstoke, England). The bla_{KPC-2} amplicon obtained from *K. pneumoniae* BAA-1705 was cloned in the plasmid pEt26a+. Primers were designed using the restriction sites for Ndel (forward 5 GGTGGTCATATGTCACTGTATCGCCGTCTAGTT 3′) and Sall (reverse ACCACCGTCGACCTGCCCGTTGACGCCCAATCCCTCGA 3′). The expression of the rKPC-2 protein was analysed using the *Escherichia coli* strain BL21 (D3E) with the addition of 0.2 mM IPTG (isopropyl-β-D-thiogalactoside) as an expression inducer for 18 h at 20 °C. The osmotic shock was performed to release rKPC-2 from the periplasmic space [19]. Purification was assessed using hydrophobic interaction chromatography with a 5 mL Resource PHE column (GE, Healthcare, Orsay, France) at a flow rate of 5 mL/min. The elution of the ligands was analysed using the linear gradient method with 0% to 100% of the elution solution (50 mM NaH_2PO_4) in 20 column volumes (CVs). The molar extinction coefficient of the rKPC-2 was calculated by applying the ProtParam tool and using λ 280 nm wavelength.

2.2. Production of Antibodies

The experiments were conducted in agreement with the Ethical Principles in Animal Research, adopted by the Brazilian College of Animal Experimentation, and they were approved by the Ethical Committee for Animal Research of the Butantan Institute (CEUA no 9599160922).

A New Zealand rabbit (60–65 days old) and two Balb/c mice (4 to 6 weeks old weighing 18–22 g) were immunised with rKPC-2 for the development of primary and secondary antibodies. The rabbit was immunised with 100 µg of rKPC-2 adsorbed to aluminium hydroxide adjuvant (10 times the amount of antigen) diluted in PBS, via intramuscular followed by two boosters at 15-day intervals. Mice were immunised subcutaneously, and the immunisation protocols consisted of one immunisation with rKPC-2 (10 µg) adsorbed to aluminium hydroxide adjuvant (10 times the amount of antigen) diluted in PBS followed, four weeks later, followed by two booster injections (rKPC-2, 10 µg) with a 15-day interval.

Once a successful immunisation was observed, the animals' blood was collected, the mice were euthanised, and the serum purified. Polyclonal antibodies (pAbs) from the rabbit and mice were filtered (0.45 µm) and purified using Protein A and Protein G Sepharose columns (GE, Healthcare, Orsay, France), respectively, both coupled to a liquid chromatography system—ÄKTA Purifier (GE Healthcare, Orsay, France). After dialysis against 20 mM sodium phosphate, they were concentrated using PEG 6000. The presence of Fab and Fc was determined via 15% polyacrylamide gel electrophoresis containing sodium dodecyl sulphate (SDS–PAGE) stained with Coomassie blue R-250. The final immunoglobulin dosage was performed at λ280 nm using a Nanodrop Lite Spectrophotometer (Thermo Scientific, Waltham, MA, USA).

2.3. Checkboard Titration

The cross-reactivity and the best concentration between mouse rKPC-2-pAb and rabbit rKPC-2-pAb for detecting specific KPC-2 in the bacterial culture supernatant were determined with a capture ELISA immunoassay using different pAb concentrations ranging from 0.781 to 50 µg/mL, with a predetermined rKPC concentration (10 µg/mL) as antigen.

2.4. Capture ELISA Immunoassay

The 96-well MaxiSorp microplates (Nunc®, Rochester, NY, USA) were incubated with 6.25 µg/mL of mouse rKPC-2-pAb in a carbonate–bicarbonate buffer, pH 9.6, at 37 °C for 2 h and then further at 4 °C for 16 h. Phosphate-buffered saline (PBS) containing 1% of bovine serum albumin (BSA) (PBS-BSA) was added as a blocking agent and incubated for 1 h at 37 °C. Thereafter, the supernatant of bacterial cultures, PBS, and 10 µg/mL rKPC-2 (negative and positive control, respectively) were incubated for 1 h at 37 °C. Next, 6.25 µg/mL of rabbit rKPC-2-pAb was added and incubated for an additional hour at 37 °C. KPC-2 from samples was then detected with goat anti-rabbit IgG peroxidase (Sigma-Aldrich, St. Louis, MO, USA) diluted 1:5000 in a 0.1% PBS-BSA solution. Reactions were developed with 0.5 mg/mL O-phenylenediamine (OPD; Sigma-Aldrich Co, St. Louis, MO, USA) plus 0.5 µL/mL hydrogen peroxide in 0.05 M citrate–phosphate buffer, pH 5.0, in the dark at room temperature. The reactions were interrupted after 15 min via the addition of 50 µL of 1 M HCl. The absorbance was measured at 492 nm using a Multiskan EX ELISA reader (Labsystems, Milford, MA, USA). At each step, the volume added was 100 µL/well, except in the washing and blocking steps, when the volume used was 200 µL/well. Between incubation periods, the plates were washed three times with PBS-Tween 0.05%. All experiments were carried out in technical duplicates, and the results correspond to three independent experiments (biological replicates).

2.5. Bacterial Sample Preparation

Thirteen isolates were selected to screen for the best condition for the isolation of bacterial colonies to be tested using an indirect ELISA (Table 1).

Initially, the isolates were subcultured using Mueller–Hinton agar supplemented with ceftazidime 2 µg/mL and incubated at 37 °C for 18 h and then submitted to four different protocols as follows: (i) direct colonies (DCs): A 10 µL loop of bacterial colonies was suspended in 500 µL of 1% PBS (140 mM NaCl; 9 µM Na_2HPO_4; 2 µM NaH_2PO_4 + H_2O; pH 7.4) in a 1.5 mL microtube; (ii) pre-inoculum (PD-OD1): Two or three colonies were inoculated in 6 mL of Mueller–Hinton broth (MHB) and incubated at 37 °C for 18 h. After the incubation period, 1% of the bacterial suspension was transferred to a new tube and incubated at 37 °C until reaching an OD of 1.0. Then, 500 µL was transferred to a 1.5 mL microtube and centrifuged at 4100× *g* for 10 min. The supernatant was discarded, and the pellet was resuspended in 500 µL 1% PBS; (iii) pre-inoculum followed by 18 h growth (PD-G): A 1 µL loop of bacterial culture was suspended in 6 mL of MHB and incubated at 37 °C for 18 h. After the incubation period, 1% of the bacterial suspension was transferred to a new tube and incubated at 37 °C for 18 h. Then, 500 µL was transferred to a 1.5 mL microtube and centrifuged at 4100× *g* for 10 min, the supernatant was discarded, and the pellet was resuspended in 500 µL 1% PBS; and (iv) growth in Mueller–Hinton broth (GMHB): A 1 µL loop of bacterial culture was suspended in 6 mL of MHB and incubated at 37 °C for 18 h. Then, 500 µL was transferred to a 1.5 mL microtube and centrifuged at 4100× *g* for 10 min, the supernatant was discarded, and the pellet was resuspended in 500 µL 1% PBS. All tubes were subjected to disruption for 3 cycles of 40 Hz in an ice bath, centrifuged at 16,200× *g* for 10 min, and employed in the ELISA test (Figure 1). The indirect ELISA was performed in duplicate, and the reading was carried out at 492 µm using an ELISA reader (Multiskan—Thermo Sci, Waltham, MA, USA).

Table 1. Isolates used for screening protocol.

Bacterial Species	Type of Carbapenemases (Number of Isolates Tested)
Carbapenemase-producing isolates	
Klebsiella pneumoniae (7)	KPC-2 (4) BKC-1 (3)
Pseudomonas aeruginosa (3)	VIM-1 (1) GES-5 (1) SPM-1 (1)
Escherichia coli (1)	NDM-1 (1)
Non-Carbapenemase-producing isolates	
Klebsiella pneumoniae (1)	SHV-18 (1)
Non-β-lactamase producing isolates	
Escherichia coli (1)	None
Total:	13 isolates

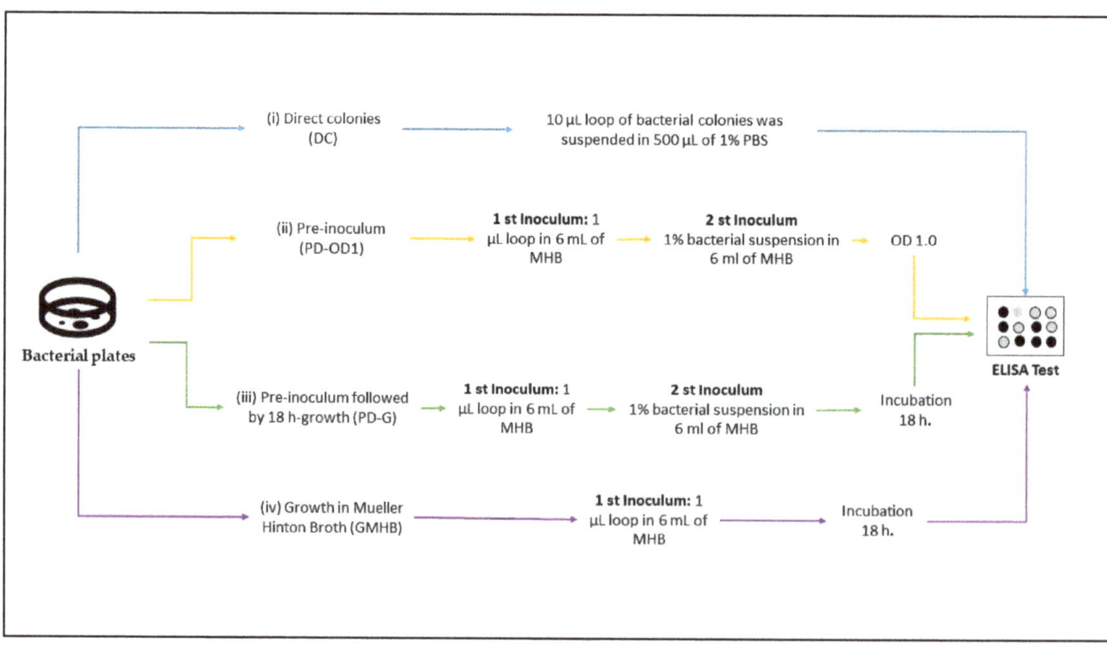

Figure 1. Graphical summary of protocols for the preparation of bacterial isolates before the ELISA test.

2.6. Validation of Capture ELISA

The final evaluation of the ELISA-KPC test was carried out using the protocol that presented the highest sensitivity, specificity, and shortest response time for performing the test. Therefore, the performance of the ELISA test was evaluated by testing 109 different isolates previously characterised via PCR and sequencing sanger as follows: 50 isolates carrying different variants of bla_{KPC}; 3 co-producer carbapenemases ($bla_{KPC\text{-}2}$/$bla_{BKC\text{-}1}$); 24 isolates other carbapenemases (non-bla_{KPC}-like); 19 isolates resistant to carbapenems but non-carbapenemase; 3 isolates producing ESBL; and 10 isolates susceptible to the antimicrobial. Isolates from different species were included in this study (*K. pneumoniae*, *K. oxytoca*, *Enterobacter* spp., *Escherichia coli*, *Proteus mirabilis*, *Shigella flexneri*, *Salmonella enterica* subsp. *Enterica*, *Pseudomonas* spp., and *Acinetobacter baumannii*) (Table 2).

Table 2. Isolates used for ELISA-KPC test verification.

Carbapenemase-Producing Isolates/Species		Type of Carbapenemase (Number of Isolates Tested)
KPC-producing isolates positive (N = 50)		
	Klebsiella pneumoniae (47)	
		KPC-2 (35)
		KPC-3 (7)
		KPC-7 (1)
		KPC-11 (1)
		KPC-33 (1)
		KPC-66 (1)
		KPC-144 (1)
	Klebsiella oxytoca (1)	
		KPC-2 (1)
	Pseudomonas aeruginosa (2)	
		KPC-2 (2)
Double-carbapenemase producer (N = 3)		
	Klebsiella pneumoniae (3)	
		KPC-2/BKC-1 (3)
Other carbapenemases (N = 24)		
	Klebsiella pneumoniae (12)	
		BKC-1 (6)
		BKC-2 (1)
		NDM-1 (1)
		OXA-48 (1)
		GES-5 (2)
		IMP-1 (1)
	Pseudomonas aeruginosa (8)	
		NDM-1 (1)
		IMP-1 (2)
		GIM-1 (2)
		SPM-1 (3)
	Pseudomonas monteilli (1)	
		VIM-1 (1)
	Enterobacter spp. (3)	
		NDM-1 (3)
Non-carbapenemase-producing isolates (N = 19)		
	Klebsiella pneumoniae (2)	
		OmpK-36 (1)
		OmpK (35/36) (1)
	Pseudomonas aeruginosa (16)	
		Non carbapenemase producer
	Serratia marcescens (1)	
		GES-16 (1)

Table 2. *Cont.*

Carbapenemase-Producing Isolates/Species	Type of Carbapenemase (Number of Isolates Tested)
ESBL-producing isolates (N = 3)	
Klebsiella pneumoniae (3)	
	CTX-M8 (2)
	SHV-18 (1)
Isolates susceptible to carbapenems (N = 10)	
Klebsiella pneumoniae (4)	
Escherichia coli (2)	
Acinetobacter baumannii (1)	
Proteus mirabilis (1)	
Shigella flexneri (1)	
Salmonella enterica subsp. *Enterica* (1)	
	Total N = 109 isolates

2.7. Data Analysis

Data were processed using RStudio to calculate the optimal cut-off, sensitivity, and specificity for each protocol. Student's *t*-test was used to assess the statistical differences between positive (KPC producers) and negative (non-KPC producers) cases for each method.

3. Results

3.1. Recombinant bla$_{KPC-2}$

Recombinant *bla*$_{KPC-2}$ was cloned into pEt-26a+. The sequencing result confirmed and demonstrated 100% similarity with the *bla*$_{KPC-2}$ deposited in the NCBI database (Accession number: CP039975.1). The production of rKPC-2 was visualised on an SDS–PAGE gel, where a component of electrophoretic mobility of 32 kDa (Figure 2) was visualised, and then nitrocefin hydrolysis was confirmed. The total yield of purified rKPC-2 protein was 14 mL at the concentration of 705 µg/mL.

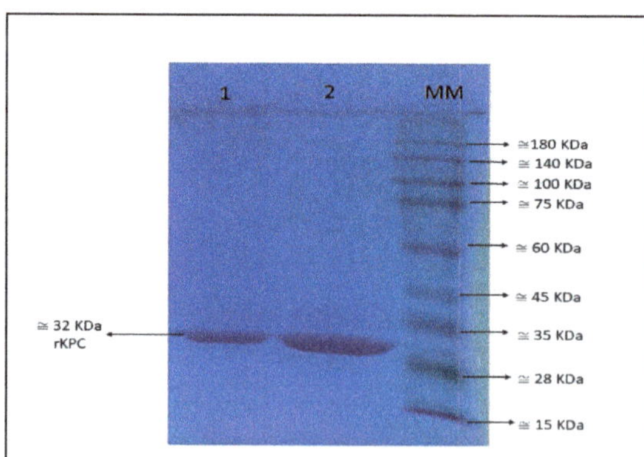

Figure 2. Presence of purified rKPC-2 on SDS–PAGE gel. The top two fractions were collected from AKTA equipment. Legend, 1: Tube 17, 2: Tube 18, MM: molecular marker.

3.2. Immunisation with rKPC-2 Induced Specific Serum Antibody Response in Rabbit and Mice and No Cross-Reactivity

The generated sera showed high titres after immunisation (at 1/100 dilution (OD 492 µm) >1.0). The titration of rabbit and mouse antibodies was performed using ELISA presenting 1/819,200 (81.92×10^4) and 1/6400 (6.4×10^3) for the rabbit and mice (Figure 3), respectively.

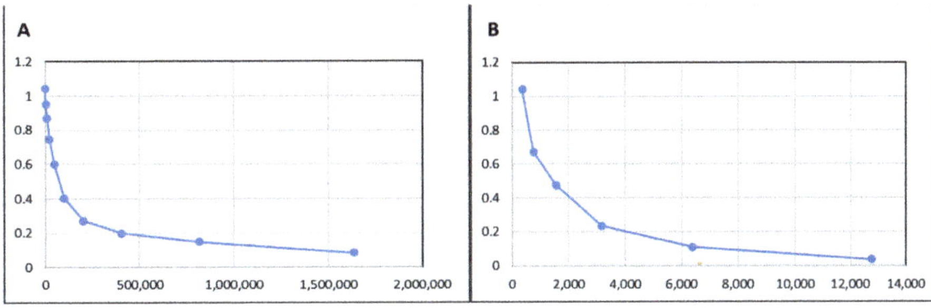

Figure 3. Titration of animal serum containing antibodies against rKPC-2. Legend: (**A**) titre of anti-rKPC-2 antibodies in rabbit serum; (**B**) titre of anti-rKPC-2 antibodies in mouse serum.

The purification of antibodies using the affinity column recovered 44 mL with a concentration of 65.7 mg/mL and 300μL with a concentration of 1.29 mg/mL for the rabbit and mice, respectively (Figure 4).

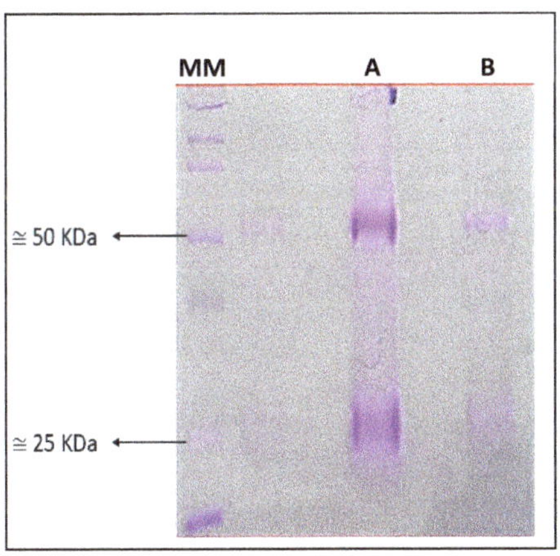

Figure 4. Purification of polyclonal antibodies from (**MM**) molecular marker; (**A**) rabbit and (**B**) mice on SDS–PAGE gel. Antibodies have two subunits, one with 50 KDa and the other with 25 KDa.

The cross-reactivity between pAbs of mouse rKPC-2-pAb and rabbit rKPC-2 for capture ELISA was investigated, and the best concentration chosen was 6.25 μg/mL for both pAbs, which presented the lowest acceptable recognition among the pAbs (cross-reacting) without losing the ability to recognise the studied antigen (rKPC) (Table 3).

Table 3. Results of the block titration. Average of three block titration assays to define optimal mouse and rabbit pAb concentrations to be used in the ELISA test. Legend: **A**: titration using PBS; **B**, titration using rKPC-2. * We used the 6.25 μg/mL mouse pAb dilution and also 6.25 μg/mL rabbit pAb.

A	Rabbit (μg/mL)			
Mice (μg/mL)	**25**	**12.5**	**6.25**	**3.125**
12.5	0.195	0.146	0.11	0.104
6.25	0.181	0.116	0.104 (*)	0.073
3.125	0.158	0.109	0.078	0.063
1.5625	0.154	0.095	0.07	0.057
0.78125	0.151	0.098	0.076	0.059
0.390625	0.146	0.091	0.066	0.057
0.1953125	0.147	0.1	0.073	0.058
0.09765625	0.149	0.096	0.078	0.06

Table 3. *Cont.*

B		Rabbit (µg/mL)		
Mice (µg/mL)	**25**	**12.5**	**6.25**	**3.125**
12.5	1.165	1.091	1.062	0.948
6.25	1.055	1.097	1.026 (*)	0.925
3.125	0.855	0.825	0.835	0.708
1.5625	0.833	0.714	0.666	0.571
0.78125	0.923	0.754	0.695	0.518
0.390625	0.936	0.695	0.722	0.57
0.1953125	0.87	0.829	0.682	0.609
0.09765625	0.988	0.871	0.737	0.576

3.3. Different Bacterial Preparation Induced Different Results in ELISA

The protocol for bacterial sample preparation influenced the ELISA results with very different optimal cut-offs obtained for the four conditions. For the DC protocol, the ROC curve demonstrated an ideal cut-off point of 0.269, with a sensitivity of 75% and a specificity of 77.8%. The pre-inoculum protocols presented cut-offs of 0.486 and 0.437 for PD-OD1 and PD-G, respectively. Both PD-OD1 and PD-G had 100% specificity and 100% sensitivity. The GMHB protocol had a cut-off of 0.159, with 100% sensitivity and 88.9% specificity. Student's *t*-test showed a statistical difference for PD-OD1, PD-G, and GMHB protocols; however, DC protocol showed $p < 0.05$, not discriminating the producers and non-producers of KPC. The turnaround time for obtaining the ELISA test results was 3–4 h, 22–24 h, 28–30 h, and 40–42 h for DC, GMHB, PD-G, and PD-OD1 protocols, respectively (Figure 5).

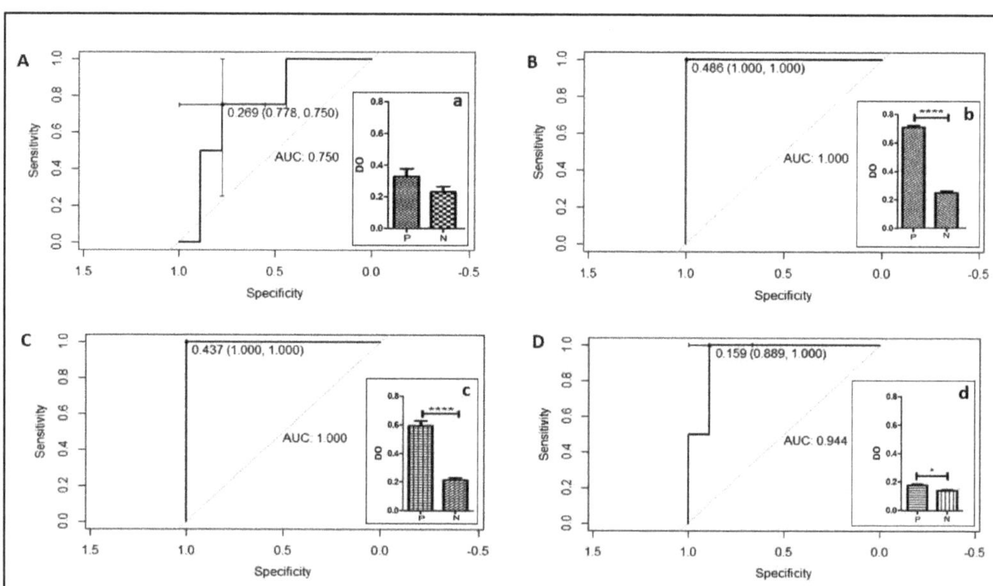

Figure 5. Statistical parameters of the four protocols of bacterial inoculum preparation: (**A**) ROC DC curve with cut-off 0.269, sensitivity: 77.8%, and specificity: 75%; (**a**) *t*-test *p*-value < 0.05; (**B**) ROC PD-OD1 curve, cut-off of 0.486, sensitivity of 100%, and specificity of 100%; (**b**) *t*-test **** *p*-value < 0.0001. (**C**) ROC PD-G curve, cut-off of 0.437; 100% sensitivity, and 100% specificity; (**c**) *t*-test with **** *p*-value < 0.0001; (**D**) ROC GMHB curve, cut-off 0.159, sensitivity 88.9%, and specificity 100%, (**d**) *t*-test * *p*-value < 0.05; legend: P: positive isolates; N: negative isolates.

3.4. Validation of KPC Detection

The GMHB was selected as the best condition for evaluating KPC production via ELISA. A total of 109 bacterial isolates, comprising 53 positives for bla_{KPC} and 56 negatives for the bla_{KPC}, were tested. The variants of KPC (KPC-33 and KPC-66) showing resistance to ceftazidime–avibactam but susceptibility to carbapenems and the KPC-144 variant resistant to ceftazidime–avibactam and carbapenems were evaluated. The KPC-ELISA-GMHB test detected all 53 KPC-producing isolates (50 *K. pneumoniae*, 1 *K. oxytoca*, and 2 *P. aeruginosa*), showing 100% sensitivity and 89.9% specificity with $p < 0.05$ (Figures 6 and 7). Of the 56 non-KPC-producing isolates, 6 (5.5%) were misclassified as KPC producers. False-negative results were observed with *K. pneumoniae* producers of BKC-1 (two isolates) and GES-5 (one isolate); NDM-1-producing *Enterobacter* spp. (one isolate); SPM-1-producing *P. aeruginosa* (one isolate); and a wild-type *K. pneumoniae*.

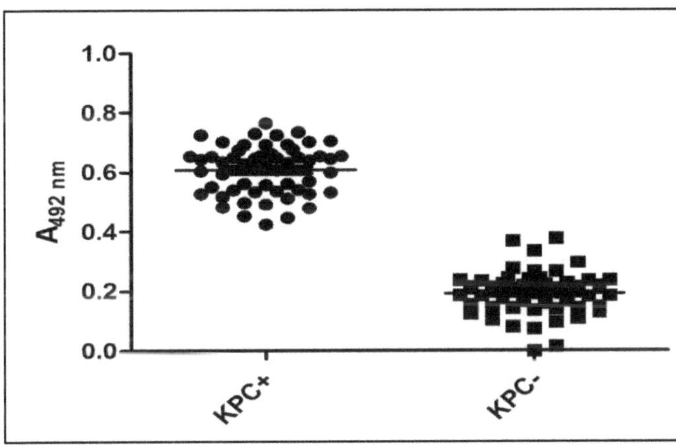

Figure 6. T-Student analysis of the KPC-ELISA-GMHB test verification. Caption: KPC+: KPC-like-producing samples. KPC-: isolates not producing KPC-like (p-value: <0.0001).

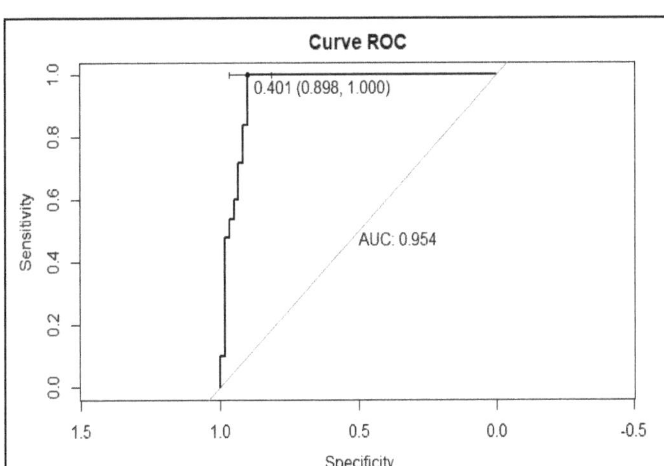

Figure 7. ROC curve of the performance of the ELISA-KPC-GMHB. This methodology obtained a sensitivity of 100% and a specificity of 89%, with 95% IC and a cut-off of 0.401.

4. Discussion

ELISA is considered the gold standard for diagnosing many infectious diseases [20]. During the last decade, different ELISA tests showing high sensitivity and specificity were developed for the detection of different pathogens such as canine parvovirus (100% sensitivity and 88.4% specificity) and *Schistosoma spindale* (86.7% sensitivity and 90.9% specificity) [21,22]. Chen and colleagues also developed an indirect ELISA test for detecting *K. pneumoniae* in animals; these authors reported that the ELISA test showed higher sensitivity (6.7% increase in positive detection rate) when compared to the agglutination test for the detection of *K. pneumoniae* [23]. Despite the wide use of ELISA tests, we were not able to find any studies conducted thus far that report the development of a similar test for β-lactamase detection. This study reports the first standardisation of an ELISA test for the detection of KPC in GNB isolates. KPC is the most frequent carbapenemase produced by GNB in many geographic regions such as Brazil.

During the standardisation process, we used mouse and rabbit polyclonal antibodies (pAbs), which showed the ability to differentiate KPC-producing isolates from non-KPC producers. pAbs can recognise different epitopes and are produced by different plasma cell clones (differentiated B cells) in response to a specific antigen. The pAbs are an important reagent that can considerably affect the test result. Bradbury (2015) reports that only 0.5% to 5% of pAbs can bind to the intended antigen. In addition, new immunisation is necessary when the amount of antibodies decreases. The production of pAbs may differ from batch to batch, making test reproducibility more difficult [24,25]. Despite these limitations, Ascoli and collaborators reported that pAbs are attractive options due to their clonal diversity and biophysical characteristics such as greater stability. In addition, the possibility of generating pAbs in different animals without the need for specialised equipment and rooms is also very attractive [26]. In agreement with other studies, we also used pAbs for this first standardisation [22,24]. In our experience, the use of pAbs proved to be a good alternative, being able to detect 100% of the KPC-producing isolates and their variants. Easy handling and having the shortest time for pAb generation were decisive in choosing pAbs as reagents. However, the lack of specificity was observed with a higher number of false positives among the isolates producing different β-lactamases, and even among non-β-lactamase producers. The development and standardisation of the KPC-ELISA-GMHB with monoclonal antibodies may be a more appropriate alternative to increase the specificity without losing the sensitivity of the KPC-ELISA-GMHB test.

The cross-reactions observed in the KPC-ELISA-GMHB test may impact clinical decisions on the management of infected patients. The false-positive reactions observed for GES-5- or BKC-1-producing isolates would not significantly lead to the change in antimicrobial prescription, as both enzymes belong to class A carbapenemases. In this case, the new beta-lactamase inhibitor–beta-lactam combinations (IBL-BL), such as ceftazidime–avibactam, are first-line therapeutic agents. In contrast, the false detection of NDM-1- and SPM-1-producing isolates as KPC producers are worrying, as this result could impact the selection of the most appropriate antimicrobial therapy, possibly leading to therapeutic failure [27]. The false detection of non-carbapenemase-producing bacteria is also concerning because it would not allow the de-escalation of antimicrobials, promoting the inappropriate use of antimicrobials and, further, the possible selection of resistant bacterial isolates. To overcome this problem, rapid tests such as CarbaNP or BlueCarba could be initially tested for excluding false-positive isolates.

Although the ELISA test had high sensitivity (100%) and specificity (84%), the incubation period for preparing the bacterial isolate before ELISA testing was an important test limitation. The use of bacterial colonies directly from the growth plate did not show the ability to differentiate KPC from other carbapenemases. Isolate preparation using an initial pre-inoculum showed better sensitivity and specificity; however, the incubation period (28–42 h) was too long. The presence of a single inoculum with incubation for 18 h showed a high sensitivity (100%) but reduced specificity (88.9%) when compared to the pre-inoculum methods (100% for both tests with the pre-inoculum). The reduction in the incubation period (22–24 h) proved to be a better alternative for the clinical use of KPC-ELISA-GMHB. The influence of bacterial sample preparation on the ELISA results could have different explanations. Bacterial growth in different culture media (broth and solid) was evaluated by Fujikawa and Morozumi in 2005. In this study, the authors noted a difference in the number of colony-forming units (CFUs), with a higher number of CFUs in agar than in broth media [28]. Studies carried out by Fortuin and collaborators demonstrated that proteomic differences might occur between the colonies obtained from agar media and those obtained from broth media. The expression of some proteins was higher in solid media, while others were more expressed in broth media [29]. Although bacterial growth was greater on agar media, Warrem and collaborators reported that bacterial colony growth occurs by pushing bacterial cells upward. This process causes the older bacterial cells to be at the top and the younger ones, which were undergoing cell division, at the bottom of the colony [30]. Older bacteria can enter a stationary phase and grow slowly, in addition to important structural changes in the bacterial cell [31]. Jaishankar and collaborators reported that in the stationary phase, there was an increase in the thickness of the bacterial membrane, a decrease in cell space, morphological change (changing from rod to spherical form), and a decrease in protein synthesis [32]. In addition to these characteristics, Kragh and collaborators also noted the presence of bacterial aggregates on agar media, which could not be removed via disruption. These same aggregates could promote tolerance to antimicrobials independent of β-lactamase production [33]. These facts could justify the poor performance of the KPC-ELISA-GMHB test when the bacterial colonies were directly obtained from the growth plate. In broth media, oxygen dispersion is also another important factor influencing bacterial growth. Changes in bacterial metabolism secondary to low oxygen dispersion in broth media have been reported [34]. In our study, the best results for the ELISA test were obtained when a pre-inoculum suspension was used for sample preparation. The inclusion of a second inoculum could have reduced the formation of aggregates and antimicrobial tolerance [33]. Studies evaluating the impact of different culture media (agar and broth) on the expression of β-lactamases are necessary for the better clarification of our results and optimisation of the sample preparation protocol.

The new beta-lactamase inhibitor–beta-lactam combinations such as ceftazidime–avibactam (CAZ-AVI) are recommended as first-line therapy for the treatment of KPC infections [35]. Resistance to CAZ-AVI, especially during or after therapy, has been increasingly reported due to the emergence of KPC-2 variants possessing mutations in the Ω loop such as KPC-33, KPC-66, and KPC-144 [36]. However, these D179Y mutants (KPC-33 and KPC-66) show a reversion of the carbapenem phenotype of resistance becoming susceptible to carbapenems (CAZ-AVI^R/Carbapenem^s phenotype) [37]. Initial studies have reported the clinical success of carbapenems in treating CAZ-AVI-resistant infections caused by such mutants [38,39]. However, sublethal meropenem concentrations may select meropenem-resistant variants in vitro [40]. The presence of a heterogeneous population of wild-type and mutated bla_{KPC-2} and the reversibility of the genotypes represent a significant challenge for managing KPC-producing isolates because phenotypic tests such as immunochromatographic, CarbaNP, and BlueCarba fail to recognise the mixed population and often report these isolates as non-carbapenemase-producing isolates, leading to the inadequate prescription of antimicrobial therapy [17,41–43]. In our study, the KPC-ELISA-GMHB test detected KPC-2 variants (KPC-33 and KPC-66,), which show CAZ-AVI^R/Carbapenem^S phenotype. It could represent an advantage of the KPC-ELISA-GMHB test over other phenotypic tests for the detection of KPC-2 with a mutation in the Ω loop. The use of polyclonal antibodies in the ELISA platform reduces the cost of the test and increases the KPC detection capacity. The evaluation of polyclonal antibodies in immunochromatographic platforms can also be a promising alternative to the detection of KPC.

5. Conclusions

The diagnosis and treatment of bacterial infections have become challenging due to the emergence and spread of multidrug-resistant pathogens harbouring multiple mechanisms of resistance. The presence of gene variants with a different spectrum of activity has become even more challenging. Fast, simple, and accurate tests are of fundamental importance to help antimicrobial stewardship activities in guiding antimicrobial therapy decisions. We proposed the first standardisation protocol for KPC detection using pAbs on an ELISA platform. The ELISA-KPC test showed high sensitivity and specificity, in addition to being the first test besides PCR to detect different KPC variants that have an ESBL profile. The test can be applied to automated systems, reducing human interference and the subjectivity of conventional tests.

Author Contributions: Conceptualisation, A.C.G., R.M.F.P. and M.M.F.M.; methodology, A.V., M.A.d.S., F.F.S., J.M.P. and M.M.F.M.; validation, A.V., M.A.d.S., F.F.S. and J.M.P.; formal analysis, A.V., M.A.d.S., F.F.S., J.M.P., M.M.F.M., A.C.G. and R.M.F.P.; investigation, A.V., M.A.d.S., F.F.S., J.M.P. and M.M.F.M.; resources, A.C.G. and R.M.F.P.; data curation, A.V., M.A.d.S. and F.F.S.; writing—original draft preparation, A.V. and A.C.G.; writing—review and editing, A.V., M.A.d.S., F.F.S., J.M.P., M.M.F.M., A.C.G. and R.M.F.P.; supervision, A.C.G., M.M.F.M. and R.M.F.P.; project administration, A.C.G. All authors have read and agreed to the published version of the manuscript.

Funding: We are grateful to the National Council for Science and Technological Development (CNPq) for providing grants to A.V. (Process Number: 830938/1999-0), A.C.G. (Process Number: 312066/2019) and to Coordination of Superior Level Staff Improvement (CAPES) for providing grants to F.F.S. (Process Number: 88887.464416/2019-00). This work was also supported by the Fundação de Amparo à Pesquisa do Estado de São Paulo (FAPESP-Process Number: 2021/01158-3).

Acknowledgments: We would like to express our gratitude to CAPES for the scholarships granted to A.V. and F.F.S., and to São Paulo Research Foundation (FAPESP) for the grant 2021/01158-3.

Conflicts of Interest: A.C.G. has received honoraria and/or consultation fees from Entasis Therapeutics, BioMerieux, Eurofarma, MSD, Pfizer, Roche, Sandoz, and United Medical, and research funding from Eurofarma.

References

1. World Health Organization Ten Threats to Global Health in 2019. Available online: https://www.who.int/news-room/spotlight/ten-threats-to-global-health-in-2019 (accessed on 20 February 2023).
2. Dadgostar, P. Antimicrobial Resistance: Implications and Costs. *Infect. Drug Resist.* **2019**, *12*, 3903–3910. [CrossRef] [PubMed]
3. Mancuso, G.; Midiri, A.; Gerace, E.; Biondo, C. Bacterial Antibiotic Resistance: The Most Critical Pathogens. *Pathogens* **2021**, *10*, 1310. [CrossRef] [PubMed]
4. Tacconelli, E.; Carrara, E.; Savoldi, A.; Harbarth, S.; Mendelson, M.; Monnet, D.L.; Pulcini, C.; Kahlmeter, G.; Kluytmans, J.; Carmeli, Y. Discovery, research, and development of new antibiotics: The WHO priority list of antibiotic-resistant bacteria and tuberculosis. *Lancet Infect. Dis.* **2018**, *18*, 318–327. [CrossRef] [PubMed]
5. Ambler, R.P. The structure of β-lactamases. *Philos. Trans. R Soc. Lond. B Biol. Sci.* **1980**, *289*, 321–331. [CrossRef] [PubMed]
6. Worthington, R.J.; Melander, C. Overcoming resistance to β-lactam antibiotics. *J. Org. Chem.* **2013**, *78*, 4207–4213. [CrossRef]
7. Lee, C.R.; Lee, J.H.; Park, K.S.; Kim, Y.B.; Jeong, B.C.; Lee, S.H. Global Dissemination of Carbapenemase-Producing *Klebsiella pneumoniae*: Epidemiology, Genetic Context, Treatment Options, and Detection Methods. *Front. Microbiol.* **2016**, *7*, 895. [CrossRef]
8. Han, R.; Shi, Q.; Wu, S.; Yin, D.; Peng, M.; Dong, D.; Zheng, Y.; Guo, Y.; Zhang, R.; Hu, F.; et al. Dissemination of Carbapenemases (KPC, NDM, OXA-48, IMP, and VIM) among Carbapenem-Resistant *Enterobacteriaceae* Isolated from Adult and Children Patients in China. *Front. Cell Infect. Microbiol.* **2020**, *10*, 314. [CrossRef]
9. Reyes, J.A.; Melano, R.; Cárdenas, P.A.; Trueba, G. Mobile genetic elements associated with carbapenemase genes in South American *Enterobacterales*. *Braz. J. Infect. Dis.* **2020**, *24*, 231–238. [CrossRef]
10. Kumar, A.; Roberts, D.; Wood, K.E.; Light, B.; Parrillo, J.E.; Sharma, S.; Suppes, R.; Feinstein, D.; Zanotti, S.; Taiberg, L.; et al. Duration of hypotension before initiation of effective antimicrobial therapy is the critical determinant of survival in human septic shock. *Crit. Care Med.* **2006**, *34*, 1589–1596. [CrossRef]
11. Ducomble, T.; Faucheux, S.; Helbig, U.; Kaisers, U.X.; König, B.; Knaust, A.; Lübbert, C.; Möller, I.; Rodloff, A.C.; Schweickert, B.; et al. Large hospital outbreak of KPC-2-producing *Klebsiella pneumoniae*: Investigating mortality and the impact of screening for KPC-2 with polymerase chain reaction. *J. Hosp. Infect.* **2015**, *89*, 179–185. [CrossRef]
12. Martin, A.; Fahrbach, K.; Zhao, Q.; Lodise, T. Association between Carbapenem Resistance and Mortality among Adult, Hospitalized Patients with Serious Infections Due to *Enterobacteriaceae*: Results of a Systematic Literature Review and Meta-analysis. *Open Forum. Infect. Dis.* **2018**, *5*, ofy150. [CrossRef] [PubMed]
13. Karaiskos, I.; Lagou, S.; Pontikis, K.; Rapti, V.; Poulakou, G. The "Old" and the "New" Antibiotics for MDR Gram-Negative Pathogens: For Whom, When, and How. *Front. Public Health* **2019**, *7*, 151. [CrossRef] [PubMed]

14. Nordmann, P.; Poirel, L.; Dortet, L. Rapid detection of carbapenemase-producing *Enterobacteriaceae*. *Emerg. Infect. Dis.* **2012**, *18*, 1503–1507. [CrossRef]
15. Nordmann, P.; Poirel, L. Epidemiology and Diagnostics of Carbapenem Resistance in Gram-negative Bacteria. *Clin. Infect. Dis.* **2019**, *69* (Suppl. S7), S521–S528. [CrossRef] [PubMed]
16. Ramos, A.C.; Gales, A.C.; Monteiro, J.; Silbert, S.; Chagas-Neto, T.; Machado, A.M.O.; Carvalhaes, C.G. Evaluation of a rapid immunochromatographic test for detection of distinct variants of *Klebsiella pneumoniae* carbapenemase (KPC) in *Enterobacteriaceae*. *J. Microbiol. Methods* **2017**, *142*, 1–3. [CrossRef] [PubMed]
17. Bianco, G.; Boattini, M.; van Asten, S.A.V.; Iannaccone, M.; Zanotto, E.; Zaccaria, T.; Bernards, A.T.; Cavallo, R.; Costa, C. RESIST-5 O.O.K.N.V. and NG-Test Carba 5 assays for the rapid detection of carbapenemase-producing cultures: A comparative study. *J. Hosp. Infect.* **2020**, *105*, 162–166. [CrossRef]
18. Castanheira, M.; Arends, S.J.R.; Davis, A.P.; Woosley, L.N.; Bhalodi, A.A.; MacVane, S.H. Analyses of a Ceftazidime-Avibactam-Resistant Citrobacter freundii Isolate Carrying bla_{KPC-2} Reveals a Heterogenous Population and Reversible Genotype. *mSphere* **2018**, *3*, e00408-18. [CrossRef]
19. Nicoletti, A.G.; Marcondes, M.F.; Martins, W.M.; Almeida, L.G.; Nicolas, M.F.; Vasconcelos, A.T.; Oliveira, V.; Gales, A.C. Characterization of BKC-1 class A carbapenemase from *Klebsiella pneumoniae* clinical isolates in Brazil. *Antimicrob. Agents Chemother.* **2015**, *59*, 5159–5164. [CrossRef]
20. Alhajj, M.; Farhana, A. Enzyme Linked Immunosorbent Assay. [Updated 6 February 2021]. In *StatPearls [Internet]*; StatPearls Publishing: Treasure Island, FL, USA, 2021. Available online: https://www.ncbi.nlm.nih.gov/books/NBK555922 (accessed on 12 June 2021).
21. Kumar, M.; Nandi, S.; Chidri, S. Development of a polyclonal antibody-based AC-ELISA and its comparison with PCR for diagnosis of Canine parvovirus infection. *Virol. Sin.* **2010**, *25*, 352–360. [CrossRef]
22. Sreenivasa Murthy, G.S.; D'Souza, P.E.; Shrikrishna Isloor, K. Evaluation of a polyclonal antibody-based sandwich ELISA for the detection of faecal antigens in *Schistosoma spindale* infection in bovines. *J. Parasit. Dis.* **2013**, *37*, 47–51. [CrossRef]
23. Chen, R.; Shang, H.; Niu, X.; Huang, J.; Miao, Y.; Sha, Z.; Zhu, R. Establishment and evaluation of an indirect ELISA for detection of antibodies to goat *Klebsiella pneumonia*. *BMC Vet. Res.* **2021**, *17*, 107. [CrossRef] [PubMed]
24. Bradbury, A.; Plückthun, A. Reproducibility: Standardize antibodies used in research. *Nature* **2015**, *518*, 27–29. [CrossRef] [PubMed]
25. Nakazawa, M.; Mukumoto, M.; Miyatake, K. Production and purification of polyclonal antibodies. *Methods Mol. Biol.* **2010**, *657*, 63–74. [CrossRef]
26. Ascoli, C.A.; Aggeler, B. Overlooked benefits of using polyclonal antibodies. *Biotechniques* **2018**, *65*, 127–136. [CrossRef] [PubMed]
27. Aitken, S.L.; Tarrand, J.J.; Deshpande, L.M.; Tverdek, F.P.; Jones, A.L.; Shelburne, S.A.; Prince, R.A.; Bhatti, M.M.; Rolston, K.V.I.; Jones, R.N.; et al. High Rates of Nonsusceptibility to Ceftazidime-avibactam and Identification of New Delhi Metallo-β-lactamase Production in *Enterobacteriaceae* Bloodstream Infections at a Major Cancer Center. *Clin. Infect. Dis.* **2016**, *63*, 954–958. [CrossRef]
28. Fujikawa, H.; Morozumi, S. Modeling surface growth of *Escherichia coli* on agar plates. *Appl. Environ. Microbiol.* **2005**, *71*, 7920–7926. [CrossRef]
29. Fortuin, S.; Nel, A.J.M.; Blackburn, J.M.; Soares, N.C. Comparison between the proteome of *Escherichia coli* single colony and during liquid culture. *J. Proteomics.* **2020**, *228*, 103929. [CrossRef]
30. Warren, M.R.; Sun, H.; Yan, Y.; Cremer, J.; Li, B.; Hwa, T. Spatiotemporal establishment of dense bacterial colonies growing on hard agar. *eLife* **2019**, *8*, e41093. [CrossRef]
31. Gray, D.A.; Dugar, G.; Gamba, P.; Strahl, H.; Jonker, M.J.; Hamoen, L.W. Extreme slow growth as alternative strategy to survive deep starvation in bacteria. *Nat. Commun.* **2019**, *10*, 890. [CrossRef]
32. Jaishankar, J.; Srivastava, P. Molecular Basis of Stationary Phase Survival and Applications. *Front. Microbiol.* **2017**, *8*, 2000. [CrossRef]
33. Kragh, K.N.; Alhede, M.; Rybtke, M.; Stavnsberg, C.; Jensen, P.Ø.; Tolker-Nielsen, T.; Whiteley, M.; Bjarnsholt, T. The Inoculation Method Could Impact the Outcome of Microbiological Experiments. *Appl. Environ. Microbiol.* **2018**, *84*, e02264-17. [CrossRef] [PubMed]
34. Somerville, G.A.; Proctor, R.A. Cultivation conditions and the diffusion of oxygen into culture media: The rationale for the flask-to-medium ratio in microbiology. *BMC Microbiol.* **2013**, *13*, 9. [CrossRef] [PubMed]
35. Tamma, P.D.; Aitken, S.L.; Bonomo, R.A.; Mathers, A.J.; van Duin, D.; Clancy, C.J. Infectious Diseases Society of America Guidance on the Treatment of Extended-Spectrum β-lactamase Producing Enterobacterales (ESBL-E), Carbapenem-Resistant Enterobacterales (CRE), and *Pseudomonas aeruginosa* with Difficult-to-Treat Resistance (DTR-P. aeruginosa). *Clin. Infect. Dis.* **2021**, *72*, e169–e183. [CrossRef] [PubMed]
36. Shields, R.K.; Nguyen, M.H.; Press, E.G.; Chen, L.; Kreiswirth, B.N.; Clancy, C.J. Emergence of Ceftazidime-Avibactam Resistance and Restoration of Carbapenem Susceptibility in *Klebsiella pneumoniae* Carbapenemase-Producing *K pneumoniae*: A Case Report and Review of Literature. *Open Forum. Infect. Dis.* **2017**, *4*, ofx101. [CrossRef] [PubMed]
37. Alsenani, T.A.; Viviani, S.L.; Kumar, V.; Taracila, M.A.; Bethel, C.R.; Barnes, M.D.; Papp-Wallace, K.M.; Shields, R.K.; Nguyen, M.H.; Clancy, C.J.; et al. Structural Characterization of the D179N and D179Y Variants of KPC-2 β-Lactamase: Ω-Loop Destabilization as a Mechanism of Resistance to Ceftazidime-Avibactam. *Antimicrob. Agents Chemother.* **2022**, *66*, e0241421. [CrossRef]

38. Tsolaki, V.; Mantzarlis, K.; Mpakalis, A.; Malli, E.; Tsimpoukas, F.; Tsirogianni, A.; Papagiannitsis, C.; Zygoulis, P.; Papadonta, M.E.; Petinaki, E.; et al. Ceftazidime-Avibactam To Treat Life-Threatening Infections by Carbapenem-Resistant Pathogens in Critically Ill Mechanically Ventilated Patients. *Antimicrob. Agents Chemother.* **2020**, *64*, e02320-19. [CrossRef]
39. Giddins, M.J.; Macesic, N.; Annavajhala, M.K.; Stump, S.; Khan, S.; McConville, T.H.; Mehta, M.; Gomez-Simmonds, A.; Uhlemann, A.C. Successive Emergence of Ceftazidime-Avibactam Resistance through Distinct Genomic Adaptations in bla_{KPC-2}-Harboring *Klebsiella pneumoniae* Sequence Type 307 Isolates. *Antimicrob. Agents Chemother.* **2018**, *62*, e02101-17. [CrossRef]
40. Gaibani, P.; Campoli, C.; Lewis, R.E.; Volpe, S.L.; Scaltriti, E.; Giannella, M.; Pongolini, S.; Berlingeri, A.; Cristini, F.; Bartoletti, M.; et al. In vivo evolution of resistant subpopulations of KPC-producing *Klebsiella pneumoniae* during ceftazidime/avibactam treatment. *J. Antimicrob. Chemother.* **2018**, *73*, 1525–1529. [CrossRef]
41. Oueslati, S.; Tlili, L.; Exilie, C.; Bernabeu, S.; Iorga, B.; Bonnin, R.A.; Dortet, L.; Naas, T. Different phenotypic expression of KPC β-lactamase variants and challenges in their detection. *J. Antimicrob. Chemother.* **2020**, *75*, 769–771. [CrossRef]
42. Tiseo, G.; Falcone, M.; Leonildi, A.; Giordano, C.; Barnini, S.; Arcari, G.; Carattoli, A.; Menichetti, F. Meropenem-Vaborbactam as Salvage Therapy for Ceftazidime-Avibactam-, Cefiderocol-Resistant ST-512 *Klebsiella pneumoniae*-Producing KPC-31, a D179Y Variant of KPC-3. *Open Forum. Infect. Dis.* **2021**, *8*, ofab141. [CrossRef]
43. Ding, L.; Shen, S.; Han, R.; Yin, D.; Guo, Y.; Hu, F. Ceftazidime-Avibactam in Combination with Imipenem as Salvage Therapy for ST11 KPC-33-Producing Klebsiella pneumoniae. *Antibiotics* **2022**, *11*, 604. [CrossRef] [PubMed]

microorganisms

Article

Assessment of New and Genome-Reduced *Pseudomonas* Strains Regarding Their Robustness as *Chassis* in Biotechnological Applications

María José Cárdenas Espinosa [1], Tabea Schmidgall [1], Jessica Pohl [1], Georg Wagner [1], Benedikt Wynands [2], Nick Wierckx [2], Hermann J. Heipieper [1] and Christian Eberlein [1,*]

[1] Department of Environmental Biotechnology, Helmholtz Centre for Environmental Research—UFZ, 04318 Leipzig, Germany
[2] Institute of Bio- and Geosciences, IBG-1: Biotechnology, Forschungszentrum Jülich, 52428 Jülich, Germany
* Correspondence: christian.eberlein@ufz.de

Abstract: Organic olvent-tolerant strains of the Gram-negative bacterial genus *Pseudomonas* are discussed as potential biocatalysts for the biotechnological production of various chemicals. However, many current strains with the highest tolerance are belonging to the species *P. putida* and are classified as biosafety level 2 strains, which makes them uninteresting for the biotechnological industry. Therefore, it is necessary to identify other biosafety level 1 *Pseudomonas* strains with high tolerance towards solvents and other forms of stress, which are suitable for establishing production platforms of biotechnological processes. In order to exploit the native potential of *Pseudomonas* as a microbial cell factory, the biosafety level 1 strain *P. taiwanensis* VLB120 and its genome-reduced *chassis* (GRC) variants as well as the plastic-degrading strain *P. capeferrum* TDA1 were assessed regarding their tolerance towards different *n*-alkanols (1-butanol, 1-hexanol, 1-octanol, 1-decanol). Toxicity of the solvents was investigated by their effects on bacterial growth rates given as the EC50 concentrations. Hereby, both toxicities as well as the adaptive responses of *P. taiwanensis* GRC3 and *P. capeferrum* TDA1 showed EC50 values up to two-fold higher than those previously detected for *P. putida* DOT-T1E (biosafety level 2), one of the best described solvent-tolerant bacteria. Furthermore, in two-phase solvent systems, all the evaluated strains were adapted to 1-decanol as a second organic phase (i.e., OD_{560} was at least 0.5 after 24 h of incubation with 1% (*v/v*) 1-decanol), which shows the potential use of these strains as platforms for the bio-production of a wide variety of chemicals at industrial level.

Keywords: *Pseudomonas*; biocatalysts; genome-reduced *chassis*; bio-production

Citation: Cárdenas Espinosa, M.J.; Schmidgall, T.; Pohl, J.; Wagner, G.; Wynands, B.; Wierckx, N.; Heipieper, H.J.; Eberlein, C. Assessment of New and Genome-Reduced *Pseudomonas* Strains Regarding Their Robustness as *Chassis* in Biotechnological Applications. *Microorganisms* **2023**, *11*, 837. https://doi.org/10.3390/microorganisms11040837

Academic Editors: Shengxi Chen and Fabio Zicker

Received: 13 February 2023
Revised: 12 March 2023
Accepted: 23 March 2023
Published: 25 March 2023

1. Introduction

Biocatalysis is one of the most promising technologies for the sustainable synthesis of complex molecules at biotechnological, pharmaceutical, and industrial scale [1]. The increased demand for bio-based chemicals is driving their growth in the global market, making them economically competitive with fossil raw materials [2]. Particularly with regard to the demand for transforming oil-based chemical production into a sustainable bio-based circular economy, modern techniques of metabolic engineering and synthetic biology have favored the biotechnological production of high-value compounds. However, next to improvements of bioconversion pathways and product yields, the biocatalysts also need to be able to tolerate high, often toxic concentrations of substrates and products, respectively [3]. Often, solvents are used as reservoirs/sinks for the substrates/products. Thus, substrate/product concentrations in the aqueous phase can be kept below inhibitory levels [4]. The reservoir function of the solvent phase represents a big advantage over batch cultures without a second phase, where the bacterial growth conditions can be adverse from the beginning or worsen during the cultivation process. However, the solvents involved in the second phase need to be tolerated by the biocatalysts as well. Biocompatibility of solvents has been widely assessed by log $P_{o/w}$ value, which is the logarithm of the partition coefficient in an octanol–water system [5,6]. Generally, log $P_{o/w}$ in the range of 1 and 4 (e.g., long chain alkanols, aromatics, esters, among others) are toxic to microorganisms at very low concentrations because these solvents accumulate in the cytoplasmic membrane. Partitioning into the membrane lipid bilayer correlates with the compound's log $P_{o/w}$ and toxicity. Solvents accumulating within the hydrophobic layer of the membrane cause an increase in fluidity [7,8], which leads to inactivation and denaturation of membrane embedded proteins, such as ion pumps and ATPases, it provokes leakage of ions and intracellular macromolecules, such as RNA, phospholipids, and proteins [7]. Among all these effects, the increase in membrane permeability is considered the main reason for cell death [7,8].

Tolerance to hydrocarbon solvents is an evolutionary trait employed by bacterial cells to overcome the stress imposed by these compounds. Tolerance mechanisms have been widely studied in Gram-negative bacteria, especially in various strains of *Pseudomonas* that exhibit an exceptional metabolic versatility and play a relevant role in biotransformation [2,9–11]. In the presence of a solvent, several strategies have been observed: (a) adaptive alterations of the membrane fatty acids and phospholipid head group composition, (b) energy-dependent active efflux pumps, and (c) membrane vesicle formation [9,10]. From these mechanisms, changes in phospholipid profile and extrusion of the solvent are considered the most efficient methods of solvent tolerance [12,13].

The resistance–nodulation–cell division (RND) efflux pumps are membrane proteins that transport multiple substrates into and out of the cell [14]. The outstanding solvent tolerance of some *Pseudomonas* species can be attributed to efflux pumps directly involved in toluene resistance called TtgABC, TtgDEF, and TtgGHI, whereby the latter was identified as the main pump causing the solvent tolerance phenotype. These pumps confer basal resistance to several solvents including styrene, xylenes, ethylbenzene, and propylbenzene [15].

The use of native solvent-tolerant strains has demonstrated several complications in industrial biotechnology. The complications reach from a possible accumulation of toxic intermediates, reduction of biomass formation, imbalances in pathway flux to unpredictable product yields [16]. During the last years, the deletion of dispensable features by metabolic engineering has optimized bioconversion pathways, production levels and strain stability [11,17].

Therefore, solvent-tolerant strains of *Pseudomonas* can be seen as perfect biocatalysts for the biotechnological production of various chemicals. However, many strains with the highest tolerance belong to the species *P. putida* that are classified as biosafety level 2 strains in Germany (*P. putida* KT2440 being an exception), which makes them less interesting for the biotechnological industry due to costly but necessary safety measures. This German classification for the *P. putida* group is highly questionable because of the absence of pathogenicity factors from the genome and the history of safe use of many strains. By the U.S. Food and Drug Administration (FDA), the *P. putida* group is considered as non-pathogenic [18]. Moreover, an occasion to review the German classification is the ongoing debate about the *P. putida* group. After a taxonomic review of the *P. putida* clade, a new species of *P. alloputida* was suggested, encompassing well-investigated strains such as *P. putida* KT2440, *P. putida* S12, and *P. putida* DOT-T1E, among others [19]. Next to the aforementioned issues in legislation and taxonomy, it is important to identify new *Pseudomonas* strains' high tolerance towards solvents and other forms of stress, which are suitable for establishing production platforms of biotechnological processes. In order to exploit the potential of other *Pseudomonas* species for biotechnological applications, solvent-tolerant features were enhanced in *Pseudomonas taiwanensis* VLB120 by successive genome reduction. The deletion of genes related to biofilm formation and flagella expression by the removal of the megaplasmid pSTY and proviral segments provided three new bacterial strains called GRC (genome-reduced *chassis*). In total, the genome was reduced by up to 10%. Strain GRC1 lacks the efflux pump TtgGHI, whereas the *ttgGHI* genes without and with regulatory genes *ttgVW* were re-integrated in GRC2 and GRC3, respectively, enhancing solvent tolerance [20]. Previously, similar genome reductions in *P. putida* KT2440 and in VLB120 considerably enhanced biomass yield coefficients and heterologous gene expression compared to the wildtype [17,20,21]. However, the main disadvantage of KT2440 is its low solvent tolerance when compared to other *Pseudomonas* strains.

Furthermore, the recently isolated strain *P. capeferrum* TDA1 has been reported as a plastic monomer and oligomer degrader [22,23], which shows its metabolic capacity and broad potential to degrade recalcitrant compounds. Together with three *P. putida* KT2440 derivatives in a defined microbial mixed culture, TDA1 was deployed in the metabolization of polyurethane hydrolysates. The mineralization of the polyurethane hydrolysates led to the subsequent production of rhamnolipids [24]. The study reflects the important role that new bacterial strains can play in the (bio)technological plastic upcycling.

In this study, the different solvent tolerance levels towards *n*-alkanols (1-butanol, 1-hexanol, 1-octanol, and 1-decanol) in *Pseudomonas taiwanensis* VLB120, GRC1, GRC2, GRC3, and in *Pseudomonas capeferrum* TDA1, and their performance in two-phase systems as potential hosts for the bio-production of chemical compounds at industrial level is described. The *n*-alkanols with different log $P_{o/w}$ values (0.89 for 1-butanol, 1.87 for 1-hexanol, 2.92 for 1-octanol, and 3.97 for 1-decanol) function as benchmarks to compare the solvent tolerance and their suitability for biotechnology, respectively, of genome-reduced *chassis* strains compared to the wildtype *P. taiwanensis* VLB120.

2. Materials and Methods

2.1. Strains

Strain *Pseudomonas taiwanensis* VLB120 and its genome-reduced strains GRC1, GRC2, and GRC3 were selected for the stress assessment because of their superior bioprocess features compared to the wildtype [20]. The genome for the three strains was reduced by about 10%. The GRC strains lack genes enabling the cells to swim and form biofilms. Moreover, the megaplasmid pSTY and large proviral segments were deleted from the genome. For *Pseudomonas taiwanensis* GRC2 and GRC3 the genes for the efflux pump TtgGHI (formerly present on the deleted megaplasmid pSTY) were reintroduced. For *Pseudomonas taiwanensis,* GRC3 as well the regulatory genes *ttgVW* were reintegrated into the bacterial chromosome (Table 1).

Table 1. Strains used in the study, their abbreviation, and the presence (+) or absence (−) of the full set of genes for the efflux pump TtgGHI and its regulators TtgVW. GRC: genome-reduced chassis, TDA: toluene diamine.

Strain	Abbreviation	*ttgTGH*	*ttgVW*
***P. taiwanensis* VLB120**	VLB120 or wildtype	+	+
***P. taiwanensis* VLB120 GRC1**	GRC1	−	−
***P.taiwanensis* VLB120 GRC2**	GRC2	+	−
***P.taiwanensis* VLB120 (GRC3)**	GRC3	+	+
***P. capeferrum* TDA1**	TDA1	−	−

Additionally, the bacterial strain *Pseudomonas capeferrum* TDA1 was included in the assessment, based on reports about its metabolic ability to degrade a polyurethane monomer and a polyurethane oligomer [22] and first applications in biotechnological approaches [24]. The genome of *Pseudomonas capeferrum* TDA1 was first published under the species name *Pseudomonas* sp. TDA1. An improved genome assembly for *Pseudomonas capeferrum* TDA1 is now available under the accession number CP116669.1.

2.2. Medium and Culture Conditions

All bacterial strains were regularly grown in "Hartmans" mineral medium [25] and Na_2-succinate (4 g/L) as a carbon and energy source at 30 °C and 180 rpm (overnight culture). A fresh culture was inoculated with an overnight culture to reach an OD_{560} of about 0.1. The OD of the new culture was measured every hour and during the exponential phase, the *n*-alkanols (1-butanol, 1-hexanol, 1-octanol, and 1-decanol) were added at different concentrations, separately, according to their log $P_{o/w}$ value. 1-Butanol was added in the range of 10 to 400 mM, 1-hexanol from 1 to 15 mM, 1-octanol from 0.1 to 2.5 mM, and 1-decanol from 0.05 to 0.4 mM. Purchased solutions of 1-octanol and 1-decanol were diluted with acetone to facilitate the addition to the cultures and the solubility in the mineral medium. Before reaching the stationary phase, the cultures were centrifuged (15 min at 10,000 rpm) (Heraeus, Hanau, Germany) and the pellets were resuspended in 1.75 mL of phosphate buffer (50 mM at pH 7.0). Finally, the bacterial cells were centrifuged (7 min at 13,000 rpm) and stored at −20 °C. Growth inhibition (Formula (2)) caused by the toxic compounds was measured by comparing the percentage difference in the growth rates μ (h^{-1}) (Formula (1)) between intoxicated cultures with that of control [26].

Formula (1): Calculation of bacterial growth rates μ (h^{-1})

$$\text{Growth rate } \mu\left(h^{-1}\right) = \frac{\text{Ln } OD_{t1} - \text{Ln } OD_{t0}}{t_1 - t_0} \tag{1}$$

Formula (2): Calculation of growth inhibition in %

$$\text{Inhibited growth } (\%) = \frac{\mu_{with\_toxin} \times 100}{\mu_{control}} \tag{2}$$

2.3. Extraction of Membrane Lipids

The resulting pellets, equivalent to a dry mass of about 15 mg, were suspended in 0.5 mL of water, 1 mL of methanol and 1.75 mL of chloroform. The solution was shaken for 3 min using a vortex (VWR, West Chester, PA, USA), and 0.5 mL of water was added to the mixture, which was agitated for 30 s. Then, the solution was centrifuged (10 min at 3000 rpm) and the chloroform phase was transferred to HPLC-flasks.

For the methylation of fatty acids, samples were incubated in BF_3-methanol (Merck, Darmstadt, Germany) for 15 min at 95 °C by applying the method of Morrison and Smith [27]. Lastly, fatty acid methyl esters (FAME) were extracted with hexane and stored at 4 °C.

2.4. Determination of Fatty Acid Composition

FAME analysis was performed using gas chromatography with flame ionization detector (GC-FID, Agilent Technologies, 6890N Network GC System, 7683B Series Injector). The instrument used a CP-Sil 88 column (Varian CP7488) in stationary phase and helium as a carrier gas. The temperature program was 40 °C, 2 min isothermal, a gradient increase up to 220 °C (8 °C × min^{-1}), and 10 min at 220 °C. The peak areas of the FAMEs were used to determine their relative amounts. The fatty acids were identified by co-injection of authentic reference compounds obtained from Supelco (Bellefonte, PA). *Trans/cis* ratio was calculated taking the sum of the FAME of palmitoleic acid (C16:1Δ9*cis*) and *cis*-vaccenic acid (C18:1Δ11*cis*) as divisor and the sum of their corresponding *trans* configuration as dividend [7].

2.5. Growth in in a Second Phase System of 1-octanol and 1-decanol

All bacterial strains were cultivated in mineral medium and Na_2-succinate as described above for 5 h. After reaching an OD_{560} of 0.5, 1 mM (0.015% v/v) of 1-octanol or 0.2 mM (0.004% v/v) of 1-decanol were added to different assays, separately. The cultures grew for 3 h and then, 10 mL of these cultures were transferred to a fresh medium to reach an OD_{560} of about 0.1. Subsequently, 1% (v/v) of 1-octanol (maximum concentration in the aqueous phase of 3.8 mM) or 1-decanol (maximum concentration in the aqueous phase of 0.23 mM) were added to the corresponding assays. These cultures grew overnight and the OD_{560} was measured after 24 h for all strains. The tested strains were considered as *adapted* at an OD_{560} of at least 0.5 or *well-adapted* at an OD_{560} of over 1 after 24 h incubation with 1% (*v*/*v*) of 1-octanol or 1-decanol.

2.6. Statistical Analysis

Data of at least threefold measurements were obtained as the mean ± one standard deviation. Student's *t*-test was used to analyze the significant differences ($p < 0.05$).

3. Results and Discussion

3.1. Growth Kinetics

In order to assess the impact of the genome reduction on the growth kinetics, all the aforementioned strains were cultivated in mineral medium with Na_2-succinate as carbon and energy source. The growth rate of *P. taiwanensis* GRC3 ($\mu = 0.38 \pm 0.02\ h^{-1}$) and GRC1 ($\mu = 0.37 \pm 0.01\ h^{-1}$) increased up to 5.2% compared to *P. taiwanensis* VLB120 ($\mu = 0.36 \pm 0.01\ h^{-1}$) (Figure 1). Growth rates of GRC2 ($\mu = 0.35 \pm 0.015\ h^{-1}$) and TDA1 ($\mu = 0.29 \pm 0.04\ h^{-1}$) were slightly below the rates of the aforementioned strains.

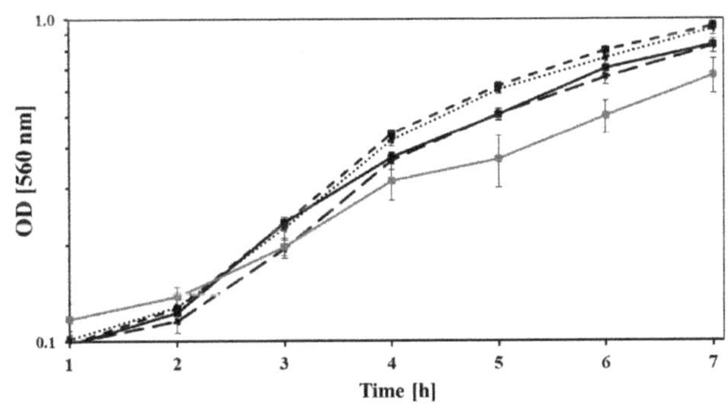

Figure 1. OD_{560} during the growth of *P. taiwanensis* VLB120 (solid black line), *P. taiwanensis* GRC1 (dotted black line), GRC2 (long dashed black line), GRC3 (short dashed black line), and *P. capeferrum* TDA1 (solid gray line) on Na_2-succinate (4 g/L). Error bars indicate the standard error of the mean (n = 3).

3.2. Effects of n-alkanols on Bacterial Growth

The effect of *n*-alkanols (1-butanol, 1-hexanol, 1-octanol, and 1-decanol) of different log $P_{o/w}$ on the growth of *P. taiwanensis* VLB120 (wildtype), GRC1, GRC2, GRC3, and *P. capeferrum* TDA1 was assessed using EC50 (effective concentration for reducing cell growth by 50% compared to the control) [28]. Previous reports have shown that higher hydrophobicity leads to a higher tendency to accumulation in the membranes [4,29]. The most important finding of this study is that the GRC strains not only show better bioprocess features compared to the wildtype VLB120, as reported before [20], but that their tolerance towards the tested *n*-alkanols is also comparable. Thus, the genome reduction process did not diminish solvent tolerance towards *n*-alkanols (Figure 2 and Table 2). It is likely that this counts also for the tolerance towards other solvents such as toluene and styrene in GRC2 and GRC3 because of the presence of the TtgGHI pump.

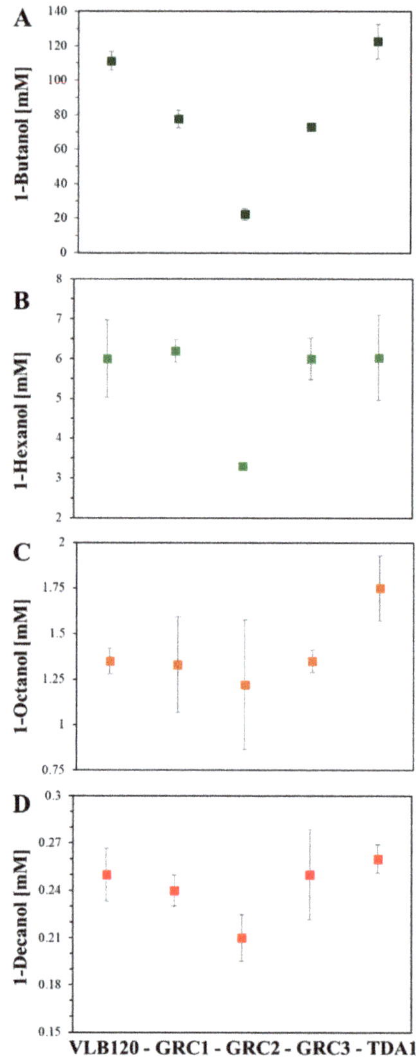

Figure 2. EC50 values for *P. taiwanensis* VLB120, GRC1, GRC2, GRC3, and *P. capeferrum* TDA1 grown on (**A**) 1-butanol (dark green squares), (**B**) 1-hexanol (green squares), (**C**) 1-octanol (orange squares), and (**D**) 1-decanol (red squares).

Table 2. EC50 values (mM) of the tested *n*-alkanols in *P. putida* DOT-T1E (data from [4]), *P. taiwanensis* GRC3 (see also Figure S1), and *P. capeferrum* TDA1.

n-alkanols	*P. putida* DOT-T1E [4]	*P. taiwanensis* GRC3	*P. capeferrum* TDA1
1-Butanol	49	73	123
1-Hexanol	6.35	6	6
1-Octanol	0.80	1.38	1.75
1-Decanol	0.11	0.25	0.26

Some Pseudomonads harbor several RND-type efflux pumps (TtgABC, TtgDEF, and TtgGHI) that are involved in the extrusion of numerous solvents. For instance, TtgABC mainly extrudes flavonoids, antibiotics, and short-chain alcohols such as *n*-butanol [30,31]. TtgGHI efflux pump seems to be induced only by aromatic hydrocarbons such as toluene, whereas TtgDEF plays an important role in tolerance to aliphatic alcohols [32] and organic solvents such as toluene, ethylbenzene, propylbenzene, and styrene [11]. The targeted elimination of the pSTY megaplasmid in tailored strain GRC1, re-integration of the pSTY-encoded efflux pump genes *ttgGHI* without (GRC2) and with (GRC3) transcriptional regulators enhanced solvent tolerance and performance indicators such as titer, yield, and product tolerance [20,33].

Previously, the TtgABC efflux pump was actively expressed in *Escherichia coli* at high concentrations of short-chain alcohols (*n*-butanol, isobutanol, isoprenol, and isopentanol) inhibiting bacterial growth. Under these conditions, the function of TtgABC was limited to improving survival, rather than supporting bacterial growth [30]. Solvent tolerance mechanisms are considered energy-intensive processes that come at the cost of growth yields to protect cells from further damage [4,34]. In the presence of sub-lethal toluene dosages, *P. putida* S12 showed a marked decrease in growth yield, which was reduced linearly with increasing toluene concentrations [35]. It is likely that the constitutive expression of the TtgGHI efflux pump in *P. taiwanensis* GRC2 has required high levels of energy and, in addition, the presence of 1-butanol and 1-hexanol might have induced other RND-type efflux systems [36], which could also increase the energy consumption, significantly reducing bacterial growth. These results showed a competitive advantage of GRC1 and GRC3 compared to GRC2 enhancing key performance indicators that can be applicable to many biotechnological processes. On the other hand, *P. taiwanensis* GRC2 showed similar EC50 values in assays containing 1-octanol and 1-decanol (1.22 $\pm$ 0.35 and 0.21 $\pm$ 0.01 mM, respectively) compared to *P. taiwanensis* VLB120 (1.35 $\pm$ 0.07 and 0.25 $\pm$ 0.01 mM), *P. taiwanensis* GRC1 (1.33 $\pm$ 0.26 and 0.24 $\pm$ 0.00 mM), and GRC3 (1.35 $\pm$ 0.06 and 0.25 $\pm$ 0.02 mM), respectively. Comparable results were observed in the tested wildtypes, GRC1 and GRC3, where the EC50 values were similar for all the strains in 1-butanol, 1-hexanol, 1-octanol, and 1-decanol (Figure 2). However, GRC1 was previously described as a sensitive strain in the presence of hydrophobic solvents such as toluene [20] which points to a preference for GRC2 and GRC3 for corresponding applications in microbial biocatalysis. In cases where aromatics are gradually accumulating such as in de novo production [31], *P. taiwanensis* GRC2 or GRC3 might provide a balance between optimal growth and solvent tolerance in high-stress conditions due to the presence of the *ttgGHI* genes. In addition, as almost exclusively typical for *Pseudomonas*, an increase in the *trans/cis* ratio, dependent on the stressor concentration, was observed (data for GRC3 shown in Figure S1). This shows the functionality of this fast stress response mechanism also in the genome-reduced *chassis* strains of *P. taiwanensis* [9,28].

P. capeferrum TDA1 showed the highest EC50 values for all the assays (1-butanol, 1-hexanol, 1-octanol, and 1-decanol) (122.66, 6.03, 1.75, and 0.26 mM, respectively). The strain TDA1 contains the *ttgABC* genes; however, none of the gene sets for TtgGHI nor for TtgDEF are complete. Several genes related to biofilm formation and induction, multidrug resistance proteins and RND transporters were identified as differentially expressed (upregulated) in *P. capeferrum* TDA1 grown on the aromatic compound 2,4-diaminotoluene [22], however, *ttgABC* were not among them. Thus, TtgABC was most probably not involved in the adaptation to 2,4-diaminotoluene. According to the EC50 values obtained in the present study, TtgABC could be rather involved in the adaptation to *n*-alkanols. However, this remains an open question for further studies. In Rojas et al. 2004, a *P. putida* DOT-T1E mutant lacking *ttgH* did not show lower tolerance to 1-octanol, 1-nonanol, and 1-decanol [32]. This is in full agreement with the data shown in Figure 2, where GRC1 is equally tolerant to 1-octanol and 1-decanol as the VLB120 wildtype. Furthermore, it was suggested by Rojas et al. 2004 that the aliphatic alcohols' extrusion was performed simultaneously by the three efflux pumps (TtgABC, TtgDEF, and TtgGHI) present in DOT-T1E because a triple mutant lacking all the pumps was much more sensitive to the *n*-alkanols tested [32]. However, further evidence is needed regarding that. Additionally, efflux pumps in *P. taiwanensis* VLB120 may have different substrate ranges than those in *P. putida* DOT-T1E. Constitutive expression of *ttgGHI* gave a clear fitness advantage to *P. taiwanensis* GRC2 compared to GRC1 or GRC3 when confronted with hydrophobic aromatics such as styrene, toluene, or 4-ethylphenol [20]. However, GRC2 showed a marked growth defect when confronted with the more hydrophilic aromatic phenol. Together, this strongly suggests that the TtgGHI pump is probably specific for hydrophobic aromatic substrates, and only poses an energetic burden for GRC2 when facing *n*-alkanols. The adaptive response of *P. taiwanensis* GRC3 and *P. capeferrum* TDA1 revealed a superior performance with EC50 values up to two-fold higher than those presented in *P. putida* DOT-T1E formerly [4] (Table 2). These results revealed the potential use of these strains as platforms for the bio-production of a wide variety of chemicals at industrial level.

3.3. Two-Phase Adaptation of 1-octanol and 1-decanol in P. taiwanensis VLB120, GRC1, GRC2, GR3, and P. capeferrum TDA1

Solvent-tolerant bacteria are well suited for biocatalytic production in two-phase systems [2]. One major problem—next to the substrate and product toxicity—concerns product recovery from the aqueous medium. A promising solution addressing both problems is the implementation of a two-phase system with an organic solvent as the second liquid phase. Since the product accumulates in the organic solvent phase, it can be isolated by routine chemical purification methods [4,10,37]. Nevertheless, due to the toxic effects, only few solvents with high hydrophobicity coefficients can be applied for this purpose, which limit their use at industrial applications involving microbial activity. To enhance tolerance to chemical stresses in bacteria, it is needed to increase yields and titers of several bioprocesses [38].

Organic solvents with log $P_{o/w} \geq 2.5$ (1-octanol, 1-nonanol, and 1-decanol) were previously tested to predict the adaptation of *Pseudomonas putida* DOT-T1E in a second-phase system. The results revealed that pre-exposed bacterial cells to the aforementioned solvents were able to adapt, whereas an increased amount of unsaturated *trans*- and saturated membrane fatty acids was crucial [4,32]. In this study, the growth of *P. taiwanensis* VLB120, GRC1, GRC2, GRC3, and *P. capeferrum* TDA1 was evaluated in a two-phase system with 1% (*v*/*v*) of 1-octanol or 1-decanol (Table 3). The cultures were preadapted with 1 mM of 1-octanol or 0.2 mM of 1-decanol, respectively, and subsequently cultivated in medium containing 1% (*v*/*v*) of 1-octanol or 1-decanol for 24 h. The tested strains were considered as adapted at an OD_{560} of at least 0.5 or well-adapted at an OD_{560} of over 1 after 24 h incubation with 1% [*v*/*v*] of 1-octanol or 1-decanol. *P. taiwanensis* VLB120 and *P. capeferrum* TDA1 adapted to these solvents in a second-phase system (OD_{560} 0.73 ± 0.02 and 0.71 ± 0.18, respectively). However, the streamlined GRC strains showed a faster adaptability (OD_{560} above 1). This can be explained most probably due to their higher growth rates compared to the wildtype strain tested.

In contrast to 1-octanol, all the tested strains were able to adapt very well to 1-decanol. One parameter to predict the toxicity of several compounds is the maximum membrane concentration (MMC) [39]. For long-chain *n*-alkanols such as 1-octanol and 1-decanol, the MMC is 588 mM and 379 mM, respectively. Values higher than 200 mM are considered extremely toxic to microorganisms [40]. The lower MMC of 1-decanol explains the better adaptability to it in contrast to 1-octanol. This was documented for *P. putida* DOT-T1E and *P. putida* KT2440 before [4,41].

Table 3. Bacterial growth (OD_{560}) adaptation to a second phase of 1-octanol and 1-decanol present in concentrations of 1% (*v*/*v*). +: adapted (OD_{560} between 0.5 and 1 after 24 h incubation with 1% (*v*/*v*)), ++: well-adapted (OD_{560} above 1 after 24 h incubation with 1% (*v*/*v*)). The errors are given as standard error of the mean (n = 3).

Strain	1-Octanol OD	1-Octanol Adaptation	1-Decanol OD	1-Decanol Adaptation
P. taiwanensis VLB120	0.73 ± 0.02	+	1.25 ± 0.31	++
P. taiwanensis GRC1	1.33 ± 0.09	++	1.36 ± 0.14	++
P. taiwanensis GRC2	1.25 ± 0.16	++	1.47 ± 0.21	++
P. taiwanensis GRC3	1.07 ± 0.11	++	1.72 ± 0.34	++
P. capeferrum TDA1	0.71 ± 0.18	+	2.28 ± 0.10	++

The addition of 1-decanol as a second phase has been widely used to alleviate product toxicity and increase productivity and product titers at industrial level [42]. Moreover, the inherent solvent tolerance of *Pseudomonas* is considered an important trait for the bio-production of chemicals. For instance, *P. putida* S12 has been used for the production of *p*-hydroxystyrene in a second phase of 1-decanol, which resulted in a four-fold increased titer compared to a standard fed-batch production [42]. *P. putida* S12 and an engineered *E.coli* TG1-326 expressing the solvent efflux pump SrpABC were employed as hosts for the production of 1-naphthol in a water-organic solvent (1-decanol) biphasic system [43]. *Pseudomonas* enhanced 1-naphthol production by approximately 42% contrasted with *E. coli* under the same conditions, which demonstrated that *Pseudomonas* is a more robust candidate for whole-cell bioprocesses.

4. Conclusions and Future Perspectives

The transition of the current fossil-based to a bio-based economy requires specialized microbial cell factories for the production of industrially relevant compounds. Despite this ongoing progress, the number of biotechnologically produced compounds at commercial scale is still very limited. The unique solvent-tolerant features of *P. taiwanensis* VLB120 in combination with tailored metabolic modifications provides several advantages including genetic stability and enhanced fitness under stress-triggering conditions. This study could show that the genome-reduced strains of *P. taiwanensis* VLB120 are in no way inferior to the wildtype in terms of tolerance. Therefore, the streamlined *P. taiwanensis* GRC3 with the TtgGHI pump to be activated on demand has all the properties required by the industry to be used as a new chassis bacterium, such as superior solvent tolerance to solvents even in two-phase systems. With the recent advances in high-throughput genome editing and evolution, more strains can be considered as attractive new host platforms, reducing host interference and increasing production yields. The potential of streamlined *chassis* strains for whole-cell biocatalytic applications might improve the development of new bio-based chemicals replacing petrochemical-based production in the near future.

Supplementary Materials: The following supporting information can be downloaded at: https://www.mdpi.com/article/10.3390/microorganisms11040837/s1, Figure S1. Growth rates (black circles) and *trans*/*cis* ratios (white diamonds) of *P. taiwanensis* GRC3 incubated with different *n*-alkanols: (A) 1-butanol, (B) 1-hexanol, (C) 1-octanol and (D) 1-decanol.

Author Contributions: Conceptualization, B.W., N.W., H.J.H. and C.E.; Methodology, H.J.H.; Investigation, M.J.C.E., T.S., J.P. and G.W.; Resources, B.W. and N.W.; Data curation, M.J.C.E., T.S., J.P. and G.W.; Writing—original draft, M.J.C.E.; Writing—review & editing, H.J.H. and C.E.; Supervision, M.J.C.E. and C.E.; Funding acquisition, N.W. and H.J.H. All authors have read and agreed to the published version of the manuscript.

Funding: This work was supported by the German Federal Ministry of Education and Research (BMBF) via the project NO-STRESS [grant numbers 031B0852A (to B.W. and N.W.) and 031B085C (to C.E. and H.J.H.)].

Data Availability Statement: The genome assembly for *Pseudomonas capeferrum* TDA1 is available under the accession number CP116669.1. The data presented in this study are available in Figure S1 (Supplemental Material) and on request from the corresponding author.

Conflicts of Interest: The authors declare no conflict of interest.

References

1. Schmid, A.; Dordick, J.; Hauer, B.; Kiener, A.; Wubbolts, M.; Witholt, B. Industrial biocatalysis today and tomorrow. *Nature* **2001**, *409*, 258–268. [CrossRef] [PubMed]
2. Kusumawardhani, H.; Hosseini, R.; de Winde, J.H. Solvent tolerance in bacteria: Fulfilling the promise of the biotech era? *Trends Biotechnol.* **2018**, *36*, 1025–1039. [CrossRef] [PubMed]
3. Bitzenhofer, N.L.; Kruse, L.; Thies, S.; Wynands, B.; Lechtenberg, T.; Ronitz, J.; Kozaeva, E.; Wirth, N.T.; Eberlein, C.; Jaeger, K.E.; et al. Towards robust Pseudomonas cell factories to harbour novel biosynthetic pathways. *Essays Biochem.* **2021**, *65*, 319–336. [CrossRef] [PubMed]
4. Neumann, G.; Kabelitz, N.; Zehnsdorf, A.; Miltner, A.; Lippold, H.; Meyer, D.; Schmid, A.; Heipieper, H.J. Prediction of the adaptability of *Pseudomonas putida* DOT-T1E to a second phase of a solvent for economically sound two-phase biotransformations. *Appl. Environ. Microbiol.* **2005**, *71*, 6606–6612. [CrossRef]
5. Blank, L.M.; Ionidis, G.; Ebert, B.E.; Bühler, B.; Schmid, A. Metabolic response of *Pseudomonas putida* during redox biocatalysis in the presence of a second octanol phase. *FEBS J.* **2008**, *275*, 5173–5190. [CrossRef]
6. Domínguez de María, P.; Hollmann, F. On the (un) greenness of biocatalysis: Some challenging figures and some promising options. *Front. Microbiol.* **2015**, *6*, 1257. [CrossRef]
7. Heipieper, H.J.; Diefenbach, R.; Keweloh, H. Conversion of *cis* unsaturated fatty acids to *trans*, a possible mechanism for the protection of phenol-degrading *Pseudomonas putida* P8 from substrate toxicity. *Appl. Environ. Microbiol.* **1992**, *58*, 1847–1852. [CrossRef]
8. Sikkema, J.; de Bont, J.A.; Poolman, B. Mechanisms of membrane toxicity of hydrocarbons. *Microbiol. Rev.* **1995**, *59*, 201–222. [CrossRef]
9. Eberlein, C.; Baumgarten, T.; Starke, S.; Heipieper, H.J. Immediate response mechanisms of Gram-negative solvent-tolerant bacteria to cope with environmental stress: Cis-*trans* isomerization of unsaturated fatty acids and outer membrane vesicle secretion. *Appl. Microbiol. Biotechnol.* **2018**, *102*, 2583–2593. [CrossRef]
10. Heipieper, H.J.; Neumann, G.; Cornelissen, S.; Meinhardt, F. Solvent-tolerant bacteria for biotransformations in two-phase fermentation systems. *Appl. Microbiol. Biotechnol.* **2007**, *74*, 961–973. [CrossRef]
11. Rojas, A.; Duque, E.; Mosqueda, G.; Golden, G.; Hurtado, A.; Ramos, J.L.; Segura, A. Three efflux pumps are required to provide efficient tolerance to toluene in *Pseudomonas putida* DOT-T1E. *J. Bacteriol.* **2001**, *183*, 3967–3973. [CrossRef] [PubMed]
12. Fillet, S.; Daniels, C.; Pini, C.; Krell, T.; Duque, E.; Bernal, P.; Segura, A.; Lu, D.; Zhang, X.; Ramos, J.L. Transcriptional control of the main aromatic hydrocarbon efflux pump in *Pseudomonas*. *Environ. Microbiol. Rep.* **2012**, *4*, 158–167. [CrossRef] [PubMed]
13. Ramos, J.L.; Duque, E.; Gallegos, M.-T.; Godoy, P. Mechanisms of solvent tolerance in gram-negative bacteria. *Annu. Rev. Microbiol.* **2002**, *56*, 743–768. [CrossRef]
14. Vasylkivska, M.; Patakova, P. Role of efflux in enhancing butanol tolerance of bacteria. *J. Biotechnol.* **2020**, *320*, 17–27. [CrossRef] [PubMed]
15. Krell, T.; Lacal, J.; Guazzaroni, M.E.; Busch, A.; Silva-Jiménez, H.; Fillet, S.; Reyes-Darías, J.A.; Muñoz-Martínez, F.; Rico-Jiménez, M.; García-Fontana, C. Responses of *Pseudomonas putida* to toxic aromatic carbon sources. *J. Biotechnol.* **2012**, *160*, 25–32. [CrossRef]
16. Mukhopadhyay, A. Tolerance engineering in bacteria for the production of advanced biofuels and chemicals. *Trends Microbiol.* **2015**, *23*, 498–508. [CrossRef] [PubMed]
17. Lieder, S.; Nikel, P.I.; de Lorenzo, V.; Takors, R. Genome reduction boosts heterologous gene expression in *Pseudomonas putida*. *Microb. Cell Factories* **2015**, *14*, 23. [CrossRef]
18. Kampers, L.F.C.; Volkers, R.J.M.; Martins Dos Santos, V.A.P. *Pseudomonas putida* KT2440 is HV1 certified, not GRAS. *Microb. Biotechnol.* **2019**, *12*, 845–848. [CrossRef]
19. Keshavarz-Tohid, V.; Vacheron, J.; Dubost, A.; Prigent-Combaret, C.; Taheri, P.; Tarighi, S.; Taghavi, S.M.; Moenne-Loccoz, Y.; Muller, D. Genomic, phylogenetic and catabolic re-assessment of the *Pseudomonas putida* clade supports the delineation of *Pseudomonas alloputida* sp. nov., *Pseudomonas inefficax* sp. nov., *Pseudomonas persica* sp. nov., and *Pseudomonas shirazica* sp. nov. *Syst. Appl. Microbiol.* **2019**, *42*, 468–480. [CrossRef]
20. Wynands, B.; Otto, M.; Runge, N.; Preckel, S.; Polen, T.; Blank, L.M.; Wierckx, N. Streamlined *Pseudomonas taiwanensis* VLB120 chassis strains with improved bioprocess features. *ACS Synth. Biol.* **2019**, *8*, 2036–2050. [CrossRef]
21. Martínez-García, E.; Nikel, P.I.; Aparicio, T.; de Lorenzo, V. *Pseudomonas* 2.0: Genetic upgrading of *P. putida* KT2440 as an enhanced host for heterologous gene expression. *Microb. Cell Factories* **2014**, *13*, 159. [CrossRef]

22. Puiggené, Ò.; Espinosa, M.J.C.; Schlosser, D.; Thies, S.; Jehmlich, N.; Kappelmeyer, U.; Schreiber, S.; Wibberg, D.; Kalinowski, J.; Harms, H. Extracellular degradation of a polyurethane oligomer involving outer membrane vesicles and further insights on the degradation of 2, 4-diaminotoluene in *Pseudomonas capeferrum* TDA1. *Sci. Rep.* **2022**, *12*, 2666. [CrossRef]
23. Espinosa, M.J.C.; Blanco, A.C.; Schmidgall, T.; Atanasoff-Kardjalieff, A.K.; Kappelmeyer, U.; Tischler, D.; Pieper, D.H.; Heipieper, H.J.; Eberlein, C. Toward biorecycling: Isolation of a soil bacterium that grows on a polyurethane oligomer and monomer. *Front. Microbiol.* **2020**, *11*, 404. [CrossRef]
24. Utomo, R.N.C.; Li, W.J.; Tiso, T.; Eberlein, C.; Doeker, M.; Heipieper, H.J.; Jupke, A.; Wierckx, N.; Blank, L.M. Defined Microbial Mixed Culture for Utilization of Polyurethane Monomers. *ACS Sustain. Chem. Eng.* **2020**, *8*, 17466–17474. [CrossRef]
25. Hartmans, S.; Smits, J.; Van der Werf, M.; Volkering, F.; De Bont, J. Metabolism of styrene oxide and 2-phenylethanol in the styrene-degrading *Xanthobacter* strain 124X. *Appl. Environ. Microbiol.* **1989**, *55*, 2850–2855. [CrossRef]
26. Heipieper, H.J.; Loffeld, B.; Keweloh, H.; Debont, J.A.M. The *Cis/Trans* Isomerization of Unsaturated Fatty-Acids in *Pseudomonas putida* S12—An Indicator for Environmental-Stress Due to Organic-Compounds. *Chemosphere* **1995**, *30*, 1041–1051. [CrossRef]
27. Morrison, W.R.; Smith, L.M. Preparation of fatty acid methyl esters and dimethylacetals from lipids with boron fluoride–methanol. *J. Lipid Res.* **1964**, *5*, 600–608. [CrossRef] [PubMed]
28. Heipieper, H.; Meinhardt, F.; Segura, A. The *cis–trans* isomerase of unsaturated fatty acids in *Pseudomonas* and *Vibrio*: Biochemistry, molecular biology and physiological function of a unique stress adaptive mechanism. *FEMS Microbiol. Lett.* **2003**, *229*, 1–7. [CrossRef]
29. Concha, E.; Heipieper, H.J.; Wick, L.Y.; Navia, R. Effects of limonene, *n*-decane and *n*-decanol on growth and membrane fatty acid composition of the microalga *Botryococcus braunii. AMB Express* **2018**, *8*, 189. [CrossRef]
30. Basler, G.; Thompson, M.; Tullman-Ercek, D.; Keasling, J. A *Pseudomonas putida* efflux pump acts on short-chain alcohols. *Biotechnol. Biofuels* **2018**, *11*, 1–10. [CrossRef] [PubMed]
31. Wynands, B.N.D.; Wierckx, N.; Blank, L.M. *Engineering of Pseudomonas taiwanensis VLB120 for the sustainable production of hydroxylated aromatics*; Fachgruppe Biologie: Aachen, Germany, 2019.
32. Rojas, A.; Duque, E.; Schmid, A.; Hurtado, A.; Ramos, J.-L.; Segura, A. Biotransformation in double-phase systems: Physiological responses of *Pseudomonas putida* DOT-T1E to a double phase made of aliphatic alcohols and biosynthesis of substituted catechols. *Appl. Environ. Microbiol.* **2004**, *70*, 3637–3643. [CrossRef]
33. Otto, M.; Wynands, B.; Marienhagen, J.; Blank, L.M.; Wierckx, N. Benzoate Synthesis from Glucose or Glycerol Using Engineered *Pseudomonas taiwanensis. Biotechnol. J.* **2020**, *15*, e2000211. [CrossRef]
34. Sayqal, A.; Xu, Y.; Trivedi, D.K.; AlMasoud, N.; Ellis, D.I.; Rattray, N.J.; Goodacre, R. Metabolomics analysis reveals the participation of efflux pumps and ornithine in the response of *Pseudomonas putida* DOT-T1E cells to challenge with propranolol. *PLoS ONE* **2016**, *11*, e0156509. [CrossRef]
35. Isken, S.; Derks, A.; Wolffs, P.F.; de Bont, J.A. Effect of organic solvents on the yield of solvent-tolerant *Pseudomonas putida* S12. *Appl. Environ. Microbiol.* **1999**, *65*, 2631–2635. [CrossRef] [PubMed]
36. Volmer, J.; Neumann, C.; Bühler, B.; Schmid, A. Engineering of *Pseudomonas taiwanensis* VLB120 for constitutive solvent tolerance and increased specific styrene epoxidation activity. *Appl. Environ. Microbiol.* **2014**, *80*, 6539–6548. [CrossRef] [PubMed]
37. Salgado, J.M.; Rodríguez-Solana, R.; Curiel, J.A.; de Las Rivas, B.; Munoz, R.; Domínguez, J.M. Bioproduction of 4-vinylphenol from corn cob alkaline hydrolyzate in two-phase extractive fermentation using free or immobilized recombinant *E. coli* expressing pad gene. *Enzym. Microb. Technol.* **2014**, *58*, 22–28. [CrossRef]
38. Calero, P.; Nikel, P.I. Chasing bacterial chassis for metabolic engineering: A perspective review from classical to non-traditional microorganisms. *Microb. Biotechnol.* **2019**, *12*, 98–124. [CrossRef]
39. De Bont, J.A. Solvent-tolerant bacteria in biocatalysis. *Trends Biotechnol.* **1998**, *16*, 493–499. [CrossRef]
40. Heipieper, H.J.; Martínez, P. Toxicity of hydrocarbons to microorganisms. In *Cellular Ecophysiology of Microbe: Hydrocarbon and Lipid Interactions. Handbook of Hydrocarbon and Lipid Microbiology*; Springer: Cham, Switzerland, 2018; Volume 335.
41. Eberlein, C.; Starke, S.; Doncel, A.E.; Scarabotti, F.; Heipieper, H.J. Quantification of outer membrane vesicles: A potential tool to compare response in *Pseudomonas putida* KT2440 to stress caused by alkanols. *Appl. Microbiol. Biotechnol.* **2019**, *103*, 4193–4201. [CrossRef]
42. Verhoef, S.; Wierckx, N.; Westerhof, R.M.; de Winde, J.H.; Ruijssenaars, H.J. Bioproduction of p-hydroxystyrene from glucose by the solvent-tolerant bacterium *Pseudomonas putida* S12 in a two-phase water-decanol fermentation. *Appl. Environ. Microbiol.* **2009**, *75*, 931–936. [CrossRef]
43. Garikipati, S.J.; Peeples, T.L. Solvent resistance pumps of *Pseudomonas putida* S12: Applications in 1-naphthol production and biocatalyst engineering. *J. Biotechnol.* **2015**, *210*, 91–99. [CrossRef] [PubMed]

microorganisms

MDPI

Article

Comparative Genomic Analyses of Virulence and Antimicrobial Resistance in *Citrobacter werkmanii*, an Emerging Opportunistic Pathogen

José R. Aguirre-Sánchez [1], Beatriz Quiñones [2], José A. Ortiz-Muñoz [3], Rogelio Prieto-Alvarado [3], Inés F. Vega-López [3], Jaime Martínez-Urtaza [4], Bertram G. Lee [2] and Cristóbal Chaidez [1,*]

1 Laboratorio Nacional para la Investigación en Inocuidad Alimentaria, Centro de Investigación en Alimentación y Desarrollo A.C. (CIAD), Coordinación Regional Culiacán, Culiacan 80110, Mexico; jose.aguirre.dc18@estudiantes.ciad.mx

2 Produce Safety and Microbiology Research Unit, Western Regional Research Center, Agricultural Research Service, U.S. Department of Agriculture, Albany, CA 94710, USA; beatriz.quinones@usda.gov (B.Q.); bertram.lee@usda.gov (B.G.L.)

3 Parque de Innovación Tecnológica de la Universidad Autónoma de Sinaloa, Culiacan 80040, Mexico; armando_3133@outlook.com (J.A.O.-M.); rogelioprieto@uas.edu.mx (R.P.-A.); ifvega@uas.edu.mx (I.F.V.-L.)

4 Departament de Genètica i de Microbiologia, Universitat Autònoma de Barcelona, 08193 Bellaterra, Spain; jaime.martinez.urtaza@uab.cat

* Correspondence: chaqui@ciad.mx; Tel.: +52-(667)-480-6950

Abstract: *Citrobacter werkmanii* is an emerging and opportunistic human pathogen found in developing countries and is a causative agent of wound, urinary tract, and blood infections. The present study conducted comparative genomic analyses of a *C. werkmanii* strain collection from diverse geographical locations and sources to identify the relevant virulence and antimicrobial resistance genes. Pangenome analyses divided the examined *C. werkmanii* strains into five distinct clades; the subsequent classification identified genes with functional roles in carbohydrate and general metabolism for the core genome and genes with a role in secretion, adherence, and the mobilome for the shell and cloud genomes. A maximum-likelihood phylogenetic tree with a heatmap, showing the virulence and antimicrobial genes' presence or absence, demonstrated the presence of genes with functional roles in secretion systems, adherence, enterobactin, and siderophore among the strains belonging to the different clades. *C. werkmanii* strains in clade V, predominantly from clinical sources, harbored genes implicated in type II and type Vb secretion systems as well as multidrug resistance to aminoglycoside, beta-lactamase, fluoroquinolone, phenicol, trimethoprim, macrolides, sulfonamide, and tetracycline. In summary, these comparative genomic analyses have demonstrated highly pathogenic and multidrug-resistant genetic profiles in *C. werkmanii* strains, indicating a virulence potential for this commensal and opportunistic human pathogen.

Keywords: *Citrobacter*; nosocomial infections; virulence; antimicrobial resistance; comparative genomics; emerging pathogens; pangenomics; whole-genome sequencing

Citation: Aguirre-Sánchez, J.R.; Quiñones, B.; Ortiz-Muñoz, J.A.; Prieto-Alvarado, R.; Vega-López, I.F.; Martínez-Urtaza, J.; Lee, B.G.; Chaidez, C. Comparative Genomic Analyses of Virulence and Antimicrobial Resistance in *Citrobacter werkmanii*, an Emerging Opportunistic Pathogen. *Microorganisms* **2023**, *11*, 2114. https://doi.org/10.3390/microorganisms11082114

Academic Editors: Shengxi Chen and Fabio Zicker

Received: 14 July 2023
Revised: 11 August 2023
Accepted: 13 August 2023
Published: 19 August 2023

1. Introduction

The genus *Citrobacter*, a member of the family *Enterobacteriaceae*, is composed of Gram-negative bacteria that are non-spore-forming bacilli and that have the ability to use citrate as a carbon source [1–3]. To date, the genus has been classified into 18 species, based on traditional and molecular techniques [4–6]. *Citrobacter* species have many reservoirs, including the human and animal gastrointestinal tracts, and these species can also be found in water, soil, and food [1–3]. Several transmission routes have been proposed for this bacterial pathogen, such as fecal-oral transmission, contaminated food, hospital equipment, and person-to-person transmission.

Citrobacter is considered an opportunistic pathogen and has been implicated as a causative agent of hospital settings (nosocomial) and community-acquired infections among immunocompromised patients and neonates [2,7–10]. Disease symptoms in humans that are caused by *Citrobacter* include urinary tract infections, bloodstream infections, brain abscesses, respiratory tract infections, and neonatal infections such as meningitis and bacteremia. When compared to other *Enterobacteriaceae* pathogens, *Citrobacter* species are considered to have low virulence since human infections are uncommon in the general population [11,12]. However, evidence revealed that *Citrobacter* species were responsible for 3–6% of all nosocomial infections attributed to the *Enterobacteriaceae* pathogens in surveys conducted in North America [3]. Among the *Citrobacter* species, *Citrobacter freundii* and *Citrobacter koseri* are the species most commonly implicated in causing the majority of opportunistic human infections [11,12]. Given that *Citrobacter* can persist in a host for long periods [13], *Citrobacter* infections can result in multidrug resistance outbreaks among neonates and immunocompromised patients who have prolonged hospital stays [8,10,14–17]. Recent evidence has demonstrated that other *Citrobacter* species, such as *Citrobacter werkmanii*, are also considered emerging opportunistic pathogens in developing countries [9,18–20].

Although considered to be commensal in humans and animals, *C. werkmanii* has previously been linked as a causative agent of wound infection, urinary tract infections, and bacteremia in humans [18,19]. More recently, the use of whole-genome sequencing has revealed that the *C. werkmanii* strain AK-8, isolated from a patient suffering chronic kidney disease, harbors genes with a functional role in virulence and antimicrobial resistance [20], highlighting the need for subsequent studies to further characterize the virulence potential of this commensal as an opportunistic human pathogen. To expand on the characterization of *C. werkmanii* as an emerging opportunistic pathogen, the present study conducted comparative genomic analyses of *C. werkmanii* strains recovered from a major agricultural region producing horticultural products in Mexico, as well as from various geographical locations and sources, to identify relevant virulence and antimicrobial resistance genes. This research identified key determinants implicated in highly pathogenic and multidrug-resistant profiles in *C. werkmanii* strains and has provided fundamental information to enable the characterization of this emerging and opportunistic pathogen in humans.

2. Materials and Methods

2.1. Bacterial Strain Isolation and Growth Conditions

The examined *Citrobacter* strains, LANIIA-031 and LANIIA-032, were isolated from a field survey study conducted in major agricultural rivers in the state of Sinaloa in Northwestern Mexico [21]. Specifically, a size-exclusion ultrafiltration method was employed to efficiently concentrate large river water volumes, resulting in a final suspension of the targeted bacterial species in the retentive volume, as recently documented in [21]. The strains named LANIIA-031 and LANIIA-032 were recovered after subjecting the concentrated river water samples to an enrichment step in tryptic soy broth (Becton Dickinson Bioxon, Mexico City, Mexico) at 37 °C for 24 h, followed by subsequent growth on xylose lysine desoxycholate selective agar, as described in a previous study [21]. The strains were subsequently preserved at −80 °C in a 50/50 mixture of glycerol and tryptic soy broth for further genomic characterization.

2.2. Genome Sequencing and Annotation

To characterize the recovered *Citrobacter* strains from river water, the LANIIA-031 and LANIIA-032 strains were subjected to whole-genome sequencing. DNA extractions were performed using the DNeasy Blood and Tissue kit (QIAGEN, Mexico City, Mexico), following the manufacturer's specifications, and the recovered genomic DNA was initially assessed and quantified with a NanoDrop 2000c spectrophotometer (Thermo Fisher Scientific, Waltham, MA, USA). For performing the sequencing reactions, genomic DNA from the *Citrobacter* strains was quantified with a Qubit™ 2.0 Fluorometer (Invitrogen,

Carlsbad, CA, USA) and adjusted to a 0.2 ng/μL concentration. The genomic DNA libraries per strain were prepared using a final amount of 1 ng with the Nextera XT DNA Library Preparation Kit (Illumina Inc., San Diego, CA, USA). They were then sequenced using a MiSeq™ Reagent Kit v2 (300-cycle format) to obtain a 2 × 150 bp paired-end read output with a MiSeq™ System (Illumina, Inc.) at the Earlham Institute (Norwich Research Park, Norwich, United Kingdom), as in previous studies [22]. To initially assess the quality of the sequencing output prior to assembly, the raw sequence data were visualized and evaluated using the FastQC program [23]. The script wrapper Trim-Galore, version 0.6.4 [24], was employed for the removal of low-quality bases (value < 30) from the 3′-end of the reads, of adapter sequences, and of reads shorter than 80 bp in length, and the Clumpify tool, version 38.75 [25], was used to eliminate the duplicated sequencing reads. For the de novo read assembly, the A5-miseq pipeline, version 20160825 [26], was used with the re-scaffolding process to reduce the contig numbers, and the results were chosen according to the final number of contigs, N50, and genome size. To perform genome annotation, Prokka software, version 1.14.5 [27] and the RAST server [28] for the *Citrobacter* genus (NCBI:txid544) were used with a total of 30 genomic sequences, downloaded from the National Center for Biotechnology Information, detailing *C. werkmanii* strains from various geographical locations and sources.

To screen the contig assemblies for the examined *C. werkmanii* strains, the virulence finder database and the comprehensive antibiotic resistance database [29,30] for genes related to virulence and antimicrobial resistance traits, respectively, were employed in conjunction with the ABRicate tool, version 1.0.1, by selecting the parameter cutoffs of 90% coverage and 95% nucleotide identity [31]. The ABRicate results were then represented on a clustered heatmap, depicting the presence or absence of the virulence and antimicrobial resistance gene profiles, and the heatmap was constructed and edited with the Interactive Tree Of Life (iTOL) Annotation Editor, version 5 [32]. Genes encoding the proteins linked to the various bacterial secretion systems were detected by employing the program MacSyFinder, version 1, with the TXSScan model, version 1.1.1 [33,34], by using a maximal E-value of 0.001 as the statistical threshold of significance for the analysis of secretion systems.

2.3. Phylogenetic, Comparative Genomics, and Pangenome Analyses

The phylogenetic analysis of the *C. werkmanii* strains, LANIIA-031 and LANIIA-032, recovered from river water was performed using the concatenated sequences of the housekeeping genes, *fusA* (protein synthesis elongation factor-G), *leuS* (leucine tRNA synthetase), *pyrG* (CTP synthetase), and *rpoB* (β-subunit of RNA polymerase), as an established sequence-based method for the identification of species within the genus *Citrobacter* [35]. The sequence of *recN* (DNA repair) was also employed as another reliable marker for molecular species identification [36]. For the phylogenetic analyses, the genome data from the *Citrobacter* type strains (Table S1 in the Supplementary Materials) were also included in the phylogenetic analyses. Sequence concatenation and alignments were performed with Geneious software, version 9.1.8 (Biomatters, Auckland, New Zealand) and were imported into MEGA-X [37] for constructing the phylogeny trees, using the neighbor-joining method [38]. The phylogenetic tree was rooted at the midpoint, then the topology was validated by performing a bootstrap test for a total of 1000 replicates. The evolutionary distances were computed using the maximum composite likelihood method, with a uniform rate for substitution [39].

Based on the results of the phylogenetic trees, the genomic sequences from closely related *Citrobacter* species were subsequently selected for further comparison with the strains LANIIA-031 and LANIIA-032, based on documented methods for phylogenetic analysis using the complete genome data [40]. Digital DNA–DNA hybridization was performed with the server located at: https://www.dsmz.de/services/online-tools/genome-to-genome-distance-calculator-ggdc (accessed on 9 December 2022), using "suggest method 2" [41]. The average nucleotide identity values between genomes and coverage were determined using JSpecies, version 1.2.1 [42], and the reported average nucleotide identity values were adjusted by the genome coverage on the JSpecies web server. To initially examine genomic variability in the recovered strains of LANIIA-031 and LANIIA-032, the genome assemblies were submitted to the IslandViewer 4 server [43] to search for genomic islands. To further characterize the genomic islands, the genomes were examined for identifying phages using PHASTER [44]. Comparison of the genomes of the strains LANIIA-031 and LANIIA-032 was performed using progressiveMauve [45], and the diagram with the genomic islands was generated with Geneious software, version 9.1.8 (Biomatters).

To examine the complete genetic composition, using the available sequence data of the *C. werkmanii* strain collection for the phylogenetic clade grouping, a pangenome was constructed using a combination of Roary, version 3.11.2 [46], and Anvi'o, version 7 [47] platforms. Phylogenetic relationships based on the core genomes of *C. werkmanii* were generated with the HarvestTools suite, version 1.2.2, by employing the multi-aligner, Parsnp, version 1.7.4, and visual platform, Gingr, version 1.3, to obtain a multi-FASTA alignment [48]. The RAxML program, version 8 [49], with a general time-reversible model, was used to construct a core genome phylogenetic tree using a gamma distribution and 1000 bootstrap replicates. The iTOL program, version 5 [32] was used to visualize and edit the core phylogenetic tree. Additionally, GFF files were generated with the Prokka tool, version 1.14.5; these files were used as input for generating a gene alignment and for identifying the presence of core and accessory genes, based on a 90% identity with the Roary program, version 3.11.2. The pangenome was visualized with Phandango, version 1.3.0 [50], and some additional plots were generated using the script entitled "roary_plots.py". Moreover, the Anvi'o platform, version 7, was used by selecting the pangenomics workflow system (available at https://merenlab.org/2016/11/08/pangenomics-v2/ and accessed on 18 November 2021) for constructing the pangenome of the *C. werkmanii* strains. The results obtained using the annotated genomic databases were subsequently visualized using the command entitled "Anvi-display-pan". The gene calls were clustered into bins, based on the following criteria: (i) core genes present in 99–100% of the genomes; (ii) shell genes present in 15–99% of the genomes; and (iii) cloud genes present in less than 15% of the genomes [46,51]. Furthermore, the data in the gene cluster summary file from the Anvi'o workflow system was used to analyze the distribution of functional clusters of orthologous gene (COG) categories in the shell and cloud pangenome in the examined *C. werkmanii* strains.

2.4. Statistical Analyses

Statistical analyses were performed by conducting Fisher's exact test, using the R Statistical Software (version 4.2.0; R Foundation for Statistical Computing, Vienna, Austria) [52]. Probability values (p-values) of lower than 0.05 were considered significant.

3. Results

3.1. Characterization of *Citrobacter* Strains from River Water

Phylogenetic analyses of *Citrobacter* reference strains (Table S1 in the Supplementary Materials), based on concatenation of the MLST genes *fusA*, *leuS*, *pyrG*, and *rpoB* (Figure 1a) and also of the housekeeping *recN* gene (Figure 1b), revealed that the river water strains LANIIA-031 and LANIIA-032 belong to the *C. werkmanii* species and were found to cluster together in a separate branch with the reference strain *C. werkmanii* FDAARGOS_364^{T} as the most closely related species. Based on the results of the phylogenetic trees (Figure 1), the genomic sequences of the *Citrobacter* river water strains, LANIIA-031 and LANIIA-032, were further examined by calculating the digital DNA–DNA hybridization, in comparison to the genome data from closely related *Citrobacter* reference strains (Table S1 in the Supplementary Materials). As shown in Table 1, the recorded values of the digital DNA–DNA hybridization analysis revealed that strains LANIIA-031 and LANIIA-032 had values of >90% when compared to the reference strain, *C. werkmanii* FDAARGOS_364^{T}. In contrast, the digital DNA–DNA hybridization values of strains LANIIA-031 and LANIIA-032 when compared to other species were below 70%, which is the established threshold for bacterial species delineation [5,53]. As an additional in silico method for corroborating the speciation of the river water strains, LANIIA-031 and LANIIA-032, the genomes were compared via the average nucleotide identity method (Table S2 in the Supplementary Materials), and the results showed values above 90% for the LANIIA-031 and LANIIA-032 strains when compared to the reference strain, *C. werkmanii* FDAARGOS_364^{T}. When compared to other *Citrobacter* species, values below 79% were observed; these findings were in agreement with the phylogenetic analysis and digital DNA–DNA hybridization, indicating that the river water strains of LANIIA-031 and LANIIA-032 are *C. werkmanii*.

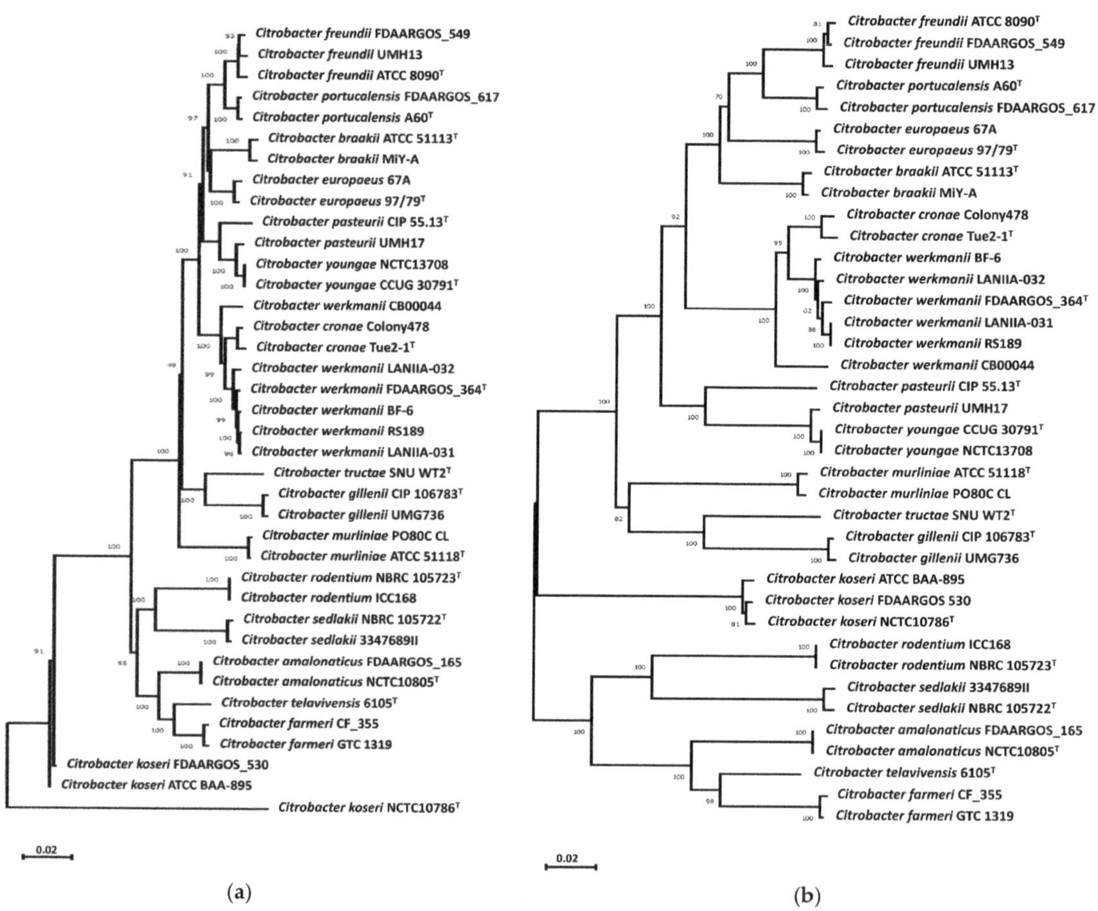

Figure 1. Phylogenetic relationships of the *Citrobacter* strains, LANIIA-031 and LANIIA-032, recovered from agricultural river water, to other *Citrobacter*-type strains. Maximum-likelihood phylogenetic trees were constructed, based on the concatenated *fusA*, *leuS*, *pyrG*, and *rpoB* genes (**a**) and the *recN* gene (**b**), by using the neighbor-joining method, rooted at the midpoint, and validated by performing a bootstrap test for a total of 1000 replicates. Bootstrap values (%) are indicated at the nodes, and the scale bar represents the expected number of nucleotide substitutions. The superscript "T" indicates a *Citrobacter* type strain.

Table 1. Digital DNA–DNA hybridization of the *Citrobacter* strains examined in the present study.

Citrobacter Species [1]	*C. werkmanii* LANIIA-032	*C. werkmanii* FDAARGOS_364 ^{T}	*C. freundii* ATCC 8090 ^{T}	*C. youngae* CCUG30791 ^{T}	*C. pasteurii* CIP55.33^{T}	*C. braakii* ATCC 51,113 ^{T}	*C. europaeus* 97/79^{T}	*C. portucalensis* A60^{T}	*C. tructae* SNU WT2 ^{T}	*C.cronae* Tue2-1^{T}
C. tructae SNU WT2 ^{T}										32.6
C. portucalensis A60 ^{T}									33.0	42.4
C. europaeus 97/79 ^{T}								50.3	32.9	42.9
C. braakii ATCC 51113 ^{T}							52.9	48.5	33.1	42.7
C. pasteurii CIP 55.13 ^{T}						38.8	38.5	38.4	32.9	36.2
C. youngae CCUG 30791 ^{T}					59.6	39.0	39.0	39.4	32.9	36.5
C. freundii ATCC 8090^{T}				39.8	35.2	45.0	45.2	52.4	29.8	37.6
C. werkmanii FDAARGOS_364 ^{T}			37.8	36.4	36.1	42.7	43.4	42.4	32.4	70.0
C. werkmanii LANIIA-032		92.2	37.7	36.5	36.1	42.6	43.3	42.5	32.4	70.0
C. werkmanii LANIIA-031	92.3	92.2	37.9	36.5	36.2	42.7	43.3	42.5	32.4	70.0

[1] The superscript "T" indicates a *Citrobacter*-type strain.

3.2. Pangenome Analyses of *C. werkmanii* Strains from Diverse Sources and Locations

To obtain a better understanding of the virulence potential in *C. werkmanii*, a pangenome analysis was conducted by employing the Roary and Anvi'o workflows with an additional thirty publicly available genomic data points from *C. werkmanii* strains (Table 2), recovered from distinct geographical locations and sources. For this pangenome analysis, the resulting gene calls were clustered into bins based on: (i) core genes present in 99–100% of the genomes; (ii) shell genes present in 15–99% of the genomes; and (iii) cloud genes present in less than 15% of the genomes. A total of 11,680 gene clusters with 153,862 genes were detected to be specific for the *C. werkmanii* pangenome, which was classified as a core, shell, and cloud genome. As indicated by the outer colored ring in the pangenome diagram (Figure 2), the core genome in the examined *C. werkmanii* strains was the most abundant and consistently detected fraction (99–100% of the strains), with 3871 gene clusters and 126,303 gene calls. Moreover, 367 gene clusters and 6832 gene calls were detected for the shell genome (15–95% of the strains). Finally, the cloud genome (0–15% of the strains) accounted for 7442 gene clusters and 20,727 gene calls. The construction of a Heap's Law chart showed that as new genomes were added to the analysis, the number of conserved genes (the core genome) decreased slightly (Figure 2, solid line), but the total number of genes increased considerably (Figure 2, dashed line). These observations were indicative of an open pangenome for *C. werkmanii* where the genomic content increased, with the number of additionally sequenced strains contributing to the species characterization.

Table 2. List of *Citrobacter werkmanii* strains examined in the present study and their characteristics.

Clade [1]	Strain [2]	GenBank Accession Number	Isolation Source Description	Sample Source Type	Date [3]	Country
Clade I	CB00011	GCA_016505505.1	Sputum	Human	2017	United States
	CB00078	GCA_016507625.1	Wound	Human	2018	United States
Clade II	FDAARGOS_364 [T]	GCA_002386385.1	Stool	Human	2014	United States
	MGYG-HGUT-02535	GCA_902388105.1	Gut	Human	2019	United States
Clade III	CB00044	GCA_016505055.1	Not collected	Human	2017	United States
	CVM 45625	GCA_015942525.1	Unknown	Environmental	2019	United States
Clade IV	BF-6	GCA_002025225.1	Industrial water	Environmental	2012	China
	Colony242	GCA_016893825.1	Food	Food	2019	Thailand
	Colony247	GCA_016893645.1	Food	Food	2019	Thailand
	CVM 45620	GCA_015943365.1	Unknown	Environmental	2019	United States
	CVM 45667	GCA_015943485.1	Unknown	Environmental	2019	United States
	CVM 45672	GCA_015943405.1	Unknown	Environmental	2020	United States
	LANIIA-031	JAUJUK000000000	River water	Environmental	2018	Mexico
	LANIIA-032	JAUJUL000000000	River water	Environmental	2018	Mexico
	RS189	GCA_015958985.1	Not collected	Human	2017	United States
	YDC730	GCA_015958865.1	Pelvic abscess	Human	2015	United States
Clade V	2580	GCA_009907085.1	Urine	Human	2015	Nigeria
	AK-8	GCA_002114305.1	Human urine	Human	2014	India
	BOX-5	GCA_009856875.1	Hospital sink	Environmental	2016	France
	C5.1	GCA_008364715.1	Sprouts	Food	2015	Germany
	CRE1173	GCA_018106225.1	Pus	Human	2015	Malaysia
	CRE1470	GCA_018106165.1	Peritoneal fluid	Human	2016	Malaysia
	CRE346	GCA_018106145.1	Foot ulcer	Human	2015	Malaysia
	CRE806	GCA_018106185.1	Foot ulcer	Human	2014	Malaysia
	CRK0001	GCA_002185305.2	Blood	Human	2014	United States
	CW_LB-887	GCA_013303045.1	Coastal water	Environmental	2014	Brazil
	FDAARGOS_616	GCA_008693645.1	Clinical isolate	Human	NA	United States
	ICR003007	GCA_004146135.1	Hospital patient	Human	2017	France
	JS97	GCA_009821535.1	Chicken	Animal	NA	Unknown
	L38	GCA_013618825.1	Chicken liver	Animal	2019	Nigeria
	UMH18	GCA_003665555.1	Bacteremia	Human	2013	United States
	YDC667-1	GCA_013336965.	Lung tissue	Human	2014	United States

[1] Clade grouping was based on whole genome analysis of *C. werkmanii* strains, as described in Section 2, Materials and Methods. [2] The superscript "T" indicates a *Citrobacter*-type strain. [3] Date refers to the sample collection date. NA refers to data that are not available.

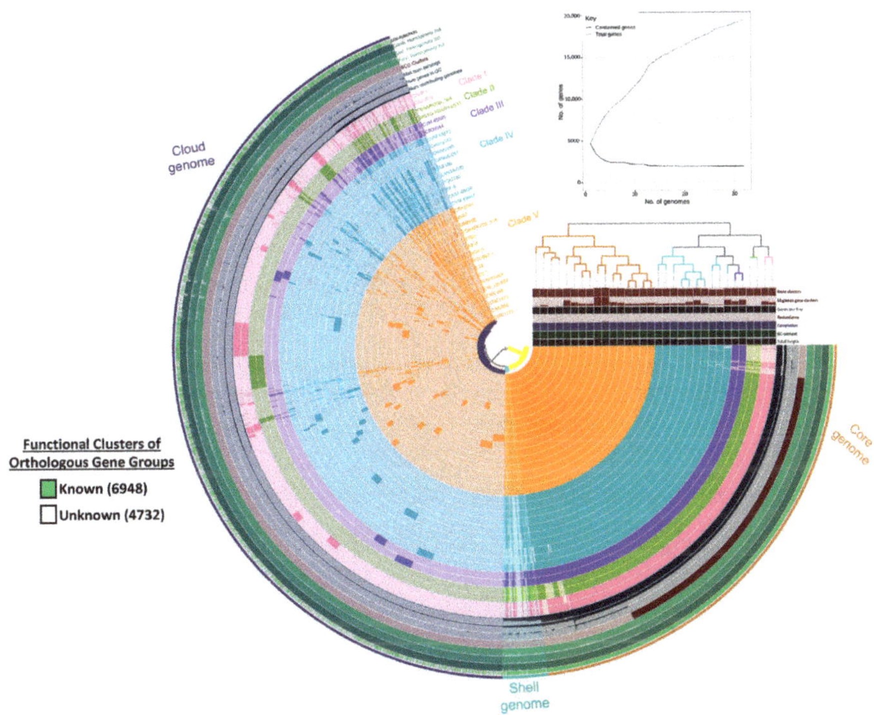

Figure 2. Pangenome analysis of the examined *Citrobacter werkmanii* strains. Pangenome analysis of over thirty *C. werkmanii* sequenced strains (Table 2) was constructed using Roary and Anvi'o platforms, and the resulting gene calls were clustered into bins, based on: (i) core genes present in 99–100% of the genomes; (ii) shell genes present in 15–99% of the genomes; and (iii) cloud genes present in less than 15% of the genomes. The construction of a phylogram, based on gene presence or absence in the examined *C. werkmanii* strains, resulted in the grouping of the strains into five distinct clades. Construction of a Heap´s Law chart (top right) showed that the number of conserved genes in the core genome decreased slightly (solid line) but the total number of genes increased considerably (dashed line), indicative of an open pangenome for *C. werkmanii*.

Phylogenetic analysis of the *C. werkmanii* strains divided the pangenome of this species into five distinct clades (Figure 2 and Table 2). The phylogram, based on gene presence or absence, revealed that most examined *C. werkmanii* strains were clustered in two major clades, namely, clades IV and V. In particular, clade IV was composed of ten strains, including the two river-derived *C. werkmanii* strains, LANIIA-031 and LANIIA-032, from this study, as well as strain BF-6 from industrial water, strains Colony242 and Colony247 from food, and the environmental strains CVM45667, CVM45620, and CVM45672. A correlation analysis revealed that this clade was significantly associated with environmental sources (Fisher's exact test p-value of < 0.01). Moreover, clade V, with the largest number of strains (Figure 2 and Table 2), a total of 16 strains, comprised *C. werkmanii* strains that had predominantly been recovered from clinical samples, associated with minor and severe human illnesses. The clinical strains included AK-8, ICR003007, CRE806, CRE1173, CRE346, 2580, CRK0001, and YDC667-1. Other isolation sources for the strains in clade V included chicken (strains JS97 and L38), sprouts (strain C5.1), and coastal water (strain CW_LB-887), and a significant correlation was identified for clade V strains with sampling sources from diverse geographical locations, excluding the United States (Fisher's exact test p-value of <0.02).

Additionally, the functional COG categories associated with the *C. werkmanii* accessory genome are presented in Figure 2. When examining the entire pangenome, the average percentage of functional COG was 85%, and subsequent analysis of the gene clusters for the core, shell, and cloud genomes revealed the average percentage of the functional COGs to be 91.5%, 26.5%, and 57.8%, respectively. Carbohydrate metabolism and transport, cell wall structure, translation, transcription, signal transduction, and general metabolism (amino acids, lipids, and nucleotides) were detected as the principal orthologous groups of proteins during the analysis of the core genome. These orthologous protein groups in the core genome were associated with *C. werkmanii* traits that are needed for growth and survival, while the shell genome was found to be mainly composed of genes that are implicated in the secretion system, pilus assembly protein, and phage structural genes. When compared to the core genome, the analysis of the cloud and shell genomes showed significant variability.

By conducting a subsequent analysis with the Anvi'o software version 7, the presence of putative genes for strains belonging to each clade was identified and they were assigned to functional categories for the shell (Figure 3a) and cloud (Figure 3b) pangenomes. The cellular process and signaling (brown-yellow) category, including genes with a proposed function in kinase signal transduction, accounted for about 26% to 39% of the genes in the shell pangenome. Approximately 25% to 32% of the genes were assigned to this functional category in the cloud pangenome. Moreover, the information storage and processing (blue) category, including genes with a proposed function in transcription, showed the lowest percentage, ranging from 14% to 18% of the genes in the shell genome and 12% to 21% of the genes in the cloud pangenome. Interestingly, the most variability was observed for metabolic functions (pink-red), such as the transport of lipids, carbohydrates, amino acids, and nucleotides. The results for this category had the highest percentages of 34 to 38% of the genes, as detected for the human strains in clade I and the environmental clade IV for the cloud and shell genomes, respectively. In addition, a high percentage of uncharacterized genes (grey-black), ranging from 23–33%, were predominantly identified in the cloud genomes for strains belonging to all clades, and the cloud genomes for the human strains in clade II and clade V were identified by having prophages, transposons, and over 20% of the genes associated with the mobilome, corresponding to those genetic elements that can confer movement within and among the different bacterial genomes [54].

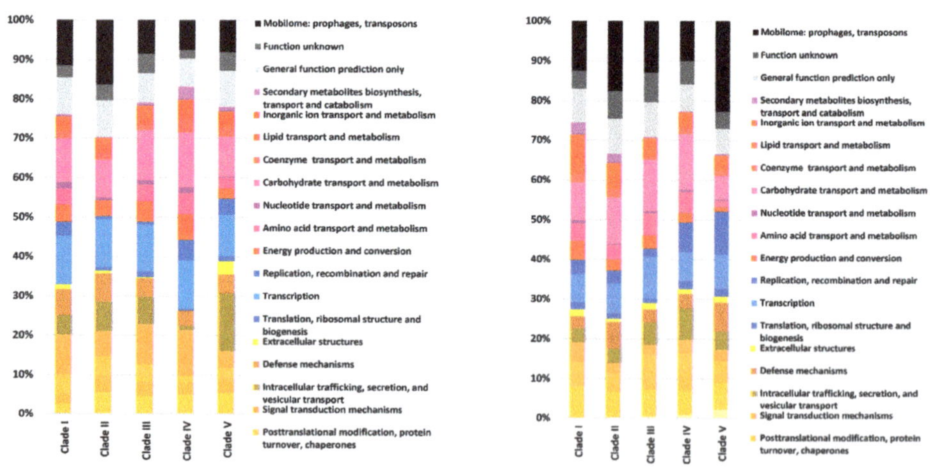

(a) (b)

Figure 3. Functional COG categories derived from the *Citrobacter werkmanii* pangenome analysis. Over thirty *C. werkmanii* genomes (Table 2) were analyzed using Anvi'o software version 7, and putative genes were assigned to functional clusters of the orthologous gene categories for each clade, as described in Figure 2, for the shell genome (**a**) and cloud genome (**b**).

3.3. Virulence and Antimicrobial Resistance Gene Identification

To gain insights about specific virulence and resistance markers in *C. werkmanii*, the available genome data from a collection of strains recovered from various geographical locations and sources were analyzed (Table 2). Additional *Citrobacter* species that are implicated in human illness, the *C. koseri* strain ATCC BAA-895 and the *C. freundii* strain FDAARGOS_549, were added to the comparative analysis. A heatmap with a phylogenetic tree was constructed by indicating the presence or absence of targeted genes that are implicated in virulence and antimicrobial resistance (Figure 4).

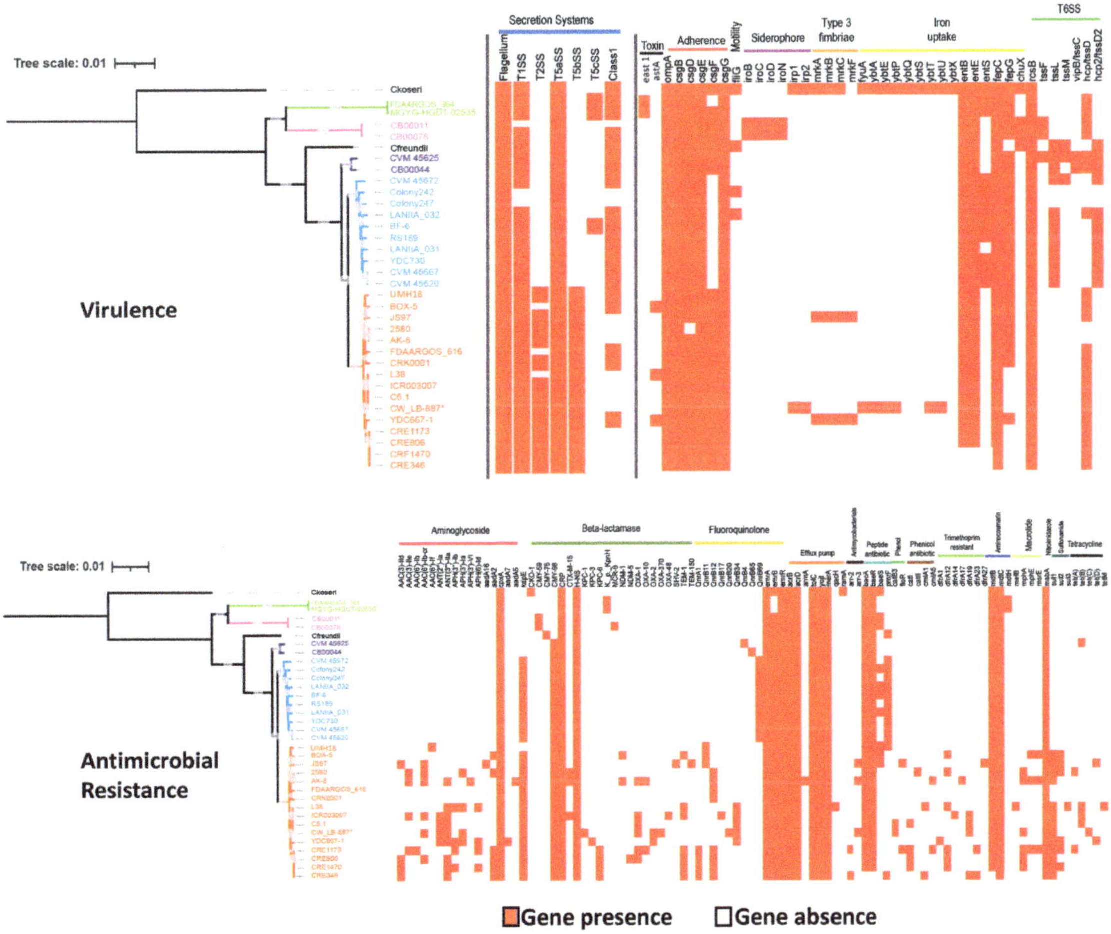

Figure 4. Virulence and antimicrobial resistance determinants in the examined *Citrobacter werkmanii* strains. A heatmap, indicating the presence or absence of virulence (**top panel**) and antimicrobial resistance genes (**bottom panel**), was constructed, coupled with a maximum-likelihood phylogenetic tree for the *C. werkmanii* strains belonging to the various clades, which are shown in distinct colors, as described in Figure 2, along with the clinical strains of *C. koseri* ATCC BAA-895 and *C. freundii* FDAARGOS_549. Gene presence in the heatmap is indicated by the red color and gene absence is indicated by the white color. Bootstrap values greater than 85% are shown with light blue circles over the phylogenetic tree nodes.

The analysis revealed that virulence traits related to the flagellum, type I, and type Va secretion systems were detected consistently in all of the examined *Citrobacter* strains. Interestingly, the phylogenetic analysis revealed the presence of mandatory secretion system-specific genes implicated in the type II and type Vb secretion system for the strains in clade V (Figure 4, top panel), comprising the largest number of examined *C. werkmanii* strains that were predominantly recovered from the clinical samples. The virulence mapping analysis also revealed that more than 55% of the examined *C. werkmanii* strains were observed to harbor genes related to the type VI-class 1 secretion system. Additionally, distinct profiles were also observed when examining the presence of the adhesin curli fimbrae *csgF* gene, the *entS* enterobactin gene, and the siderophore *iroBCDN* operon among the strains belonging to the different clades (Figure 4, top panel). In contrast, the adhesin genes *csgB* and *csgE* and the iron uptake genes *entB*, *entE*, and *fepC* were present in all examined *C. werkmanii* strains. By using the VirulenceFinder computational tool, genes implicated in the type VI secretion system were identified for *C. werkmanii* strains belonging to all five clades. Although a larger number of virulence genes were detected for the clinically relevant species, *C. koseri*, a total of 16 virulence genes were commonly detected for the *C. werkmanii* strains recovered from clinical, food, and environmental sources at various geographical locations.

Multidrug resistance was observed in the examined *C. werkmanii* strains, and resistance to beta-lactamase, efflux pumps, peptide antibiotics, fluoroquinolone, and aminocoumarin were commonly observed in this study (Figure 4, bottom panel). With the exception of the strains FDAARGOS_616 and CRK0001, the remaining *C. werkmanii* strains that were clustering in clade V, which were predominantly from clinical sources, were found to have a larger number of genes associated with antimicrobial resistance in various drug classes when compared to the other examined *C. werkmanii* strains, as well as the clinical *C. freundii* FDAARGOS_549 and *C. koseri* ATCC BAA-895 reference strains (Figure 4, bottom panel). In particular, distinct antimicrobial gene profiles were detected among the *C. werkmanii* strains belonging to clade V for aminoglycoside {*aac*(3)-*lle*, *ant*(2″)-*la*, *ant*(3″)-*lla*, and *ant*(3″)-*la*}, beta-lactamases (bla$_{OXA-10}$, bla$_{OXA-2}$, bla$_{OXA-370,}$ and bla$_{OXA-48}$), fluoroquinolone (*qnrA1*, *qnrB11*, *qnrB12*, *qnrB17*, and *qnrB20*), phenicol (*catl*, *catll*, *cmlA1*, and *cmlA5*), trimethoprim (*dfrA1*, *dfrA12*, *dfrA14*, *dfrA17*, *dfrA19*, *dfrA23*, and *dfrA27*), macrolide (*mphA* and *mphE*), sulfonamide (*sul1*, *sul2*, and *sul3*) and tetracycline (*tetA*, *tetB*, *tetC*, *tetD*, and *tetM*). The current analysis revealed that *C. werkmanii* strains belonging to clade V that were recovered from clinical sources showed the most extensive antimicrobial resistance profiles. In particular, a significantly higher number of resistance genes was observed for *C. werkmanii*, which had a total of 27 genes, when compared to the clinically relevant *C. koseri* strain ATCC BAA-895, with a total of 21 genes. In summary, these findings indicated that the *C. werkmanii* strains harbor virulence determinant genes, conferring an ability regarding host adaptation and colonization as well as multidrug resistance to various high-priority agents, indicating significant potential for *C. werkmanii* as an emerging and opportunistic pathogen in humans.

4. Discussion

In the present study, a size-exclusion ultrafiltration method, followed by genome sequencing and phylogenomic analyses, revealed the isolation of novel strains of *C. werkmanii*, the strains LANIIA-031 and LANIIA-032, which were recovered from agricultural river water in an important region for the production of fresh produce in northwestern Mexico. The examined *C. werkmanii* strains, LANIIA-031 and LANIIA-032, were isolated from major agricultural river samples as part of a field survey study aimed at the isolation of *Salmonella enterica* [21]. Preliminary typing and biochemical standardized assays falsely identified the LANIIA-031 and LANIIA-032 strains as *Salmonella*. However, the use of whole-genome sequencing led to the accurate species identification of the LANIIA-031 and LANIIA-032 strains as *Citrobacter*. In previous studies, phylogenomic analyses have indicated that *Salmonella* and *Citrobacter* have similar environmental niches and a highly related evolutionary history [55]. The genomic similarity between *Salmonella* and *Citrobacter* poses a challenge in food safety since current traditional methods may lead to the inaccurate speciation of these foodborne pathogens. These findings indicate that the use of high-resolution sequencing-based methods focused on genomic characterization enables the accurate species identification of the recovered bacterial pathogens from diverse samples relevant to food safety, environmental monitoring, and diagnostics testing.

Pangenome analyses of a large collection of sequenced *C. werkmanii* strains from diverse sources and locations resulted in the grouping of the strains into five distinct clades. Based on the identification of a gene's presence or absence, as well as single nucleotide polymorphisms, most of the strains clustered in clade IV were from environmental sources and those in clade V were predominantly from clinical sources. The pangenome analysis from this study showed that the examined *C. werkmanii* strains had genomic features and a similar number of genes that were associated with subsystem categories, a finding that was in agreement with previous reports characterizing the genomic structure for another environmental *C. werkmanii* isolate recovered from industrial spoilage samples [56]. The current analysis also revealed that *C. werkmanii* exhibited an open pangenome, and this finding was in agreement with the characteristics of other *Enterobacteriaceae* family members such as *S. enterica* and *Escherichia coli* [57,58]. The supporting evidence indicating an open pangenome for *C. werkmanii* would imply a genome in a species that is constantly exchanging genetic material and implies an ability to colonize multiple environments, as observed for other pathogenic and environmental bacterial species [59].

In agreement with previous studies examining the pangenome of other *Citrobacter* species [60], the core pangenome in *C. werkmanii* comprised the essential gene components associated with general metabolism, replication, cell wall structure, and transcription categories in *C. werkmanii* strains belonging to all clades for this opportunistic pathogen. Although small variations between the functional gene classification among the strains were observed, the functional gene category analysis revealed that the shell and cloud genomes showed the highest variability in terms of the number of genes for cellular metabolism, including carbohydrate lipid, nucleotide, and amino acid transport, as well as secondary metabolite biosynthesis. In particular, carbohydrate metabolism genes were detected as the main accessory features, resulting in additional gene copies in the shell and cloud genomes for *C. werkmanii* strains belonging to all clades. In particular, genes with an important role in galactose metabolism were identified that were coding for galactose mutarotase and beta-galactosidase, which are involved in substrate conversion and degradation [61]. Previous studies have shown that galactose is an important carbohydrate for the lipopolysaccharide composition of *Citrobacter* species [62,63].

Interestingly, the cloud genomes for the human strains in clades II and V had a large number of prophages, transposons, and mobilome-associated genes. The acquisition of mobile elements is considered an important evolutionary mechanism for bacterial pathogens to adapt and colonize many reservoirs, as well as to enhance the virulence potential [64–66]. Additionally, the presence of phage protein tail and capsid structural genes were found as part of both the shell and cloud genomes. Phages have been described as a common bacterial evolution mechanism involving horizontal gene transfer [54,66]. Moreover, the findings from this study indicated a high number of genes of unknown function or uncharacterized genes, which were also in the cloud and shell genomes. These results indicate the current challenges in the annotation of microbial genomes, supporting the need for improved computational pipelines to integrate uncharacterized genes in an evolutionary, biotechnological, and ecological framework as the volumes of genomic data continue to be deposited in public repositories for the improved characterization of a particular bacterial species [67,68].

An analysis of virulence determinants showed that *C. werkmanii* strains in this study harbor genes that are implicated in secretion, as with the flagellum, type I, and type Va secretion systems. To our knowledge, the present study is the first one to identify specific clades associated with genetic traits linked to the secretion system in this species.

Moreover, genes related to the type II and type Vb secretion systems were found among the *C. werkmanii* strains that belonged to clade V, which were recovered predominantly from clinical sources. The type II secretion system is implicated in protein effector translocation across the outer membrane in pathogenic bacteria [69], while the type Vb secretion has the characteristic structure of a two-partner secretion pathway, with a role in the adhesion to receptors in mammalian cells [70]. Moreover, the identified type II secretion system genes in this study have a known role in the translocation of virulence factors, toxins, and enzymes across the cell's outer membrane, as described for other human pathogens such as *E. coli*, *Vibrio cholerae*, *Chlamydia trachomatis*, and *Acinetobacter baumannii* [71]. Finally, genes of the type VI secretion system were identified in the *C. werkmanii* strains examined in this study. In a previous report, the type VI secretion system in *C. freundii* has been shown to contribute to adhesion and cytotoxicity in host cells [72]. Interestingly, the expression of the type VI secretion system by bacterial pathogens has been considered to be an antagonist mechanism in the colonization of the human gut by the resident microbiota [73], and the presence of the type VI secretion system could potentially provide *C. werkmanii* with a selective competitive advantage during the colonization of the human host.

Another important observation of the virulence categorization was that the curli fimbriae (Csg) and outer membrane protein A (OmpA) were detected in most of the examined *C. werkmanii* strains. In particular, curli has been well characterized in pathogenic *E. coli* and *S. enterica* and has a role in biofilm formation and in attachment to the extracellular matrix and colonization of the mammalian host cells [74]. Moreover, recent studies on the *C. werkmanii* strain BF-6 have demonstrated that OmpA plays an important role in the regulation of multiple virulence phenotypes, including biofilm formation on distinct surfaces, swimming motility, metal ion responses, and resistance to biocides [75]. Finally, the virulence typing analyses of *C. werkmanii* indicated the presence of the siderophore enterobactin genes for the acquisition of iron, an essential nutrient for bacteria growth and the increased virulence potential of *Citrobacter* [76,77].

Multidrug resistance to various antimicrobial agents was commonly observed among the examined *C. werkmanii* strains in this study. Several studies have proposed beta-lactam resistance as a common mechanism among *Citrobacter* species [78]. Resistance to the fluoroquinolone and aminocoumarin agents has previously been detected for other species of *Citrobacter* [60,77,79,80]. Interestingly, the present study demonstrated that the *C. werkmanii* strains that belong to clade V, which were predominantly from clinical sources, showed the most extensive antimicrobial resistance profiles. These observed multidrug resistances included various classes of antimicrobials, such as aminoglycoside, beta-lactamase, fluoroquinolone, phenicol, trimethoprim, macrolide, sulfonamide, and tetracycline. Due to the mortality rate associated with ineffective therapy in the treatment of *Citrobacter* infections [13], these observations highlight the importance of the resistome characterization of this opportunistic bacterial species. The global dissemination of multidrug-resistant enteric pathogens, including *Citrobacter*, has resulted in a high-priority action item under the One Health approach by integrating health solutions in various interfaces related to humans, animals, and the environment [81,82]. The spread of multidrug-resistant nosocomial infections to other niches is a critical issue that is in need of the improved identification and surveillance of bacterial virulence mechanisms and agent resistance [81]. In summary, this study demonstrated that the use of next-generation sequencing platforms in conjunction with bioinformatics allowed the identification in *C. werkmanii* of markers that are linked to virulence and antimicrobial resistance, indicating the potential of this species as an emerging and opportunistic pathogen in humans.

Supplementary Materials: The following supporting information can be downloaded at: https://www.mdpi.com/article/10.3390/microorganisms11082114/s1, Table S1: List of *Citrobacter* reference strains used in the present study; Table S2: Percentage of the average nucleotide identity for the *Citrobacter* strains examined in the present study.

Author Contributions: Conceptualization, J.R.A.-S., I.F.V.-L. and C.C.; methodology, J.R.A.-S., B.Q., R.P.-A. and B.G.L.; software, J.R.A.-S., J.A.O.-M., J.M.-U. and B.G.L.; validation, J.R.A.-S., I.F.V.-L. and B.L.; formal analysis, J.R.A.-S., B.Q., J.A.O.-M. and B.G.L.; investigation, J.R.A.-S., J.A.O.-M., R.P.-A. and B.G.L.; resources, J.R.A.-S., B.Q., J.M.-U. and C.C.; data curation, J.R.A.-S. and B.G.L.; writing—original draft preparation, J.R.A.-S., B.Q. and B.G.L.; writing—review and editing, J.R.A.-S., B.Q., B.G.L. and C.C.; visualization, J.R.A.-S., J.M.-U. and C.C.; supervision, J.R.A.-S., B.Q. and C.C.; project administration, C.C.; funding acquisition, B.Q. and C.C. All authors have read and agreed to the published version of the manuscript.

Funding: This material was based in part upon work supported by the Laboratorio Nacional para la Investigación en Inocuidad Alimentaria (LANIIA) at Centro de Investigación y Desarrollo A. C. (CIAD) in Culiacán, Sinaloa and by the United States Department of Agriculture (USDA), Agricultural Research Service (ARS), CRIS Project Number 2030-42000-055-00D.

Data Availability Statement: The complete genome sequences of the *C. werkmanii* strains of LANIIA-031 and LANIIA-032 can be found under the GenBank accession numbers JAUJUK000000000 and JAUJUL000000000, respectively, with the BioProject accession number PRJNA992637 in the National Center for Biotechnology Information (NCBI) BioProject database (https://www.ncbi.nlm.nih.gov/bioproject/ accessed on 15 July 2023).

Acknowledgments: The authors would like to thank Irvin González López (CIAD, Culiacán, Sinaloa, México) for coordinating the survey of agricultural river water in the Culiacan Valley, Sinaloa, Mexico, and Maria Brandl (USDA-ARS, Albany, CA, USA) for providing critical comments on the manuscript. Mention of trade names or commercial products in this publication is solely for the purpose of providing specific information and does not imply recommendation or endorsement by the USDA. USDA is an equal opportunity provider and employer.

Conflicts of Interest: The authors declare no conflict of interest. The funders had no role in the design of the study; in the collection, analyses, or interpretation of data; in the writing of the manuscript, or in the decision to publish the results.

References

1. Rogers, L.; Power, K.; Gaora, P.Ó.; Fanning, S. Escherichia coli and other *Enterobacteriaceae*: Occurrence and detection. In *Encyclopedia of Food and Health*; Caballero, B., Finglas, P.M., Toldrá, F., Eds.; Academic Press: Oxford, UK, 2016; pp. 545–551.
2. Kus, J.V. Infections due to *Citrobacter* and *Enterobacter*. In *Reference Module in Biomedical Sciences*; Elsevier: Amsterdam, The Netherlands, 2014.
3. Janda, J.M.; Abbott Sharon, L. The changing face of the family *Enterobacteriaceae* (Order: "*Enterobacterales*"): New members, taxonomic issues, geographic expansion, and new diseases and disease syndromes. *Clin. Microbiol. Rev.* **2021**, *34*, e00174-20. [CrossRef] [PubMed]
4. Jung, W.J.; Kim, H.J.; Giri, S.S.; Kim, S.G.; Kim, S.W.; Kang, J.W.; Kwon, J.; Lee, S.B.; Oh, W.T.; Jun, J.W.; et al. *Citrobacter tructae* sp. nov. isolated from kidney of diseased rainbow trout (*Oncorhynchus mykiss*). *Microorganisms* **2021**, *9*, 275. [CrossRef] [PubMed]
5. Oberhettinger, P.; Schüle, L.; Marschal, M.; Bezdan, D.; Ossowski, S.; Dörfel, D.; Vogel, W.; Rossen, J.W.; Willmann, M.; Peter, S. Description of *Citrobacter cronae* sp. nov., isolated from human rectal swabs and stool samples. *Int. J. Syst. Evol. Microbiol.* **2020**, *70*, 2998–3003. [CrossRef]
6. Ribeiro, T.G.; Izdebski, R.; Urbanowicz, P.; Carmeli, Y.; Gniadkowski, M.; Peixe, L. *Citrobacter telavivum* sp. nov. with chromosomal *mcr-9* from hospitalized patients. *Eur. J. Clin. Microbiol. Infect Dis.* **2021**, *40*, 123–131. [CrossRef]
7. Ong, C.C.H.; Farhanah, S.; Linn, K.Z.; Tang, Y.W.; Poon, C.Y.; Lim, A.Y.; Tan, H.R.; Binte Hamed, N.H.; Huan, X.; Puah, S.H.; et al. Nosocomial infections among COVID-19 patients: An analysis of intensive care unit surveillance data. *Antimicrob. Resist. Infect Control* **2021**, *10*, 119. [CrossRef]
8. Pletz, M.W.; Wollny, A.; Dobermann, U.-H.; Rödel, J.; Neubauer, S.; Stein, C.; Brandt, C.; Hartung, A.; Mellmann, A.; Trommer, S.; et al. A nosocomial foodborne outbreak of a VIM carbapenemase-expressing *Citrobacter freundii*. *Clin. Infect Dis.* **2018**, *67*, 58–64. [CrossRef] [PubMed]
9. Ranjan, K.P.; Ranjan, N. *Citrobacter*: An emerging health care associated urinary pathogen. *Urol. Ann.* **2013**, *5*, 313. [CrossRef]
10. Tchidjou, H.K.; Romeo, B. Infant case of co-infection with SARS-CoV-2 and *Citrobacter koseri* urinary infection. *J. Trop. Pediatr.* **2021**, *67*, fmaa032. [CrossRef]
11. Anderson, M.T.; Mitchell, L.A.; Zhao, L.; Mobley, H.L.T. *Citrobacter freundii* fitness during bloodstream infection. *Sci. Rep.* **2018**, *8*, 11792. [CrossRef] [PubMed]
12. Ariza-Prota, M.A.; Pando-Sandoval, A.; García-Clemente, M.; Fernández, R.; Casan, P. Community-acquired pneumonia and empyema caused by *Citrobacter koseri* in an immunocompetent patient. *Case Rep. Pulmonol.* **2015**, *2015*, 670373. [CrossRef]
13. Pepperell, C.; Kus, J.V.; Gardam, M.A.; Humar, A.; Burrows, L.L. Low-virulence *Citrobacter* species encode resistance to multiple antimicrobials. *Antimicrob. Agents Chemother.* **2002**, *46*, 3555–3560. [CrossRef] [PubMed]

14. Gardam, M.A.; Burrows, L.L.; Kus, J.V.; Brunton, J.; Low, D.E.; Conly, J.M.; Humar, A. Is surveillance for multidrug-resistant *Enterobacteriaceae* an effective infection control strategy in the absence of an outbreak? *J. Infect. Dis.* **2002**, *186*, 1754–1760. [CrossRef] [PubMed]
15. Hammerum, A.M.; Hansen, F.; Nielsen, H.L.; Jakobsen, L.; Stegger, M.; Andersen, P.S.; Jensen, P.; Nielsen, T.K.; Hansen, L.H.; Hasman, H.; et al. Use of WGS data for investigation of a long-term NDM-1-producing *Citrobacter freundii* outbreak and secondary in vivo spread of bla_{NDM-1} to *Escherichia coli, Klebsiella pneumoniae* and *Klebsiella oxytoca*. *J. Antimicrob. Chemother.* **2016**, *71*, 3117–3124. [CrossRef]
16. Jones, M.E.; Karlowsky, J.A.; Draghi, D.C.; Thornsberry, C.; Sahm, D.F.; Nathwani, D. Epidemiology and antibiotic susceptibility of bacteria causing skin and soft tissue infections in the USA and Europe: A guide to appropriate antimicrobial therapy. *Int. J. Antimicrob. Agents* **2003**, *22*, 406–419. [CrossRef] [PubMed]
17. Lee, R.; Choi, S.M.; Jo, S.J.; Lee, J.; Cho, S.Y.; Kim, S.H.; Lee, D.G.; Jeong, H.S. Clinical characteristics and antimicrobial susceptibility trends in *Citrobacter* bacteremia: An 11-year single-center experience. *Infect. Chemother.* **2019**, *51*, 1–9. [CrossRef]
18. Mohanty, S.; Singhal, R.; Sood, S.; Dhawan, B.; Kapil, A.; Das, B.K. *Citrobacter* infections in a tertiary care hospital in Northern India. *J. Infect.* **2007**, *54*, 58–64. [CrossRef]
19. Nayar, R.; Shukla, I.; Sultan, A. Epidemiology, prevalence and identification of *Citrobacter* species in clinical specimens in a tertiary care hospital in India. *Int. J. Sci. Res. Publ.* **2014**, *4*, 1–6.
20. Parvez, S.; Khan, A.U.; Kaur, G.; Barakat, M.; Ortet, P.; Mayilraj, S. An insight into the genome of extensively drug-resistant and uropathogenic *Citrobacter werkmanii*. *J. Glob. Antimicrob. Resist.* **2020**, *22*, 785–791. [CrossRef]
21. González-López, I.; Medrano-Félix, J.A.; Castro-Del Campo, N.; López-Cuevas, O.; González-Gómez, J.P.; Valdez-Torres, J.B.; Aguirre-Sánchez, J.R.; Martínez-Urtaza, J.; Gómez-Gil, B.; Lee, B.G.; et al. Prevalence and genomic diversity of *Salmonella enterica* recovered from river water in a major sgricultural region in Northwestern Mexico. *Microorganisms* **2022**, *10*, 1214. [CrossRef] [PubMed]
22. Aguirre-Sánchez, J.R.; Valdez-Torres, J.B.; del Campo, N.C.; Martínez-Urtaza, J.; del Campo, N.C.; Lee, B.G.; Quiñones, B.; Chaidez-Quiroz, C. Phylogenetic group and virulence profile classification in *Escherichia coli* from distinct isolation sources in Mexico. *Infect. Genet. Evol.* **2022**, *106*, 105380. [CrossRef]
23. Andrews, S. FastQC: A Quality Control Tool for High Throughput Sequence Data. Available online: http://www.bioinformatics.babraham.ac.uk/projects/fastqc (accessed on 1 May 2021).
24. Krueger, F.; Trim Galore. A Wrapper Tool around Cutadapt FastQC to Consistently Apply QUAL Adapt Trimming to FastQ Files. Available online: https://www.bioinformatics.babraham.ac.uk/projects/trim_galore/ (accessed on 1 May 2021).
25. Bushnell, B. BBTools Software Package. Available online: Sourceforge.net/projects/bbmap/ (accessed on 1 May 2021).
26. Coil, D.; Jospin, G.; Darling, A.E. A5-miseq: An updated pipeline to assemble microbial genomes from Illumina MiSeq data. *Bioinformatics* **2015**, *31*, 587–589. [CrossRef]
27. Seemann, T. Prokka: Rapid prokaryotic genome annotation. *Bioinformatics* **2014**, *30*, 2068–2069. [CrossRef]
28. Brettin, T.; Davis, J.J.; Disz, T.; Edwards, R.A.; Gerdes, S.; Olsen, G.J.; Olson, R.; Overbeek, R.; Parrello, B.; Pusch, G.D.; et al. RASTtk: A modular and extensible implementation of the RAST algorithm for building custom annotation pipelines and annotating batches of genomes. *Sci. Rep.* **2015**, *5*, 8365. [CrossRef] [PubMed]
29. Jia, B.; Raphenya, A.R.; Alcock, B.; Waglechner, N.; Guo, P.; Tsang, K.K.; Lago, B.A.; Dave, B.M.; Pereira, S.; Sharma, A.N.; et al. CARD 2017: Expansion and model-centric curation of the comprehensive antibiotic resistance database. *Nucleic Acids Res.* **2017**, *45*, D566–D573. [CrossRef] [PubMed]
30. Liu, B.; Zheng, D.; Jin, Q.; Chen, L.; Yang, J. VFDB 2019: A comparative pathogenomic platform with an interactive web interface. *Nucleic Acids Res.* **2019**, *47*, D687–D692. [CrossRef]
31. Seemann, T. ABRicate. Available online: https://github.com/tseemann/abricate (accessed on 1 May 2021).
32. Letunic, I.; Bork, P. Interactive Tree Of Life (iTOL) v4: Recent updates and new developments. *Nucleic Acids Res.* **2019**, *47*, W256–W259. [CrossRef]
33. Abby, S.S.; Néron, B.; Ménager, H.; Touchon, M.; Rocha, E.P.C. MacSyFinder: A program to mine genomes for molecular dystems with an spplication to CRISPR-Cas systems. *PLoS ONE* **2014**, *9*, e110726. [CrossRef]
34. Abby, S.S.; Rocha, E.P.C. Identification of protein secretion systems in bacterial genomes using MacSyFinder. In *Bacterial Protein Secretion Systems: Methods and Protocols*; Journet, L., Cascales, E., Eds.; Springer New York: New York, NY, USA, 2017; pp. 1–21.
35. Clermont, D.; Motreff, L.; Passet, V.; Fernandez, J.-C.; Bizet, C.; Brisse, S. Multilocus sequence analysis of the genus *Citrobacter* and description of *Citrobacter pasteurii* sp. nov. *Int. J. Syst. Evol. Microbiol.* **2015**, *65*, 1486–1490. [CrossRef] [PubMed]
36. Das, S.; Dash, H.R.; Mangwani, N.; Chakraborty, J.; Kumari, S. Understanding molecular identification and polyphasic taxonomic approaches for genetic relatedness and phylogenetic relationships of microorganisms. *J. Microbiol. Methods* **2014**, *103*, 80–100. [CrossRef] [PubMed]
37. Kumar, S.; Stecher, G.; Li, M.; Knyaz, C.; Tamura, K. MEGA X: Molecular Evolutionary Genetics Analysis across Computing Platforms. *Mol. Biol. Evol.* **2018**, *35*, 1547–1549. [CrossRef] [PubMed]
38. Saitou, N.; Nei, M. The neighbor-joining method: A new method for reconstructing phylogenetic trees. *Mol. Biol. Evol.* **1987**, *4*, 406–425. [CrossRef] [PubMed]
39. Tamura, K.; Nei, M.; Kumar, S. Prospects for inferring very large phylogenies by using the neighbor-joining method. *Proc. Natl. Acad. Sci. USA* **2004**, *101*, 11030–11035. [CrossRef]

40. Chun, J.; Oren, A.; Ventosa, A.; Christensen, H.; Arahal, D.R.; da Costa, M.S.; Rooney, A.P.; Yi, H.; Xu, X.W.; De Meyer, S.; et al. Proposed minimal standards for the use of genome data for the taxonomy of prokaryotes. *Int. J. Syst. Evol. Microbiol.* **2018**, *68*, 461–466. [CrossRef] [PubMed]
41. Meier-Kolthoff, J.P.; Carbasse, J.S.; Peinado-Olarte, R.L.; Göker, M. TYGS and LPSN: A database tandem for fast and reliable genome-based classification and nomenclature of prokaryotes. *Nucleic Acids Res.* **2022**, *50*, D801–D807. [CrossRef] [PubMed]
42. Richter, M.; Rossello-Mora, R.; Oliver Glockner, F.; Peplies, J. JSpeciesWS: A web server for prokaryotic species circumscription based on pairwise genome comparison. *Bioinformatics* **2016**, *32*, 929–931. [CrossRef] [PubMed]
43. Bertelli, C.; Laird, M.R.; Williams, K.P.; Simon Fraser University Research Computing Group; Lau, B.Y.; Hoad, G.; Winsor, G.L.; Brinkman, F.S.L. IslandViewer 4: Expanded prediction of genomic islands for larger-scale datasets. *Nucleic Acids Res.* **2017**, *45*, W30–W35. [CrossRef]
44. Arndt, D.; Grant, J.R.; Marcu, A.; Sajed, T.; Pon, A.; Liang, Y.; Wishart, D.S. PHASTER: A better, faster version of the PHAST phage search tool. *Nucleic Acids Res.* **2016**, *44*, W16–W21. [CrossRef]
45. Darling, A.E.; Mau, B.; Perna, N.T. progressiveMauve: Multiple genome alignment with gene gain, loss and rearrangement. *PLoS ONE* **2010**, *5*, e11147. [CrossRef]
46. Page, A.J.; Cummins, C.A.; Hunt, M.; Wong, V.K.; Reuter, S.; Holden, M.T.G.; Fookes, M.; Falush, D.; Keane, J.A.; Parkhill, J. Roary: Rapid large-scale prokaryote pan genome analysis. *Bioinformatics* **2015**, *31*, 3691–3693. [CrossRef]
47. Eren, A.M.; Esen, Ö.C.; Quince, C.; Vineis, J.H.; Morrison, H.G.; Sogin, M.L.; Delmont, T.O. Anvi'o: An advanced analysis and visualization platform for 'omics data. *PeerJ* **2015**, *3*, e1319. [CrossRef]
48. Treangen, T.J.; Ondov, B.D.; Koren, S.; Phillippy, A.M. The Harvest suite for rapid core-genome alignment and visualization of thousands of intraspecific microbial genomes. *Genome Biol.* **2014**, *15*, 524. [CrossRef]
49. Stamatakis, A. RAxML version 8: A tool for phylogenetic analysis and post-analysis of large phylogenies. *Bioinformatics* **2014**, *30*, 1312–1313. [CrossRef] [PubMed]
50. Hadfield, J.; Croucher, N.J.; Goater, R.J.; Abudahab, K.; Aanensen, D.M.; Harris, S.R. Phandango: An interactive viewer for bacterial population genomics. *Bioinformatics* **2018**, *34*, 292–293. [CrossRef] [PubMed]
51. Tettelin, H.; Masignani, V.; Cieslewicz, M.J.; Donati, C.; Medini, D.; Ward, N.L.; Angiuoli, S.V.; Crabtree, J.; Jones, A.L.; Durkin, A.S.; et al. Genome analysis of multiple pathogenic isolates of *Streptococcus agalactiae*: Implications for the microbial "pan-genome". *Proc. Natl. Acad. Sci. USA* **2005**, *102*, 13950–13955. [CrossRef] [PubMed]
52. Mehta, C.R.; Patel, N.R. Algorithm 643. FEXACT: A FORTRAN subroutine for Fisher's exact test on unordered r × c contingency tables. *ACM Trans. Math Softw.* **1986**, *12*, 154–161. [CrossRef]
53. Goris, J.; Konstantinidis, K.T.; Klappenbach, J.A.; Coenye, T.; Vandamme, P.; Tiedje, J.M. DNA-DNA hybridization values and their relationship to whole-genome sequence similarities. *Int. J. Syst. Evol. Microbiol.* **2007**, *57*, 81–91. [CrossRef]
54. Smets, B.F.; Barkay, T. Horizontal gene transfer: Perspectives at a crossroads of scientific disciplines. *Nat. Rev. Microbiol.* **2005**, *3*, 675–678. [CrossRef] [PubMed]
55. Pilar, A.V.C.; Petronella, N.; Dussault, F.M.; Verster, A.J.; Bekal, S.; Levesque, R.C.; Goodridge, L.; Tamber, S. Similar yet different: Phylogenomic analysis to delineate *Salmonella* and *Citrobacter* species boundaries. *BMC Genom.* **2020**, *21*, 377. [CrossRef] [PubMed]
56. Zhou, G.; Peng, H.; Wang, Y.-s.; Huang, X.-m.; Xie, X.-b.; Shi, Q.-s. Complete genome sequence of *Citrobacter werkmanii* strain BF-6 isolated from industrial putrefaction. *BMC Genom.* **2017**, *18*, 765. [CrossRef]
57. Tantoso, E.; Eisenhaber, B.; Kirsch, M.; Shitov, V.; Zhao, Z.; Eisenhaber, F. To kill or to be killed: Pangenome analysis of *Escherichia coli* strains reveals a tailocin specific for pandemic ST131. *BMC Biol.* **2022**, *20*, 146. [CrossRef]
58. Laing, C.R.; Whiteside, M.D.; Gannon, V.P.J. Pan-genome analyses of the species *Salmonella enterica*, and identification of genomic markers predictive for species, subspecies, and serovar. *Front Microbiol.* **2017**, *8*, 1345. [CrossRef] [PubMed]
59. Vernikos, G.; Medini, D.; Riley, D.R.; Tettelin, H. Ten years of pan-genome analyses. *Curr. Opin. Microbiol.* **2015**, *23*, 148–154. [CrossRef] [PubMed]
60. Yuan, C.; Yin, Z.; Wang, J.; Qian, C.; Wei, Y.; Zhang, S.; Jiang, L.; Liu, B. Comparative genomic analysis of *Citrobacter* and key genes essential for the pathogenicity of *Citrobacter koseri*. *Front Microbiol.* **2019**, *10*, 2774. [CrossRef]
61. Thoden, J.B.; Kim, J.; Raushel, F.M.; Holden, H.M. The catalytic mechanism of galactose mutarotase. *Protein Sci.* **2003**, *12*, 1051–1059. [CrossRef]
62. Kocharova, N.A.; Knirel, Y.A.; Stanislavsky, E.S.; Kholodkova, E.V.; Lugowski, C.; Jachymek, W.; Romanowska, E. Structural and serological studies of lipopolysaccharides of *Citrobacter* O35 and O38 antigenically related to *Salmonella*. *FEMS Microbiol. Immunol.* **1996**, *13*, 1–8. [CrossRef]
63. Raff, R.A.; Wheat, R.W. Carbohydrate composition of the phenol-soluble lipopolysaccharides of *Citrobacter freundii*. *J. Bacteriol.* **1968**, *95*, 2035–2043. [CrossRef] [PubMed]
64. Aguirre-Sánchez, J.R.; Ibarra-Rodriguez, J.R.; Vega-López, I.F.; Martínez-Urtaza, J.; Chaidez-Quiroz, C. Genomic signatures of adaptation to natural settings in non-typhoidal *Salmonella enterica* serovars Saintpaul, Thompson and Weltevreden. *Infect. Genet. Evol.* **2021**, *90*, 104771. [CrossRef]
65. Subedi, D.; Kohli, G.S.; Vijay, A.K.; Willcox, M.; Rice, S.A. Accessory genome of the multi-drug resistant ocular isolate of *Pseudomonas aeruginosa* PA34. *PLoS ONE* **2019**, *14*, e0215038. [CrossRef]
66. Colavecchio, A.; Cadieux, B.; Lo, A.; Goodridge, L.D. Bacteriophages contribute to the spread of antibiotic resistance genes among foodborne pathogens of the *Enterobacteriaceae* family—A review. *Front Microbiol.* **2017**, *8*, 1108. [CrossRef]

67. Miller, D.; Stern, A.; Burstein, D. Deciphering microbial gene function using natural language processing. *Nat. Commun.* **2022**, *13*, 5731. [CrossRef]
68. Vanni, C.; Schechter, M.S.; Acinas, S.G.; Barberán, A.; Buttigieg, P.L.; Casamayor, E.O.; Delmont, T.O.; Duarte, C.M.; Eren, A.M.; Finn, R.D.; et al. Unifying the known and unknown microbial coding sequence space. *eLife* **2022**, *11*, e67667. [CrossRef] [PubMed]
69. Korotkov, K.V.; Sandkvist, M.; Hol, W.G.J. The type II secretion system: Biogenesis, molecular architecture and mechanism. *Nat. Rev. Microbiol.* **2012**, *10*, 336–351. [CrossRef] [PubMed]
70. Meuskens, I.; Saragliadis, A.; Leo, J.C.; Linke, D. Type V secretion systems: An overview of passenger domain functions. *Front Microbiol.* **2019**, *10*, 1163. [CrossRef] [PubMed]
71. Chernyatina, A.A.; Low, H.H. Core architecture of a bacterial type II secretion system. *Nat. Commun.* **2019**, *10*, 5437. [CrossRef]
72. Liu, L.; Hao, S.; Lan, R.; Wang, G.; Xiao, D.; Sun, H.; Xu, J. The type VI secretion system modulates flagellar gene expression and secretion in *Citrobacter freundii* and contributes to adhesion and cytotoxicity to host cells. *Infect. Immun.* **2015**, *83*, 2596–2604. [CrossRef]
73. Chen, C.; Yang, X.; Shen, X. Confirmed and potential roles of bacterial T6SSs in the intestinal ecosystem. *Front. Microbiol.* **2019**, *10*, 1484. [CrossRef]
74. Barnhart, M.M.; Chapman, M.R. Curli biogenesis and function. *Annu. Rev. Microbiol.* **2006**, *60*, 131–147. [CrossRef]
75. Zhou, G.; Wang, Y.S.; Peng, H.; Li, S.J.; Sun, T.L.; Shen, P.F.; Xie, X.B.; Shi, Q.S. Roles of *ompA* of *Citrobacter werkmanii* in bacterial growth, biocide resistance, biofilm formation and swimming motility. *Appl. Microbiol. Biotechnol.* **2021**, *105*, 2841–2854. [CrossRef]
76. Kortman, G.A.; Boleij, A.; Swinkels, D.W.; Tjalsma, H. Iron availability increases the pathogenic potential of *Salmonella typhimurium* and other enteric pathogens at the intestinal epithelial interface. *PLoS ONE* **2012**, *7*, e29968. [CrossRef]
77. Qin, J.; Zhao, Y.; Wang, A.; Chi, X.; Wen, P.; Li, S.; Wu, L.; Bi, S.; Xu, H. Comparative genomic characterization of multidrug-resistant *Citrobacter* spp. strains in Fennec fox imported to China. *Gut Pathog.* **2021**, *13*, 59. [CrossRef]
78. Liu, L.; Lan, R.; Liu, L.; Wang, Y.; Zhang, Y.; Wang, Y.; Xu, J. Antimicrobial resistance and cytotoxicity of *Citrobacter* spp. in Maanshan Anhui Province, China. *Front Microbiol.* **2017**, *8*, 1357. [CrossRef]
79. Aoyama, H.; Fujimaki, K.; Sato, K.; Fujii, T.; Inoue, M.; Hirai, K.; Mitsuhashi, S. Clinical isolate of *Citrobacter freundii* highly resistant to new quinolones. *Antimicrob. Agents Chemother.* **1988**, *32*, 922–924. [CrossRef] [PubMed]
80. Metri, B.C.; Jyothi, P.; Peerapur, B.V. Antibiotic resistance in *Citrobacter* spp. isolated from urinary tract infection. *Urol. Ann.* **2013**, *5*, 312. [PubMed]
81. Butler Mark, S.; Gigante, V.; Sati, H.; Paulin, S.; Al-Sulaiman, L.; Rex John, H.; Fernandes, P.; Arias Cesar, A.; Paul, M.; Thwaites Guy, E.; et al. Analysis of the clinical pipeline of treatments for drug-resistant bacterial infections: Despite progress, more action is needed. *Antimicrob. Agents Chemother.* **2022**, *66*, e01991-21. [CrossRef]
82. Velázquez-Meza, M.E.; Galarde-López, M.; Carrillo-Quiróz, B.; Alpuche-Aranda, C.M. Antimicrobial resistance: One Health approach. *Vet World* **2022**, *15*, 743–749. [CrossRef] [PubMed]

microorganisms

Article

Characterization of Food Chain *Clostridioides difficile* Isolates in Terms of Ribotype and Antimicrobial Resistance

Pilar Marcos [1,2], Aoife Doyle [3,4], Paul Whyte [2], Thomas R. Rogers [3], Máire McElroy [4], Seamus Fanning [5], Jesus Frias [6] and Declan Bolton [1,*]

1 Teagasc Food Research Centre, Ashtown, Dublin 15, D15 KN3K Dublin, Ireland
2 School of Veterinary Medicine, University College Dublin, Belfield, Dublin 4, D04 V1W8 Dublin, Ireland
3 Department of Clinical Microbiology, Trinity College Dublin, Central Pathology Laboratory, St James's Hospital, Dublin 8, D08 RX0X Dublin, Ireland
4 Central Veterinary Research Laboratory, Department of Agriculture, Food and the Marine, Backweston, Celbridge, W23 X3PH Kildare, Ireland
5 UCD-Centre for Food Safety, School of Public Health, Physiotherapy and Sports Science, University College Dublin, Belfield, Dublin 4, D04 V1W8 Dublin, Ireland
6 Environmental Sustainability and Health Institute, Technological University Dublin, Grangegorman, Dublin 7, D07 H6K8 Dublin, Ireland
* Correspondence: declan.bolton@teagasc.ie; Tel.: +353-1-8059539

Abstract: The aim of this study was to characterize *C. difficile* isolates from the farm, abattoir, and retail outlets in Ireland in terms of ribotype and antibiotic resistance (vancomycin, erythromycin, metronidazole, moxifloxacin, clindamycin, and rifampicin) using PCR and E-test methods, respectively. The most common ribotype in all stages of the food chain (including retail foods) was 078 and a variant (RT078/4). Less commonly reported (014/0, 002/1, 049, and 205) and novel (RT530, 547, and 683) ribotypes were also detected, but at lower frequencies. Approximately 72% (26/36 tested) of the isolates tested were resistant to at least one antibiotic, with the majority of these (65%; 17/26) displaying a multi-drug (three to five antibiotics) resistant phenotype. It was concluded that ribotype 078, a hypervirulent strain commonly associated with *C. difficile* infection (CDI) in Ireland, was the most frequent ribotype along the food chain, resistance to clinically important antibiotics was common in *C. difficile* food chain isolates, and there was no relationship between ribotype and antibiotic resistance profile.

Keywords: *Clostridioides difficile*; ribotype; antibiotic resistance; food; farm; abattoir; retail

Citation: Marcos, P.; Doyle, A.; Whyte, P.; Rogers, T.R.; McElroy, M.; Fanning, S.; Frias, J.; Bolton, D. Characterization of Food Chain *Clostridioides difficile* Isolates in Terms of Ribotype and Antimicrobial Resistance. *Microorganisms* **2023**, *11*, 1296. https://doi.org/10.3390/microorganisms11051296

Academic Editors: Shengxi Chen and Fabio Zicker

Received: 14 March 2023
Revised: 12 May 2023
Accepted: 12 May 2023
Published: 16 May 2023

1. Introduction

Clostridioides difficile is an anaerobic, Gram-positive, spore-forming bacterium of ubiquitous nature. They have been isolated from the feces of a range of different animals that are both asymptomatic or suffering from severe enteric disease and are the leading cause of antibiotic-associated diarrhea in humans [1]. Symptoms caused by *C. difficile* infection (CDI) vary from mild to serious and fulminant colitis, which may result in mortality [2]. The factors promoting CDI include treatment with antibiotics. Although treatment with nearly all classes of antibiotics has been associated with CDI, third-generation cephalosporins, penicillins, clindamycin, and fluoroquinolones active against the commensal bowel flora are considered to present the highest risk. These anti-bacterial compounds disrupt the gut microbiota, remove competing organisms, and thus facilitate *C. difficile* outgrowth, colonization, and toxin production [3].

Ribotyping is a tool used for the definition of lineages and the study of the epidemiology in humans based on the heterogeneity of the ribosomal intergenic spacer region (ISR) [4]. This technique detects the ISR variation of the 11–12 copies within the rRNA operon in the genome. PCR of this region generates fragments of different lengths, which can be identified by capillary gel-based electrophoresis or be compared online against a database of isolates with known ribotypes [5].

There are currently over 800 *C. difficile* ribotypes reported worldwide. Only toxin-producing strains cause disease in humans, with several of these having increased virulence (hypervirulence) [6]. Diarrhea is associated with the production of toxin A (TcdA) and toxin B (TcdB), both of which cause intestinal inflammation. Toxigenic strains may also produce *C. difficile* transferase (CDT), a third unrelated binary toxin thought to play a role in epithelial adhesion that is associated with enhanced virulence and higher patient mortality [6,7]. *C. difficile*'s ribotypes can be classified into five main clades (I-V). Among these, Clade II and V are considered the most relevant as they include hypervirulent ribotypes 027 and 078, respectively, and other clinically important RTs, such as RT244 and RT176 [7].

Several different ribotypes have been detected in hospitals and associated with human illness and outbreaks. Ribotypes 014/020 (20.5%), 001 (18.2%), and 078/126 (18.2%), for example, were detected in Spanish hospitals in a study by Alcalá et al. [8]. Ribotypes RT002 (11.6%), RT015 (10.9%), and RT005 (7.2%) were isolated from CDI patients in England [9], and ribotypes 014/020 and 001/072 are considered endemic throughout Europe [10]. In recent years, ribotypes 078 (17%), 020 (9%), and 014 (8%) have been frequently detected in Ireland [11].

Although *C. difficile* was historically considered to be a hospital-associated disease, there is a high genetic relatedness between human and animal isolates which often belong to the same genotype [12]. Indeed whole genome sequencing (WGS) analysis has shown that isolates from human, animal, and environmental sources are often indistinguishable [13]. The multi-directional spread of *C. difficile* occurs between these different sources and inevitably results in the contamination of food [14].

It is, therefore, to be expected that the same *C. difficile* ribotypes occur in food-producing animals and humans [15,16]. Ribotypes 078 and 126, for example, often isolated from CDI patients, are also common in pigs [12,17], poultry [18], and cattle [19]. Although less information is available for sheep, previously detected ribotypes in humans (RT014, RT010, and RT045) have been isolated from ovine sources [20–22].

Many of these ribotypes, including 078, 126, 027, and 014/020, have also been isolated from a range of retail foods, including pork, turkey, beef [23–25], chicken [18,26], shellfish and bivalve mollusks [27–29], and vegetables [30–33].

Antibiotics have played an important role in the emergence of *C. difficile*. Tetracycline usage in animal husbandry, for example, was a key driver in the emergence of hypervirulent RT078, which acquired tetracycline resistance elements from *Streptococcus suis*, *Enterococcus faecalis*, and *Escherichia coli* when under selective pressure [34]. Treatment with antibiotics is also a risk factor for current human infections, which are often associated with the disruption of the gastrointestinal tract microflora [35].

CDI in humans is treated with vancomycin, metronidazole, or a combination of both, which has inevitably increased the prevalence of resistance to these antibiotics in *C. difficile* [36]. Plasmid-mediated resistance to metronidazole was recently identified by Boekhoud et al. [37]. Thus, the current guidelines on the management of *C. difficile* infection in adults in the USA recommend treatment using vancomycin and fidaxomicin instead of metronidazole [38]. Additionally, many *C. difficile* strains are resistant to clindamycin, erythromycin, rifamycins, fluoroquinolones, and penicillins, commonly used to treat other bacterial-associated illnesses in humans [3,39]. The prevalence of *C. difficile* resistance to metronidazole and vancomycin varies in different countries and has been reported to be 3.6–17.9% in the USA and 44.6% in Israel, respectively [40,41]. Considerably lower rates have been described in the UK and Ireland (0.7% metronidazole resistance in the former and 2% vancomycin resistance in Ireland) [42]. Resistance to rifampicin ranges from 0.7% in the UK to 63.6% in Hungary [43,44] and 65.3% in the USA [41,45].

The increasing resistance of *C. difficile* to antibiotics has driven the emergence of new ribotypes resulting in altered epidemiology in humans [46]. For example, the rapid emergence of hypervirulent *C. difficile* RT027 during the early 2000s has been attributed to the acquisition of fluoroquinolone resistance, a conjugative transposon in two distinct lineages, and the frequent use of this antibiotic for the treatment of a range of gastrointestinal infections in humans [47]. Despite this link between ribotypes and antibiotic resistance, the majority of studies to date have reported one or the other but not both.

Previous research in our group found *C. difficile* at all stages of the food chain in Ireland and obtained isolates from farms, abattoirs, and retail foods [48]. The objective of the current study was to characterize these isolates in terms of ribotype, with a representative selection being tested for antibiotic resistance.

2. Materials and Methods

2.1. *C. difficile* Isolates

The isolates were previously obtained in our laboratory as part of a surveillance study [48]. Isolates tested belonged to bovine farm water (5), bovine farm soil (6), bovine carcass (1), ovine feces (1), ovine carcasses (2), ovine farm soil (4), broiler feces (1), broiler farm water (2), broiler farm soil (11), wild rocket leaves (1), coleslaw (1), and cottage cheese (1).

2.2. PCR Ribotyping

The protocol used for PCR ribotyping was that described by ECDC [49], with minor modifications. The intergenic spacer region between the 16S and 23S rRNA genes was initially amplified by PCR using the primers described by Bidet et al. [50]. Briefly, 12.5 µL of HotStarTaq Mastermix (QIAGEN Ltd., Manchester, UK), 0.25 µL of each primer (10 µM), 10 µL RNAse free water (QIAGEN Ltd., Manchester, UK), and 2 µL of the template were added to a MicroAmp^{TM} Optical 96-well reaction plate (Applied Biosystems, Warrington, Cheshire, UK). Sealed PCR plates were then inserted into a Veriti 96-Well Thermal Cycler (Applied Biosystems, Warrington, Cheshire, UK), where the amplification protocol included a denaturation step of 15 min at 95 °C, followed by 30 cycles of 1 min at 94 °C, 1 min at 60 °C and 1 min at 72 °C, and a final extension of 30 min at 72 °C.

After the initial PCR, denaturation of the generated fragments was required before further analysis in an automated sequence and fragment analysis system. For this, 2 µL of the PCR products were added to 9.5 µL of Highly Deionized (Hi-Di) Formamide (Applied Biosystems, Warrington, Cheshire, UK) and 0.5 µL of GeneScan 1200 LIZ Size Standard (Applied Biosystems, Warrington, Cheshire, UK) and denatured for 2 min at 95 °C in a thermal cycler. The plate was cooled down for 10 min in a fridge before the PCR products were analyzed on an ABI 3500 Genetic Analyzer (Applied Biosystems, Warrington, Cheshire, UK) with default settings for POP7 and 50 cm capillary length. The raw data files obtained (*fsa files) from the ABI 3500 Genetic Analyzer were uploaded to the freely available WEBRIBO database (https://webribo.ages.at/; last accessed 1 September 2022) to compare our isolates with profiles stored in the database and confirm the PCR ribotypes [5].

2.3. Minimum Inhibitory Concentration (MIC) Testing

The antibiotics metronidazole, vancomycin, erythromycin, clindamycin, moxifloxacin, and rifampicin were selected for MIC testing. Metronidazole is the recommended first-line treatment option for mild-to-moderate CDI, while vancomycin can be used as a substitute or in combination [51]. Rifampicin is used as an adjunctive therapy to CDI [52]; however, its increased use in tuberculosis treatment has been associated with the emergence of rifampicin-resistant *C. difficile* strains [53], linked with recurrent CDI [54]. Moxifloxacin, erythromycin, and clindamycin are not frequently used to treat CDI, but are widely detected in virulent *C. difficile* strains and are important markers for treatment failure and for the spread of the disease in healthcare settings [55–58].

E-test strips (bioMérieux, Marcy-l'Étoile, France) were used following the protocol described by the manufacturer. In brief, the isolates were recovered from Protect Select Anaerobe Cryobeads (Technical Service Consultants Ltd., Lancashire, UK) and stored at −80 °C by aseptically transferring a single bead to 30 mL of Mueller–Hinton broth (Oxoid, Basingstoke, Hampshire, UK (CM0405)). Isolates were incubated anaerobically in an A35 Anaerobic workstation (Don Whitley, Victoria Works, Yorkshire, UK) at 37 °C for 24 h or until an OD_{600nm} = 0.5 (10^8 CFU mL^{-1}) was achieved as measured using a spectrophotometer DeNovix DS-C (DeNovix Inc., Wilmington, NC, USA). A sterile swab (Sparks Lab Supplies, Dublin, Ireland (SW001)) was then used to spread the broth culture on Brucella, vitamin K, and haemin agar (Sigma-Aldrich, Gillingham, UK (B2926-500G)) with 5% defibrinated horse blood (TCS Biosciences Limited, Botolph Claydon, Buckingham, UK (HB034)). Plates were air-dried for 15 min in the anaerobic cabinet before the E-test strips were aseptically placed on top of the agar using sterilized forceps and incubated at 37 °C for 48 h under anaerobic conditions (as described above). The MIC values were then obtained using the scale (μg mL^{-1}) provided by the manufacturer. MIC values for each antibiotic were compared to the epidemiological cutoff values (ECOFF) and breakpoints found in the European Committee on Antimicrobial Susceptibility Testing (EUCAST) [59] for *C. difficile* and classified as susceptible or resistant accordingly.

3. Results

The characterization of the isolates in terms of PCR ribotyping and antimicrobial resistance is presented in Table 1.

Table 1. PCR ribotyping, antibiotic susceptibility phenotype profiles, and minimum inhibitory concentration (MIC) values of the isolates tested from different stages of the food chain. ERY = Erythromycin (Resistant > 4 μg mL^{-1}), MET = Metronidazole (Resistant > 2 μg mL^{-1}), CLIN = Clindamycin (Resistant > 16 μg mL^{-1}), MOX = Moxifloxacin (Resistant > 4 μg mL^{-1}), VAN = Vancomycin (Resistant > 2 μg mL^{-1}), and RIF = Rifampicin (Resistant > 0.004 μg mL^{-1}) [59]. Resistance to each antibiotic according to the ECOFF values is highlighted in bold.

Isolate Source	RT [1]	Antimicrobial Susceptibility (S) or Resistance (R) (MIC in μg mL^{-1})					
		ERY	MET	CLIN	MOX	VAN	RIF
Bovine farm water	078	S (1)	S (0.094)	S (4)	S (<0.002)	**R (3)**	S (<0.002)
	078	S (1.5)	S (0.5)	S (<0.016)	S (<0.002)	S (<0.016)	S (<0.002)
	078/4	S (0.75)	S (0.047)	S (<0.016)	**R (>32)**	S (<0.016)	S (<0.002)
	078/4	S (0.75)	**R (>256)**	S (0.032)	S (3)	**R (>256)**	**R (0.5)**
	078/4	S (0.25)	**R (>256)**	S (8)	**R (>32)**	**R (>256)**	**R (0.5)**
Bovine farm soil	078	S (2)	S (0.125)	S (4)	S (<0.002)	**R (>256)**	S (<0.002)
	078/4	**R (8)**	S (0.19)	S (8)	**R (>32)**	**R (>256)**	**R (0.064)**
	049	**R (>256)**	**R (>256)**	**R (>256)**	S (3)	**R (>256)**	**R (1)**
	547	S (0.25)	**R (>256)**	S (0.064)	S (2)	**R (>256)**	**R (1)**
	683	S (0.75)	S (0.125)	S (4)	S (<0.002)	S (2)	S (<0.002)
	IN [2]	**R (>256)**	**R (>256)**	**R (>256)**	S (1)	**R (>256)**	**R (0.19)**
Bovine carcass	078/4	S (0.5)	**R (>256)**	S (0.047)	S (3)	**R (>256)**	**R (1)**
Ovine feces	078/4	S (<0.016)	S (<0.016)	S (6)	**R (>32)**	S (1.5)	S (<0.002)
Ovine farm soil	078	**R (>256)**	S (0.125)	**R (>256)**	**R (>32)**	S (1)	S (<0.002)
	078	S (<0.016)	S (<0.016)	S (<0.016)	S (<0.002)	S (<0.016)	S (<0.002)
	078	S (1.5)	S (0.047)	S (<0.016)	**R (>32)**	S (<0.016)	S (<0.002)
	IN [2]	S (<0.016)	S (<0.016)	S (8)	**R (>32)**	S (0.75)	S (<0.002)
Ovine carcasses	078/4	**R (24)**	**R (>256)**	S (6)	S (0.064)	**R (3)**	S (0.003)
	078/4	S (1.5)	S (0.047)	S (2)	S (0.5)	S (1.5)	S (<0.002)

Table 1. *Cont.*

Isolate Source	RT [1]	Antimicrobial Susceptibility (S) or Resistance (R) (MIC in μg mL^{-1})					
		ERY	MET	CLIN	MOX	VAN	RIF
Broiler feces	078	**R (>256)**	**R (>256)**	**R (>256)**	S (3)	**R (>256)**	**R (0.19)**
Broiler farm water	078/4	S (<0.016)	S (<0.016)	S (<0.016)	**R (>32)**	S (<0.016)	S (<0.002)
	IN [2]	S (<0.016)	S (<0.016)	S (<0.016)	S (<0.002)	S (<0.016)	S (<0.002)
Broiler farm soil	078	S (<0.016)	S (<0.016)	S (<0.016)	S (0.75)	S (<0.016)	S (<0.002)
	078	S (0.38)	**R (>256)**	S (0.064)	S (2)	**R (>256)**	**R (0.5)**
	078	S (<0.016)	S (<0.016)	S (<0.016)	S (<0.002)	S (<0.016)	S (<0.002)
	078/4	S (1.5)	**R (>256)**	S (<0.016)	S (3)	**R (>256)**	**R (1)**
	078/4	S (1)	S (0.047)	S (8)	**R (>32)**	**R (3)**	S (<0.002)
	078/4	S (1)	**R (>256)**	S (6)	**R (>32)**	S (<0.016)	S (<0.002)
	049	**R (>256)**	**R (>256)**	**R (>256)**	S (1.5)	**R (>256)**	**R (2)**
	002/1	S (0.25)	S (<0.016)	**R (>256)**	**R (>32)**	**R (>256)**	**R (0.75)**
	014/0	**R (>256)**	**R (>256)**	**R (>256)**	S (1.5)	**R (>256)**	**R (0.75)**
	205	S (0.25)	S (<0.016)	S (6)	S (<0.002)	S (<0.016)	S (<0.002)
	530	S (<0.016)	S (<0.016)	S (<0.016)	S (0.75)	S (<0.016)	S (<0.002)
Coleslaw	078	S (0.5)	**R (>256)**	S (0.5)	S (2)	**R (>256)**	**R (1)**
Cottage cheese	078	S (0.19)	S (<0.016)	S (<0.016)	S (0.5)	S (<0.016)	S (<0.002)
Wild rocket leaves	078	**R (>256)**	**R (>256)**	S (0.25)	S (1.5)	**R (>256)**	**R (0.047)**
% Resistance		25 (9/36)	41.6(15/36)	19.4 (7/36)	30.5 (11/36)	52.7 (19/36)	41.6 (15/36)

[1] RT = Ribotype; [2] IN = Inconclusive result.

In bovines, all of the five isolates from farm water belonged to RT078 (two isolates) or the variant 078/4 (3). One of these was sensitive to all of the antibiotics tested, two were resistant to VAN or MOX, while the remaining isolates displayed multi-resistant phenotypes (MET-VAN-RIF and MET-MOX-VAN-RIF). Ribotypes 078, 078/4, 049, 547, and 683 and an inconclusive result were obtained in bovine farm soil with corresponding antibiotic resistance profiles VAN, ERY-MOX-VAN-RIF, ERY-MET-CLIN-VAN-RIF, MET-VAN-RIF, and sensitive to all and ERY-MET-CLIN-VAN-RIF, respectively. A bovine carcass isolate (RT078/4) was resistant to MET-VAN-RIF.

The ovine feces isolate was ribotype 078/4 and resistant to MOX. Ovine farm soil isolates belonged to RT078 (three isolates), with one isolate being an inconclusive ribotype and demonstrating three different phenotypical resistance profiles (ERY-CLIN-MOX, MOX, and sensitive to all). Both ovine carcass isolates were RT078/4, and one showed antimicrobial resistance to ERY-MET-VAN.

In broilers, a fecal isolate (RT078) showed simultaneous resistance against ERY-MET-CLIN-VAN-RIF. In water collected from broiler farms, one of the two isolates was resistant to MOX (RT078/4), while the other was susceptible (inconclusive RT). Ribotypes detected in broiler farm soil (11 isolates) had diverse ribotypes and antibiotic resistance patterns, including RT078 (MET-VAN-RIF and two sensitive), 078/4 (MET-VAN-RIF, MOX-VAN, and MET-MOX, one each), 049 (ERY-MET-CLIN-VAN-RIF), 002/1 (CLIN-MOX-VAN-RIF), 014/0 (ERY-MET-CLIN-VAN-RIF), and 205 and 530, both of which were sensitive to all antibiotics tested.

In retail foods, all strains belonged to RT078, although antibiotic resistance profiles differed. In cottage cheese, susceptibility to all antibiotics tested was reported. In coleslaw and wild rocket leaves, resistance to MET-VAN-RIF and ERY-MET-VAN-RIF was observed, respectively.

Overall, the highest resistance was observed for VAN (52.7%), followed by RIF and MET (41.6% each), MOX (30.5%), ERY (25%), and CLIN (19.4%) and 47% of the isolates were multi-drug resistant (≥three antibiotics).

4. Discussion

4.1. *C. difficile's Ribotype Prevalence along the Food Chain*

In this study, nine different *C. difficile* ribotypes were detected, which reflects the diversity in sample type and source. RT078 and its variant (RT078/4) were the most common ribotypes. This was not unexpected as several previous studies have reported RT078 to be the predominant ribotype in meat and vegetable food chains [1,18,19,30,60–63].

The bovine isolates belonged to ribotypes 078, 078/4, 049, 547, and 683. Previous studies also reported ribotype 078 in bovine feces [1,19,60], while other authors have reported ribotype AI-82/1 [64,65], which was not detected in bovine isolates in our study.

The ovine isolates were identified as RT078 and 078/4. To the best of our knowledge, none of these ribotypes have been previously reported in sheep. Other ribotypes found in sheep and not detected in our study include 066 [21], 045 [66], 015, and 097 [67].

Broiler isolates belonged to RT-078, 078/4, 002/1, 014/0, 049, 205, and 530. Of these, ribotypes 002, 049 [68], 014 [64,65,68], and 205 [61,69] had been previously found in broilers.

The retail isolates included ribotype 078, which has only been previously reported in retail foods [30].

Moreover, to the best of our knowledge, this is the first time that ribotypes 530, 547, and 683 have been described in a *C. difficile* study.

4.2. *Antimicrobial Resistance of the Selected C. difficile Isolates*

Our study suggests there is no relationship between a specific ribotype and the antibiotic resistance phenotype. RT078, for example, had several resistance patterns, including susceptibility to all antibiotics (38.5%; 5/13), resistance to VAN (15.4%; 2/13), MET-VAN-RIF (15.4%; 2/13), MOX, ERY-MOX-CLIN, ERY-MET-VAN-RIF, and ERY-MET-CLIN-VAN-RIF (each 7.69%; 1/13). Solomon et al. [70] reported there was no correlation between clinical RT078 isolates and antimicrobial resistance patterns with diverse profiles obtained even in isolates from the same healthcare setting. These findings are consistent with the suggestion by Spigaglia et al. [39] that there are multiple drivers for antibiotic resistance in *C. difficile*, including the acquisition of genetic elements, alterations in target sites, changes in metabolic pathways, etc. Moreover, the same authors reported that antibiotic resistance is maintained in *C. difficile* regardless of the burden imposed on the cell or the absence of a selective pressure.

The observed resistance to erythromycin (25%) in isolates from bovine (three), ovine (one), and broiler (two) farm soil, ovine and broiler carcasses (one each), and wild rocket leaves (one), has been previously described in isolates from dairy and beef cattle [71,72], sheep [22], and human patients [73–76]. At the retail level, erythromycin-resistant *C. difficile* has been widely reported in beef, sheep, goat, and poultry meats [69,77], lettuce [78,79], and ready-to-eat foods [80].

Metronidazole is used to treat a range of infections in both food animals and humans, including CDI. A high percentage of isolates were resistant to metronidazole (42%), including broiler feces (one), bovine farm water (two), bovine (three), ovine (one), and broiler (five) farm soil, bovine carcass (one), wild rocket leaves (one), and coleslaw (one). Even though metronidazole resistance was reported to be rare in *C. difficile* [81], metronidazole-resistant *C. difficile* strains have been isolated from cattle [82], chicken, beef, and sheep carcasses [83,84], foods such as meat products [85], and lettuce [79] and humans [39,42,86]. This is often mediated by mutations in *gyrA/B* [81].

One-fifth (20%) of the *C. difficile* strains tested in our study displayed clindamycin resistance. These isolates came from sources such as broiler feces (one), bovines (two), and soil from ovine (one) and broiler (three) farms. *C. difficile*-resistant strains have been previously reported in cattle [87] and poultry [18,65,88], on beef and sheep carcasses [83], in raw beef and cooked pork meats [89–92], ready-to-eat salads [78–80], and human isolates [42,72,74,93]. It has been suggested that clindamycin-resistant *C. difficile* strains possess the *erm(B)* gene, which also confers resistance to erythromycin [57,94]. Our data provide further evidence of these co-resistances in *C. difficile*.

Eleven of the 36 *C. difficile* isolates tested were resistant to moxifloxacin, which included ovine feces (one), bovine (two) and broiler (one) farm water, and bovine (one), ovine (three), and broiler (three) farm soil. Similar results were previously described in *C. difficile* strains from cattle [72,95,96], soil on poultry farms [65], pig carcasses [97], ready-to-eat salads [58,78], and clinical isolates [93,98,99]. Furthermore, reduced susceptibility to moxifloxacin has been linked to increased mortality in human CDI [100].

Approximately half (53%, 19/36) of our isolates displayed vancomycin resistance. These isolates were from broiler feces (one), bovine farm water (three), bovine (five) and broiler (six) farm soil, bovine and ovine carcasses (one each), and wild rocket leaves (one) and coleslaw (one). Other authors have reported vancomycin-resistant *C. difficile* strains in calf feces [87], doner kebab and meatballs [85], and human patients [40–42,101]. Vancomycin and metronidazole are commonly used drugs to treat *C. difficile* infection [2], with Adler et al. [40] describing resistance to both antibiotics simultaneously, as observed in this study.

In addition, 15 of the isolates were resistant to rifampicin, which were isolated from broiler feces (one), bovine farm water (two), bovine (four) and broiler (five) farm soil, bovine carcass (one), wild rocket leaves (one), and coleslaw (one). Rifaximin resistance in *C. difficile* isolates is associated with point mutations in the *rpoB* gene which encodes the beta subunit of RNA polymerase [52] and has been previously reported in clinical isolates [52,81,99–103]. Indeed, in recent years there has been increasingly reported rifampicin resistance in *C. difficile*, reflecting the increased use of this antibiotic in medicine [104–106].

Multi-drug resistance, defined as resistance to three or more antibiotics simultaneously [107], was detected in 17 out of the 36 isolates tested from a range of stages along the food chain. Among these, the most relevant were two *C. difficile* strains isolated from retail foods resistant to MET-VAN-RIF in coleslaw and ERY-MET-VAN-RIF in wild rocket leaves, similar to that described in lettuce (ERY-MET-VAN) by Han et al. [79]. Multi-drug-resistant bacteria are often present in rinse water, and vegetable products may be cross-contaminated during the washing process [108]. Moreover, in recent years, *C. difficile* strains with multi-drug resistance have been associated with major outbreaks [76,109].

As humans, animals, and the environment are reservoirs for clinically important ribotypes with multi-directional spread, the health of all three is interconnected and reliant on adopting a One Health approach. Thus, a range of different stakeholders must work together to reduce the emergence and spread of antibiotic-resistant *C. difficile* strains. This should include the restricted use of antibiotics in both human and veterinary settings, but more comprehensive control will require the development of vaccines to reduce carriage in both humans and animals [13].

5. Conclusions

Overall, this study PCR-ribotyped and examined the antibiotic resistance phenotype of 36 food chain isolates. It was concluded that RT078 was the predominant ribotype, although another seven were identified, including three new ribotypes. Resistance to clinically important antibiotics such as erythromycin (25%; 9/36), metronidazole (42%; 15/36), clindamycin (19%; 7/36), moxifloxacin (31%; 11/36), vancomycin (53%; 19/36), and rifampicin (42%; 15/36) was common, with 72% (26/36) of isolates resistant to at least one antibiotic and 47% (17/36) displaying a multi-drug resistant phenotype. These data suggest that virulent ribotypes of *C. difficile* with phenotypic antibiotic resistance are present in the food chain, which could contribute to the spread of the infection to high risk groups in the community. Our data also suggested that there was no association between ribotype and antibiotic resistance profiles.

Author Contributions: P.M.: investigation, experimental design, methodology, formal analysis, writing—original draft, and writing—review and editing. A.D.: investigation, methodology, formal analysis, and writing—review and editing. P.W.: methodology, supervision, and writing—review and editing. T.R.R.: writing—review and editing. M.M.: writing—review and editing. S.F.: writing—review and editing. J.F.: writing—review and editing. D.B.: conceptualization, funding acquisition, experimental design, methodology, project administration, supervision, and writing—review and editing. All authors have read and agreed to the published version of the manuscript.

Funding: This project was funded by the Food Institutional Research Measure (FIRM) administered by the Department for Agriculture, Food and the Marine (DAFM) (Grant number 17F206). Pilar Marcos was supported by the Teagasc Walsh Scholarship Scheme (number 2018210).

Data Availability Statement: Data available upon request.

Conflicts of Interest: All authors declare that there are no conflict of interest in this work.

References

1. Rodriguez, C.; Avesani, V.; Van Broeck, J.; Taminiau, B.; Delmee, M.; Daube, G. *Clostridium difficile* in young farm animals and slaughter animals in Belgium. *Anaerobe* **2012**, *18*, 621–625. [CrossRef] [PubMed]
2. Cohen, S.H.; Gerding, D.N.; Johnson, S.; Kelly, C.P.; Loo, V.G.; McDonald, L.C.; Wilcox, M.H. Society for Healthcare Epidemiology of America, Infectious Diseases Society of America, Infectious Diseases Society of America. Clinical practice guidelines for *Clostridium difficile* in adults, 2010 update by the society for healthcare epidemiology of America (SHEA) and the infectious diseases society of America (IDSA). *Infect. Control. Hosp. Epidemiol.* **2010**, *31*, 431–455. [PubMed]
3. Owens, R.C.; Donskey, C.J.; Gaynes, R.P.; Loo, V.G.; Muto, C.A. Antimicrobial-associated risk factors for *Clostridium difficile* infection. *Clin. Infect. Dis.* **2008**, *46*, 19–31. [CrossRef]
4. Stubbs, S.L.; Brazier, J.S.; O'Neill, G.L.; Duerden, B.I. PCR targeted to the 16S-23S rRNA gene intergenic spacer region of *Clostridium difficile* and construction of a library consisting of 116 different PCR ribotypes. *J. Clin. Microbiol.* **1999**, *37*, 461–463. [CrossRef] [PubMed]
5. Indra, A.; Huhulescu, S.; Schneeweis, M.; Hasenberger, P.; Kernbichler, S.; Fiedler, A.; Wewalka, G.; Allerberger, F.; Kuijper, E.J. Characterization of *Clostridium difficile* isolates using capillary gel electrophoresis-based PCR ribotyping. *J. Med. Microbiol.* **2008**, *57 Pt 11*, 1377–1382. [CrossRef]
6. Dayananda, P.; Wilcox, M.H. A Review of Mixed Strain *Clostridium difficile* Colonization and Infection. *Front. Microbiol.* **2019**, *10*, 692. [CrossRef]
7. Knight, D.R.; Elliott, B.; Chang, B.J.; Perkins, T.T.; Riley, T.V. Diversity and Evolution in the Genome of *Clostridium difficile*. *Clin. Microbiol. Rev.* **2015**, *28*, 721–741. [CrossRef]
8. Alcalá, L.; Martin, A.; Marin, M.; Sanchez-Somolinos, M.; Catalan, P.; Pelaez, T.; Bouza, E. The undiagnosed cases of *Clostridium difficile* infection in a whole nation, where is the problem? *Clin. Microbiol. Infect.* **2012**, *18*, E204–E213. [CrossRef]
9. Herbert, R.; Hatcher, J.; Jauneikaite, E.; Gharbi, M.; d'Arc, S.; Obaray, N.; Rickards, T.; Rebec, M.; Blandy, O.; Hope, R.; et al. Two-year analysis of *Clostridium difficile* ribotypes associated with increased severity. *J. Hosp. Infect.* **2019**, *4*, 388–394. [CrossRef]
10. Couturier, J.; Davies, K.; Gateau, C.; Barbut, F. Ribotypes and New Virulent Strains Across Europe. *Adv. Exp. Med. Biol.* **2018**, *1050*, 45–58.
11. Health Service Executive (HSE)—Health Protection Surveillance Centre (HPSC) Enhanced Surveillance of *Clostridioides (Clostridium) difficile* Infection in Ireland: Q2 2021 National Report. 2021. Available online: https://www.hpsc.ie/a-z/microbiologyantimicrobialresistance/clostridioidesdifficile/enhancedsurveillance/quarterlyreports/CDINationalReport_Q2 2022_v2.0_Final.pdf (accessed on 20 January 2023).
12. Álvarez-Perez, S.; Blanco, J.L.; Harmanus, C.; Kuijper, E.; Garcia, M.E. Subtyping and antimicrobial susceptibility of *Clostridium difficile* PCR ribotype 078/126 isolates of human and animal origin. *Vet. Microbiol.* **2017**, *199*, 15–22. [CrossRef]
13. Knetsch, C.W.; Connor, T.R.; Mutreja, A.; van Dorp, S.M.; Sanders, I.M.; Browne, H.P.; Harris, D.; Lipman, L.; Keessen, E.C.; Corver, J.; et al. Whole genome sequencing reveals potential spread of Clostridium difficile between humans and farm animals in the Netherlands, 2002 to 2011. *Euro. Surveill.* **2014**, *19*, 20954. [CrossRef]
14. Lim, S.C.; Knight, D.R.; Riley, T.V. *Clostridium difficile* and One Health. *Clin. Microbiol. Infect.* **2020**, *26*, 857–863. [CrossRef]
15. Weese, J.S. *Clostridium (Clostridioides) difficile* in animals. *J. Vet. Diagn. Investig. Off. Publ. Am. Assoc. Vet. Lab. Diagn.* **2020**, *32*, 213–221. [CrossRef]
16. Bolton, D.; Marcos, P. The Environment, Farm Animals and Foods as Sources of *Clostridioides difficile* Infection in Humans. *Foods* **2023**, *12*, 1094. [CrossRef]
17. Andres-Lasheras, S.; Bolea, R.; Mainar-Jaime, R.C.; Kuijper, E.; Sevilla, E.; Martin-Burriel, I.; Chirino-Trejo, M. Presence of *Clostridium difficile* in pig faecal samples and wild animal species associated with pig farms. *J. Appl. Microbiol.* **2017**, *122*, 462–472. [CrossRef]
18. Harvey, R.B.; Norman, K.N.; Andrews, K.; Hume, M.E.; Scanlan, C.M.; Callaway, T.R.; Anderson, R.C.; Nisbet, D.J. Clostridium difficile in poultry and poultry meat. *Foodborne Pathog. Dis.* **2011**, *8*, 1321–1323. [CrossRef]

19. Costa, M.C.; Stämpfli, H.R.; Arroyo, L.G.; Pearl, D.L.; Weese, J.S. Epidemiology of *Clostridium difficile* on a veal farm, prevalence, molecular characterization and tetracycline resistance. *Vet. Microbiol.* **2011**, *152*, 379–384. [CrossRef]
20. Knight, D.R.; Riley, T.V. Prevalence of Gastrointestinal *Clostridium difficile* Carriage in Australian Sheep and Lambs. *Appl. Environ. Microbiol.* **2013**, *79*, 5689–5692. [CrossRef]
21. Romano, V.; Albanese, F.; Dumontet, S.; Krovacek, K.; Petrini, O.; Pasquale, V. Prevalence and genotypic characterization of *Clostridium difficile* from ruminants in Switzerland. *Zoonoses Public Health* **2012**, *59*, 545–548. [CrossRef]
22. Avberšek, J.; Pirš, T.; Pate, M.; Rupnik, M.; Ocepek, M. *Clostridium difficile* in goats and sheep in Slovenia, characterisation of strains and evidence of age-related shedding. *Anaerobe* **2014**, *28*, 163–167. [CrossRef]
23. Songer, J.G.; Trinh, H.T.; Killgore, G.E.; Thompson, A.D.; McDonald, L.C.; Limbago, B.M. *Clostridium difficile* in Retail Meat Products, USA, 2007. *Emerg. Infect. Dis.* **2009**, *15*, 819–821. [CrossRef] [PubMed]
24. Weese, J.S.; Avery, B.; Rousseau, J.; Reid-Smith, R. Detection and Enumeration of *Clostridium difficile* Spores in Retail Beef and Pork. *Appl. Environ. Microbiol.* **2009**, *15*, 5009–5011. [CrossRef]
25. Varshney, J.B.; Very, K.J.; Williams, J.L.; Hegarty, J.P.; Stewart, D.B.; Lumadue, J. Characterization of *Clostridium difficile* isolates from human fecal samples and retail meat from Pennsylvania. *Foodborne Pathog. Dis.* **2014**, *11*, 822–829. [CrossRef]
26. Weese, J.S.; Reid-Smith, R.J.; Avery, B.P.; Rousseau, J. Detection and characterization of *Clostridium difficile* in retail chicken. *Lett. Appl. Microbiol.* **2010**, *50*, 362–365. [CrossRef] [PubMed]
27. Pasquale, V.; Romano, V.; Rupnik, M.; Capuano, F.; Bove, D.; Aliberti, F.; Krovacek, K.; Dumontet, S. Occurrence of toxigenic *Clostridium difficile* in edible bivalve molluscs. *Food Microbiol.* **2012**, *31*, 309–312. [CrossRef]
28. Troiano, T.; Harmanus, C.; Sanders, I.M.J.G.; Pasquale, V.; Dumontet, S.; Capuano, F.; Kuijper, E.J. Toxigenic *Clostridium difficile* PCR ribotypes in edible marine bivalve molluscs in Italy. *Int. J. Food Microbiol.* **2015**, *208*, 30–34. [CrossRef] [PubMed]
29. Agnoletti, F.; Arcangeli, G.; Barbanti, F.; Barco, L.; Brunetta, R.; Cocchi, M.; Conedera, G.; D'Este, L.; Drigo, I.; Spigaglia, P.; et al. Survey, characterization and antimicrobial susceptibility of *Clostridium difficile* from marine bivalve shellfish of North Adriatic Sea. *Int. J. Food Microbiol.* **2019**, *298*, 74–80. [CrossRef]
30. Metcalf, D.S.; Costa, M.C.; Dew, W.M.; Weese, J.S. *Clostridium difficile* in vegetables, Canada. *Lett. Appl. Microbiol.* **2010**, *51*, 600–602. [CrossRef]
31. Eckert, C.; Burghoffer, B.; Barbut, F. Contamination of ready-to-eat raw vegetables with *Clostridium difficile*, France. *J. Med. Microbiol.* **2013**, *62*, 1435–1438. [CrossRef]
32. Primavilla, S.; Farneti, S.; Petruzzelli, A.; Drigo, I.; Scuota, S. Contamination of hospital food with *Clostridium difficile* in Central Italy. *Anaerobe* **2019**, *55*, 8–10. [CrossRef]
33. Tkalec, V.; Janezica, S.; Skoka, B.; Simonica, T.; Mesarica, S.; Vrabica, T.; Rupnik, M. High *Clostridium difficile* contamination rates of domestic and imported potatoes compared to some other vegetables in Slovenia. *Food Microbiol.* **2019**, *78*, 194–200. [CrossRef] [PubMed]
34. Dingle, K.E.; Didelot, X.; Quan, T.P.; Eyre, D.W.; Stoesser, N.; Marwick, C.A.; Coia, J.; Brown, D.; Buchanan, S.; Ijaz, U.Z.; et al. A Role for Tetracycline Selection in Recent Evolution of Agriculture-Associated *Clostridium difficile* PCR Ribotype 078. *mBio* **2019**, *10*, e02790-18. [CrossRef]
35. Lim, S.C.; Androga, G.O.; Knight, D.R.; Moono, P.; Foster, N.F.; Riley, T.V. Antimicrobial susceptibility of *Clostridium difficile* isolated from food and environmental sources in Western Australia. *Int. J. Antimicrob. Agents* **2018**, *52*, 411–415. [CrossRef]
36. Surawicz, C.M.; Brandt, L.J.; Binion, D.G.; Ananthakrishnan, A.N.; Curry, S.R.; Gilligan, P.H.; McFarland, L.V.; Mellow, M.; Zuckerbraun, B. Guidelines for diagnosis, treatment, and prevention of *Clostridium difficile* infections. *Am. J. Gastroenterol.* **2013**, *108*, 478–498. [CrossRef]
37. Boekhoud, I.M.; Hornung, B.V.H.; Sevilla, E.; Harmanus, C.; Bos-Sanders, I.M.J.G.; Terveer, E.M.; Bolea, R.; Corver, J.; Kuijper, E.J.; Smits, W.K. Plasmid-mediated metronidazole resistance in *Clostridioides difficile*. *Nat. Commun.* **2020**, *11*, 598. [CrossRef]
38. Johnson, S.; Lavergne, V.; Skinner, A.M.; Gonzales-Luna, A.J.; Garey, K.W.; Kelly, C.P.; Wilcox, M.H. Clinical Practice Guideline by the Infectious Diseases Society of America (IDSA) and Society for Healthcare Epidemiology of America (SHEA) Focused Update Guidelines on Management of Clostridioides difficile Infection in Adults. *Clin. Infect. Dis.* **2021**, *5*, e1029–e1044. [CrossRef]
39. Spigaglia, P.; Mastrantonio, P.; Barbanti, F. Antibiotic Resistances of *Clostridium difficile*. *Adv. Exp. Med. Biol.* **2018**, *1050*, 137–159.
40. Adler, A.; Miller-Roll, T.; Bradenstein, R.; Block, C.; Mendelson, B.; Parizade, M.; Paitan, Y.; Schwartz, D.; Peled, N.; Carmeli, Y.; et al. A national survey of the molecular epidemiology of Clostridium difficile in Israel, the dissemination of the ribotype 027 strain with reduced susceptibility to vancomycin and metronidazole. *Diagn. Microbiol. Infect. Dis.* **2015**, *83*, 21–24. [CrossRef]
41. Snydman, D.R.; McDermott, L.A.; Jacobus, N.V.; Thorpe, C.; Stone, S.; Jenkins, S.G.; Goldstein, E.J.; Patel, R.; Forbes, B.A.; Mirrett, S.; et al. U.S.-Based National Sentinel Surveillance Study for the Epidemiology of Clostridium difficile-Associated Diarrheal Isolates and Their Susceptibility to Fidaxomicin. *Antimicrob. Agents Chemother.* **2015**, *59*, 6437–6443. [CrossRef]
42. Freeman, J.; Vernon, J.; Morris, K.; Nicholson, S.; Todhunter, S.; Longshaw, C.; Wilcox, M.H. Pan-European Longitudinal Surveillance of Antibiotic Resistance among Prevalent *Clostridium difficile* Ribotypes' Study Group. *Clin. Microbiol. Infect.* **2015**, *21*, 248-e9. [CrossRef] [PubMed]
43. Huang, J.S.; Jiang, Z.-D.; Garey, K.W.; Lasco, T.; Dupont, H.L. Use of rifamycin drugs and development of infection by rifamycin-resistant strains of *Clostridium difficile*. *Antimicrob. Agents Chemother.* **2013**, *57*, 2690–2693. [CrossRef] [PubMed]

44. Terhes, G.; Maruyama, A.; Latkóczy, K.; Szikra, L.; Konkoly-Thege, M.; Princz, G.; Nagy, E.; Urbán, E. In vitro antibiotic susceptibility profile of Clostridium difficile excluding PCR ribotype 027 outbreak strain in Hungary. *Anaerobe* **2014**, *30*, 41–44. [CrossRef]
45. Goldstein, E.J.; Citron, D.M.; Sears, P.; Babakhani, F.; Sambol, S.P.; Gerding, D.N. Comparative susceptibilities to fidaxomicin (OPT-80) of isolates collected at baseline, recurrence, and failure from patients in two phase III trials of fidaxomicin against *Clostridium difficile* infection. *Antimicrob. Agents Chemother.* **2011**, *55*, 5194–5199. [CrossRef]
46. Spigaglia, P. Recent advances in the understanding of antibiotic resistance in *Clostridium difficile* infection. *Ther. Adv. Infect. Dis.* **2016**, *3*, 23–42. [CrossRef]
47. He, M.; Miyajima, F.; Roberts, P.; Ellison, L.; Pickard, D.J.; Martin, M.J.; Connor, T.R.; Harris, S.R.; Fairley, D.; Bamford, K.B.; et al. Emergence and global spread of epidemic healthcare-associated *Clostridium difficile*. *Nat. Genet.* **2013**, *45*, 109–113. [CrossRef]
48. Marcos, P.; Whyte, P.; Rogers, T.; McElroy, M.; Fanning, S.; Frias, J.; Burgess, C.M.; Bolton, D.J. The prevalence of *Clostridioides difficile* on farms, in abattoirs and in retail foods in Ireland. *Food Microbiol.* **2021**, *98*, 103781. [CrossRef]
49. European Centre for Disease Prevention and Control (ECDC). Laboratory Procedures for Diagnosis and Typing of Human *Clostridium difficile* Infection; Technical Report. 2018. Available online: www.ecdc.europa.eu (accessed on 20 January 2023).
50. Bidet, P.; Barbut, F.; Lalande, V.; Burghoffer, B.; Petit, J.C. Development of a new PCR-ribotyping method for Clostridium difficile based on ribosomal RNA gene sequencing. *FEMS Microbiol. Lett.* **1999**, *175*, 261–266. [CrossRef]
51. Vardakas, K.Z.; Polyzos, K.A.; Patouni, K.; Rafailidis, P.I.; Samonis, G.; Falagas, M.E. Treatment failure and recurrence of *Clostridium difficile* infection following treatment with vancomycin or metronidazole, a systematic review of the evidence. *Int. J. Antimicrob. Agents* **2012**, *40*, 1–8. [CrossRef]
52. Curry, S.R.; Marsh, J.W.; Shutt, K.A.; Muto, C.A.; O'Leary, M.M.; Saul, M.I.; Pasculle, A.W.; Harrison, L.H. High frequency of rifampin resistance identified in an epidemic *Clostridium difficile* clone from a large teaching hospital. *Clin. Infect. Dis.* **2009**, *48*, 425–429. [CrossRef]
53. Obuch-Woszczatyński, P.; Lachowicz, D.; Schneider, A.; Mól, A.; Pawłowska, J.; Ożdżeńska-Milke, E.; Pruszczyk, P.; Wultańska, D.; Młynarczyk, G.; Harmanus, C.; et al. Occurrence of *Clostridium difficile* PCR-ribotype 027 and it's closely related PCR-ribotype 176 in hospitals in Poland in 2008-2010. *Anaerobe* **2014**, *28*, 13–17. [CrossRef]
54. Tijerina-Rodríguez, L.; Garza-González, E.; Martínez-Meléndez, A.; Morfín-Otero, R.; Camacho-Ortiz, A.; Gonzalez-Diaz, E.; Perez-Gomez, H.R.; Villarreal-Treviño, L.; Maldonado-Garza, H.; Esparza-Ahumada, S.; et al. Clinical characteristics associated with the severity of *Clostridium [Clostridioides] difficile* infection in a tertiary teaching hospital from Mexico. *Biomed. J.* **2022**, *45*, 200–205. [CrossRef] [PubMed]
55. Ackermann, G.; Tang-Feldman, Y.J.; Schaumann, R.; Henderson, J.P.; Rodloff, A.C.; Silva Jr, J.; Cohen, S.H. Antecedent use of fluoroquinolones is associated with resistance to moxifloxacin in *Clostridium difficile*. *Clin. Microbiol. Infect.* **2003**, *9*, 526–530. [CrossRef]
56. Carroll, D.N. Moxifloxacin-induced *Clostridium difficile*-associated diarrhea. *Pharmacotherapy* **2003**, *23*, 1517–1519. [CrossRef]
57. Spigaglia, P.; Barbanti, F.; Mastrantonio, P. European Study Group on *Clostridium difficile* (ESGCD) Multidrug resistance in European *Clostridium difficile* clinical isolates. *J. Antimicrob. Chemother.* **2011**, *66*, 2227–2234. [CrossRef]
58. Rodriguez-Palacios, A.; Ilic, S.; LeJeune, J.T. *Clostridium difficile* with Moxifloxacin/Clindamycin Resistance in Vegetables in Ohio, USA, and Prevalence Meta-Analysis. *J. Pathog.* **2014**, *2014*, 158601. [CrossRef]
59. European Committee on Antimicrobial Susceptibility Testing (EUCAST). MIC Distribution for Clostridioides Difficie. 2023. Available online: https://mic.eucast.org/search/?search%5Bmethod%5D=mic&search%5Bantibiotic%5D=-1&search%5Bspecies%5D=564&search%5Bdisk_content%5D=-1&search%5Blimit%5D=50 (accessed on 1 March 2023).
60. Keel, K.; Brazier, J.S.; Post, K.W.; Weese, S.; Songer, J.G. Prevalence of PCR ribotypes among *Clostridium difficile* isolates from pigs, calves, and other species. *J. Clin. Microbiol.* **2007**, *45*, 1963–1964. [CrossRef]
61. Abdel-Glil, M.Y.; Thomas, P.; Schmoock, G.; Abou-El-Azm, K.; Wieler, L.H.; Neubauer, H.; Seyboldt, C. Presence of *Clostridium difficile* in poultry and poultry meat in Egypt. *Anaerobe* **2018**, *51*, 21–25. [CrossRef]
62. Krutova, M.; Zouharova, M.; Matejkova, J.; Tkadlec, J.; Krejci, J.; Faldyna, M.; Nyc, O.; Bernardy, J. The emergence of *Clostridium difficile* PCR ribotype 078 in piglets in the Czech Republic clusters with *Clostridium difficile* PCR ribotype 078 isolates from Germany, Japan and Taiwan. *Int. J. Med. Microbiol.* **2018**, *308*, 770–775. [CrossRef]
63. Zhang, L.J.; Yang, L.; Gu, X.X.; Chen, P.X.; Fu, J.L.; Jiang, H.X. The first isolation of *Clostridium difficile* RT078/ST11 from pigs in China. *PLoS ONE* **2019**, *14*, e0212965. [CrossRef]
64. Janezic, S.; Zidaric, V.; Pardon, B.; Indra, A.; Kokotovic, B.; Blanco, J.L.; Seyboldt, C.; Diaz, C.R.; Poxton, I.R.; Perreten, V.; et al. International *Clostridium difficile* animal strain collection and large diversity of animal associated strains. *BMC Microbiol.* **2014**, *14*, 173. [CrossRef]
65. Berger, F.K.; Mellmann, A.; Bischoff, M.; von Muller, L.; Becker, S.L.; Simango, C.; Gartner, B. Molecular epidemiology and antimicrobial resistance of *Clostridioides difficile* detected in chicken, soil and human samples from Zimbabwe. *Int. J. Infect. Dis.* **2020**, *96*, 82–87. [CrossRef]
66. de Boer, E.; Zwartkruis-Nahuis, A.; Heuvelink, A.E.; Harmanus, C.; Kuijper, E.J. Prevalence of *Clostridium difficile* in retailed meat in the Netherlands. *Int. J. Food Microbiol.* **2011**, *144*, 561–564. [CrossRef]

67. Koene, M.G.J.; Mevius, D.; Wagenaar, J.A.; Harmanus, C.; Hensgens, M.P.M.; Meetsma, A.M.; Putirulan, F.F.; van Bergen, M.A.P.; Kuijper, E. *Clostridium difficile* in Dutch animals, their presence, characteristics and similarities with human isolates. *Clin. Microbiol. Infect.* **2012**, *18*, 778–784. [CrossRef] [PubMed]
68. Frentrup, M.; Thiel, N.; Junker, V.; Behrens, W.; Münch, S.; Siller, P.; Kabelitz, T.; Faust, M.; Indra, A.; Baumgartner, S.; et al. Agricultural fertilization with poultry manure results in persistent environmental contamination with the pathogen *Clostridioides difficile*. *Environ. Microbiol.* **2021**, *23*, 7591–7602. [CrossRef] [PubMed]
69. Heise, J.; Witt, P.; Maneck, C.; Wichmann-Schauer, H.; Maurischat, S. Prevalence and phylogenetic relationship of *Clostridioides difficile* strains in fresh poultry meat samples processed in different cutting plants. *Int. J. Food Microbiol.* **2021**, *339*, 109032. [CrossRef]
70. Solomon, K.; Murray, S.; Scott, L.; McDermott, S.; Drudy, D.; Martin, A.; O'Donoghue, C.; Skally, M.; Burns, K.; Fenelon, L.; et al. An investigation of the subtype diversity of clinical isolates of Irish *Clostridium difficile* ribotypes 027 and 078 by repetitive-extragenic palindromic PCR. *J. Med. Microbiol.* **2011**, *60 Pt 8*, 1080–1087. [CrossRef]
71. Zidaric, V.; Pardon, B.; Dos Vultos, T.; Deprez, P.; Brouwer, M.S.; Roberts, A.P.; Henriques, A.O.; Rupnik, M. Different antibiotic resistance and sporulation properties within multiclonal *Clostridium difficile* PCR ribotypes 078, 126, and 033 in a single calf farm. *Appl. Environ. Microbiol.* **2012**, *78*, 8515–8522. [CrossRef]
72. Blasi, F.; Lovito, C.; Albini, E.; Bano, L.; Dalmonte, G.; Drigo, I.; Maresca, C.; Massacci, F.R.; Orsini, S.; Primavilla, S.; et al. *Clostridioides difficile* in Calves in Central Italy, Prevalence, Molecular Typing, Antimicrobial Susceptibility and Association with Antibiotic Administration. *Animals* **2021**, *11*, 515. [CrossRef]
73. Keessen, E.C.; Hensgens, M.P.; Spigaglia, P.; Barbanti, F.; Sanders, I.M.; Kuijper, E.J.; Lipman, L.J. Antimicrobial susceptibility profiles of human and piglet *Clostridium difficile* PCR-ribotype 078. *Antimicrob. Resist. Infect. Control.* **2013**, *2*, 14. [CrossRef]
74. Usui, M.; Nanbu, Y.; Oka, K.; Takahashi, M.; Inamatsu, T.; Asai, T.; Kamiya, S.; Tamura, Y. Genetic relatedness between Japanese and European isolates of *Clostridium difficile* originating from piglets and their risk associated with human health. *Front. Microbiol.* **2014**, *5*, 513. [CrossRef]
75. Chow, V.C.Y.; Kwong, T.N.Y.; So, E.W.M.; Ho, Y.I.I.; Wong, S.H.; Lai, R.W.M.; Chan, R.C.Y. Surveillance of antibiotic resistance among common *Clostridium difficile* ribotypes in Hong Kong. *Sci. Rep.* **2017**, *7*, 17218. [CrossRef] [PubMed]
76. Isidro, J.; Menezes, J.; Serrano, M.; Borges, V.; Paixão, P.; Mimoso, M.; Martins, F.; Toscano, C.; Santos, A.; Henriques, A.O.; et al. Genomic Study of a *Clostridium difficile* Multidrug Resistant Outbreak-Related Clone Reveals Novel Determinants of Resistance. *Front. Microbiol.* **2018**, *9*, 2994. [CrossRef]
77. Rahimi, E.; Jalali, M.; Weese, J.S. Prevalence of *Clostridium difficile* in raw beef, cow, sheep, goat, camel and buffalo meat in Iran. *BMC Public Health* **2014**, *14*, 119. [CrossRef]
78. Bakri, M.; Brown, D.J.; Butcher, J.P.; Sutherland, A.D. *Clostridium difficile* in ready-to-eat salads, Scotland. *Emerg. Infect. Dis.* **2009**, *15*, 817–818. [CrossRef] [PubMed]
79. Han, Y.; King, J.; Janes, M.E. Detection of antibiotic resistance toxigenic *Clostridium difficile* in processed retail lettuce. *Food Qual. Saf.* **2018**, *2*, 37–41. [CrossRef]
80. Rahimi, E.; Afzali, S.Z.; Baghbadorani, Z.T. *Clostridium difficile* in ready-to-eat foods in Isfahan and Shahrekord, Iran. *Asian Pac. J. Trop. Biomed.* **2015**, *2*, 128–131. [CrossRef]
81. Huang, H.; Weintraub, A.; Fang, H.; Nord, C.E. Antimicrobial resistance in *Clostridium difficile*. *Int. J. Antimicrob. Agents* **2009**, *34*, 516–522. [CrossRef]
82. Androga, G.O.; Knight, D.R.; Lim, S.C.; Foster, N.F.; Riley, T.V. Antimicrobial resistance in large clostridial toxin-negative, binary toxin-positive *Clostridium difficile* ribotypes. *Anaerobe* **2018**, *54*, 55–60. [CrossRef]
83. Hampikyan, H.; Bingol, E.B.; Muratoglu, K.; Akkaya, E.; Cetin, O.; Colak, H. The prevalence of *Clostridium difficile* in cattle and sheep carcasses and the antibiotic susceptibility of isolates. *Meat Sci.* **2018**, *139*, 120–124. [CrossRef]
84. Bingol, E.B.; Hampikyan, H.; Muratoglu, K.; Akkaya, E.; Cetin, O.; Colak, H. Characterisation and Antibiotic Susceptibility Profile of *Clostridioides (Clostridium) Difficile* Isolated from Chicken Carcasses. *J. Vet. Res.* **2020**, *64*, 407–412. [CrossRef] [PubMed]
85. Ersöz, Ş.Ş.; Coşansu, S. Prevalence of *Clostridium difficile* Isolated from Beef and Chicken Meat Products in Turkey. *Korean J. Food Sci. Anim. Resour.* **2018**, *38*, 759–767. [PubMed]
86. Peláez, T.; Alcalá, L.; Alonso, R.; Rodríguez-Créixems, M.; García-Lechuz, J.M.; Bouza, E. Reassessment of *Clostridium difficile* susceptibility to metronidazole and vancomycin. *Antimicrob. Agents Chemother.* **2002**, *46*, 1647–1650. [CrossRef]
87. Doosti, A.; Mokhtari-Farsani, A. Study of the frequency of *Clostridium difficile* tcdA, tcdB, cdtA and cdtB genes in feces of Calves in south west of Iran. *Ann. Clin. Microbiol. Antimicrob.* **2014**, *13*, 21. [CrossRef]
88. Dharmasena, M.; Jiang, X. Isolation of toxigenic *Clostridium difficile* from animal manure and composts being used as biological soil amendments. *Appl. Environ. Microbiol.* **2018**, *84*, e00738-18. [CrossRef]
89. Kouassi, K.A.; Dadie, A.T.; Nanga, Z.Y.; Dje, K.M.; Loukou, Y.G. *Clostridium perfringens* and *Clostridium difficile* in cooked beef sold in Cote d'Ivoire and their antimicrobial susceptibility. *Anaerobe* **2014**, *28*, 90–94. [CrossRef]
90. Rodriguez, C.; Taminiau, B.; Avesani, V.; Van Broeck, J.; Delmée, M.; Daube, G. Multilocus sequence typing analysis and antibiotic resistance of *Clostridium difficile* strains isolated from retail meat and humans in Belgium. *Food Microbiol.* **2014**, *42*, 166–171. [CrossRef]
91. Atasoy, F.; Gücükoğlu, A. Detection of *Clostridium difficile* and toxin genes in samples of modified atmosphere packaged (MAP) minced and cubed beef meat. *Ank. Üniversitesi Vet. Fakültesi Derg.* **2017**, *64*, 165–170.

92. Lee, J.Y.; Lee, D.Y.; Cho, Y.S. Prevalence of *Clostridium difficile* isolated from various raw meats in Korea. *Food Sci. Biotechnol.* **2018**, *27*, 883–889. [CrossRef]
93. Sholeh, M.; Krutova, M.; Forouzesh, M.; Mironov, S.; Sadeghifard, N.; Molaeipour, L.; Maleki, A.; Kouhsari, E. Antimicrobial resistance in *Clostridioides (Clostridium) difficile* derived from humans, a systematic review and meta-analysis. *Antimicrob. Resist. Infect. Control.* **2020**, *9*, 158. [CrossRef]
94. Bandelj, P.; Golob, M.; Ocepek, M.; Zdovc, I.; Vengust, M. Antimicrobial Susceptibility Patterns of *Clostridium difficile* Isolates from Family Dairy Farms. *Zoonoses Public Health* **2017**, *64*, 213–221. [CrossRef]
95. Rodriguez Diaz, C.; Hakimi, D.-E.; Vanleyssem, R.; Taminiau, B.; VanBroeck, J.; Delmee, M.; Kosak Koulagenko, N.; Daube, G. *Clostridium difficile* in beef cattle farms, farmers and their environment, Assessing the spread of the bacterium. *Vet. Microbiol.* **2017**, *210*, 183–187. [CrossRef] [PubMed]
96. Masarikova, M.; Simkova, I.; Plesko, M.; Eretova, V.; Krutova, M.; Cizek, A. The Colonisation of Calves in Czech Large-Scale Dairy Farms by Clonally-Related *Clostridioides difficile* of the Sequence Type 11 Represented by Ribotypes 033 and 126. *Microorganisms* **2020**, *8*, 901. [CrossRef] [PubMed]
97. Cho, A.; Byun, J.W.; Kim, J.W.; Oh, S.I.; Lee, M.H.; Kim, H.Y. Low Prevalence of *Clostridium difficile* in Slaughter Pigs in Korea. *J. Food Prot.* **2015**, *78*, 1034–1036. [CrossRef] [PubMed]
98. Spigaglia, P.; Barbanti, F.; Dionisi, A.M.; Mastrantonio, P. *Clostridium difficile* isolates resistant to fluoroquinolones in Italy, emergence of PCR ribotype 018. *J. Clin. Microbiol.* **2010**, *48*, 2892–2896. [CrossRef]
99. Tenover, F.C.; Tickler, I.A.; Persing, D.H. Antimicrobial-resistant strains of *Clostridium difficile* from North America. *Antimicrob. Agents Chemother.* **2012**, *56*, 2929–2932. [CrossRef] [PubMed]
100. Krutova, M.; Capek, V.; Nycova, E.; Vojackova, S.; Balejova, M.; Geigerova, L.; Tejkalova, R.; Havlinova, L.; Vagnerova, I.; Cermak, P.; et al. The association of a reduced susceptibility to moxifloxacin in causative *Clostridium (Clostridioides) difficile* strain with the clinical outcome of patients. *Antimicrob. Resist. Infect. Control* **2020**, *9*, 98. [CrossRef]
101. Goudarzi, M.; Goudarzi, H.; Alebouyeh, M.; Azimi Rad, M.; Shayegan Mehr, F.S.; Zali, M.R.; Aslani, M.M. Antimicrobial susceptibility of *Clostridium difficile* clinical isolates in Iran. *Iran. Red Crescent Med. J.* **2013**, *15*, 704–711. [CrossRef]
102. Hecht, D.W.; Galang, M.A.; Sambol, S.P.; Osmolski, J.R.; Johnson, S.; Gerding, D.N. In vitro activities of 15 antimicrobial agents against 110 toxigenic *Clostridium difficile* clinical isolates collected from 1983 to 2004. *Antimicrob. Agents Chemother.* **2007**, *51*, 2716–2719. [CrossRef]
103. O'Connor, J.R.; Galang, M.A.; Sambol, S.P.; Hecht, D.W.; Vedantam, G.; Gerding, D.N.; Johnson, S. Rifampin and rifaximin resistance in clinical isolates of *Clostridium difficile*. *Antimicrob. Agents Chemother.* **2008**, *52*, 2813–2817. [CrossRef]
104. Garey, K.W.; Salazar, M.; Shah, D.; Rodrigue, R.; DuPont, H.L. Rifamycin antibiotics for treatment of *Clostridium difficile*-associated diarrhea. *Ann. Pharmacother.* **2008**, *42*, 827–835. [CrossRef]
105. Zhao, Y.; Xu, S.; Wang, L.; Chin, D.P.; Wang, S.; Jiang, G.; Xia, H.; Zhou, Y.; Li, Q.; Ou, X.; et al. National survey of drug-resistant tuberculosis in China. *N. Engl. J. Med.* **2012**, *366*, 2161–2170. [CrossRef]
106. Johanesen, P.A.; Mackin, K.E.; Hutton, M.L.; Awad, M.M.; Larcombe, S.; Amy, J.M.; Lyras, D. Disruption of the Gut Microbiome, *Clostridium difficile* Infection and the Threat of Antibiotic Resistance. *Genes* **2015**, *6*, 1347–1360. [CrossRef]
107. Magiorakos, A.P.; Srinivasan, A.; Carey, R.B.; Carmeli, Y.; Falagas, M.E.; Giske, C.G.; Harbarth, S.; Hindler, J.F.; Kahlmeter, G.; Olsson-Liljequist, B. Multidrug-resistant, extensively drug-resistant and pandrug-resistant bacteria, An international expert proposal for interim standard definitions for acquired resistance. *Clin. Microbiol. Infect.* **2012**, *18*, 268–281. [CrossRef]
108. Doyle, M.E. Multidrug-resistant pathogens in the food supply. *Foodborne Pathog. Dis.* **2015**, *12*, 261–279. [CrossRef]
109. Mutai, W.C.; Mureithi, M.W.; Anzala, O.; Revathi, G.; Kullin, B.; Brugu, M.; Kany'a, C.; Odoyo, E.; Otieno, P.; Musila, L. High Prevalence of Multidrug-Resistant *Clostridioides difficile* Following Extensive Use of Antimicrobials in Hospitalized Patients in Kenya. *Front. Cell. Infect. Microbiol.* **2021**, *10*, 604986. [CrossRef]

microorganisms MDPI

Article

PreS1 Containing HBc VLPs for the Development of a Combined Therapeutic/Prophylactic Hepatitis B Vaccine

Andris Dishlers, Ivars Petrovskis, Dace Skrastina, Ieva Zarina, Ilva Lieknina, Juris Jansons, Inara Akopjana, Jelena Zakova, Velta Ose and Irina Sominskaya *

Latvian Biomedical Research and Study Centre, Ratsupites Str. 1, 1067 Riga, Latvia
* Correspondence: irina@biomed.lu.lv; Tel.: +37-167-808-212

Abstract: The available HBV vaccines based on the HBV surface protein are manufactured in yeasts and demonstrate excellent prophylactic but no therapeutic activity and are thus ineffective against chronic HBV infection. Five different HBV core proteins (HBc)—full length and C-terminally truncated—were used for the insertion of the short, preS1,aa 20–47 and long, preS1phil, aa 12–60 + 89–119 fragments. Modified virus-like particles (VLPs) were compared for their biotechnological and immunological properties. The expression level of HBc-preS1 proteins was high for all investigated proteins, allowing us to obtain 10–20 mg of purified VLPs from a gram of biomass with the combination of gel filtration and ion-exchange chromatography to reach approximately 90% purity of target proteins. The immunogenicity of chimeric VLPs was tested in BALB/c mice, showing a high anti-preS1 response and substantial T-cell proliferation after stimulation with HBc protein. Targeted incorporation of oligonucleotide ODN 1668 in modified HBc-preS1 VLPs was demonstrated.

Keywords: HBV; HBc; virus like particles; immunogenicity; packaging

Citation: Dishlers, A.; Petrovskis, I.; Skrastina, D.; Zarina, I.; Lieknina, I.; Jansons, J.; Akopjana, I.; Zakova, J.; Ose, V.; Sominskaya, I. PreS1 Containing HBc VLPs for the Development of a Combined Therapeutic/Prophylactic Hepatitis B Vaccine. *Microorganisms* **2023**, *11*, 972. https://doi.org/10.3390/microorganisms11040972

Academic Editors: Isabelle Chemin, Shengxi Chen and Fabio Zicker

Received: 20 December 2022
Revised: 31 March 2023
Accepted: 5 April 2023
Published: 8 April 2023

1. Introduction

Despite the effective immunization of newborns with the marketed HBV surface protein (HBs)-based vaccines, some completely vaccinated adults do not reach dependable protection and 5 to 10% of infants of highly viremic pregnant women become chronically infected despite immediate after-birth immunization [1]. The rate of HB non-responders to existing vaccines remains high in groups of aged people, overweight people, medical workers, patients with renal insufficiency and on dialysis, patients after transplantation, patients with HIV as well as travelers to HBV-endemic regions [2]. Interferon therapy or treatment with nucleos(t)ide analogs suppresses HBV replication and decreases the development of cirrhosis, liver failure, hepatocellular carcinoma (HCC), and death, but is not able to eliminate the virus [3]. Thus, the remaining problems with hepatitis B include the inability to achieve a complete cure for chronic HBV infections, with the potential reactivation of HBV, as well as limited protection against HBV escape mutants.

Impressive efforts have been directed to improve existing HBV vaccines (1) by including preS1 and/or preS2 epitopes in the HB-based vaccine to enhance its prophylactic effect, and (2) through the combination of both structural proteins of HBV—the HBs and the HBc—in one formulation to aid the therapeutic effect of the vaccine. The combined protein–DNA vaccines have been developed in parallel as an alternative immunization strategy to achieve the goal of a universal HBV vaccine [4,5].

With epitope mapping of the HBV preS1 region, virus-neutralizing epitopes within preS1 sequences 19–26 and 37–45 were identified [6]. Within the preS1 region of HBV, aa - 13-59 can induce virus-neutralizing antibodies in mice [7]. The pre-S1 monoclonal antibody (mAb) used in this study (MA18/7) recognizes preS1 epitope DPAFR (aa 31–35) [8–10], and can inhibit the infection of primary Tupaia hepatocytes with HBV [11].

Available in several countries, Sci-B-Vac™ vaccine (SciVac Israel, Rehovot, Israel) containing the preS1 and preS2 fragments in addition to the HBs protein and produced in the mammalian cell line enabled a robust immune response against HBV infection during the phase III trial [12,13] and remarkably high anti-HB response (>100 mIU/mL) in 20 of 21 non- or low responders [14].

Due to its unique T- and B-cell immunogenicity, recombinant HBc was investigated as a promising therapeutic antigen against HBV—the first real trials to protect chimpanzees against HBV with the HBc VLPs were performed more than 30 years ago [15–17]. In a model with woodchuck hepatitis virus (WHV), it was shown that core protein (WHc) is a necessary and sufficient agent to protect woodchucks against WHV infection [18–20]; remarkably, the woodchucks were protected not only with the WHc, but also with the HBc vaccination [19]. Moreover, immunization of woodchucks with plasmids expressing both WHc and WHs efficiently suppressed WHV infection [21,22]; and later, the differences featured by the immunization of mice with the DNA or protein prototypes of the WHc vaccine were evaluated [23]. Furthermore, vaccination with whole cells expressing duck hepatitis B virus (DHBV) core (DHBc) protein was able to resolve chronic DHBV infection [24]. Recently, Boudewijns et al. proposed a novel therapeutic HBc vaccine that induced a strong polyfunctional cytotoxic T-cell response in mice using yellow fever vaccine as a live-attenuated vector for the expression of the HBc gene [25]. The plant-produced HBc VLPs were also presented as real prototypes of therapeutic vaccines [26,27]. An exhaustive study in a Tupaia model dealt with possible formulations of the intranasal HBc/HBs-based therapeutic vaccine [28].

Ulrike Protzer's team, together with Rhein Biotech (Düsseldorf, Germany), contributed to the use of HBc as a therapeutic component in the HBV vaccine [4,29,30]. These activities resulted in the therapeutic vaccines TherVacB [4] and DV-601 (Dynavax Technologies, Emeryville, CA, USA) which combined HBs and HBc [29]. In parallel, a combination of the prospective combined HBc/HBs vaccine with an immune-stimulating CpG adjuvant was presented [30]. NASVAC, the therapeutic nasal HBc/HBs vaccine, was designed in Cuban laboratories [31–34] and underwent a phase III clinical trial for chronically infected hepatitis B patients [34].

HBc consists of 183 or 185 amino acid residues (aa), depending on the genotype [35]. The primary structure of the core protein can be divided into two domains, namely, the N-terminal self-assembly (SA) domain (aa 1–140) and the C-terminal RNA-binding protamine-like arginine-rich domain (CTD) (aa 150–183) [36] (Figure 1A). These domains are separated by the hinge peptide 141-STLPETTVV-149, which performs morphogenic functions [37]. The SA domain possesses a set of variable and conserved stretches that correspond to B-cell epitopes and structural elements, respectively, whereas the CTD and hinge peptide are the most conserved HBc regions without immunological importance (for reviews, see [38,39]). Four arginine blocks function as nucleic acid binding sites within the CTD [40]. HBc capsids that have been self-assembled from the truncated HBc proteins lacking the CTD, so-called HBcΔ variants, are not able to incorporate nucleic acids [41].

The HBc monomer is formed by five alpha helices connected by loops, where the hairpin of helices 3 and 4 dimerizes with the next monomer to form the central helical bundle. The major immunodominant region of the HBc antigen (MIR) is located within the SA domain (aa 78–82) on the tips of the HBc spikes [42] within B-cell epitopes *c* (HBc epitope) and *e1* (HBe epitope 1) [43,44] (Figure 1B). The MIR has been extensively used for the exposure of foreign immunogenic sequences (epitopes) on the HBc VLP surface and therefore provides the most efficient immunogenic activity (for reviews, see [45,46]).

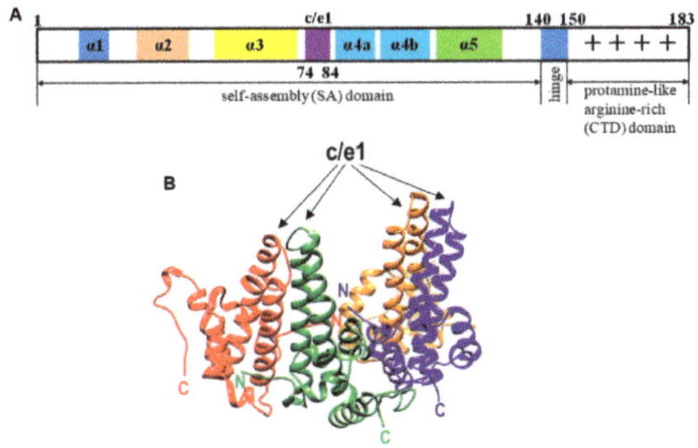

Figure 1. The structure of the HBc protein. (**A**) HBV HBc monomer represented by 183 aa-long HBc, genotype D1, subtype ayw2 (GenBank accession number XO2496) [47]. (**B**) The structure of two HBc dimers. Monomer chains A, B, C, and D are marked in red, green, blue, and orange, respectively. Surface-exposed loops with c and e1 epitopes are shown with arrows, N- and C- termini are labeled with N and C letters, respectively. Data were downloaded from the VIPERdb v3.0 (http://viperdb.scripps.edu, accessed on 1 May 2022) and visualized using UCSF Chimera Version 1.16 software [48].

Many laboratories around the world have successfully produced HBc VLPs in plants, insect, and mammal cells (for a review, see [38,39,49,50]). HBc variants from different genotypes have been produced and their corresponding VLPs purified by our group using *E. coli* [51–54] and yeast *P. pastoris* as expression systems [55].

It was found that the presence of nucleic acids is essential for the enhanced immunogenicity of the VLPs. The adjuvant effect of the bacterial RNA bound to the arginine-rich C-terminal domain of HBc was shown first by the immunization of mice with the HBc VLPs together with other proteins including HBs by the Reinhold Schirmbeck's team [56,57]. Later, this group clearly demonstrated that particle-bound mammalian RNA functioned as TLR7 ligand and induced a Th1-biased humoral immunity in B6 but not in TLR7−/− mice [58,59]. Therefore, both endogenous bacterial and mammalian RNAs functioned as a natural adjuvant, facilitating the priming of Th1-biased immune responses. Later, the use of VLP as nanocontainers for genes and/or immunostimulatory oligonucleotide sequences was developed in detail by Strods et al. [53].

In the current study, we modified the surface of HBc VLPs by exposure of selected HBV preS1 epitopes to combine in one platform efficient anti-preS1 immunogenicity and T-cell immunogenicity provided by the T-cell epitope-rich HBc antigen. In parallel, the impact of the length of the CTD domain of HBc on recombinant HBc-preS1 VLP formation, immunogenicity, and their biotechnological properties was investigated. This approach was realized by our group before, first, by insertion of the hydrophilic stretch of the preS1 sequence into the MIR of unmodified HBc [60,61] or into the C-terminus of modified HBc HBc [62]; in both cases, the surface exposure of the inserts was achieved. It is worth mentioning that both of these modified VLPs provided highly efficient induction of adequate anti-HBV B- and T-cell responses.

2. Materials and Methods

2.1. Bacterial E. coli Strains

Escherichia coli (*E. coli*) strain RR1 [F^- · r_B^- m_B^- leuB6 proA2 thi-1 araC14 lacY1 galK2 xyl-5 mtl-1 rpsL20 (Str^r) glnV44 Δ (mcrC-mrr)] was used for the cloning and selection of recombinant plasmids. For the expression of HBc-preS1 genes, *E. coli* strains K802 (F^- r_K^- m_K^+ e14 McrA metB1 lacY1 [or lacI-Y6] galK2 galT22 glnV44 mcrB) and BL21 [F^- ompT hsdS $_B$ (r_B^- m_B^-) gal dcm lon] were used.

2.2. Design of HBc-preS1 Proteins and Expression Plasmids

Plasmids for the expression of chimeric HBcΔ-preS1(20–47) and HBcΔ-preS1phil proteins were described earlier [61,62], and these plasmids served here as a source for the amplification of corresponding preS1 containing sequences. HBcΔ refers here to C-terminally truncated, 144 aa-long HBc as a carrier protein, preS1(20–47) for the 28 aa long, and preS1phil for the 79 aa-long fragment linking together two hydrophilic fragments—aa 12–60 and aa 89–119—of the preS1 region of the LHBs protein of HBV genotype D [47]; aa numeration is according to HBV genotype A (Figure 2A). The source of HBV genes was the plasmid pHB320 with the full HBV genome (genotype D1, subtype ayw2) cloned in our lab (GenBank accession number X70185) [47]. In all our constructs the MIR of HBc was a target site for preS1 insertions (between aa D78 and P79). Corresponding PCR fragments were obtained using upstream primer 5′-GCCTCTAGATAACGCCTCAGCTCTGTATCG-3′ and downstream primer 5′-ACGAACAACAGTAGTCTCCGGAAGTGTTGATAAGATAG-3′. Primers were designed for the amplification of the HBc part (30–150 aa) that contained preS1 insertions in MIR and restriction sites *XbaI* and *Kpn2I* for further cloning. A gel extraction kit (Fermentas UAB, Vilnius, Lithuania) was used for the isolation of PCR fragments from agarose followed by incubation with *Xba*I and *Kpn*21 endonucleases at 37 °C for 1 h. For cloning of preS1 fragments, *Xba*I/*Kpn*21-treated appropriate HBc expression vectors [52] based on the pBR327 plasmid [63] were used. The expression of HBc-preS1 proteins in our constructs is under the control of the *E. coli* pTrp promoter [64]. Several HBc variants—full-length HBc (aa 1–183) and four truncated HBc proteins ending at aa positions 163, 167, 171, 178—were used as VLP carriers for the described preS sequences. Altogether, ten HBc-preS1 fusion proteins were designed (Figure 2B). In the designation of fusion proteins, the HBc vector is shown first, followed by the appropriate preS1 fragment—for example, 183preS1phil refers to the full-length HBc carrying the long preS1phil fragment. After *E. coli* transformation, three clones for each construct were analyzed for the presence of designed recombinant plasmid and plasmid structures were verified by sequencing (Figure 2).

2.3. Monoclonal Antibodies

Mouse anti-HBc mAb 13C9, recognizing epitope 134-PPNAPIL-140 [65] of HBc protein, and mouse anti-preS1 mAb MA18/7, recognizing the DPAFR subtype-independent linear 31-DPAFR-35 epitope [8–10], were used in this study.

2.4. Cultivation of Recombinant E. coli Cells and Purification of HBc-preS1 Proteins

Cultivation of *E. coli* cells for the expression of HBc-preS1 proteins and purification of target VLPs was performed essentially as described before [54] using Trp-rich phosphate-buffered 2xTY (2TYP), or Trp-deficient standard minimal M9 medium supplemented with casamino acid (M9Cas) medium; both media contained glucose (0.2%, *w*/*v*). Briefly, the cultivation was performed in flasks (200 mL of culture in a 750 mL flask) on a shaker at 37 °C for 20 h. Purification of HBc-preS1 VLPs was performed with a combination of ion-exchange chromatography (IEX) and gel filtration (GF). A standard 2 g portion of frozen (−20 °C) cells was used for ultrasonic disintegration and obtained crude lysate was clarified at 10,000 rpm (13,000× *g*) for 30 min at +4 °C. The obtained suspension was first loaded on a Fractogel®EMD DEAE (M) (Merck, Burlington, Massachusetts, USA) column and fractions containing target protein were concentrated by tangential filtration using a 500 kD cartridge and obtained concentrate was further loaded on a Sepharose 4 Fast Flow (FF) (Cytiva Europe GmbH, Turku, Finland) column. SDS-polyacrylamide gel electrophoresis (SDS-PAGE, shortly PAGE) was used throughout the purification process to follow the presence of the target protein in fractions. The presence of VLPs in fractions was followed by native agarose gel electrophoresis (NAGE, see Section 2.5). The purity of protein samples was evaluated by PAGE.

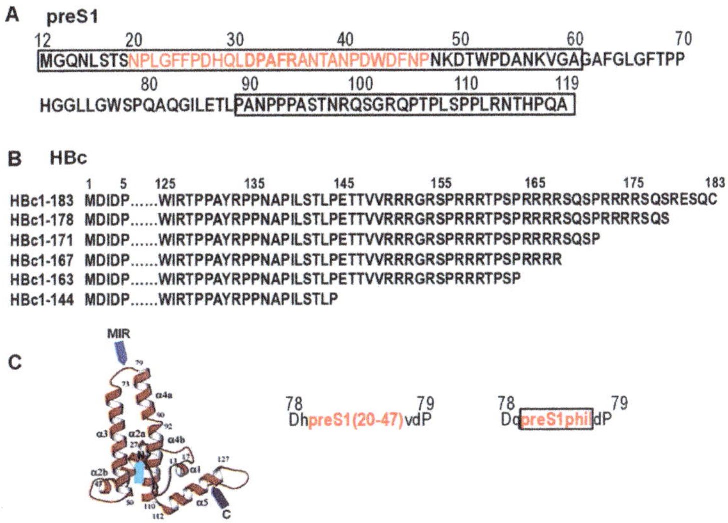

Figure 2. Structural elements used to design HBc-preS1 VLPs. (**A**) Structure of the preS1 region of HBV genotype D, and numeration of preS1 aa is according to the preS1 of HBV genotype A (GenBank accession number X70185); note: preS1 of genotype D is shorter by 11 aa than pres1 of genotype A. Short insertion fragment 20–47 is shown in red, and hydrophilic parts of the preS1 sequence spliced to create the long preS1phil fragment are shown in two boxes. Epitope DPAFR recognized by mAb MA18/7 is highlighted in bold red. (**B**) Full-length and truncated HBc vector variants were used for the insertion of preS1 fragments. Not shown aa of HBc are dotted. (**C**) Localization of surface-exposed MIR of HBc and aa surrounding the preS1 insertions: capital letters show aa of the HBc vector and small letters—linker aa added at the construction of expression plasmids. The numeration of aa in (**B**,**C**) is according to HBV genotype D (GenBank accession number XO2496).

2.5. Characterization of HBc-preS1 VLPs

For VLP detection, samples were subjected to (NAGE) using 1% UltraPure agarose (Thermo Fisher Scientific, Waltham, MA, USA) in TBE buffer. Ethidium bromide (EtBr, 5 μL of a 10 mg/mL stock in 100 mL of PBS) was used for staining NAGE gels and Coomassie Brilliant Blue R-250 (60 mg/L in 10% acetic acid) was used for the staining of PAGE and NAGE gels. All chemicals were from Sigma-Aldrich (St. Louis, MO, USA).

The purity of final HBc-preS1 preparations as VLPs was evaluated by PAGE gels (15%) stained with Coomassie Brilliant Blue R-250. Anti-HBc mAb 13C9 [65] and/or anti-preS1 mAb MA18/7 [9] at 1:1000 dilution were used for the Western blot of HBc-preS1 VLPs after PAGE. The morphology of VLP preparations was analyzed by transmission electron microscopy (EM) and the homogeneity of particles in VLP preparation by dynamic light scattering (DLS), as described earlier [54].

2.6. The Antigenicity of the HBc-preS1 VLPs

For the competitive ELISA, 96-well microplates were coated with 100 μL of preS1(20–47) peptide solution (10 μg/mL) in 50 mM sodium carbonate buffer, pH 9.6 per well and incubated overnight at 4 °C. After blocking with phosphate-buffered saline (PBS) containing 1% BSA for 1 h at RT, 50 μL aliquots of serial dilutions of test proteins and 50 μL of the anti-preS1 mAb MA18/7 (dilution 1:500) [9] were added to the wells simultaneously. Plates were incubated at 37 °C for 1 h, then washed four times with Tween-20 containing (0.05%) PBS. Thereafter, 100 μL of horseradish peroxidase-conjugated anti-mouse antibody (Sigma-Aldrich St. Louis, MO, USA) was added to wells at a 1:10,000 dilution and incubated at 37 °C for 1 h. After washing the plates four times as before, OPD substrate (Sigma–Aldrich St. Louis, MO, USA) was added to develop the color. The percent inhibition (I%) of antibody binding by the competing protein was calculated as follows: I% = [(OD492 test sample—OD492 of negative control)/(OD492 of positive control—OD492 negative control)] × 100. The molar amount of the protein necessary for 50% inhibition (I50) was calculated.

2.7. Immunogenicity of the HBc-preS1 VLPs

Immunization of BALB/c mice with HBc-preS1 VLPs was performed as described before in [52] with the permission of the Latvian Animal Protection Ethics Committee (Permission No. 61/12.05.2014). Five animals in each group were immunized subcutaneously with 25 µg of VLPs in PBS formulated with 250 µg of Alhydrogel in a total volume of 0.2 mL per mouse at days 0, 14, and 28. Sera injected with Alhydrogel only animals were used as negative controls. Anti-HBc and anti-preS1 titers in the sera were detected with direct ELISA. The recombinant full-length HBc protein (as VLPs) or preS1 peptide (20–47 aa), both at 10 µg/mL, was used for plate coating. The end-point titers were defined as the highest mAb dilution that resulted in an absorbance value three times greater than that of the negative control.

T-cell proliferation tests were performed as described in [52] in the lymphocytes of the mice immunized with the HBc-preS1 VLPs. Spleens were collected on day 42 post-immunization and splenocytes from the mice of each group were pooled. In vitro stimulation was performed using full-length HBc1-183 protein at 1.0 µg/mL and 10 µg/mL concentration. Concanavalin A (ConA) at 4 µg/mL was used as a positive control. Results of T-cell proliferation were presented as stimulation indexes (SI), which were calculated as a ratio of mean cpm obtained in the presence and absence of HBc.

2.8. CpG Oligodeoxynucleotide Packaging

Synthetic oligodeoxynucleotide (ODN) 1668 (5′-tccatgacgttcctgatgct-3′) is the B-class unmethylated CpG dinucleotide specific for mouse Toll-like receptor 9 (TLR9), strongly activates B cells but weakly IFN-α secretion. ODN 1668 was obtained from InvivoGen (Toulouse, France) and it was tested here with 183preS1(20–47) VLPs for the packaging. ODN packaging was performed according to the method described in [66] with the use of RNase and urea. For packaging the following mix was prepared: 50 µg/12,5 µL of VLPs in PBS + 100 µg/10 µL RNaseA (Thermo Fisher Scientific, cat. No R1253) + 50 µL 1 M urea in water + 7 µg/30 µL ODN 1668 in water. The packaging mix was incubated at room temperature overnight.

2.9. The 3D Modeling of HBc-preS1 VLPs

The VIPERdb v3.0 (http://viperdb.scripps.edu, accessed on 1 May 2022) [67] was used to create maps of HBc VLPs and 3DJIGSAW protein modeling program [68] was applied for the prediction of three-dimensional structures of HBc-preS1 VLPs. The UCSF Chimera Version 1.16 package from the Resource for Biocomputing, Visualization, and Informatics at the University of California, San Francisco [49] was utilized for the production of molecular graphics images.

3. Results

3.1. Modeling of HBc-preS1 Structures

The externally exposed region MIR within the c/e1 epitope located on the tip of the spikes of HBc (Figure 1) was chosen for the insertion of two preS1 fragments: one, short (aa 20–47) of the preS1, and a second long, preS1phil fragment, (aa 12–60 + 89–119) representing the preS1 sequence with the deleted hydrophobic region (aa 61–88) [60] (Figure 2A). Five HBc gene variants ending at aa positions 163, 167, 171, 178, and 183 of HBV were used for the preS1 insertions, generating in total ten HBc-preS1 chimeric proteins (Figure 2B).

The possible three-dimensional organization of the preS1 fragments within chimeric 183HBc-preS1 VLPs as predicted by 3D-JIGSAW is shown in Figure 3. This program creates three-dimensional maps grounded on homologies of highly resolved structures [64].

Figure 3 depicts the modeling results of the modified MIR region for all four possible HBc monomers (conformers [42]) A–D exposing the long preS1phil fragment (Figure 3A–D) and for two monomers A and D exposing short preS1(20–47) (Figure 3E,F).

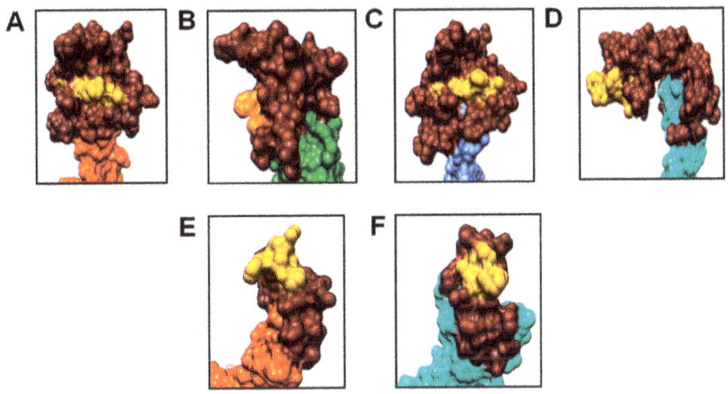

Figure 3. Structure modeling of 183preS1 VLPs. (**A–D**) preS1phil insertion in four possible HBc monomers A, B, C, and D, and (**E,F**) preS1(20–47) insertion in monomers A and D. HBc monomers are colored: A—red-orange, B—green, C—blue, and F—(cyan). PreS1 insertions are colored dark red with the DPAFR shown in yellow. Note: inserted preS1 fragments are enlarged and in orientation to highlight the position of the DPAFR within the insert. Data were downloaded from the VIPERdb v3.0 (http://viperdb.scripps.edu, accessed on 1 May 2022) and visualized using Chimera Version 1.16 software [48].

3.2. *The Expression of HBc-preS1 Proteins*

The expression level of the target protein in 10 individual transformed cell clones was compared before the selection of the best clone of transformed cells for further use. The combination of host strain (*E. coli* K802 or *E. coli* BL21) and cultivation medium (2xTYP or M9+Cas [54]) was experimentally found for each of ten HBc-preS1 proteins by the cultivation of transformed cells first in tubes (with 5 mL of culture in 15 mL tubes). The 5 mL culture served later as the seed culture for scale-up cultivation in flasks to obtain several grams of biomass (see Materials and Methods, Section 2.4). *E. coli* K802 was found optimal for HBc-preS1constructs based on the full-length HBc protein but *E. coli* BL21—for other constructs based on shortened HBc variants as a carrier protein. The lowest cell density was found for cultures expressing HBc-preS1 proteins based on 183 aa-long full-length HBc. Thus, the typical OD_{540} for cell cultures expressing 183preS1 proteins reached: approximately 6 OD units in 2TYP medium, and approximately 4 OD units in M9Cas medium, approximately 8 for cell cultures expressing 161-preS1, 163preS1, 171preS1, and 178preS1 proteins in 2xTYP medium, and approximately 6 in M9Cas medium. The expression level of different HBc-preS1 proteins varied within the range of 7–10% of total cellular protein with an overall higher level for fusion proteins based on shortened HBc and with shorter preS1(20–47) insert. It was found that in all investigated cases, HBc-preS1 proteins were able to form capsid-like structures (VLPs) within expressing cells.

3.3. *Purification of HBc-preS1 Proteins as VLP*

Target proteins were purified as VLPs from the soluble protein fraction of disintegrated cells after extraction with urea, followed by ammonium sulfate fractionation, and two-step chromatography, using IEX on Fracto DEAE as a first step followed by GF on 4 FF Sepharose. Purification of 183preS1phil as an example is shown in Figure 4.

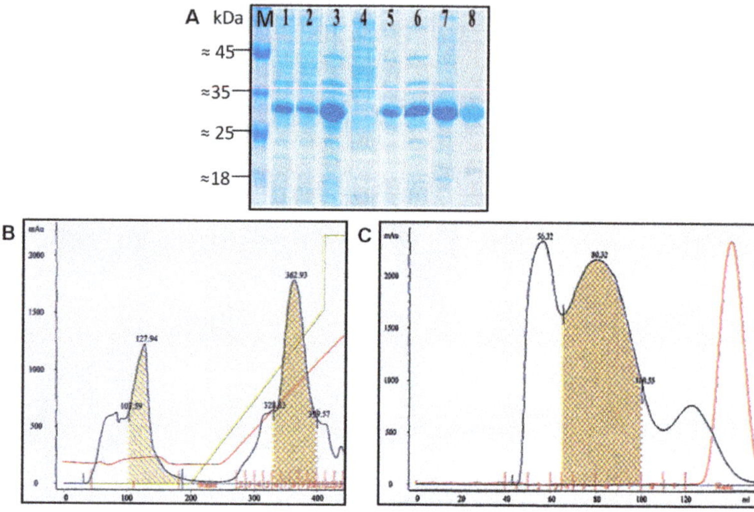

Figure 4. The purification of 183preS1phil VLPs. (**A**) SDS-PAGE of fractions from the entire purification process, staining with Coomassie Brilliant Blue G-250: M - Pierce™ Prestained Protein MW Marker (cat N 26612, Thermo Fisher Scientific Baltics UAB, Vilnius, Lithuania), lane 1—crude cell lysate (after French press), lanes 2, 3—extraction with 1M urea: lane 2—supernatant, lane 3—debris; lane 4—supernatant after precipitation with 35% ammonium sulfate (AS); lanes 5, 6—extraction with 1M urea of the precipitate after 35% AS: lane 5—soluble part, lane 6—insoluble part; lane 7—central peak fractions from Fractogel EMD DEAE (see second peak on (**B**)pooled fractions for GF are shown cross-striped and colored in brown); lane 8—central peak fractions from GF on Sepharose 4 FF (see (**C**), pooled fractions are shown crossed-striped and colored in brown. (**B**) Fractogel EMD DEAE chromatography; (**C**) GF on Sepharose 4 FF. In (**B**,**C**) X-axis show mL, Y-axis—optical density (OD, A_{254}) in adsorption units (mAu); analyzed fractions are shown under the peaks and are red colored; blue line shows OD, red line shows conductivity and green line shows NaCl gradient at elution study. Left part in (**B**) shows that the capacity of the column was exceeded and part of target protein was not loaded (shown striped).

3.4. Characterization of HBc-preS VLP

The identity of HBc-preS1 proteins after their purification as VLPs was verified by Western blot (WB) using anti-HBc mAb 13C9 [65] and anti-preS1 mAb MA18/7 [8]. Representative WB for 183preS1(20–47) and 183preS1phil proteins purified as VLPs is shown in Figure 5.

NAGE gels revealed the presence of nucleic acids in VLPs of all kinds of HBc-preS1 VLPs. Figure 6 shows representative PAGE and NAGE for the part of HBc-preS1 proteins, with the short preS1(20–47) as an insert.

DLS was used as a method to characterize the medium size of VLPs in the preparations (Z-average) along with homogeneity of particles (presence of aggregates), and polydispersity, expressed as polydispersion index (Pdi). The Z-average in pooled central GF peak fraction from GF as the final step in VLP purification (see Figure 4C) among all 10 different HBc-preS1 VLP preparations was, with the exception of 178preS1phil variant, in the range of 40–45 nm, (Figure 7A). The presence of aggregates was not revealed in any of the ten preparations of investigated VLP variants (see Figure 7B,C for two representative constructs—183preS1(20–47) and 183preS1phil). Pdi for all investigated VLP preparations was below 0.2 (0.08–0.168). Pooled central GF peak fractions contained at least 0.5 mg/mL of protein in all cases, with the content of target protein over 90% as estimated by SDS-PAGE. The yield of VLPs for all investigated HBc-preS1 protein variants was in the range of 20–30 mg/g of fresh biomass.

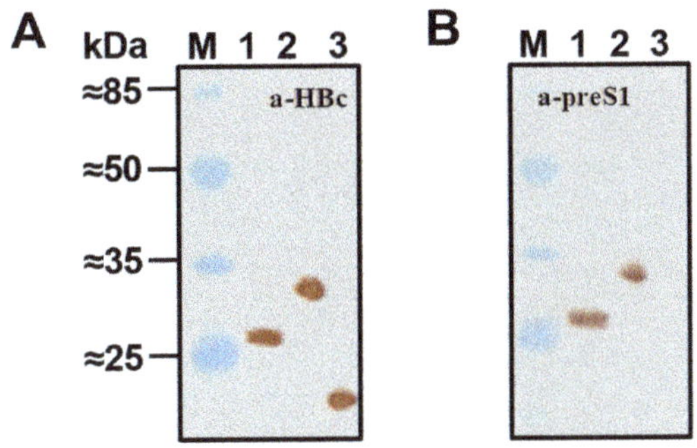

Figure 5. WB of 183preS1(20–47) and 183preS1phil proteins. (**A**) Blot with anti-HBc mAb13C9 [66]; (**B**) blot with anti-preS1 mAb MA18/7 [9]. Lane 1—183preS1(20–47), lane 2—183preS1phil, lane 3—HBc183. M—Prestained Protein Molecular Weight Marker (26612, Pierce™). Normalized amounts of VLPs (10 μg) were used for SDS-PAGE.

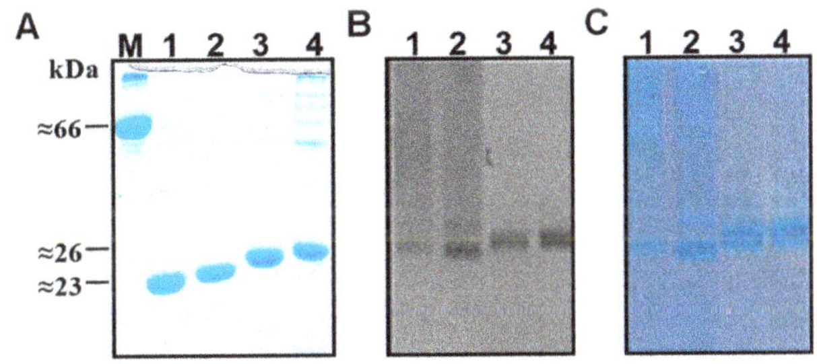

Figure 6. SDS-PAGE and NAGE of HBc-preS1(20–47) VLPs preparations. Loaded proteins: M—BSA, lane 1—163preS1(20–47), lane 2—167preS1(20–47), lane 3—178preS1(20–47), and lane 4—183preS1(20–47). (**A**) 15% SDS-PAGE, (**B**,**C**) 1% NAGE in TBE buffer of the same VLP samples: (**B**) EtBr staining, (**A**,**C**) Coomassie Brilliant Blue R-250 staining. Note: 171preS1(20–47) not shown.

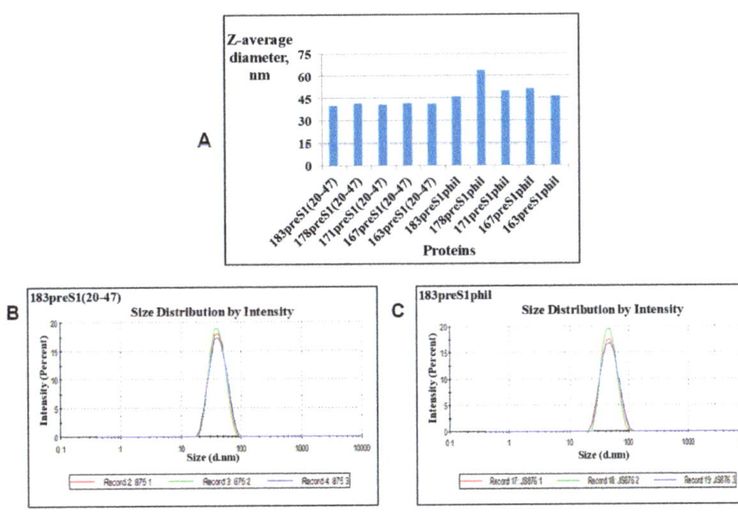

Figure 7. DLS analysis of the HBc-preS1 VLPs preparations. (**A**) Z-average of VLPs for all investigated VLP preparations. (**B**,**C**) Size distribution for VLPs of 183preS1(20–47) and 183preS1phil proteins, respectively.

VLP quality evaluation was performed by visualizing the particles by transmission EM using a negative staining protocol. VLPs formed by preS1-HBc proteins with a short preS1(20–47) insert were of the size similar to the VLPs of non-chimeric HBc and slightly bigger with a longer preS1phil insert used for the construction of HBc-preS1 proteins (Figure 8). We did not observe the instability of VLPs after storage in 50% glycerol at −18 °C for several years (Figure 8B).

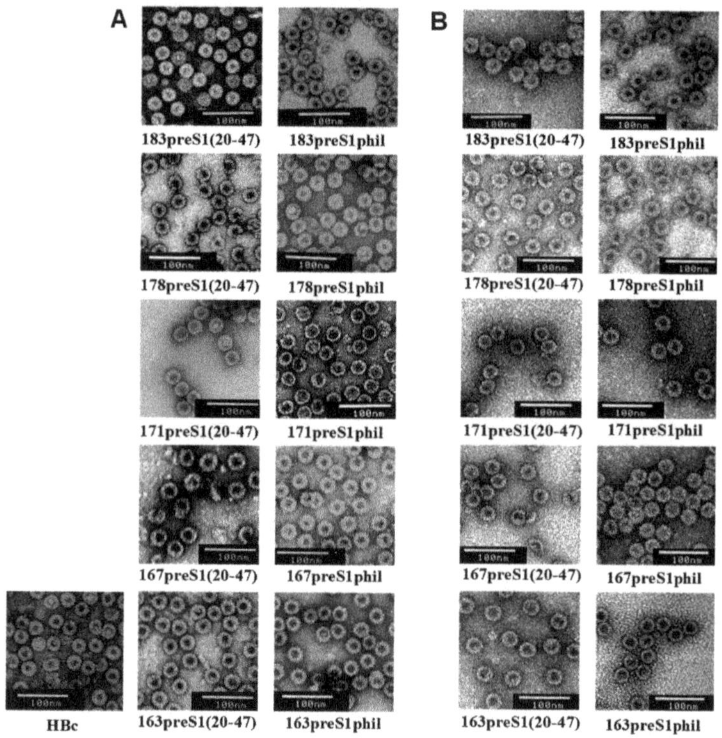

Figure 8. EM of HBc-preS1 VLP preparations. (**A**) Original preparations (years 2016—2018); (**B**) the same samples in the year 2021. Scale bar, 100 nm.

3.5. Accessibility of the preS1 Epitope in the HBc-preS1 VLPs to the mAb MA18/7

Accessibility of the inserted preS1 epitope to a specific antibody was characterized by competitive ELISA using mAb MA18/7 [9]. The 50% inhibitory concentration (Figure 9) was in the range of 48–185 nM for different HBc-preS1 constructs. The competition lag of preS1phil-containing constructs compared to constructs with preS1(20–47) is possibly due to the remarkably longer inserted preS1 sequence (80 aa).

3.6. Immunogenicity of HBc-preS1 VLPs

Anti-HBc response in the control group immunized with unmodified HBc VLPs (without any insertion) reached the titer of anti-HBc to the level 1:164,025. However, the anti-HBc titers for HBc constructs with MIR insertions were substantially decreased, with some exception for construct 178preS1(20–47) (Figure 10B). As for anti-preS1(20–47) response, there was a tendency of higher response for constructs with short preS1(20–47) insertions rather than with preS1phil insertions; however, the differences were not always significant (Figure 10C).

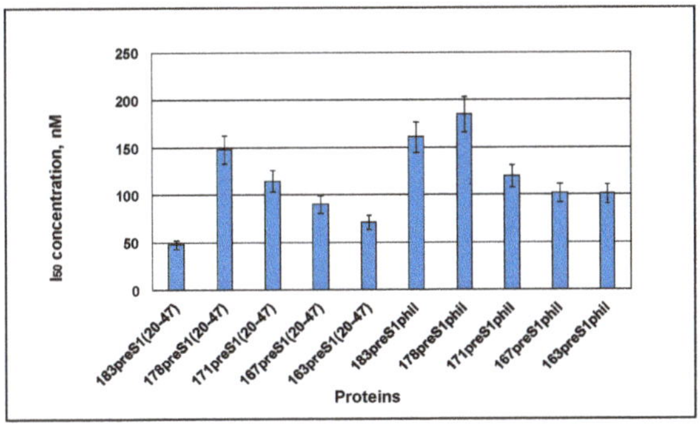

Figure 9. The preS1 antigenicity of the HBc-preS1 VLPs. The VLP concentration necessary and sufficient to inhibit 50% of the binding of mAb MA18/7 [9] to the 20–47 peptide on the support during the competitive ELISA is shown. Tests were done in triplicate.

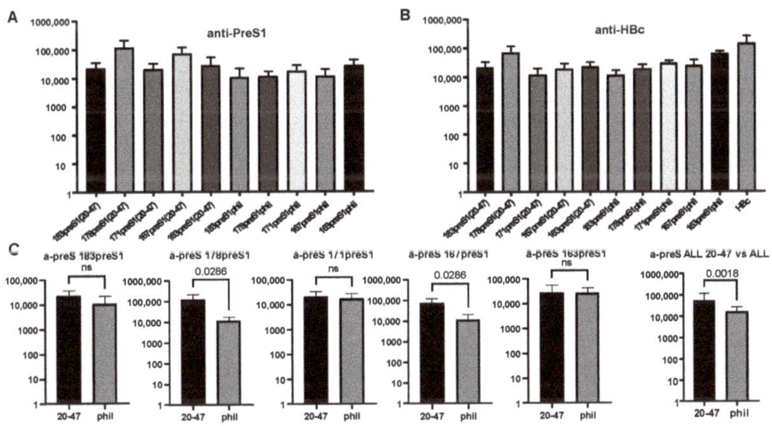

Figure 10. Immunogenicity of HBc-preS1 proteins. BALB/c mice were immunized with VLPs of HBc-preS1 proteins shown under bars in (**A**,**B**). (**A**) Anti-preS1 response and (**B**) anti-HBc response. (**C**) Statistical analysis of antibody response (Mann–Whitney U test). The type of HBc vector (with 183, 178, 171, 167, 163 aa of HBc) used in fusion constructs is shown on the top of the bars and preS1 fragment (20–47 or preS1phil) used in fusion proteins is shown under the bars. The last two bars show the comparison of all five constructs with preS1(20–47) insertion to all five constructs with preS1phil insertion. p-value is indicated on the top of the bars; ns = not significant.

T cells were stimulated in vitro using two doses of full-length 183HBc (1 µg and 10 µg), and appropriate SI were determined. SI values of 2.0 and above were considered positive. Clear proliferation was detected for all immunization cases (Figure 11). For the constructs with a preS1(20–47) insert, the best proliferation effect was found in the group of animals immunized with the VLPs formed by a 178 aa-long HBc vector (178preS1(20–47)), and for constructs with preS1phil insert—in the group of animals immunized with a 163 aa-long HBc vector (163preS1phil).

3.7. Packaging of Oligonucleotide ODN 1668 in Chimeric HBc-preS1 VLPs

The ability of chimeric HBc-preS1 VLPs to incorporate immunostimulating CpG sequences was tested with the use of synthetic ODN 1668 and VLPs formed by 183preS1phil protein. To eliminate from VLPs incorporated RNA of host origin, treatment of VLPs with RNaseA was performed in the presence of urea and ODN 1668 (Materials and Methods, Section 2.7). As shown in Figure 12, the packaging was successful (Figure 12A, line 6). It was observed that although RNA can be eliminated with RNase treatment alone, effective packaging is ensured only when also in the presence of urea (compare lanes 4 and 6 in Figure 11A); however, without the urea, some packaging was observed in VLPs of unmodified HBc (Figure 12, lane 2). Treated VLPs (with RNaseA and with RNaseA + urea) lose their ability to move in gel and treated VLP material stays on the start position (see some staining signal at the start of lanes 2, 4, and 5 in Figure 12). RNaseA in NAGE moves in the opposite direction as seen in Figure 12B.

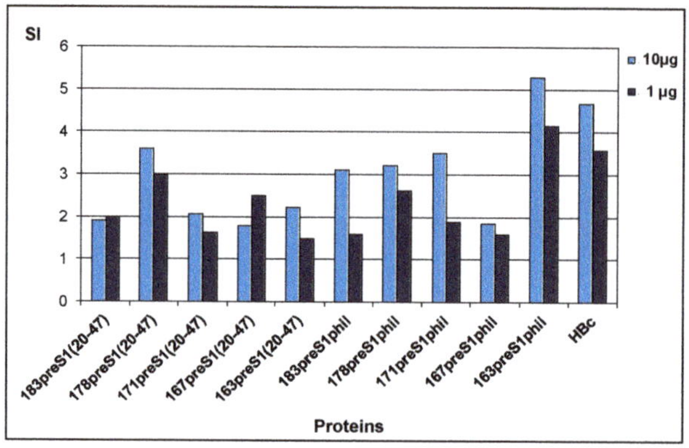

Figure 11. T-cell proliferation in splenocytes of BALB/c mice immunized with HBc-preS1 VLPs. Stimulation indexes (SI) were calculated after stimulation with HBc183 (1 μg and 10 μg) on day 42 after the first immunization. Types of VLPs used for immunization are shown under the bars.

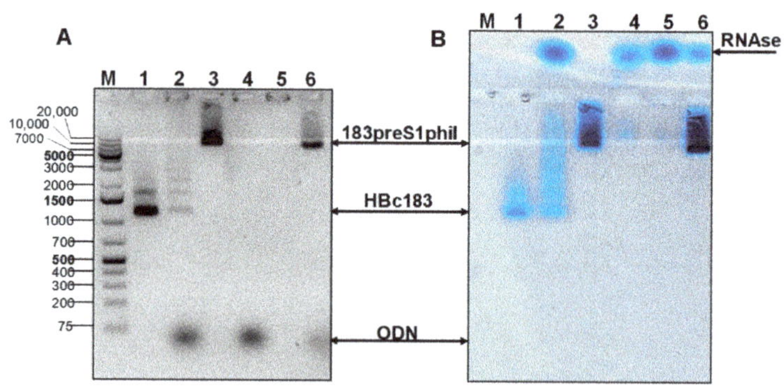

Figure 12. Packaging of ODN 1668 in the VLPs of 183pS1phil protein in the conditions of treatment with RNase and urea. (**A**) EtBr stained NAGE, of HBc and 183preS1phil VLPs before and after treatment: M—marker (GeneRuller 1 kb DNA Ladder (SM1331, Thermo Fisher Scientific™, Waltham, MA, USA), 1—full-length HBc VLPs (HBc183) without treatment, 2—HBc183 treated with RNaseA and added ODN 1668, 3—183preS1phil VLPs without treatment, 4—183preS1phil VLPs treated with RNaseA and added ODN 1668, 5—183preS1phil VLPs treated with RNaseA and urea, without ODN 1668, and 6—183preS1phil VLPs treated with RNaseA and urea with added ODN 1668. (**B**) The same NAGE gel, stained with Coomassie Brilliant Blue G250. Note: lanes 4 and 5 show the absence of protein which is explained by the precipitation of empty VLPs after treatment with RNase; the supernatants of centrifuged (10 min at 13,000× *g*) samples of treated VLPs were loaded on a gel.

4. Discussion

Currently available HBV vaccines are for prophylactic use, being ineffective for the treatment of chronic HBV carriers. These vaccines are based on the pure S protein of the HB antigen or are the combination of the S with the M and/or L forms of HBs. Although these vaccines have demonstrated their effectiveness in the vaccination of newborns, they are less effective in certain groups of people such as aged and obese people or are ineffective in chronic HBV carriers.

This work aimed to generate the recombinant VLPs presenting in two HBV antigens—preS1 for B-cell immunogenicity and HBc for T-cell immunogenicity: preS1 contains the virus-neutralizing epitope and HBc is the source of HBV-specific CTL epitopes. Thus, the rationale of our study lies within the findings that the combination of preS1 sequences and HBc in the particular HBV vaccine candidate not only extends its protective effect but also, due to the presence of HBc, compensates for the lack of T-cell immunogenicity.

The exposure of the preS1 epitopes on the HBc VLPs has a long history and was generally performed by our group [9,69–72], Ken Murray's group [73–75], David Milich and Florian Schödel's group [76–78], and Xinchun Chen's group [79]. Later, Matti Sällberg's team added additional reasons for the expected efficiency of the HBc-preS vaccines [80]. In this resumptive study, we summarized our research on the construction of HBc-preS1 VLPs.

Here, two fragments of the preS1 region of the hepatitis B virus, genotype D1, subtype ayw2 [47] have been used for insertion in HBc protein: (i) the "pure" preS1 epitope corresponding to aa 20–47 of the preS1, and (ii) a preS1phil fragment, aa 12–60 + 89–119, representing the preS1 sequence with a deleted hydrophobic (aa 61–88) region [60]. Both preS1 fragments contain a linear preS1 epitope 31-DPAFR-35 [9,10] which is recognized by highly specific mAb MA18/7 [8]. Selected preS1 sequences were inserted in the MIR of HBc protein using a set of full-length and C-terminally truncated variants of HBc, and the immunogenicity of ten different HBc-preS1 VLP constructs was compared in mice. The production level of the different HBc-preS1 fusion proteins in *E. coli* cells was remarkable (reaching 7–10% of total cell protein in crude cell lysate), allowing us the development of non-sophistical purification protocol and to obtain high-quality VLPs suitable for further immunological evaluation.

As for B-cell immunogenicity, both preS1(20–47) and preS1phil containing VLPs induced a significant anti-preS1 response (Figure 10A) along with the decreased response to the carrier HBc (Figure 10B), caused by damaged MIR within HBc formed VLPs. All variants of the HBc-preS1 VLPs competed well in ELISA with preS1 peptide coated on the plate for the MA18/7 antibodies (Figure 9). These data confirm that the major preS1 epitope DPAFR is exposed and localized correctly on the surface of the chimeric particles, and that conformation of preS1 in our constructs is native like. Additionally, it can be concluded that the N- and C-terminals surrounding of the major immunodominant DPAFR epitope are not important for the correct exposure of the preS1 epitope for induction of the preS1-specific humoral response.

As for T-cell immunogenicity, the T-cell proliferation index was found to be high for all investigated constructs; however, the SI varied significantly among the different constructs with the highest SI for 163preS1phil as a representative of the constructs with long preS1phil insert, and 178preS1(20–47) as a representative of constructs with a short preS1(20–47) insert (Figure 11).

As the antibodies elicited against the preS1 epitope should be strongly virus neutralizing [7], we suggest that recombinant HBc VLPs bearing preS1 sequences may serve as real prototypes for the creation of a combined therapeutic/prophylactic HBV vaccine according to the criteria formulated by Gerlich [81]. The sound biotechnological background of the production and purification of HBc-preS1 VLPs allowed us to obtain tens of mg of highly purified chimeric VLPs from the amount of fresh biomass by standardized protocol and this factor underlines the possible practical application of the elaborated HBc-preS1 VLPs as the candidates for further immunological investigations aimed to the development of the universal prophylactic/therapeutic vaccine.

Packaging of a selected oligonucleotide inside the VLPs has also been demonstrated in the case of full-length HBc bearing a long preS1phil insert after removal of intact RNA (Figure 12).

5. Patents

A. Dišlers, I. Petrovskis, I. Lieknina, I. Berza, J. Bogans, I. Akopjana, I. Sominska, P. Pumpens Latvian Patent C12N15/71 21.06.2013. Expression system for obtaining of HBc-pres1 virus-like particles.

I. Lieknina, I. Petrovskis, I. Sominska, J. Bogans, I. Akopjana, P. Pumpens, A. Dišlers Latvian Patent C12N15/70 20.12.2017. Method for obtaining empty and packed with nucleic acids capsids of virus-like particles of HBc-protein.

Author Contributions: Conceptualization, A.D.; methodology, I.P.; investigation, I.S., D.S., I.L., I.Z. and J.J.; visualization V.O. and J.J; technical assistance, I.A. and J.Z., writing—original draft preparation, I.S.; writing—review and editing, A.D.; supervision, I.S. All authors have read and agreed to the published version of the manuscript.

Funding: This research was funded by ERDF grant Nr. 2010/0211/2DP/2.1.1.1.0/10/APIA/VIAA/048.

Data Availability Statement: Not applicable.

Acknowledgments: We thank Paul Pumpens for helpful discussions, constant support and significant contribution in preparing of this manuscript. We wish to thank our colleagues Juris Ozols and Irina Stahovska for excellent technical assistance.

Conflicts of Interest: The authors declare no conflict of interest. The funders had no role in the design of the study; in the collection, analyses, or interpretation of data; in the writing of the manuscript; or in the decision to publish the results.

References

1. Chen, H.L.; Lin, L.H.; Hu, F.C.; Lee, J.T.; Lin, W.T.; Yang, Y.J.; Huang, F.C.; Wu, S.F.; Chen, S.C.; Wen, W.H.; et al. Effects of maternal screening and universal immunization to prevent mother-to-infant transmission of HBV. *Gastroenterology* **2012**, *142*, 773–781. [CrossRef]
2. Shouval, D.; Roggendorf, H.; Roggendorf, M. Enhanced immune response to hepatitis B vaccination through immunization with a Pre-S1/Pre-S2/S vaccine. *Med. Microbiol. Immunol.* **2015**, *204*, 57–68. [CrossRef] [PubMed]
3. Suk-Fong Lok, A. Hepatitis B Treatment: What We Know Now and What Remains to Be Researched. *Hepatol. Commun.* **2018**, *3*, 8–19. [CrossRef]
4. Backes, S.; Jäger, C.; Dembek, C.J.; Kosinska, A.D.; Bauer, T.; Stephan, A.S.; Dišlers, A.; Mutwiri, G.; Busch, D.H.; Babiuk, L.A.; et al. Protein-prime/modified vaccinia virus Ankara vector-boost vaccination overcomes tolerance in high-antigenemic HBV-transgenic mice. *Vaccine* **2016**, *34*, 923–932. [CrossRef] [PubMed]
5. Whitacre, D.C.; Peters, C.J.; Sureau, C.; Nio, K.; Li, F.; Su, L.; Jones, J.E.; Isogawa, M.; Sallberg, M.; Frelin, L.; et al. Designing a therapeutic hepatitis B vaccine to circumvent immune tolerance. *Hum. Vaccines Immunother.* **2020**, *16*, 251–268. [CrossRef] [PubMed]
6. Maeng, C.Y.; Ryu, C.J.; Gripon, P.; Guguen-Guillouzo, C.; Hong, H.J. Fine mapping of virus-neutralizing epitopes on hepatitis B virus PreS1. *Virology* **2000**, *270*, 9–16. [CrossRef] [PubMed]
7. Bremer, C.M.; Sominskaya, I.; Skrastina, D.; Pumpens, P.; El Wahed, A.A.; Beutling, U.; Frank, R.; Fritz, H.J.; Hunsmann, G.; Gerlich, W.H.; et al. N-terminal myristoylation-dependent masking of neutralizing epitopes in the preS1 attachment site of hepatitis B virus. *J. Hepatol.* **2011**, *55*, 29–37. [CrossRef] [PubMed]
8. Heermann, K.H.; Goldmann, U.; Schwartz, W.; Seyffarth, T.; Baumgarten, H.; Gerlich, W.H. Large surface proteins of hepatitis B virus containing the pre-S sequence. *J. Virol.* **1984**, *52*, 396–402. [CrossRef]
9. Sominskaya, I.; Pushko, P.; Dreilina, D.; Kozlovskaya, T.; Pumpen, P. Determination of the minimal length of preS1 epitope recognized by a monoclonal antibody which inhibits attachment of hepatitis B virus to hepatocytes. *Med. Microbiol. Immunol.* **1992**, *181*, 215–226. [CrossRef]
10. Germaschewski, V.; Murray, K. Screening a monoclonal antibody with a fusion-phage display library shows a discontinuity in a linear epitope within PreS1 of hepatitis B virus. *J. Med. Virol.* **1995**, *45*, 300–305. [CrossRef]
11. Glebe, D.; Aliakbari, M.; Krass, P.; Knoop, E.V.; Valerius, K.P.; Gerlich, W.H. Pre-s1 antigen-dependent infection of Tupaia hepatocyte cultures with human hepatitis B virus. *J. Virol.* **2003**, *77*, 9511–9521. [CrossRef] [PubMed]
12. Brian, P. Dunleavy Public Health Watch: New HBV Vaccine Shows Promise in Phase 3 Trial, 25 May 2021. Available online: https://www.contagionlive.com/view/public-health-watch-new-hbv-vaccine-shows-promise-in-phase-3-trial (accessed on 20 December 2022).
13. Hellström, U.B.; Madalinski, K.; Sylvan, S.P. PreS1 epitope recognition in newborns after vaccination with the third-generation Sci-B-Vac vaccine and their relation to the antibody response to hepatitis B surface antigen. *Virol. J.* **2009**, *6*, 7. [CrossRef] [PubMed]

14. Krawczyk, A.; Ludwig, C.; Jochum, C.; Fiedler, M.; Heinemann, F.M.; Shouval, D.; Roggendorf, M.; Roggendorf, H.; Lindemann, M. Induction of a robust T- and B-cell immune response in non- and low-responders to conventional vaccination against hepatitis B by using a third generation PreS/S vaccine. *Vaccine* **2014**, *32*, 5077–5082. [CrossRef] [PubMed]
15. Murray, K.; Bruce, S.A.; Hinnen, A.; Wingfield, P.; van Erd, P.M.; de Reus, A.; Schellekens, H. Hepatitis B virus antigens made in microbial cells immunise against viral infection. *EMBO J.* **1984**, *3*, 645–650. [CrossRef] [PubMed]
16. Iwarson, S.; Tabor, E.; Thomas, H.C.; Goodall, A.; Waters, J.; Snoy, P.; Shih, J.W.; Gerety, R.J. Neutralization of hepatitis B virus infectivity by a murine monoclonal antibody: An experimental study in the chimpanzee. *J. Med. Virol.* **1985**, *16*, 89–96. [CrossRef] [PubMed]
17. Murray, K.; Bruce, S.A.; Wingfield, P.; van Eerd, P.; de Reus, A.; Schellekens, H. Protective immunisation against hepatitis B with an internal antigen of the virus. *J. Med. Virol.* **1987**, *23*, 101–107. [CrossRef]
18. Roos, S.; Fuchs, K.; Roggendorf, M. Protection of woodchucks from infection with woodchuck hepatitis virus by immunization with recombinant core protein. *J. Gen. Virol.* **1989**, *70 Pt 8*, 2087–2095. [CrossRef]
19. Schödel, F.; Neckermann, G.; Peterson, D.; Fuchs, K.; Fuller, S.; Will, H.; Roggendorf, M. Immunization with recombinant woodchuck hepatitis virus nucleocapsid antigen or hepatitis B virus nucleocapsid antigen protects woodchucks from woodchuck hepatitis virus infection. *Vaccine* **1993**, *11*, 624–628. [CrossRef]
20. Menne, S.; Maschke, J.; Tolle, T.K.; Lu, M.; Roggendorf, M. Characterization of T-cell response to woodchuck hepatitis virus core protein and protection of woodchucks from infection by immunization with peptides containing a T-cell epitope. *J. Virol.* **1997**, *71*, 65–74. [CrossRef]
21. Lu, M.; Hilken, G.; Kruppenbacher, J.; Kemper, T.; Schirmbeck, R.; Reimann, J.; Roggendorf, M. Immunization of woodchucks with plasmids expressing woodchuck hepatitis virus (WHV) core antigen and surface antigen suppresses WHV infection. *J. Virol.* **1999**, *73*, 281–289. [CrossRef]
22. Lu, M.; Roggendorf, M. Evaluation of new approaches to prophylactic and therapeutic vaccinations against hepatitis B viruses in the woodchuck model. *Intervirology* **2001**, *44*, 124–131. [CrossRef]
23. Zhang, E.; Kosinska, A.D.; Ma, Z.; Dietze, K.K.; Xu, Y.; Meng, Z.; Zhang, X.; Wang, J.; Wang, B.; Dittmer, U.; et al. Woodchuck hepatitis virus core antigen-based DNA and protein vaccines induce qualitatively different immune responses that affect T cell recall responses and antiviral effects. *Virology* **2015**, *475*, 56–65. [CrossRef] [PubMed]
24. Miller, D.S.; Halpern, M.; Kotlarski, I.; Jilbert, A.R. Vaccination of ducks with a whole-cell vaccine expressing duck hepatitis B virus core antigen elicits antiviral immune responses that enable rapid resolution of de novo infection. *Virology* **2006**, *348*, 297–308. [CrossRef] [PubMed]
25. Boudewijns, R.; Ma, J.; Neyts, J.; Dallmeier, K. A novel therapeutic HBV vaccine candidate induces strong polyfunctional cytotoxic T cell responses in mice. *JHEP Rep.* **2021**, *3*, 100295. [CrossRef]
26. Pyrski, M.; Rugowska, A.; Wierzbiński, K.R.; Kasprzyk, A.; Bogusiewicz, M.; Bociąg, P.; Samardakiewicz, S.; Czyż, M.; Kurpisz, M.; Pniewski, T. HBcAg produced in transgenic tobacco triggers Th1 and Th2 response when intramuscularly delivered. *Vaccine* **2017**, *35*, 5714–5721. [CrossRef]
27. Pyrski, M.; Mieloch, A.A.; Plewiński, A.; Basińska-Barczak, A.; Gryciuk, A.; Bociąg, P.; Murias, M.; Rybka, J.D.; Pniewski, T. Immunization with Plant-Derived HBcAg as a Potential Therapeutic Vaccine against Chronic Hepatitis B. *Vaccines* **2019**, *7*, 211. [CrossRef] [PubMed]
28. Sanada, T.; Yamamoto, N.; Kayesh, M.E.H.; Tsukiyama-Kohara, K.; Hasegawa, H.; Miyazaki, T.; Takano, J.I.; Shiogama, Y.; Yasutomi, Y.; Goh, Y.; et al. Intranasal vaccination with HBs and HBc protein combined with carboxyl vinyl polymer induces strong neutralizing antibody, anti-HBs IgA, and IFNG response. *Biochem. Biophys. Res. Commun.* **2019**, *520*, 86–92. [CrossRef] [PubMed]
29. Spellman, M.; Martin, J.T. Treatment of chronic hepatitis B infection with DV-601, a therapeutic vaccine. *J. Hepatol.* **2011**, *54*, S302. [CrossRef]
30. Li, J.; Ge, J.; Ren, S.; Zhou, T.; Sun, Y.; Sun, H.; Gu, Y.; Huang, H.; Xu, Z.; Chen, X.; et al. Hepatitis B surface antigen (HBsAg) and core antigen (HBcAg) combine CpG oligodeoxynucletides as a novel therapeutic vaccine for chronic hepatitis B infection. *Vaccine* **2015**, *33*, 4247–4254. [CrossRef]
31. Betancourt, A.A.; Delgado, C.A.; Estévez, Z.C.; Martínez, J.C.; Ríos, G.V.; Aureoles-Roselló, S.R.; Zaldívar, R.A.; Guzmán, M.A.; Baile, N.F.; Reyes, P.A.; et al. Phase I clinical trial in healthy adults of a nasal vaccine candidate containing recombinant hepatitis B surface and core antigens. *Int. J. Infect. Dis.* **2007**, *11*, 394–401. [CrossRef]
32. Lobaina, Y.; Michel, M.L. Chronic hepatitis B: Immunological profile and current therapeutic vaccines in clinical trials. *Vaccine* **2017**, *35*, 2308–2314. [CrossRef] [PubMed]
33. Lopez, M.; Rodriguez, E.N.; Lobaina, Y.; Musacchio, A.; Falcon, V.; Guillen, G.; Aguilar, J.C. Characterization of the size distribution and aggregation of virus-like nanoparticles used as active ingredients of the HeberNasvac therapeutic vaccine against chronic hepatitis B. *Adv. Nat. Sci. Nanosci. Nanotechnol.* **2017**, *8*, 025009. [CrossRef]
34. Al Mahtab, M.; Akbar, S.M.F.; Aguilar, J.C.; Guillen, G.; Penton, E.; Tuero, A.; Yoshida, O.; Hiasa, Y.; Onji, M. Treatment of chronic hepatitis B naïve patients with a therapeutic vaccine containing HBs and HBc antigens (a randomized, open and treatment controlled phase III clinical trial). *PLoS ONE* **2018**, *13*, e0201236. [CrossRef]
35. Chain, B.M.; Myers, R. Variability and conservation in hepatitis B virus core protein. *BMC Microbiol.* **2005**, *5*, 33. [CrossRef]

36. Birnbaum, F.; Nassal, M. Hepatitis B virus nucleocapsid assembly: Primary structure requirements in the core protein. *J. Virol.* **1990**, *64*, 3319–3330. [CrossRef]
37. Seifer, M.; Standring, D.N. A protease-sensitive hinge linking the two domains of the hepatitis B virus core protein is exposed on the viral capsid surface. *J. Virol.* **1994**, *68*, 5548–5555. [CrossRef]
38. Pumpens, P.; Grens, E. The true story and advantages of the famous Hepatitis B virus core particles: Outlook 2016. *Mol. Biol.* **2016**, *50*, 558–576. [CrossRef]
39. Pumpens, P.; Pushko, P. Order Blubervirales. In *Virus-like Particles: A Comprehensive Guide*; CRC Press/Taylor & Francis Group: Boca Raton, FL, USA; London, UK; New York, NY, USA, 2022; pp. 563–602.
40. Hatton, T.; Zhou, S.; Standring, D.N. RNA- and DNA-binding activities in hepatitis B virus capsid protein: A model for their roles in viral replication. *J Virol.* **1992**, *66*, 5232–5241. [CrossRef]
41. Gallina, A.; Bonelli, F.; Zentilin, L.; Rindi, G.; Muttini, M.; Milanesi, G. A recombinant hepatitis B core antigen polypeptide with the protamine-like domain deleted self-assembles into capsid particles but fails to bind nucleic acids. *J. Virol.* **1989**, *63*, 4645–4652. [CrossRef]
42. Wynne, S.A.; Crowther, R.A.; Leslie, A.G. The crystal structure of the human hepatitis B virus capsid. *Mol. Cell* **1999**, *3*, 771–780. [CrossRef]
43. Salfeld, J.; Pfaff, E.; Noah, M.; Schaller, H. Antigenic determinants and functional domains in core antigen and e antigen from hepatitis B virus. *J. Virol.* **1989**, *63*, 798–808. [CrossRef] [PubMed]
44. Sällberg, M.; Rudén, U.; Wahren, B.; Noah, M.; Magnius, L.O. Human and murine B-cells recognize the HBeAg/beta (or HBe2) epitope as a linear determinant. *Mol. Immunol.* **1991**, *28*, 719–726. [CrossRef] [PubMed]
45. Pumpens, P.; Grens, E. HBV core particles as a carrier for B cell/T cell epitopes. *Intervirology* **2001**, *44*, 98–114. [CrossRef] [PubMed]
46. Pumpens, P.; Ulrich, R.; Sasnauskas, K.; Kazaks, A.; Ose, V.; Grens, E. Construction of novel vaccines on the basis of the virus-like particles: Hepatitis B virus proteins as vaccine carriers. In *Medicinal Protein Engineering*; Khudyakov, Y., Ed.; CRC Press, Taylor & Francis Group: Boca Raton, FL, USA; London, UK; New York, NY, USA, 2008; pp. 205–248.
47. Bichko, V.; Pushko, P.; Dreilina, D.; Pumpen, P.; Gren, E. Subtype ayw variant of hepatitis B virus. DNA primary structure analysis. *FEBS Lett.* **1985**, *185*, 208–212. [CrossRef] [PubMed]
48. Pettersen, E.F.; Goddard, T.D.; Huang, C.C.; Couch, G.S.; Greenblatt, D.M.; Meng, E.C.; Ferrin, T.E. UCSF Chimera—A visualization system for exploratory research and analysis. *J. Comput. Chem.* **2004**, *25*, 1605–1612. [CrossRef]
49. Pushko, P.; Pumpens, P.; Grens, E. Development of Virus-Like Particle Technology from Small Highly Symmetric to Large Complex Virus-Like Particle Structures. *Intervirology* **2013**, *56*, 141–165. [CrossRef] [PubMed]
50. Zeltins, A. Construction and characterization of virus-like particles: A review. *Mol. Biotechnol.* **2013**, *53*, 92–107. [CrossRef]
51. Borisova, G.P.; Kalis, J.V.; Dishler, A.V.; Pumpen, P.P.; Gren, E.J.; Tsibinogin, V.V.; Kukaine, R.A. Expression of Human Hepatitis B-virus Core Antigen Gene Variants in *Escherichia coli*. *Biopolym. Cell* **1985**, *1*, 99–105. [CrossRef]
52. Sominskaya, I.; Skrastina, D.; Petrovskis, I.; Dishlers, A.; Berza, I.; Mihailova, M.; Jansons, J.; Akopjana, I.; Stahovska, I.; Dreilina, D.; et al. A VLP library of C-terminally truncated Hepatitis B core proteins: Correlation of RNA encapsidation with a Th1/Th2 switch in the immune responses of mice. *PLoS ONE* **2013**, *8*, e75938. [CrossRef]
53. Strods, A.; Ose, V.; Bogans, J.; Cielens, I.; Kalnins, G.; Radovica, I.; Kazaks, A.; Pumpens, P.; Renhofa, R. Preparation by alkaline treatment and detailed characterisation of empty hepatitis B virus core particles for vaccine and gene therapy applications. *Sci. Rep.* **2015**, *5*, 11639. [CrossRef]
54. Petrovskis, I.; Lieknina, I.; Dislers, A.; Jansons, J.; Bogans, J.; Akopjana, I.; Zakova, J.; Sominskaya, I. Production of the HBc Protein from Different HBV Genotypes in *E. coli*. Use of Reassociated HBc VLPs for Packaging of ss- and dsRNA. *Microorganisms* **2021**, *9*, 283. [CrossRef] [PubMed]
55. Freivalds, J.; Dislers, A.; Ose, V.; Pumpens, P.; Tars, K.; Kazaks, A. Highly efficient production of phosphorylated hepatitis B core particles in yeast Pichia pastoris. *Protein Expr. Purif.* **2011**, *75*, 218–224. [CrossRef] [PubMed]
56. Krieger, J.; Stifter, K.; Riedl, P.; Schirmbeck, R. Cationic domains in particle-forming and assembly-deficient HBV core antigens capture mammalian RNA that stimulates Th1-biased antibody responses by DNA vaccination. *Sci. Rep.* **2018**, *8*, 14660. [CrossRef] [PubMed]
57. Krieger, J.; Riedl, P.; Stifter, K.; Roman-Sosa, G.; Seufferlein, T.; Wagner, M.; Schirmbeck, R. Endogenously Expressed Antigens Bind Mammalian RNA via Cationic Domains that Enhance Priming of Effector CD8 T Cells by DNA Vaccination. *Mol. Ther.* **2019**, *27*, 661–672. [CrossRef]
58. Riedl, P.; Buschle, M.; Reimann, J.; Schirmbeck, R. Binding immune-stimulating oligonucleotides to cationic peptides from viral core antigen enhances their potency as adjuvants. *Eur. J. Immunol.* **2002**, *32*, 1709–1716. [CrossRef]
59. Riedl, P.; Stober, D.; Oehninger, C.; Melber, K.; Reimann, J.; Schirmbeck, R. Priming Th1 immunity to viral core particles is facilitated by trace amounts of RNA bound to its arginine-rich domain. *J. Immunol.* **2002**, *168*, 4951–4959. [CrossRef]
60. Skrastina, D.; Bulavaite, A.; Sominskaya, I.; Kovalevska, L.; Ose, V.; Priede, D.; Pumpens, P.; Sasnauskas, K. High immunogenicity of a hydrophilic component of the hepatitis B virus preS1 sequence exposed on the surface of three virus-like particle carriers. *Vaccine* **2008**, *26*, 1972–1981. [CrossRef]
61. Sominskaya, I.; Skrastina, D.; Dislers, A.; Vasiljev, D.; Mihailova, M.; Ose, V.; Dreilina, D.; Pumpens, P. Construction and immunological evaluation of multivalent hepatitis B virus (HBV) core virus-like particles carrying HBV and HCV epitopes. *Clin. Vaccine Immunol.* **2010**, *17*, 1027–1033. [CrossRef]

62. Dishlers, A.; Skrastina, D.; Renhofa, R.; Petrovskis, I.; Ose, V.; Lieknina, I.; Jansons, J.; Pumpens, P.; Sominskaya, I. The hepatitis B virus core variants that expose foreign C-terminal insertions on the outer surface of virus-like particles. *Mol. Biotechnol.* **2015**, *57*, 1038–1049. [CrossRef]
63. Soberón, X.; Covarrubias, L.; Bolivar, F. Construction and characterization of new cloning vehicles, IV. Deletion derivatives of pBR322 and pBR325. *Gene* **1980**, *9*, 287–305. [CrossRef]
64. Ovchinnikov, I.A.; Sverdlov, E.D.; Tsarev, S.A.; Khodkova, E.M.; Monastyrskaia, G.S. Direct expression of the gene of human leukocyte interferon F in *Escherichia coli* cells. *Dokl. Akad. Nauk SSSR* **1982**, *265*, 238–242. (In Russian)
65. Bichko, V.; Schodel, F.; Nassal, M.; Gren, E.; Berzinsh, I.; Borisova, G.; Miska, S.; Peterson, D.L.; Gren, E.; Pushko, P.; et al. Epitopes recognized by antibodies to denatured core protein of hepatitis B virus. *Mol. Immunol.* **1993**, *30*, 221–231. [CrossRef] [PubMed]
66. Kazaks, A.; Balmaks, R.; Voronkova, T.; Ose, V.; Pumpens, P. Melanoma vaccine candidates from chimeric hepatitis B core virus-like particles carrying a tumor-associated MAGE-3 epitope. *J. Biotechnol.* **2008**, *3*, 1429–1436. [CrossRef] [PubMed]
67. Reddy, V.S.; Natarajan, P.; Okerberg, B.; Li, K.; Damodaran, K.V.; Morton, R.T.; Brooks, C.L., 3rd; Johnson, J.E. Virus particle explorer (VIPER), a website for virus capsid structures and their computational analyses. *J. Virol.* **2001**, *75*, 11943–11947. [CrossRef]
68. Bates, P.A.; Kelley, L.A.; MacCallum, R.M.; Sternberg, M.J. Enhancement of protein modeling by human intervention in applying the automatic programs 3D-JIGSAW and 3D-PSSM. *Proteins* **2001**, *45* (Suppl. S5), 39–46. [CrossRef] [PubMed]
69. Borisova, G.; Arya, B.; Dislers, A.; Borschukova, O.; Tsibinogin, V.; Skrastina, D.; Eldarov, M.A.; Pumpens, P.; Skryabin, K.G.; Grens, E. Hybrid hepatitis B virus nucleocapsid bearing an immunodominant region from hepatitis B virus surface antigen. *J. Virol.* **1993**, *67*, 3696–3701. [CrossRef]
70. Borisova, G.; Borschukova, O.; Skrastina, D.; Dislers, A.; Ose, V.; Pumpens, P.; Grens, E. Behavior of a short preS1 epitope on the surface of hepatitis B core particles. *Biol. Chem.* **1999**, *380*, 315–324. [CrossRef]
71. Fehr, T.; Skrastina, D.; Pumpens, P.; Zinkernagel, R.M. T cell-independent type I antibody response against B cell epitopes expressed repetitively on recombinant virus particles. *Proc. Natl. Acad. Sci. USA* **1998**, *95*, 9477–9481. [CrossRef]
72. Kazaks, A.; Borisova, G.; Cvetkova, S.; Kovalevska, L.; Ose, V.; Sominskaya, I.; Pumpens, P.; Skrastina, D.; Dislers, A. Mosaic hepatitis B virus core particles presenting the complete preS sequence of the viral envelope on their surface. *J. Gen. Virol.* **2004**, *85 Pt 9*, 2665–2670. [CrossRef] [PubMed]
73. Stahl, S.J.; Murray, K. Immunogenicity of peptide fusions to hepatitis B virus core antigen. *Proc. Natl. Acad. Sci. USA* **1989**, *86*, 6283–6287. [CrossRef]
74. Shiau, A.L.; Murray, K. Mutated epitopes of hepatitis B surface antigen fused to the core antigen of the virus induce antibodies that react with the native surface antigen. *J. Med. Virol.* **1997**, *51*, 159–166. [CrossRef]
75. Murray, K.; Shiau, A.L. The core antigen of hepatitis B virus as a carrier for immunogenic peptides. *Biol. Chem.* **1999**, *380*, 277–283. [CrossRef] [PubMed]
76. Schödel, F.; Will, H.; Milich, D.R. Hybrid hepatitis-B virus core/pre-S particles expressed in live attenuated *Salmonellae* for oral immunization. In *Vaccines 91*; Brown, F., Chanock, R.M., Ginsberg, H.S., Lerner, R.A., Eds.; Cold Spring Harbor Laboratory Press: Cold Spring Harbor, NY, USA, 1991; pp. 319–325.
77. Schödel, F.; Peterson, D.; Hughes, J.; Milich, D.R. A virulent Salmonella expressing hybrid hepatitis B virus core/pre-S genes for oral vaccination. *Vaccine* **1993**, *11*, 143–148. [CrossRef] [PubMed]
78. Schödel, F.; Kelly, S.M.; Peterson, D.L.; Milich, D.R.; Curtiss, R., 3rd. Hybrid hepatitis B virus core-pre-S proteins synthesized in avirulent Salmonella typhimurium and Salmonella typhi for oral vaccination. *Infect. Immun.* **1994**, *62*, 1669–1676. [CrossRef] [PubMed]
79. Chen, X.; Li, M.; Le, X.; Ma, W.; Zhou, B. Recombinant hepatitis B core antigen carrying preS1 epitopes induce immune response against chronic HBV infection. *Vaccine* **2004**, *22*, 439–446. [CrossRef] [PubMed]
80. Malik, I.R.; Chen, A.; Brass, A.; Ahlén, G.; Rahman, M.; Sällberg, M.; Qureshi, J.A.; Frelin, L. A bi-functional hepatitis B virus core antigen (HBcAg) chimera activates HBcAg-specific T cells and preS1-specific antibodies. *Scand. J. Infect. Dis.* **2012**, *44*, 55–59. [CrossRef] [PubMed]
81. Gerlich, W.H. Do we need better hepatitis B vaccines? *Indian J. Med. Res.* **2017**, *145*, 414–419.

microorganisms

Article

Knocking Down Gm16685 Decreases Liver Granuloma in Murine Schistosomiasis Japonica

Ruyu Zhao †, Xiaoxue Tang †, Huiyao Lin, Chen Xing, Na Xu, Bingxin Dai, Pingping Wang, Wei Shao, Miao Liu, Jijia Shen, Shengqun Deng * and Cuiping Ren *

Department of Microbiology and Parasitology, Anhui Provincial Laboratory of Pathogen Biology, Anhui Key Laboratory of Zoonosis of High Institution, Laboratory of Tropical and Parasitic Diseases Control, School of Basic Medical Sciences, Anhui Medical University, Hefei 230032, China
* Correspondence: dengshengqun@ahmu.edu.cn (S.D.); cuipingren@ahmu.edu.cn (C.R.)
† These authors contributed equally to this work.

Abstract: Long noncoding RNAs (lncRNAs) can regulate key genes and pathways in liver disease development. Moreover, macrophages are speculated to play an important role in regulating granulomatous inflammation during schistosomiasis. However, the role of lncRNAs in the formation of liver granulomas by influencing the polarization of macrophages in *Schistosoma japonicum* infection is unclear. Our study aimed to determine whether lncRNAs can play a role in *S. japonicum*-induced hepatic egg granulomas and elucidate their effect on macrophages. We established *S. japonicum* infection models and screened the target lncRNA Gm16685 highly expressed in schistosomiasis mice using high-throughput sequencing. Hematoxylin and eosin staining revealed that the knockdown of Gm16685 reduced the area of egg granulomas. Moreover, M1 macrophage factor genes were significantly downregulated in Gm16685 knockdown livers. Meanwhile, M2 macrophage factor genes were significantly upregulated, which was consistent with the protein detection results. Hepatocytes, hepatic stellate cells, and macrophages were isolated from mouse models infected with *S. japonicum*, with Gm16685 being significantly upregulated in macrophages. Moreover, the knockdown of Gm16685 in RAW264.7 cells revealed similar results to in liver tissue. RNA fluorescence *in situ* hybridization (FISH) and nucleocytoplasmic separation experiments revealed that Gm16685 was predominantly localized in the cytoplasm of cells. We found that miR-205-5p was upregulated after Gm16685 was knocked down. After overexpression of miR-205-5p, the expression of Gm16685 and inflammatory factors was significantly downregulated. These results indicate that Gm16685 can participate in the pathogenesis of hepatic disease in schistosomiasis and promote M1 macrophage polarization by regulating miR-205-5p. Thus, our study may provide a new target for schistosomiasis japonica treatment.

Keywords: *Schistosoma japonicum*; long noncoding RNAs; liver granuloma; macrophage polarization

Citation: Zhao, R.; Tang, X.; Lin, H.; Xing, C.; Xu, N.; Dai, B.; Wang, P.; Shao, W.; Liu, M.; Shen, J.; et al. Knocking Down Gm16685 Decreases Liver Granuloma in Murine Schistosomiasis Japonica. *Microorganisms* **2023**, *11*, 796. https://doi.org/10.3390/microorganisms11030796

Academic Editors: Shengxi Chen and Fabio Zicker

Received: 2 March 2023
Revised: 16 March 2023
Accepted: 17 March 2023
Published: 21 March 2023

1. Introduction

Schistosomiasis is an important but often neglected tropical disease that poses a serious threat to the health of more than 250 million people worldwide [1–3]. Schistosomiasis is mainly caused by *Schistosoma hematobium*, *S. mansoni*, and *S. japonicum* [4,5]. As the pathogen of hepatointestinal schistosomiasis, *S. japonicum* is primarily prevalent in China, with a low prevalence in the Philippines and Indonesia [4,5]. Female and male *S. japonicum* worms parasitize human veins and mate to produce fertilized eggs [6]. These eggs secrete antigenic glycoproteins, which promote their transfer from blood vessels (spawning sites) into the intestinal cavity or bladder by inducing inflammatory reactions. Moreover, these soluble egg antigens (SEAs) can also induce granulomas, which are collections of inflammatory cells, around the eggs and in the surrounding tissues [7]. The main cellular components of granulomas include macrophages, which play a significant role in egg granulomatous inflammation caused by *S. japonicum* [8–10]. Macrophages are divided into two types: M1 macrophages and M2 macrophages, which have opposite functions in the inflammatory response [11]. M1 macrophages promote the progression of inflammatory responses by secreting proinflammatory cytokines such as IL-1β, whereas M2 macrophages exhibit anti-inflammatory effects by inducing the high expression of IL-10 and TGFβ [12,13]. Additionally, macrophages can govern inflammation caused by schistosomiasis [14]. However, the regulatory mechanism of macrophage polarization remains unclear.

Long noncoding RNAs (lncRNAs) are defined as a group of RNA transcripts that are longer than 200 nucleotides without coding potential [15,16]. They may be in the cytoplasm or nucleus and play a crucial role in cellular processes [17]. Notably, lncRNA is an essential potential regulatory molecule of tumor cells, playing a pivotal role in tumor metastasis, immune escape, metabolism, and angiogenesis, becoming a core in tumor-related signaling pathways [18]. Increasing evidence also suggests that the aberrant expression of lncRNAs is closely related to the initiation and progression of viral infection [19], cancer [20], liver disease, and other diseases [21]. Currently, lncRNA-related research in the liver mainly focuses on nonalcoholic steatohepatitis, hepatocellular carcinoma, and cholestatic liver disease [22–24]. RNA sequencing analysis revealed the differential expression of many lncRNAs in the liver after 25 days of infection with *S. japonicum*, suggesting that lncRNAs could participate in the pathogenesis of the liver after infection with *S. japonicum*. [25].

MicroRNA (miRNA) is a small noncoding RNA molecule that can target mRNA and regulate translation inhibition. miRNA participates in many biological processes and can also induce messenger RNA (mRNA) degradation or prevent its translation [26]. lncRNA can play its biological function as competitive endogenous RNA (ceRNA), thus reducing the inhibition of miRNA-mediated downstream transcripts of miRNA [27].

However, it is unclear how the expression profile and functions of these lncRNAs change during schistosomiasis. This study aims to establish the expression profile of lncRNAs in the liver of *S. japonicum*-infected mice at various time points and explore the potential regulatory mechanism of lncRNAs in schistosomiasis.

2. Materials and Methods

2.1. Schistosomiasis Mouse Model

Six-week-old female C57BL/6 mice were acquired from the Animal Center of Anhui Medical University and housed in a special pathogen-free animal room. Animal-related experiments were conducted according to the regulations on management and animal ethical standards. Mice were assigned into four groups stochastically, with 6 mice in each group. For the *S. japonicum* infection mouse model, the abdominal hair of mice was removed, and three groups of mice were selected to infect 18 $\pm$ 2 *S. cercariae*. The key period of liver granuloma formation is 45 days after schistosomiasis infection [28]. Therefore, after 45 days of infection with *S. japonicum*, the mice were divided into the model, LV-shRNA, and LV-sh-Gm16685 groups, and the corresponding lentivirus was injected into the tail vein of mice in the LV-shRNA and LV-sh-Gm16685 groups. Three weeks later, the mice were dissected for further experimentation.

2.2. Cell Culture

RAW 264.7 cells were maintained in DMEM (Gibco, New York, NY, USA) supplemented with 10% fetal bovine serum at 5% CO_2 and 37 °C. For lipopolysaccharide (LPS) stimulation, the cells were placed into a six-well plate and incubated overnight. Subsequently, diluted LPS was added, and the final concentration was 2 ng/mL. Then, the cells were incubated for 12 h.

2.3. Quantitative Reverse Transcription-PCR (qRT-PCR)

Total RNA from RAW264.7 cells or liver tissues was extracted using TRIzol (Ambion, Austin, TX, USA). According to the manufacturer's instructions, extracted RNA was reverse transcribed with Evo M-MLV RT Premix. Subsequently, the SYBR® Green Premix

Pro Taq HS qRT-PCR Kit (Accurate Biotechnology Co., Ltd., Changsha, China) was used for subsequent qRT-PCR experiments. Actin was chosen as an internal control for the experiments, and gene expression was computed using the $2^{-\Delta\Delta Ct}$ method. The primer sequences are displayed in Table S1.

2.4. Cell Transfection

To silence and overexpress Gm16685, sh-Gm16685, Gm16685-expressing plasmid, and the negative control (sh-NC) were synthesized by Wuhan Miaoling Biotechnology Co., Ltd. (Wuhan, China). miR-205-5p mimics were synthesized by General Biol Co., Ltd. (Chuzhou, China). For transfection, cells were placed in a 6-well culture plate and cultured overnight until 70–80% confluence was reached. EndoFectin™ MAX transfection reagent (GeneCopoeia, Maryland, NY, USA) was also used for transfection according to the instructions. After thirty-six hours, RAW264.7 cells were subjected to treatment with LPS for 12 h. Finally, transfection efficiency was detected.

2.5. Histological Examination

The liver tissues were washed with phosphate-buffered saline (PBS) and then placed in 4% paraformaldehyde. Then, 4 μm thick paraffin slices were obtained. The pathological changes in the liver tissues were observed using hematoxylin and eosin (H&E) staining. The proportion of egg granuloma area in the total area of the H&E slice was determined using ImageJ-win64.

2.6. Adult Counts and Egg Counts

The adults were extracted from the hepatic portal vein and mesentery of the mice infected with *S. japonicum* for counting. Next, 2 mL of 5% KOH was added to a 5 mL Eppendorf (EP) tube. Then, 0.2 g of mouse liver tissue was separated and cut into pieces in an EP tube. This was incubated at 37 °C for 3 h. Then, 50 μL was isolated from the liquid to count the eggs of *S. japonicum* under the microscope. Counting was performed three times, and the average value was considered. Finally, the total number of eggs of each mouse was calculated.

2.7. Isolation of Cells

After anesthesia, the mouse liver tissues were digested with Streptomyces griseus (Sigma, Saint Louis, MO, USA) and collagenase D (Sigma, USA). The digested liver was cut into pieces and then filtered through a sieve. The filtered cell fluid was centrifuged for 5 min to obtain hepatocyte sediment. Then, we centrifuged the supernatant obtained in the previous step at 600× *g* for 10 min, and then took the sediment. We added 5 mL of 20% Nycodenz into the centrifuge tube, mixed it with the sediment, and carefully added 5 mL of 12% Nycodenz and 3mL DMEM on the suspension surface. Then it underwent 1400× *g* centrifugation for 17 min. After centrifugation, the liquid in the centrifuge tube was in a stratified state. One layer of cells in the middle layer was hepatic stellate cells [29].

Other mice were selected to isolate macrophages. The mice were killed and then soaked in 75% alcohol for 5 min. The abdomen was lifted with tweezers, and a small wound was inflicted. To this end, 8 mL PBS was injected, and the peritoneal fluid was removed using a pasteurized straw. Finally, centrifugation was performed to obtain peritoneal macrophage precipitation.

2.8. RNA Fluorescence in Situ Hybridization (FISH)

Gm16685 probes were synthesized by Shanghai GenePharma Co., Ltd. (Shanghai, China). Then, RAW264.7 cells were resuspended in complete medium and mixed evenly. The cells were counted, and then the diluted cell suspension was added to the hole where the climbing tablets were placed. The paved 24-well plates were then incubated overnight. Then, FISH was performed following the instructions of Shanghai GenePharma Co., Ltd. Finally, the cells were observed under a 63 × oil lens of a laser confocal microscope.

2.9. Nucleocytoplasmic Separation

The NE-PER Nuclear Cytoplasmic Extraction Reagent kit (Thermo Scientific, USA) was used for nucleocytoplasmic separation. RAW264.7 cells were washed and then placed in a 1.5 mL EP tube and centrifuged at 500× *g*. Precooled 500 μL cytoplasmic extraction reagent I was then added to the cell precipitate and vortexed for 15 s to suspend the cell precipitate. This was placed on ice for 10 min. Then, 11 mL of a second cytoplasmic extraction reagent II was added to the sample. It was then placed on ice and centrifuged for 5 min at 16,000× *g*. The supernatant was placed into a new precooled EP tube. Precooled 250 μL NER (Nuclear Extraction Agent) was used to resuspend the sediment (which contained crude nuclei), and then the sample was incubated on ice after vortexing for 15 s. After centrifuging at 4 °C for 10 min, the supernatant (nuclear extract) was transferred into a new precooled EP tube. The RNA of the nuclear extract and cytoplasmic extract were extracted, and the expression of Gm16685 in the nucleus and cytoplasm was further analyzed using qRT-PCR.

2.10. Protein Determination Method

Approximately 0.1 g of the same liver part of the three groups was isolated. The liver tissues were homogenized on ice using RIPA buffer (Beyotime Inst. Biotech, Shanghai, China), which was supplemented with phenylmethylsulfonyl fluoride (PMSF), and then centrifuged at 4 °C for 15 min at 12,000 rpm. The supernatant, after cell transfection, was collected and stored. The protein concentrations of liver tissue were detected using the enhanced BCA protein assay kit (Beyotime Inst. Biotech, Shanghai, China). M1 and M2 cytokine determination was performed by Shanghai Universal Biotech Co., Ltd. (Shanghai, China). Murine IL-1β, IL-12A, IL-4, and IL-10 were analyzed using Luminex technology and reagents (R&D Systems, Minneapolis, MN, USA).

2.11. Luciferase Reporter Assay

The wild-type (wt) and mutant-type (mut) sequence fragments of Gm16685 (containing the binding sites of miR-205-5p or the mutated binding sites of miR-205-5p) were cloned into the GV272 vector (Shanghai Genechem, Shanghai, China). In 293T cells, the vectors were transfected with miR-205-5p or miR-NC. After 48 h, the activity of luciferase in cells was determined.

2.12. miRDB

The NCBI gene database (https://www.ncbi.nlm.nih.gov/, accessed on 1 March 2023) was used to query the sequence of lncRNA, accessed on 1 September 2020. The miRDB database (http://mirdb.org, accessed on 1 March 2023) was used to predict the miRNAs of lncRNA interaction, accessed on 1 March 2022. The lncRNA sequence was used for prediction. The specific binding sites can be viewed through the "Details" on the left side of each result.

2.13. Statistical Analysis

GraphPad Prism 8 was used to analyze the data. Data from at least three biological replicates are expressed as the mean ± standard deviation. The difference between the two groups was evaluated using Student's *t*-test, and $p < 0.05$ was considered statistically significant.

3. Results

3.1. LncRNA Expression in the Liver of Mice with Schistosomiasis

To explore the characteristics of lncRNAs in the livers of mice infected with Schistosoma at different time points, *S. japonicum*-infected livers and normal livers were collected and sent to Shanghai Biotechnology Corporation for lncRNA expression profiling (Figure 1a). Compared with normal mice at the same time point, 1231 lncRNAs were upregulated and 1247 lncRNAs were downregulated in the livers of mice infected for 15 days, while 1685 lncRNAs were upregulated and 1883 lncRNAs were downregulated in those infected for 24 days. Moreover, 3029 lncRNAs were upregulated and 4574 lncRNAs were downregulated in mice infected for 45 days (Figure 1b, $p < 0.05$, fold-change > 2). A total of 157 lncRNAs were upregulated and 96 lncRNAs were downregulated in the liver tissues of mice infected with S. japonicum for 15, 24, and 45 days (Figure 1c,d, $p < 0.05$, fold-change > 2). Gene ontology enrichment analysis revealed that at 45 days after infection, the differentially expressed lncRNAs were mainly involved in the biological process of the assembly of the spindle body during mitosis and the morphogenesis of the optic nerve. Furthermore, differentially expressed genes were enriched in the cellular components of the IPAF inflammatory body complex, whereas in terms of molecular function, the genes were highly enriched in coenzyme A ligase activity (Figure 1e). Kyoto Encyclopedia of Genes and Genomes analysis showed that the differentially expressed lncRNAs 45 days after infection were mainly involved in the interaction between cytokines and cytokine receptors, the formation of hematopoietic cells, the differentiation and development of osteoclasts, the biosynthesis of steroid hormones, the metabolism of cytochrome P450, and the metabolism of arachidonic acid (Figure 1f).

3.2. The Expression of lncRNA-Gm16685 Was Significantly Increased in the Livers of Mice with Schistosomiasis

A total of 12 lncRNAs that were differentially expressed in mouse livers after 45 days of *S. japonicum* infection were identified. These lncRNAs were related or unknown to inflammation, and there are more than two exons in the sequence. They were further verified using qRT-PCR. The results were consistent with the high-throughput sequencing analysis, indicating the feasibility of the screening method (Figure 2a).

Three lncRNAs with unknown function (1700025B11Rik, Cdkn1B) and H19 with the highest differential expression were further selected for qRT-PCR analysis to detect their expression levels in the livers of mice 15, 24, and 45 days after *S. japonicum* infection. The results showed that the expression level of lncRNA-Gm16685 was increased 6-fold ($t_{(4)} = 37.03$, $p < 0.0001$) and 61-fold ($t_{(4)} = 26.18$, $p < 0.001$) after 24 and 45 days of infection, respectively (Figure 2b). Combining the results of high-throughput sequencing and qRT-PCR revealed that Gm16685 was significantly increased in the liver of mice infected with *S. japonicum* after 45 days, except for H19. Therefore, we speculate that Gm16685 could be involved in the progression of schistosomiasis.

3.3. Knockdown of Gm16685 Alleviates S. japonicum-Induced Hepatic Granulomas

To investigate the potential impact of Gm16685 on egg-induced liver granuloma formation in schistosomiasis, an *S. japonicum* infection model in mice was constructed. Hepatic granuloma areas were significantly decreased ($t_{(10)} = 3.19$, $p = 0.005$) in the Gm16685 knockdown groups. However, there was no significant difference in the parasite and egg count in each group (Figure 3a–c). H&E staining revealed that egg granuloma reaction, inflammatory cell infiltration, and hepatocyte necrosis were alleviated in the Gm16685 knockdown group (Figure 3c). Compared with the control group, M1 macrophage factor genes, including IL-12A ($t_{(10)} = 4.53$, $p < 0.001$) and CCL-1 ($t_{(10)} = 8.92$, $p < 0.001$), were significantly downregulated in Gm16685 knockdown livers. Meanwhile, M2 macrophage factor genes, including IL-4 ($t_{(10)} = 3.99$, $p = 0.0012$), IL-10 ($t_{(10)} = 6.14$, $p < 0.001$), IL-13 ($t_{(10)} = 9.55$, $p < 0.001$), TGF-β ($t_{(10)} = 2.86$, $p = 0.008$), and Fizzl ($t_{(10)} = 6.74$, $p < 0.001$), were significantly upregulated (Figure 3d). Moreover, IL-12A ($t_{(10)} = 2.26$, $p = 0.023$) and IL-1β ($t_{(10)} = 2.51$, $p = 0.015$) protein expression levels significantly decreased, while IL-4 ($t_{(10)} = 4.31$, $p < 0.001$) and IL-10 ($t_{(10)} = 2.44$, $p = 0.017$) protein levels significantly increased in the treatment group (Figure 3e).

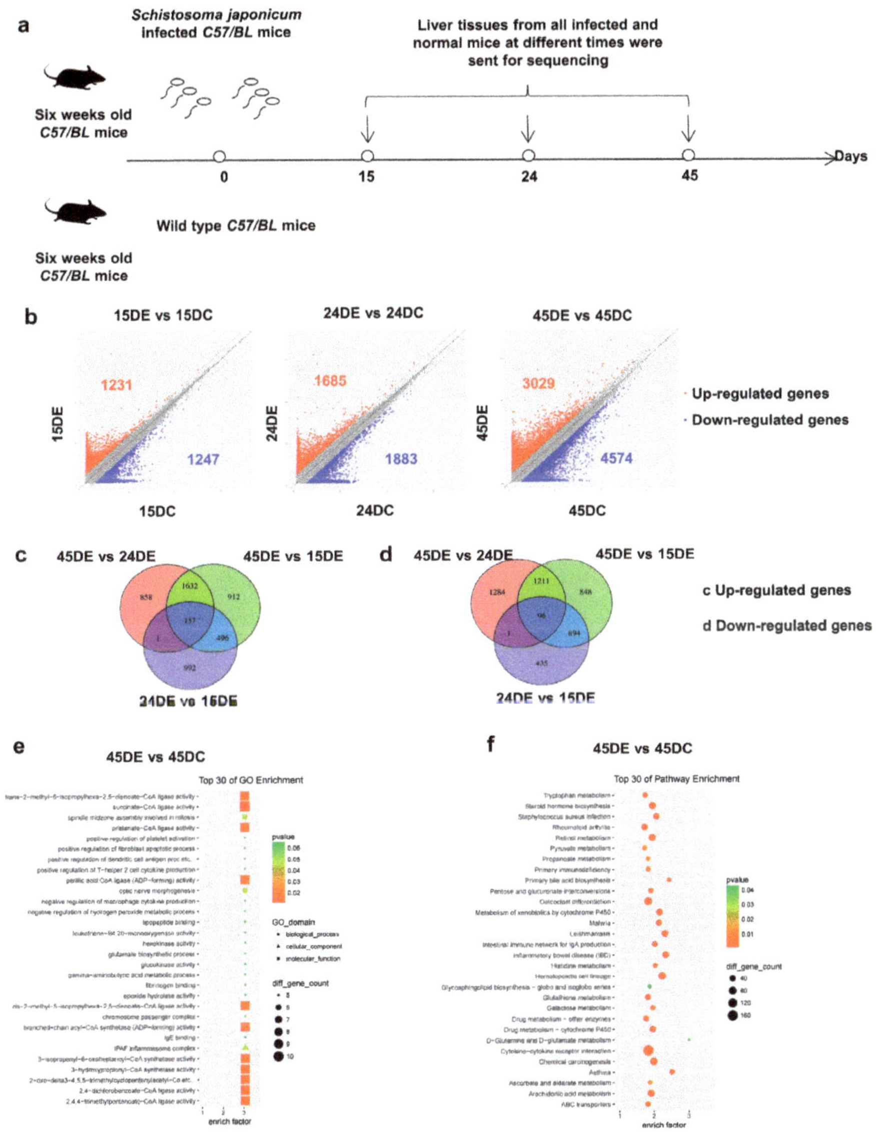

Figure 1. Different expression profiles of lncRNAs in mice at 15, 24, and 45 days after infection with *Schistosoma japonicum*: (**a**) The schematic timeline shows that mice were infected with *S. japonicum*, and the infected mice were sacrificed together with the normal mice on the 15th, 24th, and 45th days. (**b**) Volcano plots showing differentially expressed lncRNAs in liver tissues of the mice on the 15th, 24th, and 45th days after *S. japonicum* infection. (**c**,**d**) The pie chart shows the number of upregulated and downregulated lncRNAs in the liver tissues of mice infected with *S. japonicum* for 15, 24, and 45 days. (15DE, 24DE, and 45DE represent 15, 24, and 45 days after *S. japonicum* infection; 15DC, 24DC, and 45DC denote normal mice raised simultaneously). (**e**,**f**) Gene Ontology and Kyoto Encyclopedia of Genes and Genomes pathway analyses were used to analyze differentially expressed lncRNAs.

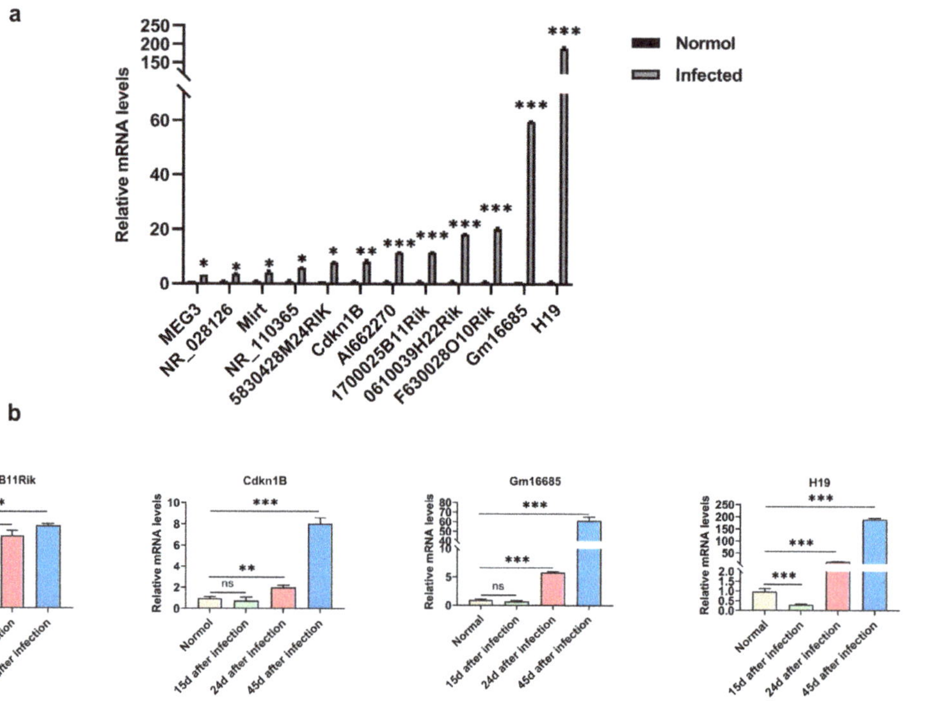

Figure 2. Validation of relative lncRNA expression using quantitative real-time PCR (qRT-PCR): (**a**) The differentially expressed lncRNAs in the liver tissues of model mice (mice infected with *Schistosoma japonicum* for 45 days) and normal mice were verified using qRT-PCR. The relative expression level of lncRNA was consistent with the high-throughput sequencing results. (**b**) The mRNA levels of 1700025B11Rik, Cdkn1B, Gm16685, and H19 in the liver tissues of the mice infected with *S. japonicum* for 15, 24, and 45 days and the liver tissues of the normal group at the same time point were detected using qRT-PCR. Data are presented as the mean ± standard deviation, n = 3. ns, not significant, * $p < 0.05$, ** $p < 0.01$, *** $p < 0.001$.

3.4. Knockdown of Gm16685 Promotes the Polarization of Macrophages to the M2 Phenotype In Vitro

Hepatocytes, hepatic stellate cells, and macrophages were isolated from *S. japonicum*-infected mice and healthy mice. Notably, Gm16685 was highly expressed in activated macrophages ($t_{(4)} = 3.82$, $p = 0.009$) (Figure 4a). Additionally, Gm16685 was significantly upregulated in the activated RAW264.7 cell line ($t_{(4)} = 11.03$, $p < 0.001$) (Figure 4b).

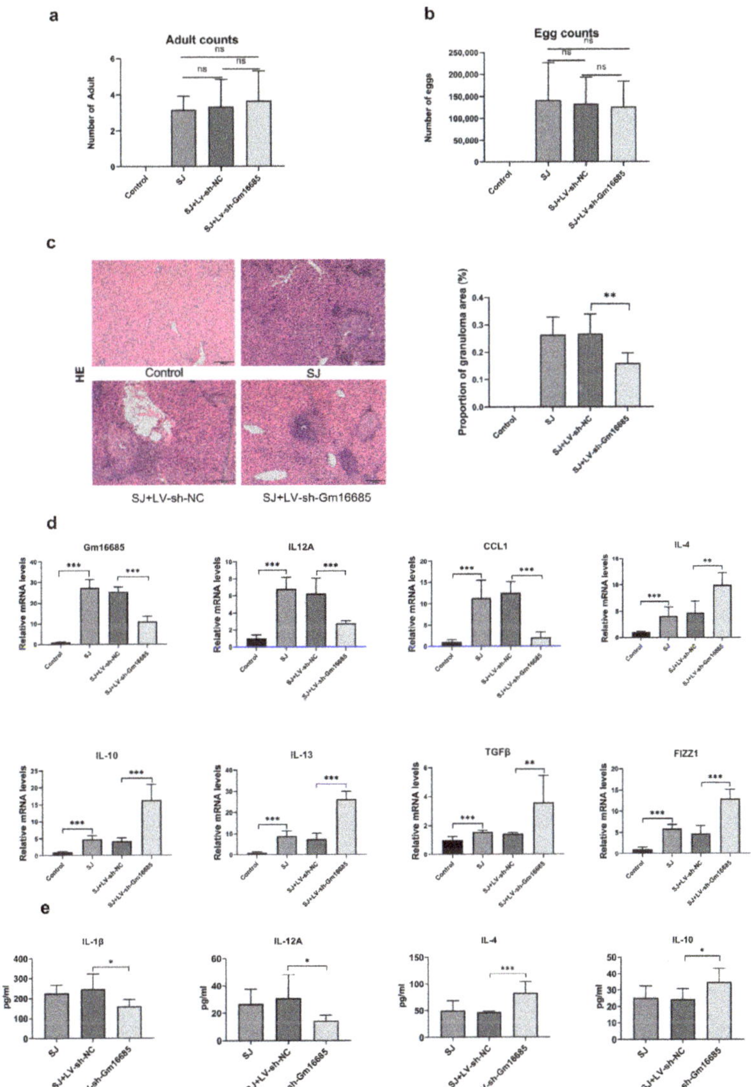

Figure 3. Functional analysis *in vivo*: (**a**) Number of *Schistosoma japonicum* eggs in the liver tissue of the three groups. (**b**) Number of *S. japonicum* adults in the liver tissues of the three groups. (**c**) Hematoxylin and eosin staining was performed to measure the pathological changes in liver tissues (original magnification × 100, scale bar = 200 μm). The area of egg granulomas of *S. japonicum* in the total area of the liver was counted. (**d**) The expression levels of Gm16685, IL-12A, CCL1, IL-4, IL-10, IL-13, TGFβ, and FIZZ1 mRNAs were detected using qRT-PCR in mouse tissues of each group. Control: normal mice; SJ: mice infected with *S. japonicum* for 45 days; SJ+LV-NC: model mice treated with blank vector; SJ+LV-Gm16685: model mice treated with lentivirus that can silence target gene expression. (**e**) Cytokine analysis: Murine IL-1β, IL-12A, IL-4, and IL-10 were analyzed using Luminex technology and reagents in mouse tissues of each group. Data are presented as the mean ± standard deviation, n = 6 mice per group. Ns, not significant, * $p < 0.05$, ** $p < 0.01$, *** $p < 0.001$.

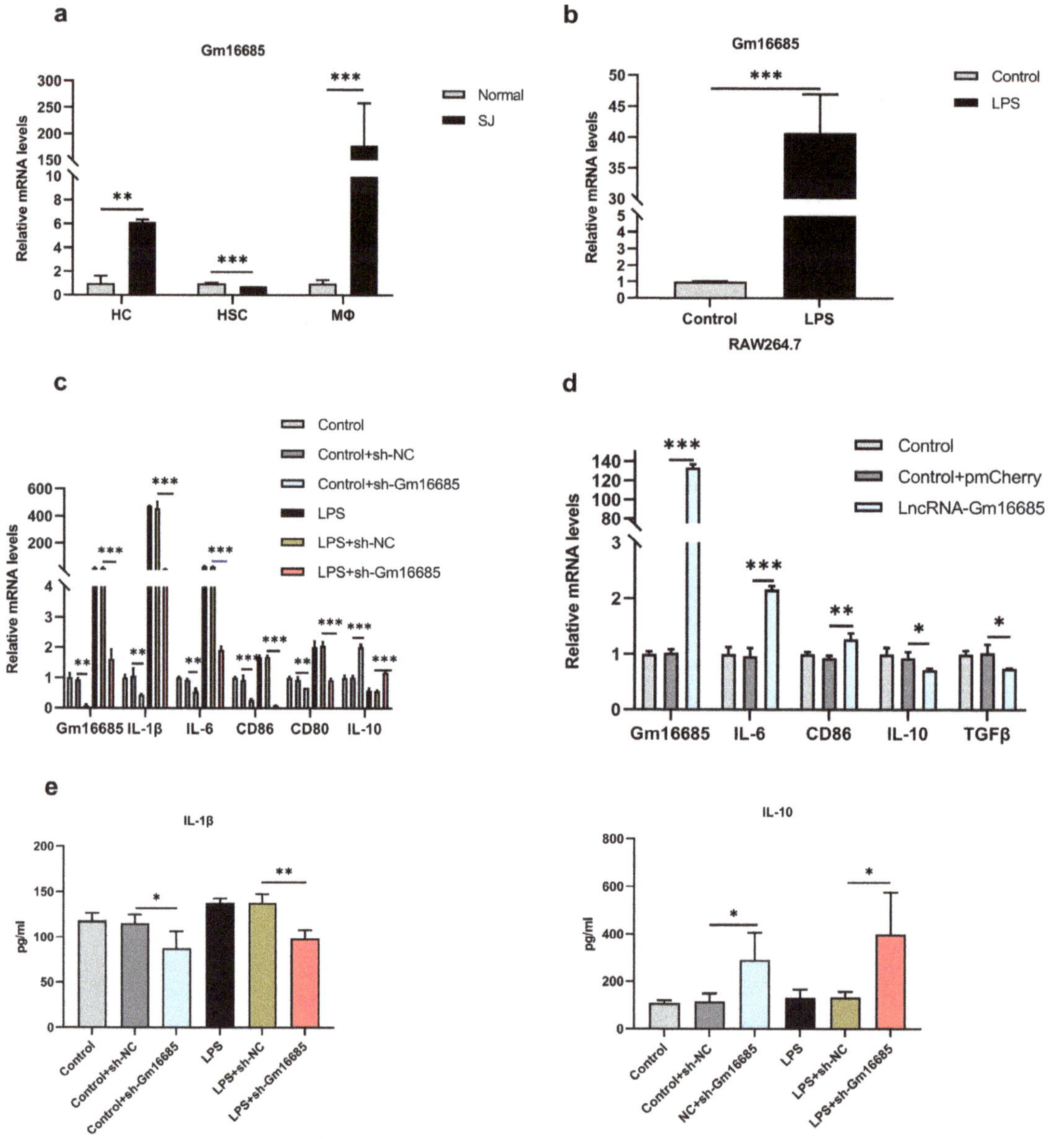

Figure 4. Functional analysis *in vitro*: (**a**) The expression of Gm16685 mRNAs in three primary cells of mice infected with *Schistosoma japonicum* for 45 days. (**b**) The expression of Gm16685 mRNAs in RAW264.7 cells stimulated with lipopolysaccharide was detected using quantitative real-time PCR. (**c**) Knockdown of lncRNA-Gm16685 in RAW 264.7 cells. The expression levels of Gm16685, IL-1β, CD80, CD86, IL-6, and IL-10 mRNAs were detected using qRT-PCR in RAW264.7 cells. (**d**) Overexpression of lncRNA-Gm16685 in RAW 264.7 cells. The expression levels of Gm16685, IL-6, CD86, IL-10, and TGFβ mRNAs were detected using qRT-PCR in RAW264.7 cells. (**e**) Cytokine analysis: Cell supernatant IL-1β and IL-10 were analyzed using Luminex technology and reagents. Data are presented as the mean ± standard deviation, n = 3. ns, not significant, * $p < 0.05$, ** $p < 0.01$, *** $p < 0.001$.

To further evaluate the role of Gm16685 in regulating macrophage polarization, Gm16685 was knocked down with Gm16685-shRNAs in RAW264.7 cells and then treated with LPS or without LPS. The knockdown of Gm16685 decreased the expression levels of IL-1β ($t_{(4)} = 4.22$, $p = 0.007$), IL-6 ($t_{(4)} = 6.20$, $p = 0.002$), CD86 ($t_{(4)} = 7.39$, $p = 0.0008$), and CD80 ($t_{(4)} = 5.02$, $p = 0.004$) compared with the control cells, while a higher level of IL-10 ($t_{(4)}$ $= 29.64$, $p < 0.001$) was observed. Moreover, the downregulation of Gm16685 blocked the LPS-induced upregulation of these M1-phenotype genes and reversed the downregulation of M2-phenotype genes in RAW264.7 cells (Figure 4c). However, the overexpression of Gm16685 increased M1-phenotype gene expression (such as IL-6 ($t_{(4)} = 13.01$, $p = 0.0001$) and CD86 ($t_{(4)} = 4.84$, $p = 0.004$)) and decreased M2-phenotype gene expression (Figure 4d). Simultaneously, treatment with shRNA-Gm16685 caused significant increases in IL-10 (LPS, $t_{(4)} = 2.63$, $p = 0.029$; control, $t_{(4)} = 2.48$, $p = 0.033$) in the supernatant of the cell cultures. In contrast, IL-1β (LPS, $t_{(4)} = 4.99$, $p = 0.004$; control, $t_{(4)} = 2.24$, $p = 0.044$) was decreased in the shRNA-Gm16685-treated group compared to the control groups (Figure 4e). Thus, these data strongly suggest that the knockdown of Gm16685 promoted the polarization of macrophages to the M2 phenotype.

3.5. Further Study on the Mechanism of Action of Gm16685

FISH revealed that Gm16685 was predominantly localized in the cytoplasm of normal cultured RAW264.7 cells. Actin, used as a positive control, was mainly located in the cytoplasm (Figure 5a). Moreover, the nucleocytoplasmic separation experiment showed consistent results with FISH analyses (Figure 5b). miRDB predicted the miRNA that might interact with Gm16685 (Figure 5c). Through qRT-PCR analyses, it was found that miR-205-5p was upregulated after Gm16685 was knocked down in RAW264.7 cells ($t_{(4)} = 10.95$, $p < 0.001$) (Figure 5d). The interaction between Gm16685 and miR-205-5p was confirmed by luciferase reporter assay (Figure 5e,f). After overexpression of miR-205-5p, the expression of Gm16685 was significantly downregulated ($t_{(4)} = 4.24$, $p = 0.007$) (Figure 5g). The expression of inflammatory factors was also downregulated (Figure 5h).

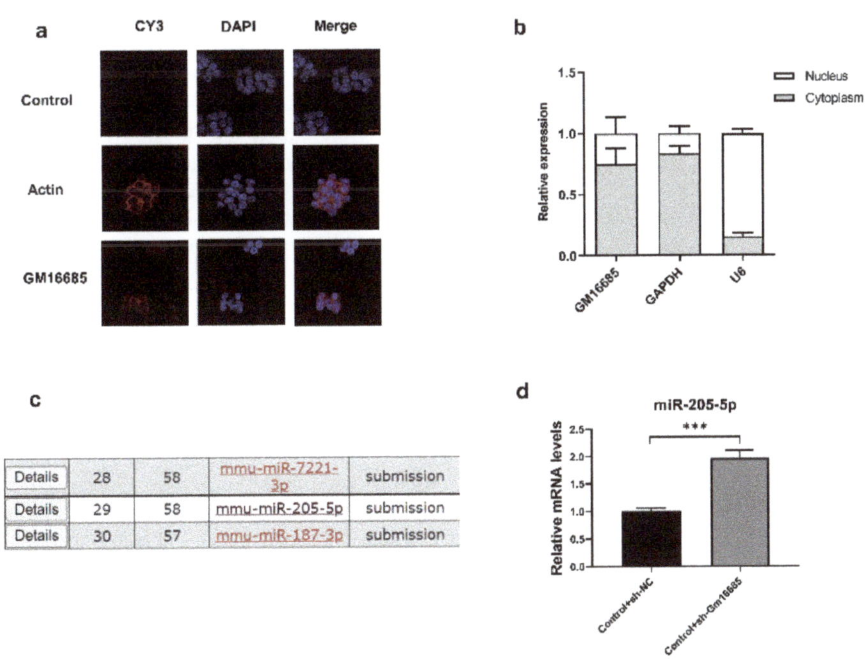

Figure 5. *Cont.*

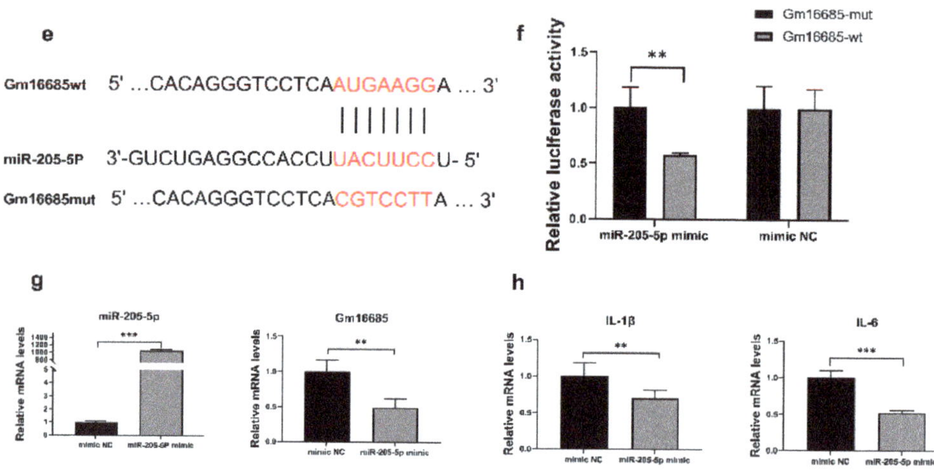

Figure 5. Further study on the mechanism of action of Gm16685: (**a**) Representative images of RNA fluorescence *in situ* hybridization (FISH) present Gm16685 (in red) localization in the macrophage cytoplasm. Actin mRNA is used as a reference and is present in the cytoplasm, scale bar = 20 μm. (**b**) qRT-PCR analysis of Gm16685 abundance in the nuclear and cytoplasmic fractions of RAW264.7 cells. (**c**) miRDB predicted the miRNAs that might interact with Gm16685. (**d**) qRT-PCR analysis of miR-205-5p expression after knockdown of Gm16685 in RAW264.7 cells. (**e**) The sequence fragments of wild type (wt) and mutant type (mut) of Gm16685 contained the binding site of miR-205-5p or the mutation binding site of miR-205-5p. (**f**) A luciferase reporter assay was used to determine the interaction between Gm16685 and miR-205-5p. (**g**) The expression of Gm16685 after overexpression of miR-205-5p was measured by qRT-PCR. (**h**) The expression of proinflammatory cytokine mRNAs after overexpression of miR-205-5p was measured by qRT-PCR. Data are presented as the mean ± standard deviation, n = 3. ** $p < 0.01$, *** $p < 0.001$.

4. Discussion

The regulatory functions of lncRNAs associated with liver disease have garnered increasing attention [30]. Regarding the regulation of lncRNAs on the biological function of liver cells, previous studies have focused on the progression of liver fibrosis and liver cancer. For instance, silencing lnc-LFAR1 decreases TGFβ-induced hepatocyte apoptosis, impairs HSC activation *in vitro*, and alleviates liver fibrosis induced by CCL4 [31]. lnc-LFAR1 plays a crucial role in regulating the activation and pyroptosis of macrophages, thereby providing an underlying target against inflammation-related diseases, including hepatic fibrosis [32]. lnc-TLNC1 promotes hepatocellular carcinoma progression and metastasis through the TLNC1-TPR-p53 axis [33]. However, the role of lncRNAs in liver lesions caused by schistosomiasis has rarely been studied.

One of the most serious clinicopathological features is liver egg granuloma caused by egg deposition during schistosomiasis [34]. Many lncRNAs are differentially expressed in the liver of *S. japonicum*-infected mice [25]. Hence, we first detected the expression profile of lncRNAs in the liver samples of mice infected with *S. japonicum* at different time points using RNA-seq. Many important lncRNA molecules related to the pathogenesis and progression of egg granulomas of *S. japonicum* were identified. Among the 12 genes selected, the expression levels of 0610039H22Rik, 5830428M24Rik, NR028126, H19, MEG3, and Mirt were upregulated after 45 days of schistosomiasis infection, which is the same as the study by Xia et al. [25]. According to the literature, H19 was also significantly induced by bile acids in mouse cholangiocytes [35]. Gm16685, also known as NAIL, was significantly upregulated in the colitis site of patients with ulcerative colitis. It can directly regulate the initiation and progression of colitis [36]. In this study, qRT-PCR results showed that the expression level of Gm16685 was increased 6-fold and 61-fold after 24 and 45 days of infection, respectively. Combining the results of high-throughput sequencing and qRT-PCR experiments, we found that Gm16685 was highly expressed in the liver of mice 45 days after *S. japonicum* infection. The expression level of Gm16685 was highly correlated with NF-κB activity. Genistein attenuates schistosomiasis liver granulomas by inhibiting the activity of NF-κB [34]. Therefore, our studies suggested that Gm16685 may exert effects on the pathogenesis of hepatic disease in schistosomiasis.

Interestingly, the knockdown of Gm16685 in mice infected with *S. japonicum* reduced the symptoms of egg granulomas, suggesting that Gm16685 participated in the pathogenesis of hepatic disease in schistosomiasis. Granuloma is an organized aggregation of macrophages and other immune cells, and the granuloma response is characterized by macrophage activation and transformation [37]. M1 macrophages can induce a chronic inflammatory state, whereas M2 macrophages can reduce this state in diseased tissues [38]. Our results suggest that the knockdown of Gm16685 in mouse liver can reduce the expression of M1 macrophage-related indicators and increase the expression of M2 macrophage-related indicators after *S. japonicum* infection, thus indicating that knocking down lncRNA-Gm16685 can inhibit M1 macrophage polarization and promote M2 macrophage polarization.

To further study the effects of Gm16685 on macrophages, hepatocytes, hepatic stellate cells, and macrophages were isolated from mouse models infected with *S. japonicum*. Gm16685 was significantly upregulated in macrophages. We also found that LPS treatment promoted Gm16685 expression in mouse macrophage RAW264.7 cells. The findings showed the proinflammatory effects of Gm16685 in macrophages. Meanwhile, in mouse macrophage lines, Gm16685 knockdown was consistent with the changing trend in tissues, with the downregulation of proinflammatory cytokines and upregulation of anti-inflammatory cytokines.

To date, a series of studies proved that subcellular localization of lncRNAs was critical to their function [39–41]. lncRNAs located in the cytoplasm can interfere with the posttranslational modification of the protein, serve as bait for miRNA, or regulate the translation, stabilization, and degradation of mRNA [17]. In our study, FISH and nucleocytoplasmic separation experiments revealed that Gm16685 was predominantly localized in the cytoplasm of normal cultured RAW264.7 cells. Thus, we performed a preliminary study on the mechanism of action of Gm16685. After the knockdown of Gm16685, the expression of miRNA-205-5p was upregulated. Previous research reported that the function of miRNA-205-5p was related to inflammation, and overexpression of miRNA-205-5p could inhibit inflammation [42]. Therefore, Gm16685 could repress conversion by acting as an miRNA-205-5p sponge.

5. Conclusions

This study established an expression profile of lncRNAs in the livers of mice with schistosomiasis at different time points. Notably, lncRNA-Gm16685 expression was upregulated in the livers of schistosomiasis mice. Gm16685 knockdown promoted M2 macrophage polarization and alleviated *S. japonicum*-induced hepatic granulomas. Additionally, Gm16685 promoted M1 macrophage polarization by regulating miR-205-5p. Thus, our study may provide a new target for schistosomiasis japonica treatment.

Supplementary Materials: The following supporting information can be downloaded at https://www.mdpi.com/article/10.3390/microorganisms11030796/s1, Table S1: Primer sequences of genes.

Author Contributions: Conceptualization, J.S. and C.R.; methodology and investigation, R.Z. and X.T.; software, H.L.; validation, C.X.; resources, W.S. and M.L.; writing—original draft preparation, R.Z.; writing—review and editing, C.R. and S.D.; supervision, J.S. and S.D.; data curation, B.D., N.X. and P.W.; funding acquisition, C.R. and J.S. All authors have read and agreed to the published version of the manuscript.

Funding: This research was funded by National Natural Science Foundation of China (81802027 and 82072305) and the Scientific Research of BSKY from Anhui Medical University (XJ201723).

Institutional Review Board Statement: The animal study protocol was approved by the Institutional Animal Care and Use Committee at Anhui Medical University (approval number: LLSC20180442).

Data Availability Statement: The datasets generated and/or analyzed during the current study are available from the corresponding author on reasonable request.

Conflicts of Interest: The authors declare no conflict of interest.

Sample Availability: Samples are available from the authors.

References

1. Boniatti, J.; Januskaite, P.; Fonseca, L.B.D.; Vicosa, A.L.; Amendoeira, F.C.; Tuleu, C.; Basit, A.W.; Goyanes, A.; Re, M.I. Direct Powder Extrusion 3D Printing of Praziquantel to Overcome Neglected Disease Formulation Challenges in Paediatric Populations. *Pharmaceutics* **2021**, *13*, 1114. [CrossRef]
2. Cardoso, I.A.; de Souza, A.K.L.; Burgess, A.M.G.; Chalmers, I.W.; Hoffmann, K.F.; Nonato, M.C. Characterization of class II fumarase from Schistosoma mansoni provides the molecular basis for selective inhibition. *Int. J. Biol. Macromol.* **2021**, *175*, 406–421. [CrossRef]
3. Abdel Aziz, N.; Musaigwa, F.; Mosala, P.; Berkiks, I.; Brombacher, F. Type 2 immunity: A two-edged sword in schistosomiasis immunopathology. *Trends Immunol.* **2022**, *43*, 657–673. [CrossRef]
4. Liang, S.; Abe, E.M.; Zhou, X.N. Integrating ecological approaches to interrupt schistosomiasis transmission: Opportunities and challenges. *Infect. Dis. Poverty* **2018**, *7*, 124. [CrossRef]
5. Lam, H.Y.P.; Liang, T.R.; Peng, S.Y. Ameliorative effects of Schisandrin B on Schistosoma mansoni-induced hepatic fibrosis *in vivo*. *PLoS Negl. Trop. Dis.* **2021**, *15*, e0009554. [CrossRef]
6. Colley, D.G.; Bustinduy, A.L.; Secor, W.E.; King, C.H. Human schistosomiasis. *Lancet* **2014**, *383*, 2253–2264. [CrossRef]
7. Fung, H.S.; Siu, J.C.; Fan, T.W.; Lai, K.C.; Kwan, T.L. Hepatic schistosomiasis. *Hong Kong Med. J.* **2009**, *15*, 75–76.
8. Pesce, J.T.; Ramalingam, T.R.; Mentink-Kane, M.M.; Wilson, M.S.; El Kasmi, K.C.; Smith, A.M.; Thompson, R.W.; Cheever, A.W.; Murray, P.J.; Wynn, T.A. Arginase-1-expressing macrophages suppress Th2 cytokine-driven inflammation and fibrosis. *PLoS Pathog.* **2009**, *5*, e1000371. [CrossRef]
9. Vannella, K.M.; Barron, L.; Borthwick, L.A.; Kindrachuk, K.N.; Narasimhan, P.B.; Hart, K.M.; Thompson, R.W.; White, S.; Cheever, A.W.; Ramalingam, T.R.; et al. Incomplete deletion of IL-4Ralpha by LysM(Cre) reveals distinct subsets of M2 macrophages controlling inflammation and fibrosis in chronic schistosomiasis. *PLoS Pathog.* **2014**, *10*, e1004372. [CrossRef]
10. Takaki, K.K.; Rinaldi, G.; Berriman, M.; Pagan, A.J.; Ramakrishnan, L. Schistosoma mansoni Eggs Modulate the Timing of Granuloma Formation to Promote Transmission. *Cell Host Microbe* **2021**, *29*, 58–67.e55. [CrossRef]
11. Zhao, J.; Ling, L.; Zhu, W.; Ying, T.; Yu, T.; Sun, M.; Zhu, X.; Du, Y.; Zhang, L. M1/M2 re-polarization of kaempferol biomimetic NPs in anti-inflammatory therapy of atherosclerosis. *J. Control Release* **2022**, *353*, 1068–1083. [CrossRef]
12. Rhee, I. Diverse macrophages polarization in tumor microenvironment. *Arch. Pharm. Res.* **2016**, *39*, 1588–1596. [CrossRef]
13. Faure-Dupuy, S.; Delphin, M.; Aillot, L.; Dimier, L.; Lebosse, F.; Fresquet, J.; Parent, R.; Matter, M.S.; Rivoire, M.; Bendriss-Vermare, N.; et al. Hepatitis B virus-induced modulation of liver macrophage function promotes hepatocyte infection. *J. Hepatol.* **2019**, *71*, 1086–1098. [CrossRef]
14. Barron, L.; Wynn, T.A. Macrophage activation governs schistosomiasis-induced inflammation and fibrosis. *Eur. J. Immunol.* **2011**, *41*, 2509–2514. [CrossRef]
15. Feyder, M.; Goff, L.A. Investigating long noncoding RNAs using animal models. *J. Clin. Investig.* **2016**, *126*, 2783–2791. [CrossRef]
16. Begolli, R.; Sideris, N.; Giakountis, A. LncRNAs as Chromatin Regulators in Cancer: From Molecular Function to Clinical Potential. *Cancers* **2019**, *11*, 1524. [CrossRef]
17. Bridges, M.C.; Daulagala, A.C.; Kourtidis, A. LNCcation: lncRNA localization and function. *J. Cell Biol.* **2021**, *220*, e202009045. [CrossRef]
18. Carlevaro-Fita, J.; Lanzos, A.; Feuerbach, L.; Hong, C.; Mas-Ponte, D.; Pedersen, J.S.; Drivers, P.; Functional Interpretation, G.; Johnson, R.; Consortium, P. Cancer LncRNA Census reveals evidence for deep functional conservation of long noncoding RNAs in tumorigenesis. *Commun. Biol.* **2020**, *3*, 56. [CrossRef]
19. Wang, Y.L.; Huang, L.Z.; Wang, Y.; Luo, W.S.; Li, F.; Xiao, J.; Qin, S.R.; Wang, Z.Y.; Song, X.W.; Wang, Y.; et al. Single-cell RNA-sequencing analysis identifies host long noncoding RNA MAMDC2-AS1 as a co-factor for HSV-1 nuclear transport. *Int. J. Biol. Sci.* **2020**, *16*, 1586–1603. [CrossRef]
20. Ma, Y.; Zhang, J.; Wen, L.; Lin, A. Membrane-lipid associated lncRNA: A new regulator in cancer signaling. *Cancer Lett.* **2018**, *419*, 27–29. [CrossRef]
21. Radhakrishnan, R.; Kowluru, R.A. Long Noncoding RNA MALAT1 and Regulation of the Antioxidant Defense System in Diabetic Retinopathy. *Diabetes* **2021**, *70*, 227–239. [CrossRef]
22. Yang, Z.; Jiang, S.; Shang, J.; Jiang, Y.; Dai, Y.; Xu, B.; Yu, Y.; Liang, Z.; Yang, Y. LncRNA: Shedding light on mechanisms and opportunities in fibrosis and aging. *Ageing Res. Rev.* **2019**, *52*, 17–31. [CrossRef]

23. Hu, B.; Ma, X.; Fu, P.; Sun, Q.; Tang, W.; Sun, H.; Yang, Z.; Yu, M.; Zhou, J.; Fan, J.; et al. The mRNA-miRNA-lncRNA Regulatory Network and Factors Associated with Prognosis Prediction of Hepatocellular Carcinoma. *Genom. Proteom. Bioinform.* **2021**, *19*, 913–925. [CrossRef]
24. DiStefano, J.K.; Gerhard, G.S. Long Noncoding RNAs and Human Liver Disease. *Annu. Rev. Pathol.* **2022**, *17*, 1–21. [CrossRef]
25. Xia, T.; Giri, B.R.; Liu, J.; Du, P.; Li, X.; Li, X.; Li, S.; Cheng, G. RNA sequencing analysis of altered expression of long noncoding RNAs associated with Schistosoma japonicum infection in the murine liver and spleen. *Parasites Vectors* **2020**, *13*, 601. [CrossRef]
26. Kho, C. Tough Decoy-Mediated Cardiac Gene Suppression. *Methods Mol. Biol.* **2022**, *2573*, 13–30. [CrossRef]
27. Guo, K.; Qian, K.; Shi, Y.; Sun, T.; Wang, Z. LncRNA-MIAT promotes thyroid cancer progression and function as ceRNA to target EZH2 by sponging miR-150-5p. *Cell Death Dis.* **2021**, *12*, 1097. [CrossRef]
28. Chuah, C.; Jones, M.K.; Burke, M.L.; Owen, H.C.; Anthony, B.J.; McManus, D.P.; Ramm, G.A.; Gobert, G.N. Spatial and temporal transcriptomics of Schistosoma japonicum-induced hepatic granuloma formation reveals novel roles for neutrophils. *J. Leukoc. Biol.* **2013**, *94*, 353–365. [CrossRef]
29. Mederacke, I.; Dapito, D.H.; Affo, S.; Uchinami, H.; Schwabe, R.F. High-yield and high-purity isolation of hepatic stellate cells from normal and fibrotic mouse livers. *Nature Protocols* **2015**, *10*, 305–315. [CrossRef]
30. He, Z.; Yang, D.Y.; Fan, X.L.; Zhang, M.W.; Li, Y.; Gu, X.B.; Yang, M.Y. The Roles and Mechanisms of lncRNAs in Liver Fibrosis. *Int. J. Mol. Sci.* **2020**, *21*, 1482. [CrossRef]
31. Zhang, K.; Han, X.H.; Zhang, Z.; Zheng, L.N.; Hu, Z.M.; Yao, Q.B.; Cui, H.M.; Shu, G.M.; Si, M.J.; Li, C.; et al. The liver-enriched lnc-LFAR1 promotes liver fibrosis by activating TGF beta and Notch pathways. *Nature Commun.* **2017**, *8*, 44. [CrossRef]
32. Zhang, K.; Shi, Z.M.; Zhang, M.X.; Dong, X.Y.; Zheng, L.N.; Li, G.T.; Han, X.H.; Yao, Z.; Han, T.; Hong, W. Silencing lncRNA Lfar1 alleviates the classical activation and pyoptosis of macrophage in hepatic fibrosis. *Cell Death Dis.* **2020**, *11*, 132. [CrossRef] [PubMed]
33. Yuan, K.; Lan, J.; Xu, L.; Feng, X.; Liao, H.; Xie, K.; Wu, H.; Zeng, Y. Long noncoding RNA TLNC1 promotes the growth and metastasis of liver cancer via inhibition of p53 signaling. *Mol. Cancer* **2022**, *21*, 105. [CrossRef]
34. Wan, C.; Jin, F.; Du, Y.; Yang, K.; Yao, L.; Mei, Z.; Huang, W. Genistein improves schistosomiasis liver granuloma and fibrosis via dampening NF-kB signaling in mice. *Parasitol. Res.* **2017**, *116*, 1165–1174. [CrossRef]
35. Song, Y.; Liu, C.; Liu, X.; Trottier, J.; Beaudoin, M.; Zhang, L.; Pope, C.; Peng, G.; Barbier, O.; Zhong, X.; et al. H19 promotes cholestatic liver fibrosis by preventing ZEB1-mediated inhibition of epithelial cell adhesion molecule. *Hepatology* **2017**, *66*, 1183–1196. [CrossRef] [PubMed]
36. Akincilar, S.C.; Wu, L.L.; Ng, Q.F.; Chua, J.Y.H.; Unal, B.; Noda, T.; Chor, W.H.J.; Ikawa, M.; Tergaonkar, V. NAIL: An evolutionarily conserved lncRNA essential for licensing coordinated activation of p38 and NF kappa B in colitis. *Gut* **2021**, *70*, 1857–1871. [CrossRef] [PubMed]
37. Pagan, A.J.; Ramakrishnan, L. The Formation and Function of Granulomas. *Annu. Rev. Immunol.* **2018**, *36*, 639–665. [CrossRef]
38. Yunna, C.; Mengru, H.; Lei, W.; Weidong, C. Macrophage M1/M2 polarization. *Eur. J. Pharmacol.* **2020**, *877*, 173090. [CrossRef]
39. Chen, L.L. Linking Long Noncoding RNA Localization and Function. *Trends Biochem. Sci.* **2016**, *41*, 761–772. [CrossRef]
40. Guo, C.J.; Ma, X.K.; Xing, Y.H.; Zheng, C.C.; Xu, Y.F.; Shan, L.; Zhang, J.; Wang, S.; Wang, Y.; Carmichael, G.G.; et al. Distinct Processing of lncRNAs Contributes to Non-conserved Functions in Stem Cells. *Cell* **2020**, *181*, 621–636.e622. [CrossRef]
41. Sang, L.; Yang, L.; Ge, Q.; Xie, S.; Zhou, T.; Lin, A. Subcellular distribution, localization, and function of noncoding RNAs. *Wiley Interdiscip. Rev. RNA* **2022**, *13*, e1729. [CrossRef] [PubMed]
42. Yang, J.J.; Zhao, Y.H.; Yin, K.W.; Zhang, X.Q.; Liu, J. Dexmedetomidine inhibits inflammatory response and oxidative stress through regulating miR-205-5p by targeting HMGB1 in cerebral ischemic/reperfusion. *Immunopharmacol. Immunotoxicol.* **2021**, *43*, 478–486. [CrossRef] [PubMed]

microorganisms

MDPI

Article

IL-27 Signaling Promotes Th1 Response by Downregulating IL-10 Production in DCs during Chlamydial Respiratory Infection

Jiajia Zeng †, Shuaini Yang †, Yuqing Tuo, Xiaoyu Zha, Ruoyuan Sun, Tingsha Lu, Hong Zhang, Lu Tan, Sai Qiao and Hong Bai *

Key Laboratory of Immune Microenvironment and Disease (Ministry of Education), Department of Immunology, School of Basic Medical Sciences, Tianjin Medical University, Tianjin 300070, China
* Correspondence: hongbai25@tmu.edu.cn
† These authors have contributed equally to this work.

Abstract: *Chlamydia trachomatis* usually causes mucosal infections, bringing considerable morbidity and socioeconomic burden worldwide. We previously revealed that IL-27/IL-27R mediates protection against chlamydial invasion by promoting a protective Th1 response and suppressing neutrophilic inflammation. Here, we used the mouse model of *Chlamydia muridarum* (*C. muridarum*) respiratory infections to further investigate the impact of IL-27 signaling in the DCs-regulated immune response, since an elevated IL-27/IL-27R expression in DCs was identified following chlamydial infection. An adoptive transfer of *Chlamydia muridarum*-stimulated DCs to wild-type mice approach was subsequently used, and the donor-DCs-promoted resistance with a higher Th1 response against chlamydial infection was attenuated when DCs lacking IL-27R were used as donor cells. Flow cytometry analysis revealed the suppression of IL-27 signaling on DCs phenotypic maturation. A further functional maturation analysis of DCs revealed that IL-27 signaling restricted the protein and mRNA expression of IL-10 from DCs following infection. Thus, these findings suggest that IL-27 signaling could support the Th1 response via inhibiting IL-10 production in DCs, thus mediating the protective host defense against chlamydial respiratory infection.

Keywords: chlamydial infection; IL-27; dendritic cells; maturation; Th1 response; IL-10

Citation: Zeng, J.; Yang, S.; Tuo, Y.; Zha, X.; Sun, R.; Lu, T.; Zhang, H.; Tan, L.; Qiao, S.; Bai, H. IL-27 Signaling Promotes Th1 Response by Downregulating IL-10 Production in DCs during Chlamydial Respiratory Infection. *Microorganisms* **2023**, *11*, 604. https://doi.org/10.3390/microorganisms11030604

Academic Editors: Shengxi Chen and Fabio Zicker

Received: 2 February 2023
Revised: 22 February 2023
Accepted: 23 February 2023
Published: 27 February 2023

1. Introduction

As an obligate intracellular pathogen, *Chlamydia trachomatis* (*C. trachomatis*) frequently causes mucosal infection through the eye, genital tract, and respiratory system, resulting in trachoma, urogenital tract infection, and neonatal pneumonia [1–3]. Moreover, *C. trachomatis* infection can induce immunopathological damage by delayed-type hypersensitivities, such as lymphogranuloma of sexually transmitted diseases [4]. The host typically displays a weak defense with persistent and asymptomatic infection; therefore, recognizing and treating infected hosts is key to preventing irreparable tissue damage and controlling chlamydial illnesses. *C. trachomatis* mouse pneumonitis biovar (*C. muridarum*) is a natural pathogen of mouse pneumonia, and the mouse model of *C. muridarum* infection is widely used for chlamydial infection in vivo [5,6]. The *C. muridarum*-induced Th1 response exerts a protective effect through the production of interferon-gamma (IFN-γ), which was reported to enhance the phagocytic capacity of macrophages and directly inhibit the growth of chlamydia by promoting the synthesis of inducible nitric oxide synthase (iNOS) and indoleamine 2, 3-dioxygenase (IDO) [7,8]. By modifying the DC function, IL-17/Th17 was found to boost the Th1 response in our earlier research [3], but excessive IL-17 could cause immunopathological effects by eliciting neutrophilic inflammation [9].

Dendritic cells (DCs), as innate immune cells and professional antigen-presenting cells, are crucial for resisting pathogen invasion and initiating and regulating adaptive immunity [10,11]. As reported, DCs actively participate in the regulation of the chlamydia-induced host response. Chlamydia might remain alive and infectious in DCs [12]. Chlamydia inclusion bodies were demonstrated to stimulate the maturation and generation of the IL-12 and TNF of DCs; adoptive transferring DCs enhanced resistance against chlamydia [13]. We previously showed that IL-17 neutralization inhibits the chlamydia-specific Th1 response and decreases DC's phenotypic maturation and IL-12 secretion. Furthermore, adoptively transferring DCs isolated from IL-17-neutralized mice induces a diminished protective effect on recipient mice, suggesting that the IL-17/Th17's promoting effect on the Th1 response is achieved by regulating the DC activity and function [3]. Our partner Dr. Yang further compared the contributions of DC subsets in chlamydia defense. Following the adoption of CD8α^+ splenic DCs, Th1-related cytokines were considerably higher than those following the adoption of CD8α^- DCs [14]. Similarly, mice-receiving CD103$^+$ LDCs exhibited better protection than CD11b^{hi} LDCs recipients, correlated with more robust Th1/Th17 responses [15]. The above documents indicate that DCs pose significant regulatory effects in *C. trachomatis* infection.

Known as the "Immunological Playmakers", the Interleukin (IL)-12 cytokines family is the only family of heterodimeric cytokines, which endows them with several unique connections and functional interactions [16]. IL-27, a member of the IL-12 family, is composed of a unique IL-12p35-like protein, IL-27p28, and a protein related to IL-12p40 encoded by the Epstein–Barr-virus-induced gene 3 (EBI3) [17,18]. The dominant cellular sources of IL-27 are reported to be DCs and macrophages, although other myeloid cell populations such as monocytes, microglia, and epithelial cells also express IL-27 [19]. IL-27 signals through a heterodimeric surface receptor consist of glycoprotein 130 (GP130) and IL-27Rα (also known as WSX-1) [20]. Being implicated in a variety of infectious diseases, IL-27 is recognized as both immunosuppressive and immunostimulatory in innate and adaptive immunity. In research on post-influenza pneumococcal pneumonia, IL-27 sensitized mice to secondary pneumococcal infection, attributable to the inhibition of IL-17A production in $\gamma\delta$ T cells in a STAT1-dependent manner [21]. In *Clostridium difficile* colitis infection, recombinant IL-27 administration caused an increased IFN-γ with a decreased mortality in mice [22]. In secondary *Staphylococcus aureus* pneumonia following influenza infection, IL-27 caused an enhanced susceptibility by inducing IL-10 and suppressing IL-17 [23]. In *C. muridarum*-induced respiratory infection, our previous study found that IL-27/IL-27R protects the host by promoting the Th1 response and suppressing excessive IL-17-induced neutrophilic inflammation [9]. As DCs are involved in chlamydial infection and its regulation on T cell responses, we hypothesized that DCs participate in the IL-27/IL-27R's protective effect during chlamydial respiratory infection.

Using the *C. muridarum*-infected mouse model, we were able to identify how IL-27/IL-27R affected DCs maturation as well as the associated immunological effect and mechanism. Though IL-27 signaling inhibited the accumulation and phenotypic maturation of DCs, the adoptive transfer experiment revealed that WSX-1$^{-/-}$-DC recipients experienced a worsened disease with weaker Th1 levels. Further exploration found an inhibitory effect of IL-27/IL-27R on IL-10 production by DCs. Our findings thus indicate that IL-27 signaling could promote the protective Th1 response by regulating the cytokine production by DCs in *C. muridarum* infection, which may provide novel insights into the treatment of intracellular bacterial infection.

2. Materials and Methods

2.1. Animals

Wild-type (WT) female C57BL/6 mice (6–8 weeks old, 18–20 g) were purchased from Huafukang Biotechnology Co., Ltd. (Beijing, China). Female IL-27Rα-deficient (WSX-1$^{-/-}$) mice on the C57BL/6 background were granted by Professor Yin Zhinan (Nankai University, China). All mice were fed in specific pathogen-free (SPF) circumstances at Tianjin Medical University (Tianjin, China). All procedures involving animals were approved by the Animal

Ethical and Welfare Committee (AEWC) of Tianjin Medical University (number of animal permit: SYXK: 2016-0012, approval date: 7 March 2018).

2.2. Bacterial Strains, Chlamydial Infection and Administration of rIL-27

Chlamydia muridarum (*C. muridarum*) strains were first gifted from Dr. Xi Yang (the University of Manitoba, Canada) and cultured, and then expanded in our laboratory as previously described [15]. To induce respiratory infection, WT and WSX-$1^{-/-}$ mice were anesthetized with isoflurane, intranasally (i.n.) inoculated with 1×10^3 inclusion-forming units (IFUs) of *C. muridarum* in 40 μL sucrose–phosphate–glutamic acid buffer (SPG), with the uninfected mice (0 d) inoculated 40 μL SPG buffer as a control. For administration of recombinant murine IL-27 (rIL-27) (BioLegend, San Diego, CA, USA), WT mice were anesthetized, i.n. inoculated with 0.2 μg rIL-27 in 30 μL PBS the day before and days 0, 2, 4, and 6 after *C. muridarum* infection or 30 μL aseptic protein-free PBS on the same schedules as the control. Following infection, body weight was recorded daily, and mice were sacrificed at the indicated time points.

2.3. Lung, Spleen, and Lymph Nodes (LNs) Single-Cell Preparation

C. muridarum-infected lung tissues were minced and digested with 2 mg/mL collagenase XI (Sigma-Aldrich, St. Louis, MO, USA) in PRMI-1640 for 55 min at 37 °C. Tissue fibers and erythrocytes were successively removed by 35% Percoll (GE Healthcare, Chicago, London, UK) and ACK Lysis buffer (Tris-NH_4Cl). *C. muridarum*-infected spleens and LNs were directly ground and filtered through 70 μm cell strainers, and erythrocytes were lysed using an ACK lysing buffer. These single cells were resuspended in the complete RPMI-1640 medium (RPMI-1640 supplemented with 10% heat-inactivated FBS, 0.05 mmol/L 2-mercaptoethanol, 100 U/mL penicillin, and 0.1 mg/mL streptomycin), stained with trypan blue, and then counted under light microscopy for further analysis.

2.4. Antibodies and Flow Cytometry

All of the antibodies were indicated in Table S1 in the Supplementary Material. Single-cell suspensions were resuspended in FACS buffer (PBS with 2% FBS) and incubated with Fc receptor block Abs (anti-CD16/CD32 monoclonal Abs; eBioscience, San Diego, CA, USA) at 4 °C for 20 min in the dark. For cell surface staining, cells were stained with conjugated antibodies specific for cell surface markers at 4 °C for 30 min in the dark. For intracellular cytokine staining, cells were firstly stimulated with cocktails including PMA (50 ng/mL, Solarbio, Beijing, China), Ionomycin (1 μg/mL, MCE, Monmouth Junction, NJ, USA), and brefeldin A (10 μg/mL, BioLegend) at 37 °C for 5–6 h. Following a blocking step similar to cell surface staining, cells were stained for surface antigens and fixed with the Fixation buffer (Biolegend). Fixed cells were washed and permeabilized with 1 × Intracellular Staining Perm Wash Buffer (Biolegend) followed by incubation with anti-IFN-γ, anti-IL-4, or anti-IL-17A mAb for 30 min at room temperature. Finally, cells were suspended with FACS buffer, detected using a FACSCanto II flow cytometer (BD Biosciences, Franklin, NJ, USA), and analyzed by the FlowJo V10 software.

2.5. Generation and Stimulation of Bone-Marrow-Derived DCs (BMDCs)

Bone marrows were flushed from the femurs and tibias of naïve C57BL/6 mice, clusters were dispersed by vigorous pipetting, and erythrocytes were lysed by ACK buffer. Cells were washed and cultured at 1–2 × 10^6 /mL in the complete RPMI-1640 medium supplemented with 20 ng/mL murine GM-CSF (R&D Systems, Madrid, Spain) and 10 ng/mL murine IL-4 (PeproTech, Cranbury, NJ, USA) in 6-well cell culture plates. A total of 20 ng/mL rIL-27 was added or not added to the BMDC culture media to compare the impact of IL-27/IL-27R. On day 3, the entire medium was discarded and replaced with fresh medium. On day 5, half of the culture supernatant was removed and replaced with 2 mL fresh culture medium. On day 7, non-adherent cells were collected and transferred to 10 mm cell culture dishes. On day 9, suspending cells were harvested and the purity of CD11c$^+$ BMDCs was >80% as determined by flow cytometry. For maturation, BMDCs were stimulated with 1 μg/mL LPS (Sigma-Aldrich) for 48 h. With the presence of LPS, 1×10^6 IFUs of *C. muridarum* was added to the medium. Finally, cells were harvested for flow cytometry and PCR, and the supernatants were harvested for ELISA.

2.6. Splenic DCs Isolation and Adoptive Transfer

Mice were i.n. infected with *C. muridarum* and spleens were harvested on day 7 following infection. The spleens were minced and digested with 1 mg/mL of collagenase D (Sigma-Aldrich) in RPMI-1640 for 25 min at 37 °C. The cell suspensions were filtered through 70 μm cell strainers, and erythrocytes were lysed by ACK lysing buffer. After the lysis, the cell suspension was incubated with CD11c microbeads (Miltenyi Biotec, Auburn, CA, USA) for 15 min at 4 °C and resuspended in MACS buffer (PBS with 2% FBS and 2 mM EDTA). After incubation, the splenic cells passed through magnetic columns for positive selection of CD11c$^+$ cells. The purity of the sorted splenic DCs was more than 90% as detected by flow cytometry. Freshly isolated CD11c$^+$ spleen cells were adoptively transferred to syngeneic naïve C57BL/6 mice through tail vein injection at 1×10^6 DCs/mouse in 200 μL PBS, with mice injected 200 μL PBS taken as control. Two hours later, the recipient mice were intranasally inoculated with 40 μL SPG containing 1×10^3 IFUs *C. muridarum*. Following infection, body weight was recorded daily, and recipient mice were sacrificed on day 14 post-infection (p.i.).

2.7. Co-Culture of Splenic DCs and CD4$^+$ T Cells

CD11c$^+$ DCs were isolated from the spleen of *C. muridarum*-immunized mice on day 7 p.i. or naïve C57BL/6 mice using magnetic beads and MACS columns through positive selection (Miltenyi Biotec, Auburn, CA, USA) as described above. CD4$^+$ T cells were isolated from the spleen of *C. muridarum*-immunized mice on day 7 p.i. or naïve C57BL/6 mice using magnetic beads and MACS columns through negative selection (Miltenyi Biotec, Auburn, CA, USA) as described previously [24]. Briefly, spleen single-cell suspensions were mixed with Biotin-Antibody Cocktail, then incubated with Anti-Biotin MicroBeads. Labeled cells passed through magnetic columns for negative selection of CD3$^+$ CD4$^+$ T cells. As shown by flow cytometric analysis, the purity of the CD4$^+$ T cells was more than 96%. CD4$^+$ T cells were co-cultured with CD11c$^+$ DCs (DC/T cell ratio, 1:5; CD11c$^+$ DCs: 1×10^5, CD4$^+$ T cells: 5×10^5) in 96-well plates in the presence or absence of UV-sterilized *C. muridarum* (UV-Cm) for 48h. Cell supernatants were collected for analyzing IFN-γ, IL-4, and IL-17 production by ELISA.

2.8. Pulmonary Chlamydial Loads

Chlamydia IFUs were detected to determine the growth of *C. muridarum* in infected lungs as described previously [9,25]. Briefly, the lung tissues were homogenized aseptically in SPG buffer, acquired lung homogenates were centrifuged at 3000 rpm for 30 min, and diluted supernatants were added into a confluent monolayer of HeLa cells and incubated at 37 °C for 2 h. After being cultured in complete RPMI-1640 medium for 24 h, media were removed and infected cells were fixed with methanol for 10 min. The cells were then stained with anti-Chlamydia LPS antibody (Invitrogen, Carlsbad, CA, USA) and HRP-conjugated goat anti-mouse IgG secondary Abs (Solarbio) and developed with the substrate (4-chloro-1-naphthol; Solarbio). The inclusion bodies were counted under the microscope [100X] to calculate IFUs per sample.

2.9. Histology Analysis and Semi-Quantitative Pathological Scoring

For histopathological analyses, *C. muridarum*-infected lung tissues were fixed in 10% formalin and routinely embedded in paraffin, sectioned (5 μm), and stained with hematoxylin and eosin (H&E). The pathological changes were evaluated by the semi-quantitative histology score in a blinded manner as described previously [9] and introduced in Table S2 in the electronic Supplementary Material.

2.10. ELISA

The spleen DCs isolated from *C. muridarum*-infected WT and WSX-$1^{-/-}$ mice on day 7 p.i., T cells co-cultured with DC, and BMDCs under different treatments were cultured to test *C. muridarum*-driven cytokine production. Briefly, these cell suspensions were cultured at a concentration of 5×10^5 splenic DCs/well, 5×10^5 $CD4^+$ T cells with 1×10^5 DCs/well, and 1×10^6 BMDCs/well, respectively. After 48/72 h culture, the supernatants were collected for cytokine detection. IFN-γ, IL-4, IL-17, IL-12p40, IL-10, and IL-6 productions were measured by ELISA according to the manufacturer's instructions (Invitrogen). The OD values were read at 450 nm on the Epoch microplate reader (BioTeK, Winooski, VT, USA).

2.11. RNA Extraction and Quantitative Real-Time PCR (qPCR)

Total RNA of sorted spleen DCs and BMDCs was extracted by TRIzol reagent (Invitrogen) and reverse transcription was performed using the cDNA Synthesis SuperMix (TransGen Biotech, Beijing, China) according to the manufacturer's instructions. qPCR was further performed using RealStar Fast SYBR qPCR Mix (GenStar, Beijing, China) on Light Cycler 96 (Roche, Basel, Switzerland). The mRNA expression of target genes was presented as the "fold change" relative to that of control samples, with fold changes calculated by the $2^{-\Delta\Delta Ct}$ method using the mouse β-actin gene as an endogenous control. The PCR primer sequences used are shown in Table S3 in the electronic Supplementary Material.

2.12. Statistical Analysis

Data are represented as means ± SD and were assessed with GraphPad Prism 9. The significance of differences in two different groups was analyzed by unpaired Student's *t*-test or two-way ANOVA followed by Šidák's multiple comparisons test; differences between multiple groups were analyzed by one-way ANOVA followed by Dunnett's multiple comparisons test. *p* values < 0.05 were considered significant (* or # $p < 0.05$, ** or ## $p < 0.01$, *** or ### $p < 0.001$, **** or #### $p < 0.0001$).

3. Results

3.1. C. muridarum Respiratory Infection Induces IL-27/IL-27R Expression of DC

We recently discovered that *C. muridarum* infection aggravated the disease in WSX-$1^{-/-}$ mice, with a reduced Th1 response and IL-17-induced neutrophilic inflammation [9]. Given that DCs boosted the Th1 and Th17 responses against *C. muridarum* invasion, it was conceivable that DCs contributed to the protective effects of IL-27 signaling. Using flow cytometry, we identified pulmonary and splenic DCs as $CD45^+$ $CD11c^+$ $MHCII^+$ cells (Figure S1 in the Supplementary Material). The expressions of WSX-1 on pulmonary and splenic DCs gradually increased from day 3 post-infection (p.i.) and peaked on day 7 p.i. (Figure 1A,B). After sorting the splenic DCs, qPCR was conducted to detect the mRNA expression of IL-27 (composed of p28 and EBI3) and IL-27R (composed of WSX-1 and GP130) (Figure 1C,D). The mRNA level of p28 was higher on day 3 p.i. and WSX-1 increased on day 7 p.i. compared with uninfected controls. To further confirm these findings, we cultured BMDCs and identified an increased WSX-1 expression after the *C. muridarum* challenge, with the LPS treatment taken as the positive control (Figure 1E). These data indicated an increased expression of IL-27 and IL-27R on DCs following infection, suggesting that DCs may participate in the IL-27/IL-27R-modulated host defense against *C. muridarum* infection.

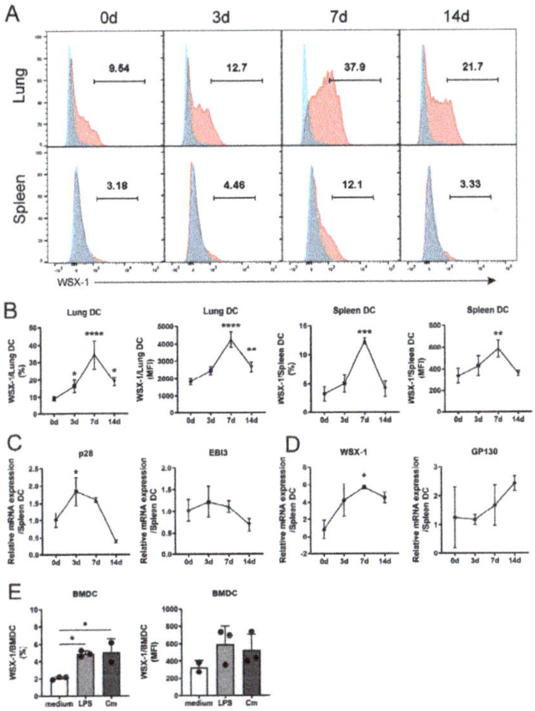

Figure 1. The expression of IL-27/IL-27R in DCs following *Chlamydia muridarum* (*C. muridarum* or Cm) respiratory infection. The wild-type (WT) mice were intranasally infected with 1×10^3 inclusion-forming units (IFUs) *C. muridarum* and euthanized at days 0, 3, 7, and 14 post-infection (p.i.). (**A**,**B**) Lung and spleen single cells were prepared, and the WSX-1 expressions (red) with fluorescence minus one (FMO) control (blue) were analyzed by flow cytometry based on gated DCs as described in Figure S1 (**A**). The percentages and mean fluorescence intensity (MFI) of WSX-1 positive cells were indicated (**B**). (**C**,**D**) Spleen $CD11c^+$ cells were purified using MACS $CD11c^+$ cell isolation column, total RNA was prepared, and the mRNA expression of p28, EBI3, WSX-1, and GP130 was detected by quantitative real-time PCR (qPCR). (**E**) The mouse bone marrow dendritic cells (BMDCs) were induced from naïve C57BL/6 mice, grouped as noninfected BMDCs (white), LPS-stimulated BMDCs (light grey), or Cm-infected BMDCs (deep grey) as described in the methods, stained with anti WSX-1, and analyzed by flow cytometry. Data are shown as means ± SD, representing one of three independent experiments (n = 2–4/group/experiments). Statistical significances of differences are determined by one-way ANOVA. * $p < 0.05$, ** $p < 0.01$, *** $p < 0.001$, **** $p < 0.0001$. Abbreviations: DCs, dendritic cells; WSX-1, interleukin 27 receptor subunit alpha (IL-27Rα); *C. muridarum*/Cm, *Chlamydia muridarum*; IFUs, inclusion-forming units; p.i., post-infection; MFI, mean fluorescence intensity; EBI3, Epstein–Barr virus-induced gene 3; GP130, glycoprotein 130; LPS, lipopolysaccharide; BMDC, bone marrow dendritic cell.

3.2. Adoptive Transfer of WSX-1$^{-/-}$ DC Reduces Protection against C. muridarum Respiratory Infection

To assess the impact of IL-27/IL-27R-modulated DCs, we compared the immunopathological consequences of recipient mice adoptively transferred with splenic DCs derived from WT (WT-DC) and WSX-1$^{-/-}$ (WSX-1$^{-/-}$-DC) mice. As shown in Figure 2A,B, the WT-DCs and WSX-1$^{-/-}$-DCs were sorted with purity validated (Figure 2B) and adoptively transferred into recipients, with mice receiving PBS used as control. As reported, the adoptive transfer of DCs results in a protective effect, as WT-DC recipients developed fewer body weight losses and decreased IFUs (indicating chlamydial burden) than the PBS group

(Figure 2C,D). Compared with WT-DC recipients, WSX-1 deficiency caused greater body weight losses and higher IFUs, suggesting the negative regulation of WSX-1$^{-/-}$-DCs on the host defense. In agreement with the above conditions, *C. muridarum* respiratory infection induced a pulmonary inflammatory pathology, while WSX-1$^{-/-}$-DC recipients displayed graver inflammatory grades with more massive inflammatory cell infiltration than WT-DC recipients (Figure 2E,F). These results show that the adoptive transfer of WSX-1$^{-/-}$-DCs failed to protect the recipients from *C. muridarum* infection more effectively than WT-DCs.

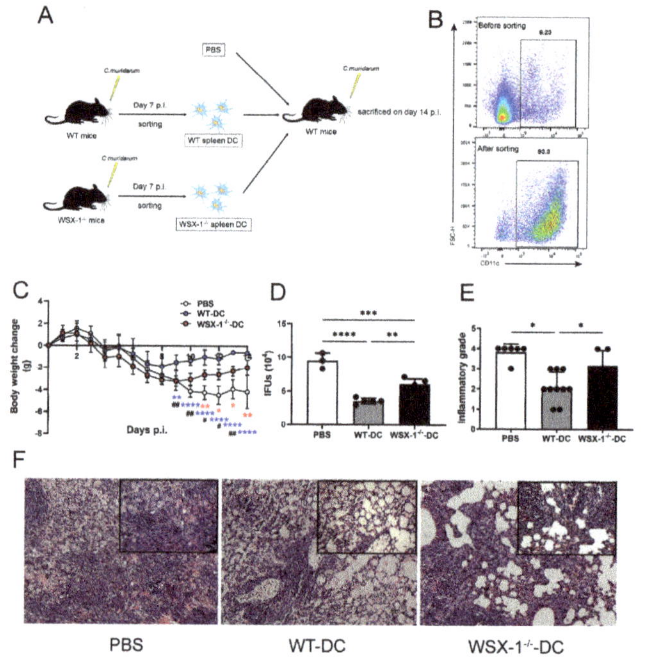

Figure 2. Disease progression of recipient mice after receiving DCs in *C. muridarum* lung infection. (**A**,**B**) WT mice and WSX-1-deficient (WSX-1$^{-/-}$) mice were intranasally infected and sacrificed on day 7 p.i., spleen cells were isolated, and DCs were sorted with purity higher than 90% (**B**). Either 1×10^6 WT-DCs or WSX-1$^{-/-}$-DCs in 200 μL PBS were adoptively transferred to naïve C57BL/6 recipient mice by tail vein injection (the control group was given the same dose of PBS). Two hours after the adoptive transfer, the recipient mice were intranasally infected with 1×10^3 IFUs *C. muridarum*. (**C**) The body weight changes were monitored daily and the mice were sacrificed on day 14 p.i. The red asterisks (*) represent statistical significances between the PBS and WSX-1$^{-/-}$-DC recipients, blue asterisks (*) represent statistical significances between the PBS and WT-DC recipients, and the pounds (#) represent statistical significances between the WT-DC and WSX-1$^{-/-}$-DC recipients. (**D**) The lung homogenates were prepared for determining chlamydia IFUs. (**E**,**F**) Lung sections were stained by H&E, the inflammatory grades were scored with the semi-quantitative pathological scoring method (**E**), and the representative pathological changes were captured under light microscopy [100X and 200X] (**F**). Data are shown as means ± SD, representing one (**C**,**D**) or two (**E**) of three independent experiments (n = 3–5/group/experiments). Statistical significances of differences are determined by two-way ANOVA (**C**) and one-way ANOVA (**C–E**). * p or # < 0.05, ** p or ## < 0.01, *** $p < 0.001$, **** $p < 0.0001$. Abbreviations: *C. muridarum*, *Chlamydia muridarum*; WSX-1$^{-/-}$, IL-27Rα-deficient; p.i., post-infection; DC, dendritic cells; IFUs, inclusion-forming units; WT-DC, DC derived from wild-type mice; WSX-1$^{-/-}$-DC, DC derived from WSX-1$^{-/-}$ mice.

3.3. Adoptive Transfer of WSX-1$^{-/-}$ DC Reduces Th1 Responses following C. muridarum Respiratory Infection

The Th1 and Th17 responses were reported to be protective against *C. muridarum* infection, whereas the Th2 response causes pathological effects [3,26]. Importantly, we reported that IL-27/IL-27R suppresses neutrophilic inflammation [9]. Thus, Th responses and neutrophil levels of recipient mice were detected to clarify the regulating target of WSX-1$^{-/-}$-DC. Consistent with disease conditions, the percentage of lung Th1 cells was higher in WSX-1$^{-/-}$-DC recipients than in PBS recipients but lower in WT-DC recipients (Figure 3A). Similar changes were observed in the spleen. Figure 3B–D illustrate that WSX-1 deficiency elicited an increased neutrophil infiltration in the lung, whereas the Th2 and Th17 responses were comparable between the WT-DC and WSX-1$^{-/-}$-DC recipients.

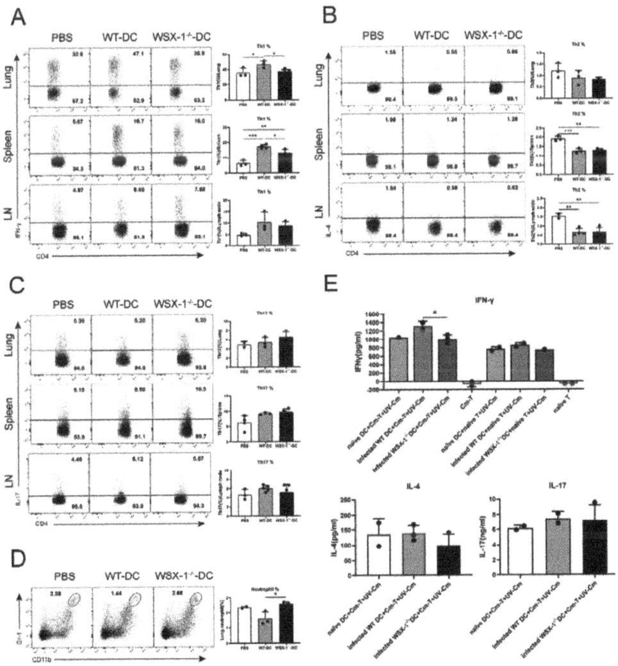

Figure 3. Th responses of recipient mice after receiving DCs in *C. muridarum* lung infection. The recipient mice were sacrificed on day 14 p.i. Lung, spleen, and lymph node single cells were prepared, and the Th responses were analyzed by flow cytometry and ELISA. (**A–D**) Representative flow cytometric images of IFN-γ-producing CD4^{+} T cells (Th1 cells, (**A**)), IL-4-producing CD4^{+} T cells (Th2 cells, (**B**)), IL-17-producing CD4^{+} T cells (Th17 cells, (**C**)), CD11b^{+} Ly-6G^{+} cells (neutrophils, (**D**)), and summaries of their percentages in the lung, spleen, and lymph node are shown. (**E**) Spleen DCs were sorted from naïve C57BL/6 mice, *C. muridarum*-infected C57BL/6 mice, and *C. muridarum*-infected WSX-1$^{-/-}$ mice (naïve DC, infected WT-DC, and infected WSX-1$^{-/-}$-DC) by magnetic microbeads. Spleen CD4^{+} T cells were sorted from naïve C57BL/6 mice and *C. muridarum*-infected mice (naïve T and Cm-T) by magnetic microbeads. The sorted DCs and CD4^{+} T cells were co-cultured for 48 h in the presence or absence of UV-sterilized *C. muridarum* (UV-Cm). IFN-γ, IL-4, and IL-17 in the culture supernatants were determined by ELISA. Data are shown as means ± SD, representing one of three independent experiments (n = 2–5/group/experiments). Statistical significances of differences are determined by one-way ANOVA. * $p < 0.05$, ** $p < 0.01$, *** $p < 0.001$. Abbreviations: *C. muridarum*/Cm, *Chlamydia muridarum*; WSX-1$^{-/-}$, IL-27Rα-deficient; p.i., post-infection; LN, lymph node; UV-Cm, UV-sterilized *Chlamydia muridarum*; WT-DC, DC derived from wild-type mice; WSX-1$^{-/-}$-DC, DC derived from WSX-1$^{-/-}$ mice.

To further confirm the impact of IL-27/IL-27R-modulated DCs on Th responses, splenic DCs and CD4$^+$ T cells were sorted and co-cultured. As indicated in Figure 3E, in the absence of UV-Cm, neither the naïve T cells nor Cm-T cells alone induced IFN-γ secretion. Under UV-Cm stimulation, there were no differences in the amount of IFN-γ when naïve CD4$^+$ T cells and DCs were co-cultured. When *C. muridarum*-immunized T cells were co-cultured with DCs, infected WSX-1$^{-/-}$-DCs secreted fewer IFN-γ than infected WT-DCs did. The IL-4 and IL-17 production in the co-culture system does not differ significantly from the flow cytometry analysis in Figure 3B,C. All of these results suggested that DCs derived from WSX-1$^{-/-}$ mice could weaken the host immunity against *C. muridarum* respiratory infection by suppressing the Th1 response.

3.4. Deficiency of WSX-1 Promotes the Accumulation and Phenotypic Maturation of DCs following Chlamydial Respiratory Infection

Next, we compared DCs infiltration and maturation between *C. muridarum*-infected WT and WSX-1$^{-/-}$ mice for clarification of the mechanism by which WSX-1$^{-/-}$-DCs inhibit the Th1 response. Following *C. muridarum* invasion, DCs accumulated to the lung and spleen, while WSX-1$^{-/-}$ mice exhibited more pulmonary DCs on days 3 and 7 p.i. and increased splenic DCs in both the percentage and absolute number on day 7 p.i. (Figure 4A,B).

DC maturation is determined by changes at the phenotypic and functional levels [27–29]. When DCs increase the expression of costimulatory molecules CD40, CD80, and CD86, as well as the MHC class II molecule, phenotypic maturation occurs [30]. DCs at an elevated functional level typically secrete cytokines, including inflammatory cytokines and immunosuppressive cytokines [31]. When WSX-1$^{-/-}$ mice were compared to WT mice on day 7 p.i., flow cytometry analysis showed a higher percentage and mean fluorescence intensity (MFI) of CD80 on pulmonary DCs in WSX-1$^{-/-}$ mice (Figure 4C,D). Similarly, on day 7 p.i., splenic DCs from WSX-1$^{-/-}$ mice displayed increased percentages and MFIs of CD40, CD80, and CD86 (Figure 4E,F). These findings indicated that the accumulation and phenotypic maturation of DCs during *C. muridarum* infection may be constrained by IL-27 signaling.

3.5. rIL-27 Treatment Suppresses the Accumulation and Phenotypic Maturation of DC following Chlamydial Respiratory Infection

To further confirm the impact of IL-27 signaling on DCs accumulation and phenotypic maturation, recombinant murine IL-27 (rIL-27) or an equal volume of PBS was given to naïve C57BL/6 mice that were subsequently challenged with *C. muridarum*. On day 7 following infection, the rIL-27-treated mice developed a lower percentage of DCs than control mice in the spleen (Figure 5A). On day 14 following infection, a reduced CD40 expression on pulmonary DCs was observed in rIL-27-treated mice when compared to PBS-treated mice (Figure 5B). Similar patterns of costimulatory molecules, such as CD40, CD80, and CD86 on splenic DCs and CD40 on BMDCs, were seen in the groups treated with rIL-27 (Figure 5C,D). Collectively, these data further support the role of IL-27 signaling in limiting DCs accumulation and phenotypic maturation during *C. muridarum* infection.

3.6. IL-27 Signaling Inhibits IL-10 Production in DCs following C. muridarum Respiratory Infection

DCs-derived cytokines are essential for modulating the adaptive immune response. Next, we investigated the DC cytokines profile to identify its functional maturation [31,32]. Among them, DCs-derived IL-12 is essential for the generation of the Th1 cell response, whereas IL-10 and IL-6 can inhibit it [33]. The uninfected and infected spleen DCs were sorted from the wild-type and WSX-1$^{-/-}$ mice for qPCR and ELISA tests. While mRNA expressions of IL-12p40 and IL-12p35 are comparable, the qPCR results demonstrated an increased IL-10 and IL-6 expression in WSX-1$^{-/-}$-DCs compared to WT-DCs on day 7 p.i. (Figure 6A). Further, in the ELISA analysis, the amount of IL-10 in the culture supernatant of WSX-1$^{-/-}$-DCs was significantly higher than WT-DCs (Figure 6B). To further confirm the cytokines production, BMDCs were treated or not treated with rIL-27, and the qPCR results show that *C. muridarum* infection induced an increased IL-12p40, IL-12p35, IL-10, and IL-6 mRNA expression, and that rIL-27 treatment inhibited the level of IL-10 and IL-6 (Figure 6C). ELISA further supported the inhibition of rIL-27 to IL-10 and IL-6 production on the protein level (Figure 6D). Together, these data demonstrate a mechanistic connection between IL-27 signaling and DCs-secreted IL-10, which suppress the protective Th1 response during *C. muridarum* lung infection.

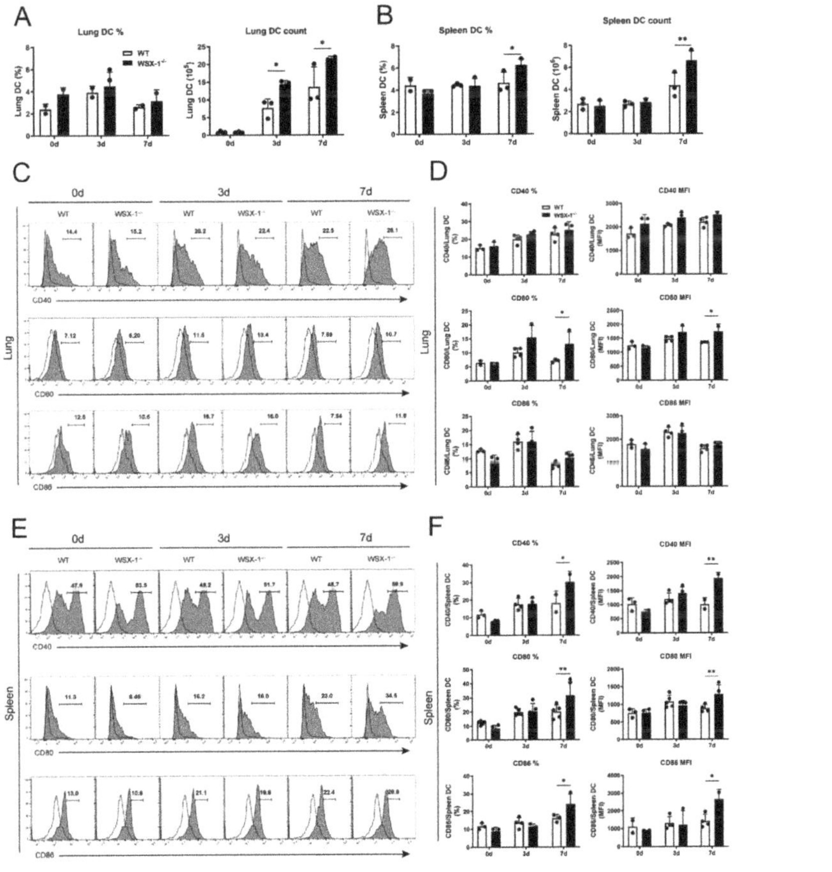

Figure 4. DCs accumulation and phenotypic maturation in WSX-1$^{-/-}$ mice following *C. muridarum* respiratory infection. The lung and spleen single cells of WT and WSX-1$^{-/-}$ mice on days 0, 3, and 7 p.i. were prepared. (**A**,**B**) The frequencies and total numbers of pulmonary DCs (**A**) and splenic DCs (**B**) were detected by flow cytometry. (**C**–**F**) The CD40, CD80, and CD86 expressions (shaded histogram) with fluorescence minus one (FMO) control (solid lines) on pulmonary DCs (**C**) and splenic DCs (**E**) were analyzed by flow cytometry. The percentages and MFI of positive cells at indicated times after infection were indicated (**D**,**F**). Data are shown as means ± SD, representing one of three independent experiments (n = 3–5/group/experiments). Statistical significances of differences are determined by two-way ANOVA. * $p < 0.05$, ** $p < 0.01$. Abbreviations: *C. muridarum*/Cm, *Chlamydia muridarum*; WSX-1$^{-/-}$, IL-27Rα-deficient; p.i., post-infection; FMO, fluorescence minus one; MFI: mean fluorescence intensity.

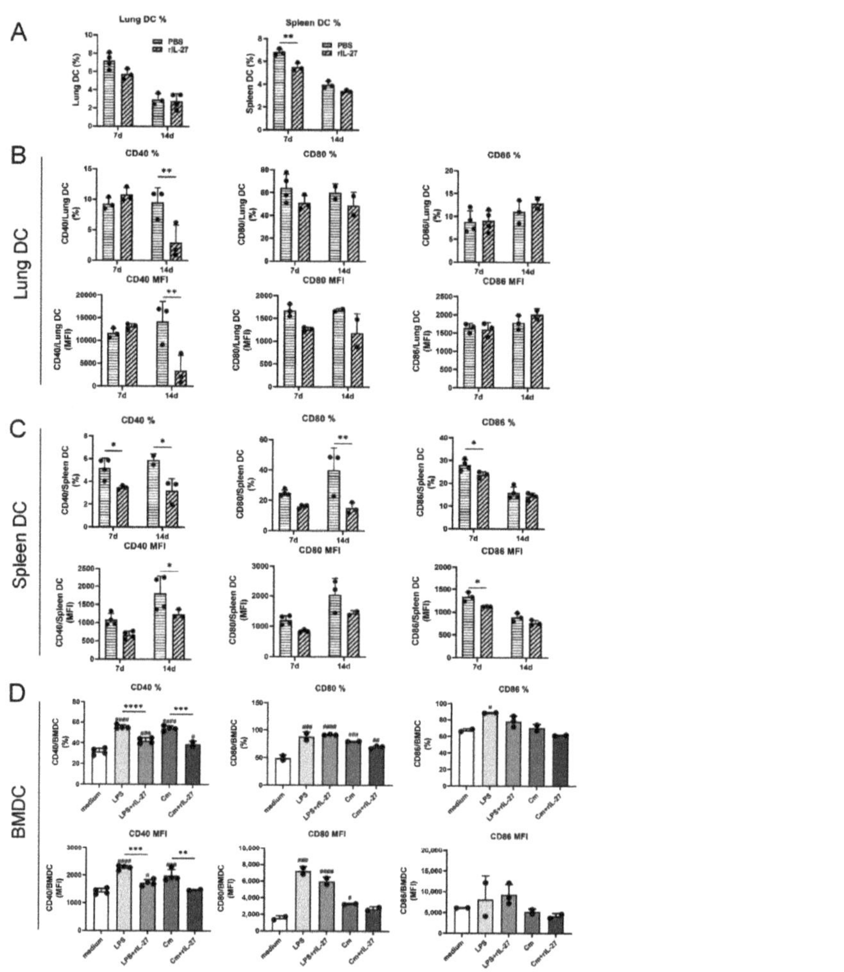

Figure 5. DCs accumulation and phenotypic maturation after rIL-27 administration against *C. muridarum* respiratory infection. (**A**–**C**) For recombinant murine IL-27 (rIL-27) administration, WT mice were inoculated intranasally with 0.2 μg rIL-27 in 30 μL PBS on the day before and days 0, 2, 4, and 6 after *C. muridarum* infection, and the control group was given 30 μL sterile PBS in the same schedule. The lung and spleen single cells on days 7 and 14 p.i. were prepared. (**A**) The percentages of pulmonary DCs and splenic DCs were detected by flow cytometry. Costimulatory molecules of DCs were further detected and altered expression of CD40, CD80, and CD86 on pulmonary DCs (**B**) and splenic DCs (**C**) was shown. (**D**) BMDCs were induced from naïve C57BL/6 mice and grouped as noninfected BMDCs, LPS-stimulated BMDCs, LPS-stimulated BMDCs with rIL-27 administration, Cm-infected BMDCs, and Cm-infected BMDCs with rIL-27 administration. Analyzed by flow cytometry, the percentages and MFI of CD40, CD80, and CD86 on BMDCs were indicated. Asterisks (*) represent statistical significances between two groups indicated by the lines, and the pounds (#) represent statistical significances compared to noninfected BMDCs. Data are shown as means ± SD, representing one of three independent experiments (n = 2–4/group/experiments). Statistical significances of differences are determined by two-way ANOVA (**A**–**C**) and one-way ANOVA (**D**). * or # $p < 0.05$, ** or ## $p < 0.01$, *** or ### $p < 0.001$, **** or #### $p < 0.0001$. Abbreviations: *C. muridarum*/Cm, *Chlamydia muridarum*; rIL-27, recombinant IL-27; p.i., post-infection; MFI: mean fluorescence intensity; BMDC, bone marrow dendritic cell.

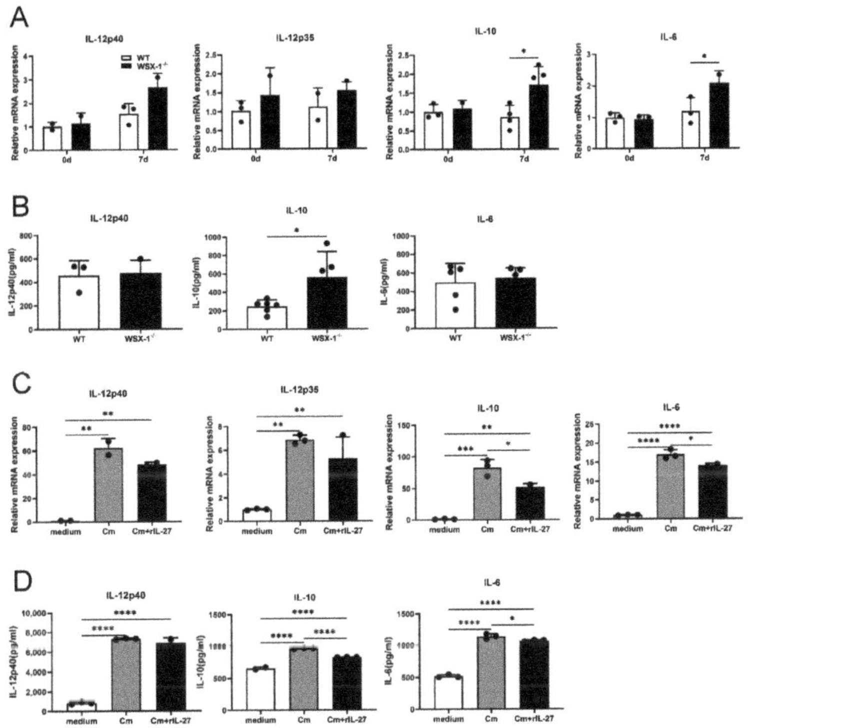

Figure 6. DCs cytokines production after WSX-1 deficiency or rIL-27 administration against *C. muridarum* infection. (**A**,**B**) The spleen DCs of WT and WSX-1$^{-/-}$ mice on days 0 and 7 p.i. were sorted as described in the methods. The total RNA of sorted DCs was prepared for qPCR, and the mRNA expressions of cytokines IL-12p40, IL-12p35, IL-10, and IL-6 were shown (**A**). The culture supernatants of sorted DCs were prepared for ELISA, and IL-12p40, IL-10, and IL-6 productions in the culture supernatant were determined on day 7 p.i. (**B**). BMDCs were induced from naïve C57BL/6 mice, grouped as noninfected control BMDCs (white), Cm-infected BMDCs (grey), and Cm-infected BMDCs with rIL-27 administration (black), analyzed by qPCR (**C**) and ELISA (**D**) for cytokine production. Data are shown as means ± SD, representing one of three independent experiments (n = 3–6 /group/experiments). Statistical significances of differences are determined by two-way ANOVA (**A**), unpaired Student's *t*-test (**B**), and one-way ANOVA (**C**,**D**). * $p < 0.05$, ** $p < 0.01$, *** $p < 0.001$, **** $p < 0.0001$. Abbreviations: *C. muridarum*/Cm, *Chlamydia muridarum*; WSX-1$^{-/-}$, IL-27Rα-deficient; p.i., post-infection; BMDC, bone marrow dendritic cell; rIL-27, recombinant IL-27.

4. Discussion

IL-27 signaling was recently demonstrated to mediate protective immunity against chlamydial infection [9]. Here, we further determined that *C. muridarum* infection induces the WSX-1 expression on DCs, and the adoptive transfer of WSX-1$^{-/-}$ mice-derived DCs failed to mediate effective protection, attributable to a diminished Th1 response. Further investigation revealed that WSX-1$^{-/-}$-DCs increased the IL-10 production compared to WT-DCs, and that the rIL-27 administration reduced IL-10 secretion, suggesting the inhibition of IL-27 on the Th1 response through raising IL-10 production from DCs. This study firstly reveals IL-27/IL-27R's encouraging role in the Th1 response by regulating DCs cytokines during chlamydial infection.

The WSX-1$^{-/-}$ mice displayed a lower Th1 response and a higher Th17 response with excessive neutrophilic inflammation [9], so we compared the Th1/2/17 response and neutrophil infiltration between WSX-1$^{-/-}$-DC and WT-DC recipient mice. Consistently with a reduced protection, the WSX-1$^{-/-}$ DC recipient conferred a lower Th1 response (Figure 3A) and increased neutrophil infiltration, as well as no difference with the PBS recipient (Figure 3D) and comparable Th2 and Th17 levels. These data suggest that though IL-27 signaling has a regulatory role on the Th1 and Th17 response, IL-27/IL-27R-modulated DCs primarily affect the Th1 response. These data potently indicate the link between DC and T-cell-mediated responses, and, more importantly, also explain why we focus on measuring Th1-related cytokines in subsequent experiments.

Identified as an anti-inflammatory factor and a Th2 cytokine, IL-10 was engaged in inflammatory responses and controlling the Th1 response [34,35]. During acute *Clostridioides difficile* infection, IL-10$^{-/-}$ mice exhibited an elevated IL-22 response and decreased mortality [36]. Sarah E. Clark et al. reported that, in *Streptococcus pneumoniae (S. pneumoniae)* infection, NK-dependent IL-10 restricts myeloid cell recruitment, causing an increased bacterial growth and exacerbated infection [37]. In *Mycobacterium tuberculosis* (*M. tuberculosis*) infection, early IL-10 overexpression reduces the T cells' capacity for parenchymal migration and interacting with infected phagocytic cells, impeding the limitation of *M. tuberculosis* growth [38]. In *Schistosomiasis mansoni* infection, a lack of IL-10 caused dysregulated Th1 responses, resulting in aggravated granulomatous inflammation and an increased mortality in the acute phase [39]. In *C. trachomatis* respiratory infection, IL-10 was reported to inhibit the priming and expansion of Th1 responses and contribute to the fibrotic reaction following infection [40]. In this study, in line with the weaker Th1 response of WSX-1$^{-/-}$-DC recipients in the adoptive transfer experiments, IL-10 expression and secretion in WSX-1$^{-/-}$ splenic DCs were significantly higher than in WT mice, suggesting that IL-27/IL-27R may mediate protective Th1 immunity by suppressing the IL-10 production of DCs in *C. muridarum* respiratory infection.

During *C. muridarum* respiratory infection, previous studies showed that Th1-promoting DCs display a more mature phenotype [3,41]; however, this study prompted more phenotypically mature WSX-1$^{-/-}$-DCs to pose a reduced potency to trigger a powerful Th1 response. Indeed, one of the crucial features of DC biology is its maturation. The immature DCs recognize and capture antigens whereas the mature ones are ineffective; in contrast, the mature DCs are potent stimulators for T-cell-mediated responses, making DCs key regulators of both protective immune responses and tolerance to autoantigens [42,43]. As detailed in the *Results*, the phenotypic maturation is not identical to functional maturation, and the capacity of DCs to produce cytokines is a critical step, as DCs participate in regulating adaptive immunity through producing IL-12, IL-10, and IL-6 [32,44,45]. Combined with the consistency between the higher IL-10 level and weaker Th1 response in the WSX-1$^{-/-}$ mice, we tend to draw the conclusion that, though impacting the mature phenotype, IL-27/IL-27R primarily regulates the Th1 response through modulating the DC cytokines profile in chlamydial infection.

Two major subsets of murine DCs, cDC1 (characterized as CD8α^+ splenic DCs or CD103$^+$ CD11b$^-$ pulmonary DCs) and cDC2 (characterized as CD8α^- splenic DCs or CD103$^-$ CD11b$^+$ pulmonary DCs), reported posing distinct potentials for inducing protective immunity against *C. muridarum* respiratory infection [14,15], were detected. As shown in Figures S2–S4 in the electronic Supplementary Material, comparable infiltrations of CD8α^+ DCs and CD8α^- DCs were identified in the spleens between infected WT and WSX-1$^{-/-}$ mice. CD8α^+ DCs and CD8α^- DCs from WSX-1$^{-/-}$ mice both express higher CD40, CD80, and CD86 than WT mice, indicating that both subsets have a more mature phenotype. These data might imply that WSX-1 depletion cannot induce a significant difference in the infiltration or phenotypic maturation of splenic DC subsets. Further research is needed for the infiltration and maturation of pulmonary DC subsets, the impacts of IL-27/IL-27R on the functional maturation of DC subsets, and, more significantly, the DC-subsets-driven immune regulations to improve this work.

5. Conclusions

Collectively, we demonstrate that IL-27 signaling promotes the protective Th1 response by regulating cytokine production from DCs in chlamydial infection. Based on the IL-27/IL-27R-mediated protection by regulating the Th response proved in the recent work, this study further clarified the potential mechanism, contributing to the in-depth understanding of the pathogenesis of chlamydial infection, which will provide new insights into an effective immunotherapy for controlling bacterial respiratory infections.

Supplementary Materials: The following supporting information can be downloaded at: https://www.mdpi.com/article/10.3390/microorganisms11030604/s1. Table S1: Sequences of primers used in this study; Table S2: Antibodies used for flow cytometry in this study; Table S3: The scoring of semi-quantitative pathological; Figure S1: The flow cytometry gate strategy for the analysis of dendritic cells (DCs); Figure S2: Infiltration of DC subsets in the spleen of WSX-1$^{-/-}$ mice following *Chlamydia muridarum* (*C. muridarum*) respiratory infection; Figure S3: Costimulatory molecules expression on CD8α^+ DCs in the spleen of WSX-1$^{-/-}$ mice following *C. muridarum* respiratory infection; Figure S4: Costimulatory molecules expression on CD8α^- DCs in the spleen of WSX-1$^{-/-}$ mice following *C. muridarum* respiratory infection.

Author Contributions: Conceptualization, J.Z.; methodology, J.Z. and S.Y.; formal analysis, J.Z.; investigation, J.Z., S.Y., Y.T., and X.Z.; data curation, R.S. and T.L.; writing—original draft preparation, J.Z. and Y.T.; writing—review and editing, J.Z. and S.Y.; visualization, H.Z., L.T., and S.Q.; supervision, project administration, H.B.; funding acquisition, H.B., S.Q. All authors have read and agreed to the published version of the manuscript.

Funding: This research was funded by the National Natural Science Foundation of China, grant number 31870889 and 31070797 to H.B., the Key Program of Tianjin Science and Technology Commission, grant number 19JCZDJC35800 and 15JCZDJC34900 to H.B., the National Natural Science Foundation of China (NSFC) Youth Program, grant number 81901596 to S.Q., and the National Natural Science Foundation of China, grant number 81900041 to J.Y.

Data Availability Statement: The raw data used to support the findings of this study are available from the corresponding author upon request.

Acknowledgments: We thank Zhinan Yin (Nankai University, China) for the generous provision of WSX-1$^{-/-}$ mice, Xi Yang (the University of Manitoba, Canada) for *Chlamydial muridarum* strains, and Jing Yang (Tianjin Nankai Hospital, China) for the funding.

Conflicts of Interest: The authors declare no conflict of interest.

References

1. Murthy, A.K.; Li, W.; Ramsey, K.H. Immunopathogenesis of Chlamydial Infections. *Curr. Top. Microbiol. Immunol.* **2018**, *412*, 183–215. [CrossRef] [PubMed]
2. Klockner, A.; Buhl, H.; Viollier, P.; Henrichfreise, B. Deconstructing the Chlamydial Cell Wall. *Curr. Top. Microbiol. Immunol.* **2018**, *412*, 1–33. [CrossRef] [PubMed]
3. Bai, H.; Cheng, J.; Gao, X.; Joyee, A.G.; Fan, Y.; Wang, S.; Jiao, L.; Yao, Z.; Yang, X. IL-17/Th17 promotes type 1 T cell immunity against pulmonary intracellular bacterial infection through modulating dendritic cell function. *J. Immunol.* **2009**, *183*, 5886–5895. [CrossRef] [PubMed]
4. Ziklo, N.; Huston, W.M.; Hocking, J.S.; Timms, P. Chlamydia trachomatis Genital Tract Infections: When Host Immune Response and the Microbiome Collide. *Trends Microbiol.* **2016**, *24*, 750–765. [CrossRef]
5. De Clercq, E.; Kalmar, I.; Vanrompay, D. Animal models for studying female genital tract infection with Chlamydia trachomatis. *Infect. Immun.* **2013**, *81*, 3060–3067. [CrossRef]
6. Giebel, A.M.; Hu, S.; Rajaram, K.; Finethy, R.; Toh, E.; Brothwell, J.A.; Morrison, S.G.; Suchland, R.J.; Stein, B.D.; Coers, J.; et al. Genetic Screen in Chlamydia muridarum Reveals Role for an Interferon-Induced Host Cell Death Program in Antimicrobial Inclusion Rupture. *mBio* **2019**, *10*, e00385-19. [CrossRef]
7. Roan, N.R.; Starnbach, M.N. Immune-mediated control of Chlamydia infection. *Cell Microbiol.* **2008**, *10*, 9–19. [CrossRef]
8. Rottenberg, M.E.; Gigliotti-Rothfuchs, A.; Wigzell, H. The role of IFN-gamma in the outcome of chlamydial infection. *Curr. Opin. Immunol.* **2002**, *14*, 444–451. [CrossRef]
9. Zha, X.; Yang, S.; Niu, W.; Tan, L.; Xu, Y.; Zeng, J.; Tang, Y.; Sun, L.; Pang, G.; Qiao, S.; et al. IL-27/IL-27R Mediates Protective Immunity against Chlamydial Infection by Suppressing Excessive Th17 Responses and Reducing Neutrophil Inflammation. *J. Immunol.* **2021**, *206*, 2160–2169. [CrossRef]

10. Qian, C.; Cao, X. Dendritic cells in the regulation of immunity and inflammation. *Semin. Immunol.* **2018**, *35*, 3–11. [CrossRef]
11. Liu, J.; Zhang, X.; Cheng, Y.; Cao, X. Dendritic cell migration in inflammation and immunity. *Cell Mol. Immunol.* **2021**, *18*, 2461–2471. [CrossRef]
12. Rey-Ladino, J.; Jiang, X.; Gabel, B.R.; Shen, C.; Brunham, R.C. Survival of Chlamydia muridarum within dendritic cells. *Infect. Immun.* **2007**, *75*, 3707–3714. [CrossRef]
13. Rey-Ladino, J.; Koochesfahani, K.M.; Zaharik, M.L.; Shen, C.; Brunham, R.C. A live and inactivated Chlamydia trachomatis mouse pneumonitis strain induces the maturation of dendritic cells that are phenotypically and immunologically distinct. *Infect. Immun.* **2005**, *73*, 1568–1577. [CrossRef]
14. Bilenki, L.; Wang, S.; Yang, J.; Fan, Y.; Jiao, L.; Joyee, A.G.; Han, X.; Yang, X. Adoptive transfer of CD8alpha+ dendritic cells (DC) isolated from mice infected with Chlamydia muridarum are more potent in inducing protective immunity than CD8alpha- DC. *J. Immunol.* **2006**, *177*, 7067–7075. [CrossRef]
15. Shekhar, S.; Peng, Y.; Wang, S.; Yang, X. CD103+ lung dendritic cells (LDCs) induce stronger Th1/Th17 immunity to a bacterial lung infection than CD11b(hi) LDCs. *Cell Mol. Immunol.* **2018**, *15*, 377–387. [CrossRef]
16. Vignali, D.A.; Kuchroo, V.K. IL-12 family cytokines: Immunological playmakers. *Nat. Immunol.* **2012**, *13*, 722–728. [CrossRef]
17. Hunter, C.A.; Kastelein, R. Interleukin-27: Balancing protective and pathological immunity. *Immunity* **2012**, *37*, 960–969. [CrossRef]
18. Morita, Y.; Masters, E.A.; Schwarz, E.M.; Muthukrishnan, G. Interleukin-27 and Its Diverse Effects on Bacterial Infections. *Front. Immunol.* **2021**, *12*, 678515. [CrossRef]
19. Yoshida, H.; Hunter, C.A. The immunobiology of interleukin-27. *Annu. Rev. Immunol.* **2015**, *33*, 417–443. [CrossRef]
20. Harker, J.A.; Wong, K.A.; Dallari, S.; Bao, P.; Dolgoter, A.; Jo, Y.; Wehrens, E.J.; Macal, M.; Zuniga, E.I. Interleukin-27R Signaling Mediates Early Viral Containment and Impacts Innate and Adaptive Immunity after Chronic Lymphocytic Choriomeningitis Virus Infection. *J. Virol.* **2018**, *92*, e02196-17. [CrossRef]
21. Cao, J.; Wang, D.; Xu, F.; Gong, Y.; Wang, H.; Song, Z.; Li, D.; Zhang, H.; Li, D.; Zhang, L.; et al. Activation of IL-27 signalling promotes development of postinfluenza pneumococcal pneumonia. *EMBO Mol. Med.* **2014**, *6*, 120–140. [CrossRef] [PubMed]
22. Wang, L.; Cao, J.; Li, C.; Zhang, L. IL-27/IL-27 Receptor Signaling Provides Protection in Clostridium difficile-Induced Colitis. *J. Infect. Dis.* **2018**, *217*, 198–207. [CrossRef] [PubMed]
23. Robinson, K.M.; Lee, B.; Scheller, E.V.; Mandalapu, S.; Enelow, R.I.; Kolls, J.K.; Alcorn, J.F. The role of IL-27 in susceptibility to post-influenza Staphylococcus aureus pneumonia. *Respir. Res.* **2015**, *16*, 10. [CrossRef] [PubMed]
24. Shekhar, S.; Peng, Y.; Gao, X.; Joyee, A.G.; Wang, S.; Bai, H.; Zhao, L.; Yang, J.; Yang, X. NK cells modulate the lung dendritic cell-mediated Th1/Th17 immunity during intracellular bacterial infection. *Eur. J. Immunol.* **2015**, *45*, 2810–2820. [CrossRef] [PubMed]
25. Zheng, N.; Sun, L.; Pang, G.; Zha, X.; Niu, W.; Tan, L.; Zhang, H.; Bai, H. Chlamydia muridarum infection induces CD4+ T cells apoptosis via PI3K/AKT signal pathway. *Pathog. Dis.* **2019**, *77*, ftz029. [CrossRef] [PubMed]
26. Perry, L.L.; Feilzer, K.; Caldwell, H.D. Immunity to Chlamydia trachomatis is mediated by T helper 1 cells through IFN-gamma-dependent and -independent pathways. *J. Immunol.* **1997**, *158*, 3344–3352. [CrossRef]
27. Steinman, R.M.; Idoyaga, J. Features of the dendritic cell lineage. *Immunol. Rev.* **2010**, *234*, 5–17. [CrossRef]
28. Reis e Sousa, C. Dendritic cells in a mature age. *Nat. Rev. Immunol.* **2006**, *6*, 476–483. [CrossRef]
29. Steinman, R.M. Decisions about dendritic cells: Past, present, and future. *Annu. Rev. Immunol.* **2012**, *30*, 1–22. [CrossRef]
30. Fujii, S.; Liu, K.; Smith, C.; Bonito, A.J.; Steinman, R.M. The linkage of innate to adaptive immunity via maturing dendritic cells in vivo requires CD40 ligation in addition to antigen presentation and CD80/86 costimulation. *J. Exp. Med.* **2004**, *199*, 1607–1618. [CrossRef]
31. Ugur, M.; Mueller, S.N. T cell and dendritic cell interactions in lymphoid organs: More than just being in the right place at the right time. *Immunol. Rev.* **2019**, *289*, 115–128. [CrossRef]
32. Eisenbarth, S.C. Dendritic cell subsets in T cell programming: Location dictates function. *Nat. Rev. Immunol.* **2019**, *19*, 89–103. [CrossRef]
33. Zhu, X.; Zhu, J. CD4 T Helper Cell Subsets and Related Human Immunological Disorders. *Int. J. Mol. Sci.* **2020**, *21*, 8011. [CrossRef]
34. Perona-Wright, G.; Jenkins, S.J.; Crawford, A.; Gray, D.; Pearce, E.J.; MacDonald, A.S. Distinct sources and targets of IL-10 during dendritic cell-driven Th1 and Th2 responses in vivo. *Eur. J. Immunol.* **2006**, *36*, 2367–2375. [CrossRef]
35. Rasquinha, M.T.; Sur, M.; Lasrado, N.; Reddy, J. IL-10 as a Th2 Cytokine: Differences between Mice and Humans. *J. Immunol.* **2021**, *207*, 2205–2215. [CrossRef]
36. Cribas, E.S.; Denny, J.E.; Maslanka, J.R.; Abt, M.C. Loss of Interleukin-10 (IL-10) Signaling Promotes IL-22-Dependent Host Defenses against Acute Clostridioides difficile Infection. *Infect. Immunol.* **2021**, *89*, e00730-20. [CrossRef]
37. Clark, S.E.; Schmidt, R.L.; Aguilera, E.R.; Lenz, L.L. IL-10-producing NK cells exacerbate sublethal Streptococcus pneumoniae infection in the lung. *Transl. Res.* **2020**, *226*, 70–82. [CrossRef]
38. Ferreira, C.M.; Barbosa, A.M.; Barreira-Silva, P.; Silvestre, R.; Cunha, C.; Carvalho, A.; Rodrigues, F.; Correia-Neves, M.; Castro, A.G.; Torrado, E. Early IL-10 promotes vasculature-associated CD4+ T cells unable to control Mycobacterium tuberculosis infection. *JCI Insight* **2021**, *6*, e150060. [CrossRef]

39. Wynn, T.A.; Cheever, A.W.; Williams, M.E.; Hieny, S.; Caspar, P.; Kühn, R.; Müller, W.; Sher, A. IL-10 regulates liver pathology in acute murine Schistosomiasis mansoni but is not required for immune down-modulation of chronic disease. *J. Immunol.* **1998**, *160*, 4473–4480. [CrossRef]
40. Yang, X.; Gartner, J.; Zhu, L.; Wang, S.; Brunham, R.C. IL-10 gene knockout mice show enhanced Th1-like protective immunity and absent granuloma formation following Chlamydia trachomatis lung infection. *J. Immunol.* **1999**, *162*, 1010–1017. [CrossRef]
41. Jiao, L.; Gao, X.; Joyee, A.G.; Zhao, L.; Qiu, H.; Yang, M.; Fan, Y.; Wang, S.; Yang, X. NK cells promote type 1 T cell immunity through modulating the function of dendritic cells during intracellular bacterial infection. *J. Immunol.* **2011**, *187*, 401–411. [CrossRef] [PubMed]
42. Dalod, M.; Chelbi, R.; Malissen, B.; Lawrence, T. Dendritic cell maturation: Functional specialization through signaling specificity and transcriptional programming. *EMBO J.* **2014**, *33*, 1104–1116. [CrossRef] [PubMed]
43. Mellman, I.; Steinman, R.M. Dendritic cells: Specialized and regulated antigen processing machines. *Cell* **2001**, *106*, 255–258. [CrossRef] [PubMed]
44. Hilligan, K.L.; Ronchese, F. Antigen presentation by dendritic cells and their instruction of CD4+ T helper cell responses. *Cell Mol. Immunol.* **2020**, *17*, 587–599. [CrossRef] [PubMed]
45. Curato, C.; Bernshtein, B.; Zupancic, E.; Dufner, A.; Jaitin, D.; Giladi, A.; David, E.; Chappell-Maor, L.; Leshkowitz, D.; Knobeloch, K.P.; et al. DC Respond to Cognate T Cell Interaction in the Antigen-Challenged Lymph Node. *Front. Immunol.* **2019**, *10*, 863. [CrossRef]

microorganisms

Article

LncRNA 8244-ssc-miR-320-CCR7 Regulates IFN-β during SVA Infecting PK-15 Cells

Xiaoyu Tang [1,2], Ruiyu Zhang [1,2], Long Gao [1,2], Xiaocheng Lv [1,2], Yuan Sun [1,2,*] and Jingyun Ma [1,2,*]

1 Guangdong Provincial Key Laboratory of Agro-Animal Genomics and Molecular Breeding, College of Animal Science, South China Agricultural University, Guangzhou 510642, China
2 Guangdong Laboratory for Lingnan Modern Agriculture, Guangzhou 510642, China
* Correspondence: sunyuan@scau.com.cn (Y.S.); majy2400@scau.edu.cn (J.M.)

Abstract: Seneca Valley virus (SVV), a member of the Picornaviridae family, is an oncolytic RNA virus that can cause idiopathic vesicular disease and increase mortality in newborn piglets. Although research on the pathogenic characteristics, epidemiology, pathogenic mechanism, and clinical diagnosis of SVA has increased due to its emergence and prevalence, the interaction between SVA and its host lncRNA has not been fully studied. This study used qualcomm sequencing to analyze differentially expressed lncRNAs and found that during SVA infection, lncRNA 8244 was significantly down-regulated in both PK-15 cells and piglets. Further analysis through quantitative real-time PCR and dual luciferase experiments demonstrated that lncRNA8244 could compete with ssc-miR-320 to regulate the expression of CCR7. The lncRNA824-ssc-miR-320-CCR7 axis activated the TLR-mediated signaling pathway, which recognized viral molecules and induced the expression of IFN-β. These findings provide new insight into the interaction between lncRNA and SVA infection, which could lead to a better understanding of SVA pathogenesis and contribute to the prevention and control of SVA disease.

Keywords: Senecavirus A; lncRNA 8244; ssc-miR-320; CCR7; TLR signaling pathway; IFN-β

Citation: Tang, X.; Zhang, R.; Gao, L.; Lv, X.; Sun, Y.; Ma, J. LncRNA 8244-ssc-miR-320-CCR7 Regulates IFN-β during SVA Infecting PK-15 Cells. *Microorganisms* **2023**, *11*, 688. https://doi.org/10.3390/microorganisms11030688

Academic Editors: Shengxi Chen and Fabio Zicker

Received: 2 February 2023
Revised: 3 March 2023
Accepted: 3 March 2023
Published: 8 March 2023

1. Introduction

Seneca Valley virus (SVV), referred to as Seneca virus (Senecavirus A, SVA), and classified as a small RNA virus within the Picornaviridae family's Senecavirus genus, was first reported in Brazil in late 2014 [1,2]. Since 2015, it has rapidly spread across various regions and countries, infecting pigs of different ages [3–5]. In 2017, both the United States and China reported new cases of SVA infection, and the year of 2016 was seen as a pivotal moment in China's SVA epidemic. Prior to 2016, the SVA strains that surfaced in China had a higher nucleotide homology with those isolated in Brazil and Canada, whereas strains reported after 2016 had greater similarity with those from the United States [6,7]. Using TaqMan-based quantitative reverse transcription PCR (qRT-PCR) analysis, researchers have identified SVA in various tissues of newborn piglets, including the lung, heart, bladder, kidney, spleen, tonsils, and small intestine, indicating that these tissues can be examined for SVA [2,8,9]. Studies have also shown that SVA causes multiple organ diseases in piglets and leads to multi-system diseases in infected piglets. Despite considerable progress in SVA-related research, there is still a dearth of knowledge, particularly about the interaction between the virus and host lncRNAs.

The lncRNA is a non-coding RNA molecule that is longer than 200 nucleotides and cannot produce proteins [10]. In mammals, thousands of lncRNAs have been identified to be involved in various cellular processes, including metabolism, apoptosis, proliferation, and innate immunity [11–14]. Specifically, lncRNA acts as a regulator on a variety of key molecules or links, such as pathogen-recognition-receptor-related signals, transcription factor translocation and activation, interferon (IFN) and cytokine production, interferon-activated JAK-STAT signal and antiviral ISG transcription, thus playing an important role in the natural immune response [15]. For example, lnc-Lsm3b induced by IFN can competitively bind to the RIG-I monomer with viral RNA, and feedback inactivation of the innate function of RIG-I in the late stage of the innate reaction. This binding restricts the conformational shift of RIG-1 protein, prevents downstream signal transduction, and terminates the production of type I IFN [16]. Our study focuses on ssc-miR-320 and CCR7, which have been reported to regulate various cellular functions, including cell proliferation, invasion, apoptosis, and glucose and lipid metabolism disorder-associated diseases [17,18]. CCR7 plays a pivotal role in regulating tissue immunity and inflammation through the CCL19-CCR7 axis. It can regulate DC morphologic change, thymic T cell development, and suppress DC apoptosis, leading to the regulation of adaptive immunity and tolerance [17,18]. This study explored the role of the lncRNA 8244-ssc-miR-320-CCR7 axis.

SVA is an emerging porcine blistering pathogen that causes porcine blistering disease outbreaks in different regions. However, knowledge about SVA immunity remains limited. To address this gap, we conducted a study using high-throughput sequencing to identify differentially expressed lncRNAs following SVA infection of PK-15 cells. Our objective was to investigate the targeting relationship between lncRNA 8244-ssc-miR-320-CCR7 and its regulatory role on key molecules of the TLR signaling pathway and IFN-β, with the aim of revealing how SVA regulates host immune responses by manipulating lncRNA expression. The results of our research will provide important theoretical and practical guidance for establishing disease prevention and control strategies for SVA. Specifically, our findings will help to establish a molecular strategy for immune escape through the regulation of lncRNA expression.

2. Materials and Methods

2.1. Cells, Viruses and Clinical Samples of Piglets

The cell line (Porcine kidney cells (PK-15)) used in this experiment was obtained from the Poultry Research Laboratory, School of Animal Science, South China Agricultural University, while the SVA CH-01-2015 isolate was isolated and preserved by our laboratory.

Huanong Wen's Co., Ltd. (Yunfu, China) provided clinical samples of piglets, including two samples each of small intestine, kidney, lung, and brain.

2.2. Plasmids and Antibodies

The pmirGLO vector (genepharma) for dual luciferase experiments is kept in our laboratory. PmirGLO-wild-CCR7 and pmirGLO-mut-CCR7 were constructed based on the binding site of CCR7 and miRNA-320. PmirGLO-wild-8244 and pmirGLO-mut-8244 were constructed based on the binding site of lncRNA 8244 and miRNA-320. These plasmids are utilized for verifying targeting relationships. The plasmid expressing the CCR7 gene (pmCherry-CCR7) was purchased from the miaoling plasmid platform (Miaolingbio, China). Sus scrofa_NONSUSG008244.1 (http://www.noncode.org, accessed on 1 February 2023) overexpression lentiviral vector was constructed by Jiman Biotechnology Co., Ltd. (Zhaoqing, China). Additionally, miRNA-320 mimics and inhibitor were purchased from Jiman Biotechnology Co., Ltd., while anti-β-actin (rabbit) monoclonal antibody (abcam), anti-TLR3 (pig) monoclonal antibody (abcam), mouse IFN-beta antibody (R&D). Anti-SVA-VP1 are kept in our laboratory.

2.3. Quantitative Real-Time PCR (qPCR)

Total RNA was isolated from cells with TRIzol reagent (Invitrogen, Waltham, MA, USA). Portions (1 μg) of each RNA sample were reverse-transcribed by using a PrimeScript RT reagent kit (TaKaRa, Kusatsu, Japan), and cDNA was quantified with these genes' specific primer pairs (Table 1). Primers were designed to target conserved regions of genes using primer-Premier 5.0 (Premier Biosoft Interpairs, Palo Alto, CA, USA). qPCRs were conducted in the ABI7900HT (Applied Biosystems, Waltham, MA, USA) using a SYBR

Premix Ex Taq kit (TaKaRa), and reactions were denatured at 95 °C for 30 s, followed by 40 two-step cycles of 95 °C for 5 s and 60 °C for 30 s. The relative gene expression levels were normalized against that of GAPDH (glyceraldehyde-3-phosphate dehydrogenase). The specific primer sequence is shown in Table 1.

Table 1. Primers used for real-time RT-PCR.

Primers	Sequence (5′-3′)	Amplification Length
IFN-β-F	GCTAACAAGTGCATCCTCCAAA	77 bp
IFN-β-R	AGCACATCATAGCTCATGGAAAGA	
IRF3-F	CCAGTGGTGCCTACACTCCT	191 bp
IRF3-R	AGAGGTGTCTGGCTCAGGAA	
TLR3-F	CGCCTCCTGGAAAACCAA	76 bp
TLR3-R	CCCTGAGTTGTCCTGCAACA	
TBK1-F	GCCTTTCTCGGGGTCTTCAA	74 bp
TBK1-R	ACACTTTTCCTGATCCGCCT	
TLR7-F	CGGTGTTTGTGATGACAGAC	134 bp
TLR7-R	AACTCCCACAGAGCCTCTTC	
TLR8-F	CACATTTGCCCGGTATCAAG	145 bp
TLR8-R	TGTGTCACTCCTGCTATTCG	
TLR9-F	GGCCTTCAGCTTCACCTTGG	151 bp
TLR9-R	GGTCAGCGGCACAAACTGAG	
MyD88-F	TGATGAACCGCAGGAT	438 bp
MyD88-R	ACTGTGCTACGGGCTGGATT	
TRIF-F	TGGGACATCCTTAGGGACATG	158 bp
TRIF-R	CCAGTGGACCTCAGGGAATG	
TRAF3-F	GTGTCAAGAAGGCATCG	164 bp
TRAF3-R	CCTCAAACTGGCAATCA	
CCR7-F	TGCTGGTGGTGGCTCTCCTTG	87 bp
CCR7-R	CCGTGGTGTTGTCGCCGATG	
LncRNA 8244-F	CAGAGGCAGGAACTGTGATGGC	145 bp
LncRNA 8244-R	GTGGTAGGTGAATCTGCGGAAGG	

2.4. Cell Infection

Cells with 80–90% confluence in good growth condition were aspirated, washed twice with PBS, spliced with SVACH-01-2015 (MOI = 1.5), and incubated in an incubator for 1 h. Finally, culture medium containing 2% fetal bovine serum was added and placed in the incubator at 37 °C with 5% CO_2.

2.5. Transduction and Western Blotting

The total cell protein was extracted 24 h after transfection. Cells were collected and treated with lysis buffer (Roche, UK). Protein concentrations of whole cell lysates were determined using a bicinchoninic acid protein assay kit (Thermo Scientific, Waltham, MA, USA) to assess protein expression. Equal amounts of proteins were separated using 12% sodium dodecyl sulfate-polyacrylamide gels (SDS-PAGE) and transferred to polyvinylidene difluoride (PVDF) membranes (Roche, UK). These membranes were blocked with 5% skim milk in 1× tris-buffered saline (TBS) plus for 2 h at room temperature. Subsequently incubated with diluted primary antibody for 2 at room temperature. Anti-rabbit or anti-pig IgG antibodies conjugated to horseradish peroxidase (HRP) were used as secondary antibodies. Enhanced chemiluminescent substrates were used for detection using the HRP kit (Thermo Scientific, USA).

2.6. Dual Luciferase Activity Detection

When the cell confluence reached 60–80%, the transfection mixture was added dropwise to a cell culture plate containing 2% fetal calf serum, shaken gently, and the medium was changed 4 h after transfection as needed.

PK-15 cells (1×10^5) were inoculated in 24-well plates 24 h prior to transfection. Cells were cotransfected with the reporter plasmids encoding lncRNA 8244 and CCR7 and the desired expression plasmid. The empty vector was used as a negative control to adjust the total amount of transfected DNA. A dual luciferase reporter assay system (E1910; Promega, Madison, WI, USA) was used, following the manufacturer's instructions. All reporter gene assays were repeated six times.

2.7. Statistical Analysis

The various treatments were compared using an unpaired, two-tailed Student t test with an assumption of unequal variance. Data are expressed as the mean ± standard deviation from at least three independent experiments. SPSS 17.0 software package (SPSS, Chicago, IL, USA) was used to analyze the qRT-PCR data. $p < 0.05$ was considered statistically significant.

3. Results

3.1. SVA Infection Inhibits the Expression of lncRNA 8244

To investigate whether the differential expression of lncRNA 8244 is related to SVA infection concentration, PK-15 cells were exposed to various titers of SVA. As the MOI increased, the expression of lncRNA 8244 was suppressed compared to uninfected cells, with an MOI of about 0.3 PFU/cell resulting in a significant decrease in lncRNA 8244 expression ($p < 0.01$) (Figure 1A). At an MOI of 1.5, PK-15 cells were infected with SVA, and quantitative real-time PCR (qPCR) analysis indicated a significant down-regulation of lncRNA 8244 expression at 12 h postinfection (hpi). At 24 hpi, lncRNA 8244 exhibited the lowest expression level ($p < 0.05$) (Figure 1B). In addition, PK-15 cells were infected with pseudorabies virus (PRV) and swine acute diarrhea syndrome coronavirus (SADS-CoV) at an MOI of approximately 1.5 PFU/cell, and the expression of lncRNA 8244 was significantly down-regulated, suggesting that the differential expression of lncRNA 8244 is related to virus infection (Figure 1C). We also detected the expression of lncRNA 8244 in clinical samples from SVA-infected piglets and cell, and the results showed a significant decrease in lncRNA 8244 expression in the small intestine, kidney, lung, brain and PK-15 cell of infected piglets compared to uninfected piglets, suggesting that SVA infection may inhibit the expression of lncRNA 8244 (Figures 1D and S1A).

3.2. LncRNA 8244 Promotes the Production of IFN-β after SVA Infection in PK-15 Cells

To investigate the regulatory effect of lncRNA 8244 on IFN-β expression, PK-15 cells were transfected with 2.0 μg of plasmid 8244/plasmid NC and 80 pmol of iNC/si-lncRNA 8244. After 24 h, the cells were infected with SVA at a 1.5 MOI and harvested at 24 hpi. The results demonstrated that both mRNA and protein levels of IFN-β expression were significantly higher in the plasmid 8244 group compared to the control group, while IFN-β expression was significantly suppressed in the si-lncRNA 8244 group ($p < 0.01$) (Figures 2A,B and S1B). These findings suggest that lncRNA 8244 positively regulates IFN-β expression in SVA-infected PK-15 cells.

3.3. LncRNA8244 Promotes the Expression of Key Signal Molecules in the TLR Signaling Pathway

To explore the impact of lncRNA 8244 on the signal pathways related to natural immunity, nucleic acids were extracted for quantitative real-time PCR analysis. We found that the expression of key molecules in the TLR signaling pathway was significantly up-regulated after the cell transfection with plasmids, with TLR3 molecules showing an extremely significant up-regulation ($p < 0.01$) (Figure 3A). Further analysis of the regulatory effect of lncRNA 8244 on the TLR3 signaling pathway revealed a significant up-regulation of TRIF expression in PK-15 cells transfected with plasmid 8244 at 24 h, 48 h, and 72 h ($p < 0.01$). Similarly, TBK1 expression was significantly up-regulated after 72 h of inoculation ($p < 0.01$). However, the expression of TRAF3 and IRF3 did not show significant differences from those in the control group ($p \geq 0.05$) (Figure 3B). These findings suggest that lncRNA

8244 affects the signal pathway involved in TLR3. To explore the regulatory effect of lncRNA 8244 on TLR3 during SVA infection, fluorescence quantitative experiments and Western Blot experiments were performed. The results revealed that transfection with plasmid 8244 significantly promoted the expression of TLR3 in PK-15 cells infected with SVA ($p < 0.01$). TLR3 expression was also significantly upregulated in clinical samples (Figure S1C). Conversely, transfection with si-lncRNA 8244 significantly inhibited TLR3 expression compared to the control group ($p < 0.05$) (Figures 3C,D and S1D). Taken together, these findings indicate that lncRNA 8244 promotes the expression of TLR3.

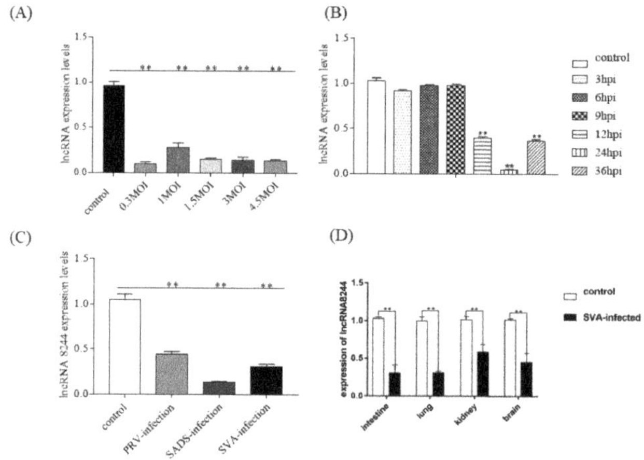

Figure 1. The expression of lncRNA 8244 in cells or clinical samples. (**A**) PK-15 cells were infected or no-infected with SVA (MOI = 0.3, 1, 1.5, 3, 4.5) for 24 h. Total RNA was extracted to determine the relative mRNA expression of lncRNA 8244 by real-time RT-PCR assay. (**B**) PK-15 cells were no-infected or infected with SVA (MOI = 1.5). The mRNA expression of lncRNA 8244 at 3, 6, 9, 12, 24, 36 hpi. (**C**) The expression level of lncRNA 8244. (**D**) Expression of lncRNA 8244 in clinical samples of SVA-infected piglets. Total RNA was extracted to determine the relative mRNA expression of lncRNA 8244 with real-time RT-PCR assay. All data are represented as mean ± SD with three replicates. ** $p < 0.01$.

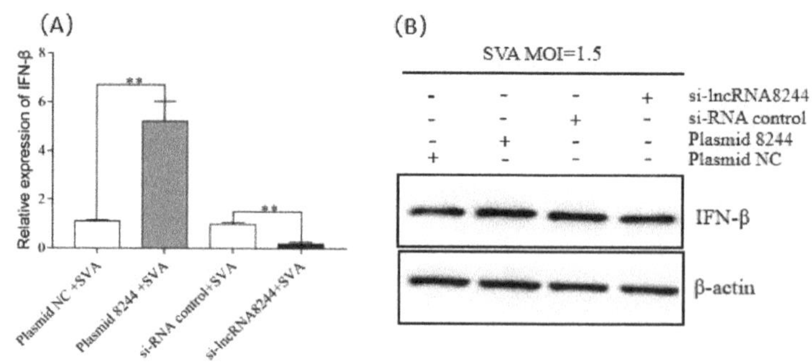

Figure 2. LncRNA 8244 can promote the expression of IFN-β. (**A**) PK-15 cells were treated with plasmid NC/plasmid 8244/iNC/si-lncRNA 8244 for 24 h, and then infected with SVA (MOI = 1.5) for an additional 24 h. The expression level of IFN-β was determined by real-time RT-PCR or Western blot (**B**). The mRNA level of IFN-β was normalized to the mRNA level of GAPDH, and anti-βactin was used as a control for Western blot sample loading. All data are represented as mean ± SD with three replicates. ** $p < 0.01$.

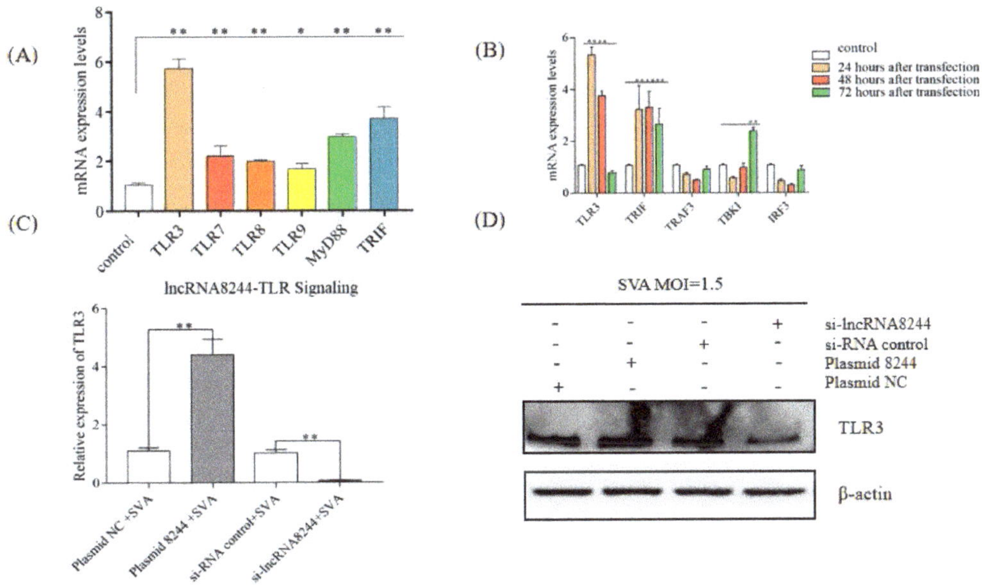

Figure 3. LncRNA 8244 can up-regulate the expression of key molecules in the TLR signaling pathway. (**A**) The expression of TLR3/TLR7/TLR8/TLR9/MyD88/TRIF. (**B**) The expression level of TLR3/TRIF/TRAF3/TBK1 and IRF3 was normalized to the mRNA level of GAPDH. (**C**) The expression level of TLR3 was determined by real-time RT-PCR assay or Western blot (**D**). The mRNA level of TLR3 was normalized to the mRNA level of GAPDH, and anti-βactin was used as a control for Western blot sample loading. All data are represented as mean ± SD with three replicates. * $p < 0.05$; ** $p < 0.01$.

3.4. LncRNA 8244 Targets miR-320

The software RegRNA2.0 [19] and the Miranda algorithm were utilized to predict miRNAs that could target lncRNA 8244. After evaluating the predicted scores and complementary matching free energy, it was determined that ssc-miR-320 was the target miRNA of lncRNA 8244 (Figure 4A). To validate this prediction, nucleic acid samples were collected, and a fluorescence quantitative experiment was conducted to measure the expression of lncRNA 8244. The pmirGLO-8244 and pmirGLO-mut-8244 plasmids were constructed according to the binding sites. The results showed that ssc-miR-320 mimics and inhibitors had the expected effects (Figure 4B). The expression of lncRNA 8244 was significantly reduced in the ssc-miR-320 mimics group compared to the NC group ($p < 0.05$). Conversely, the expression of lncRNA 8244 was significantly increased in the ssc-miR-320 inhibitor group compared to the iNC group in SVA-infected cells ($p < 0.05$) (Figure 4C). The targeting relationship was further confirmed through a dual luciferase experiment. The relative luciferase activity of the ssc-miR-320 mimics group was significantly lower than that of the NC group after transfection of the lncRNA 8244 wild-type plasmid ($p < 0.05$). However, there was no significant difference in the relative dual luciferase activity of the ssc-miR-320 mimics group after transfection with the lncRNA 8244 mutant plasmid compared to the NC group ($p \geq 0.05$) (Figure 4D). Furthermore, the expression of ssc-miR-320 was significantly upregulated in SVA-infected cells and clinical samples ($p < 0.05$). In clinical samples and cell, ssc-miR-320 expression was significantly increased in the kidney, lung, brain and PK-15 cell of SVA-infected piglets compared to normal piglets ($p < 0.05$) (Figures 4E and S1A). Overall, the results confirm that ssc-miR-320 is the target miRNA of lncRNA 8244, and there is a negative target relationship between them.

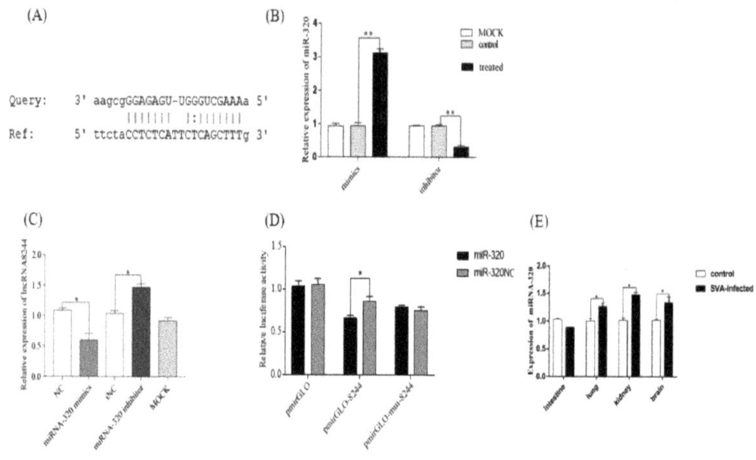

Figure 4. MiR-320 is the target gene of lncRNA 8244. (**A**) Predicted binding sites for miR-320 on the lncRNA 8244 transcript. (**B**) The expression level of miR-320 (**C**) The expression level of lncRNA 8244 after infected with SVA (MOI = 1.5) (**D**) Luciferase activity in PK-15 cells co-transfected with luciferase reporter containing pmirGLO-8244 or pmirGLO-mut-8244 and the mimics of miR-320. Data are presented as the relative ratio of Renilla luciferase activity and firefly luciferase activity. (**E**) The expression level of miR-320 in clinical samples of SVA-infected piglets. The mRNA level of TLR3 was normalized to the mRNA level of GAPDH, and anti-β-actin was used as a control for Western blot sample loading. All data are represented as mean ± SD with three replicates. * $p < 0.05$; ** $p < 0.01$.

3.5. LncRNA 8244 Affects IFN-β Secretion by Regulating the Expression of ssc-miR-320 in PK-15 Cells Infected by SVA

PK-15 cells were subjected to transfection with NC, ssc-miR-320 mimics, iNC, or ssc-miR-320 inhibitor, along with the co-transfection of plasmids NC and NC, plasmid 8244 and ssc-miR-320 mimics, or si-lncRNA 8244 and ssc-miR-320 inhibitor. The cells were then infected with SVA at a 1.5 MOI. The results indicated that compared to the control group, the expression of IFN-β was significantly down-regulated in the ssc-miR-320 mimics transfection group ($p < 0.05$), while it was significantly up-regulated in the ssc-miR-320 inhibitor transfection group ($p < 0.05$) (Figure 5A). Additionally, the co-transfection group with plasmid 8244 and ssc-miR-320 mimics or si-lncRNA 8244 and ssc-miR-320 inhibitor showed no significant difference in IFN-β expression levels compared to the control group ($p \geq 0.05$) (Figures 5B,C and S1E,F). This suggests that ssc-miR-320 counteracted the effect of lncRNA 8244 on IFN-β expression. The Western blot experiment results corroborated the fluorescence quantification results, indicating that lncRNA 8244 modulates ssc-miR-320, thereby regulating the expression of IFN-β.

3.6. LncRNA 8244 Affects the Expression of Key Signal Molecules in TLR3-Related Signaling Pathways by Regulating the Expression of ssc-miR-320

LncRNA 8244 has been shown to impact the expression of key signaling molecules in the TLR3 signaling pathway, with ssc-miR-320 being its target gene. To explore whether lncRNA 8244 acts through ssc-miR-320, the study examined the expression of key signaling molecules in the TLR3 pathway. The results showed that compared to the control group, transfection with ssc-miR-320 mimics resulted in significant down-regulation of TRIF and TBK1 expression 24 h after inoculation ($p < 0.05$), and TRAF3 expression was significantly reduced 48 h after inoculation ($p < 0.01$) (Figure 6A,B). Building on the previous findings that lncRNA 8244 affects IFN-β expression through ssc-miR-320, the study investigated whether lncRNA 8244 regulates TLR3 by modulating ssc-miR-320 expression. Co-transfection of plasmid 8244 and ssc-miR-320 mimics did not significantly differ from the control group in TLR3 expression. Similarly, co-transfection of si-lncRNA 8244 and ssc-miR-320 inhibitor did not significantly differ from the control group in TLR3 expression. The expression of ssc-miR-320 counteracted the effect of lncRNA 8244, indicating that lncRNA 8244 regulates TLR3 expression by modulating ssc-miR-320 expression (Figures 6C,D and S1G).

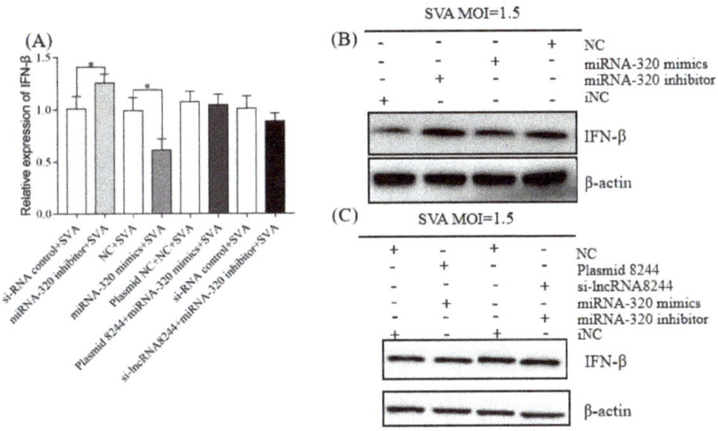

Figure 5. MiR-320 can suppress the expression of IFN-β (**A**) PK-15 cells were transfected or co-transfected with iNC/miRNA-320 inhibitor/miRNA-320 mimics/NC/plasmid NC and iNC/plasmid 8244 and ssc-miR-320 mimics/iNC and iNC/si-lncRNA 8244 and miR-320 inhibitor for 24 h, and infected with SVA (MOI = 1.5) for additional 24 h. The expression of IFN-βwas determined by real-time RT-PCR assay or Western blot (**B**,**C**). The mRNA level of IFN-β was normalized to mRNA level of GAPDH, and anti-βactin was used as a control for Western blot sample loading. All data are represented as mean ± SD with three replicates. *, $p < 0.05$.

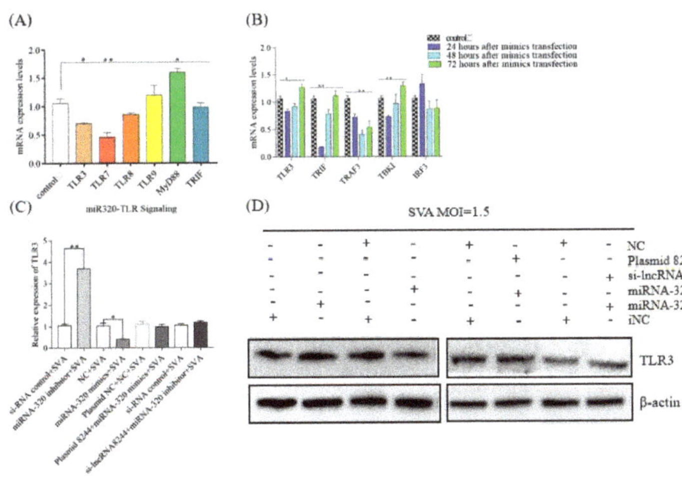

Figure 6. MiR-320 can down-regulate the expression of key molecules in the TLR signaling pathway. (**A**) The expression of TLR3/TLR7/TLR8/TLR9/MyD88/TRIF. (**B**) The expression of TLR3/TRIF/TRAF3/TBK1 and IRF3. (**C**) The expression of TLR3 was determined by real-time RT-PCR assay or Western blot (**D**). The mRNA level of TLR3 was normalized to the mRNA level of GAPDH, and anti-βactin was used as a control for Western blot sample loading. All data are represented as mean ± SD with three replicates. * $p < 0.05$; ** $p < 0.01$.

3.7. CCR7 Was the Target Gene of ssc-miR-320

RegRNA2.0 and the Miranda algorithm, two online target gene prediction software packages, predicted that CCR7 is a target gene of ssc-miR-320. Subsequent experiments confirmed the negative regulatory relationship between ssc-miR-320 and CCR7 ($p < 0.01$) (Figure 7A,B). To further validate this targeting relationship, a dual luciferase experiment was performed. After transfecting the pmirGLO-wild-CCR7 plasmid for 24 h, the ssc-miR-320 mimics group showed significantly lower relative luciferase activity than the NC group ($p < 0.01$). When the pmirGLO-mut-CCR7 plasmid was transfected, the relative dual luciferase activity of the ssc-miR-320 mimics group was not significantly different from the NC group. These results confirm that CCR7 is the targeted mRNA of ssc-miR-320 (Figure 7C).

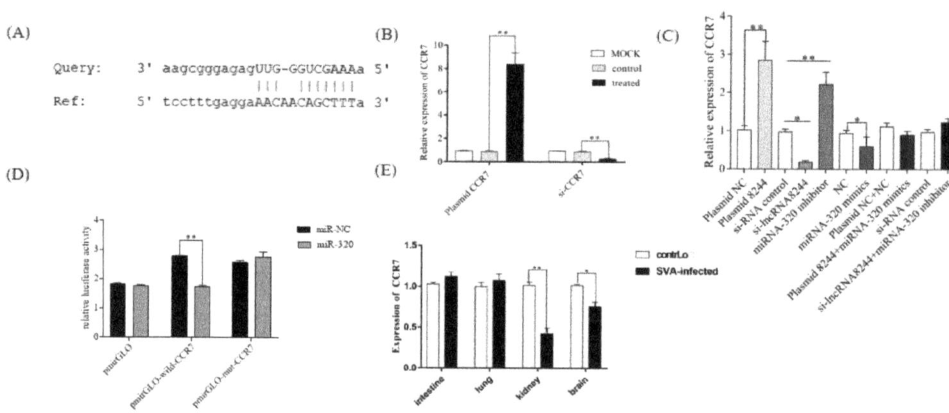

Figure 7. CCR7 is the target gene of miR-320. (**A**) Predicted binding sites for CCR7 on the miR-320. (**B**) PK-15 cells were treated with plasmid NC/plasmid CCR7/iNC/si-CCR7 for 24 h. The expression of CCR7 was determined by real-time RT-PCR assay and the mRNA level of CCR7 was normalized to the mRNA level of GAPDH. (**C**) The expression level of CCR7 was determined by real-time RT-PCR assay and the mRNA level of CCR7 was normalized to the mRNA level of GAPDH. (**D**) Luciferase activity in PK-15 cells co-transfected with luciferase reporter containing CCR7 or mutant and the mimics of miR-320. Data are presented as the relative ratio of Renilla luciferase activity and firefly luciferase activity. (**E**) Expression of CCR7 in clinical samples of SVA-infected piglets. Total RNA was extracted to determine the relative mRNA expression of CCR7 by real-time RT-PCR assay. All data are represented as mean ± SD with three replicates. * $p < 0.05$; ** $p < 0.01$.

Simultaneously, PK-15 cells were transfected with iNC/s-lncRNA8244, plasmid NC/plasmid 8244, and co-transfection of plasmid 8244 and ssc-miR-320 mimics, si-lncRNA8244 and ssc-miR-320 inhibitor, and the expression of CCR7 was examined. It was observed that CCR7 expression was significantly increased in the plasma 8244 transfection group compared to the control group ($p < 0.01$), while the si-lncRNA8244 transfection group exhibited a significant decrease in CCR7 expression compared to the control group ($p \geq 0.05$). The co-transfection group showed no significant difference in CCR7 expression compared to the control group ($p \geq 0.05$) (Figure 7D). Furthermore, the expression of CCR7 was detected in SVA-infected cells and clinical samples, and it was found to be significantly up-regulated compared to the control group ($p < 0.05$) (Figures 7E and S1A). These results indicate that lncRNA 8244 functions as a competitive endogenous RNA by competitively binding to ssc-miR-320, and indirectly regulates the expression of CCR7.

3.8. *After SVA Infection, the lncRNA 8244-ssc-miR-320-CCR7 Axis Can Regulate the Production of IFN-β by PK-15 Cells*

This study investigated whether the ssc-miR-320-mediated regulation of CCR7 affects the expression of IFN-β during SVA infection in PK-15 cells. The efficiency of plasma CCR7 expression and si-CCR7 interference was confirmed through experiments, with plasmid CCR7 and si-CCR7 performing their respective functions (Figure 8A). After transfecting iNC/si-CCR7/plasmid NC/plasmid CCR7 into PK-15 cells, the cells were infected with SVA at a 1 MOI and harvested 24 h post-infection. The results revealed a positive regulatory relationship between CCR7 and IFN-β.

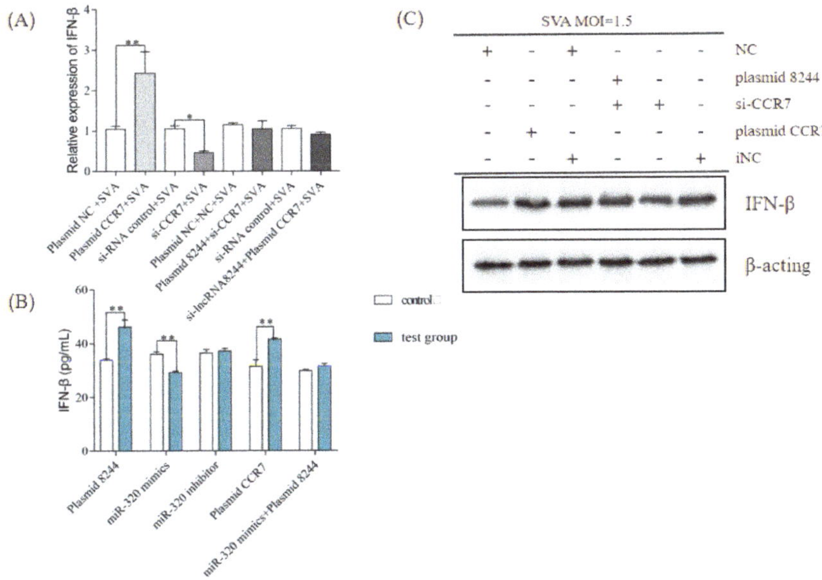

Figure 8. CCR7 can promote the expression of IFN-β (**A**) PK-15 cells were transfected or co-transfected with plasmid NC/plasmid CCR7/iNC/si-CCR7/plasmid NC and NC/plasmid 8244 and si-CCR7/iNC and iNC/si-lncRNA 8244 and plasmid CCR7 for 24 h. The expression level of IFN-β was determined by real-time RT-PCR assay and Western blot (**B**). The mRNA level of IFN-β was normalized to mRNA level of GAPDH or ELISA (**C**). All data are represented as mean ± SD with three replicates. * $p < 0.05$; ** $p < 0.01$.

This study investigated the regulatory relationship between ssc-miR-320, CCR7, and IFN-β during SVA infection in PK-15 cells. The results confirmed that CCR7 is a target gene of ssc-miR-320 and can positively regulate IFN-β expression. The study identified a regulatory axis of lncRNA 8244-ssc-miR-320-CCR7 and demonstrated that lncRNA 8244 indirectly regulates IFN-β expression through competitive binding to ssc-miR-320 and the subsequent regulation of CCR7 expression. Further experiments were conducted to verify the regulatory effect of each component of the axis on IFN-β expression. Si-CCR7 reversed the promotion of lncRNA 8244 on IFN-β, and CCR7 reversed the down-regulation of si-lncRNA 8244 on IFN-β. The results from Western blot and ELISA assays were consistent with quantitative real-time PCR (Figures 8B,C and S1H). These findings suggest that lncRNA 8244, ssc-miR-320, and CCR7 have a significant impact on IFN-β expression during SVA infection.

3.9. The lncRNA 8244-ssc-miR-320-CCR7 Axis Regulates the Expression of Key Signal Molecules in TLR3 Involved the Signaling Pathway

This study revealed that lncRNA 8244 modulates the expression of IFN-β by competitively binding to ssc-miR-320, which indirectly impacts the expression of CCR7 and alters the expression of key molecules in the TLR signaling pathway following transfection with plasmid CCR7. Notably, the levels of TLR3 and TLR8 were up-regulated (Figure 9A). Furthermore, the expression of TLR3 signaling pathway molecules was found to be up-regulated in the transfection group compared to the control group at 24 hpi ($p \geq 0.05$), including TLR3, TBK1, and TRIF (Figure 9B). LncRNA 8244 and ssc-miR-320 were found to regulate the expression of TLR3. Additionally, a positive regulatory relationship between CCR7 and TLR3 was observed. However, there was no significant difference in TLR3 expression between the co-transfected plasmid 8244 and si-CCR7 group and the control group ($p \geq 0.05$) (Figures 9C,D and S1I). LncRNA 8244 indirectly modulates the expression of CCR7, inhibiting the biological activity of ssc-miR-320, which ultimately affects the regulation of CCR7 on TLR3.

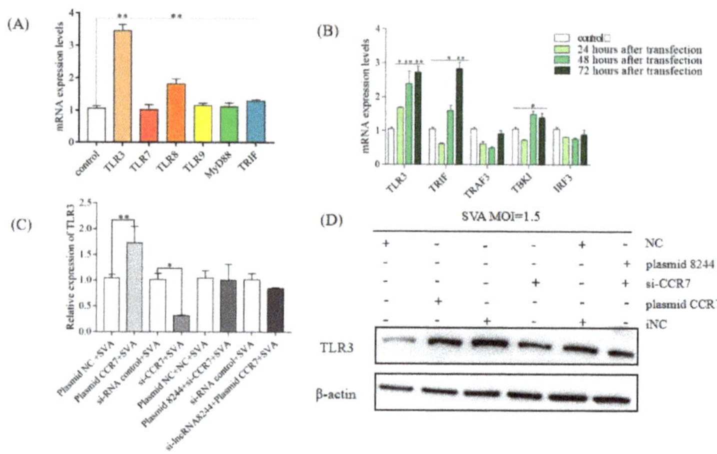

Figure 9. CCR7 can up-regulate the expression of key molecules in the TLR signaling pathway. (**A**) The expression level of TLR3/TLR7/TLR8/TLR9/MyD88/TRIF. (**B**) The expression level of TLR3/TRIF/TRAF3/TBK1 and IRF3. (**C**) The expression level of TLR3 was normalized to the mRNA level of GAPDH, and anti-βactin was used as a control for Western blot sample loading (**D**). All data are represented as mean ± SD with three replicates. * $p < 0.05$; ** $p < 0.01$.

4. Discussion

SVA is known to cause vesicular disease, which presents similar clinical symptoms to other vesicular virus diseases such as foot-and-mouth disease (FMD), vesicular stomatitis (VS), and swine vesicular disease (SVD) [20]. Recent research has shown that long non-coding RNAs (lncRNAs) play a significant role in the interaction between viruses and hosts [21]. In this study, we investigated the interaction between SVA infection and host lncRNA for the first time. Our results demonstrate that lncRNA 8244 can up-regulate the expression of IFN-β through the LncRNA 8244-ssc-miR-320-CCR7 axis following SVA infection. Furthermore, we found that lncRNA 8244 modulates the expression of IFN-β through the TLR signaling pathway.

LncRNA has been shown to participate in IFN-mediated innate immune regulation through histone modulation and the regulation of IFN or ISG expression [22,23]. At the transcriptional level, lnc-Lsm3b is a host-derived lncRNA that induces IFN-β production and competes with viral RNA to inactivate RIG-I function [16]. Our previous research showed that IFN-β expression increased after SVA infection, and this study aimed to investigate why. We found that the expression of lncRNA 8244 was down-regulated after

SVA infection, and the overexpression of lncRNA 8244 resulted in increased IFN-β mRNA and protein expression levels at 24 hpi. Recent studies have shown that lncRNAs act as competitive endogenous RNA by binding miRNA with mRNA, reducing free miRNA content and regulating mRNA function [24,25]. Dual luciferase experiments showed a negative regulatory relationship between lncRNA 8244 and ssc-miR-320. Our reversal test showed that miR-320 expression can reverse the regulation of IFN-β by lncRNA 8244. The downstream target gene of ssc-miR-320 involved in the regulation of IFN-β was found to be CCR7, with a negative regulatory relationship. In the clinical sample assay, we found that the expression of lncRNA 8244 was significantly down-regulated in the small intestine, but the expression of miRNA-320 and CCR7 was close to that of the control group. The reason is that the expression profiles of CCR7 as well as miR-320 were initially determined by high-throughput sequencing results of SVA-infected PK-15 cell, so there is some tissue variability. We discovered a positive regulatory relationship between CCR7 and IFN-β, and CCR7 can reverse the regulation of IFN-β by lncRNA 8244. Previous studies have shown that CCR7 down-regulates IFN-γ and IFN-γ response genes, enabling B16 cells to evade immunity [26].

The innate immune response serves as the primary defense mechanism against viral invasion, relying on pattern recognition receptors (PRRs) to detect pathogen-associated molecular patterns (PAMPs) [27]. Activation of the signal pathway via PRRs leads to the production of type I interferons and pro-inflammatory factors, with RLR, TLR, and NLR-mediated pathways being the most common [28]. This study focuses on the lncRNA 8244-ssc-miR-320-CCR7 axis and its regulation of IFN-β expression, examining how the axis influences the TLR signaling pathway to modulate IFN-β production. Results revealed that lncRNA 8244, CCR7, and ssc-miR-320 can affect the expression of signal transduction molecules within the TLR pathway. Host cell response to viral infections hinges on PRRs, with Toll-like receptors (TLRs) being the most prominent. LncRNA 8244 and CCR7 significantly up-regulate the expression of TLR3, whereas ssc-miR-320 down-regulates its expression. TLR3-mediated signaling pathways play a crucial role in the antiviral response to poliovirus (PV) infection, as TLR3-deficient mice fail to produce IFN in their serum [29]. Results showed that CCR7 reverses the regulation of TLR3 by lncRNA 8244, indicating that the lncRNA 8244-ssc-miR-320-CCR7 axis activates the TLR-mediated signaling pathway, with TLR3 recognizing viral molecules and inducing the expression of IFN-β. TLR3 induces IFN-β expression via a MyD88-independent pathway, activating IRF3, a key transcription factor responsible for IFN gene induction [30]. The protease activity of SVA 3Cpro can mediate the degradation of IRF3 and IRF7, reducing IFN-β, IFN-α1, IFN-α4, and ISG54 mRNA expression induced by these factors [31,32]. The current study highlights how the lncRNA 8244-ssc-miR-320-CCR7 axis regulates IFN-β expression by modulating the TLR3-mediated signaling pathway in response to SVA infection. Further studies are needed to gain a better understanding of the mechanisms underlying the axis's effect on IFN-β expression from TLR3 to IFN-β.

Supplementary Materials: The following supporting information can be downloaded at: https://www.mdpi.com/article/10.3390/microorganisms11030688/s1, Figure S1: (A) The expression level of lncRNA 8244, miRNA-320, and CCR7 in the PK-15 cell. (B) The ratio of IFN-β/β-actin in Figure 2B. (C) The expression level of TLR3 in the tissues. (D) The ratio of TLR3/β-actin in Figure 3D. (E) The ratio of IFN-β/β-actin in Figure 5B. (F) The ratio of IFN-β/β-actin in Figure 5C. (G) The ratio of TLR3/β-actin in Figure 6D. (H) The ratio of IFN-β/β-actin in Figure 8C. (I) The ratio of TLR3/β-actin in Figure 9D. All data are represented as mean ± SD with three replicates. * $p < 0.05$; ** $p < 0.01$.

Author Contributions: Conceptualization, Y.S.; Data curation, X.L.; Funding acquisition, J.M.; Investigation, X.T.; Methodology, X.T.; Project administration, Y.S. and J.M.; Software, R.Z.; Validation, R.Z. and L.G.; Writing—original draft, X.T. All authors have read and agreed to the published version of the manuscript.

Funding: The authors would like to acknowledge the Natural Science Foundation of Guangdong Province (2022A1515012473; 2020A1515010295).

Data Availability Statement: The original contributions presented in the study are included in the article. Further inquiries can be directed to the corresponding authors.

Conflicts of Interest: The authors declare no conflict of interest.

References

1. Hales, L.; Knowles, N.; Reddy, S.; Xu, L.; Hallenbeck, P. Complete genome sequence analysis of Seneca Valley virus-001, a novel oncolytic picornavirus. *J. Gen. Virol.* **2008**, *89 Pt 5*, 1265. [CrossRef] [PubMed]
2. Canning, P.; Canon, A.; Bates, J.; Gerardy, K.; Linhares, D.; Piñeyro, P.; Schwartz, K.; Yoon, K.; Rademacher, C.; Holtkamp, D. Neonatal Mortality, Vesicular Lesions and Lameness Associated with Senecavirus A in a U.S. Sow Farm. *Transbound. Emerg. Dis.* **2016**, *63*, 373–378. [CrossRef] [PubMed]
3. Leme, R.A.; Zotti, E.; Alcantara, B.K.; Oliveira, M.V.; Freitas, L.A.; Alfieri, A.F.; Alfieri, A.A. Clinical Manifestations of Senecavirus A Infection in Neonatal Pigs, Brazil. *Emerg. Infect. Dis. J.* **2015**, *22*, 1238–1241. [CrossRef]
4. Leme, R.; Oliveira, T.; Alfieri, A.; Headley, S.; Alfieri, A. Pathological, Immunohistochemical and Molecular Findings Associated with Senecavirus A-Induced Lesions in Neonatal Piglets. *J. Comp. Pathol.* **2016**, *155*, 145–155. [CrossRef] [PubMed]
5. Leme, R.; Zotti, E.; Alcantara, B.; Oliveira, M.; Freitas, L.; Alfieri, A.; Alfieri, A. Senecavirus A: An Emerging Vesicular Infection in Brazilian Pig Herds. *Transbound. Emerg. Dis.* **2015**, *62*, 603–611. [CrossRef]
6. Yang, M.; Van Bruggen, R.; Xu, W. Generation and diagnostic application of monoclonal antibodies against Seneca Valley virus. *J. Vet. Diagn. Investig.* **2012**, *24*, 42. [CrossRef] [PubMed]
7. Yang, F.; Zhu, Z.; Cao, W.; Liu, H.; Zhang, K.; Tian, H.; Liu, X.; Zheng, H. Immunogenicity and protective efficacy of an inactivated cell culture-derived Seneca Valley virus vaccine in pigs. *Vaccine* **2018**, *36*, 841–846. [CrossRef] [PubMed]
8. Vannucci, F.; Linhares, D.; Barcellos, D.; Lam, H.; Collins, J.; Marthaler, D. Identification and Complete Genome of Seneca Valley Virus in Vesicular Fluid and Sera of Pigs Affected with Idiopathic Vesicular Disease, Brazil. *Transbound. Emerg. Dis.* **2015**, *62*, 589–593. [CrossRef]
9. Segales, J.; Barcellos, D.; Alfieri, A.; Burrough, E.; Marthaler, D. Senecavirus A: An Emerging Pathogen Causing Vesicular Disease and Mortality in Pigs? *Vet. Pathol.* **2016**, *11*. [CrossRef]
10. Rapicavoli, N.; Kun, Q.; Zhang, J.; Megan, M.; Remi-Martin, L.; Chang, H. A mammalian pseudogene lncRNA at the interface of inflammation and anti-inflammatory therapeutics. *eLife* **2013**, *2*, e00762. [CrossRef]
11. Xiong, Y.; Yuan, J.; Zhang, C.; Zhu, Y.; Kuang, X. The STAT3-regulated long non-coding RNA Lethe promote the HCV replication. *Biomed. Pharmacother.* **2015**, *72*, 165–171. [CrossRef] [PubMed]
12. Zhu, M.; Liu, J.; Xiao, J.; Yang, L.; Wang, X. Lnc-mg is a long non-coding RNA that promotes myogenesis. *Nat. Commun.* **2017**, *8*, 14718. [CrossRef]
13. Mercer, T.R.; Dinger, M.E.; Mattick, J.S. Long non-coding RNAs: Insights into functions. *Nat. Rev. Genet.* **2009**, *10*, 155–159. [CrossRef] [PubMed]
14. Wang, P.; Xu, J.; Wang, Y.; Cao, X. An interferon-independent lncRNA promotes viral replication by modulating cellular metabolism. *Science* **2017**, *358*, 1051–1055. [CrossRef] [PubMed]
15. Ouyang, J.; Hu, J.; Chen, J.L. lncRNAs regulate the innate immune response to viral infection. Wiley Interdisciplinary Reviews. *Comput. Mol. Sci.* **2017**, *7*, 129–143.
16. Jiang, M.; Zhang, S.; Yang, Z.; Lin, H.; Zhu, J.; Liu, L.; Wang, W.; Liu, S.; Liu, W.; Ma, Y.; et al. Self-Recognition of an Inducible Host lncRNA by RIG-I Feedback Restricts Innate Immune Response. *Cell* **2018**, *173*, 906–919.e13. [CrossRef]
17. Du, H.; Zhao, Y.; Yin, Z.; Wang, D.W.; Chen, C. The role of miR-320 in glucose and lipid metabolism disorder-associated diseases. *Int. J. Biol. Sci.* **2021**, *17*, 402–416. [CrossRef]
18. Zhang, Z.; Zhang, J.; Li, J.; Geng, H.; Zhou, B.; Zhang, B.; Chen, H. miR320/ELF3 axis inhibits the progression of breast cancer via the PI3K/AKT pathway. *Oncol. Lett.* **2019**, *19*, 3239–3248.
19. Chang, T.H.; Huang, H.-Y.; Hsu, J.B.-K.; Weng, S.-L.; Horng, J.-T. An enhanced computational platform for investigating the roles of regulatory RNA and for identifying functional RNA motifs. *BMC Bioinform.* **2013**, *14* (Suppl. 2), S4. [CrossRef]
20. Pasma, T.; Davidson, S.; Shaw, S.L. Idiopathic vesicular disease in swine in Manitoba. *Can. Vet. J.-Rev. Vet. Can.* **2008**, *49*, 84–85.
21. Meng, X.; Luo, Y.; Anwar, M.; Yao, G.; Qiu, H. Long Non-Coding RNAs: Emerging and Versatile Regulators in Host–Virus Interactions. *Front. Immunol.* **2017**, *8*, 1663. [CrossRef]
22. Gomez, J.A.; Wapinski, O.; Yang, Y.; Bureau, J.F.; Gopinath, S.; Monack, D.; Chang, H.; Brahic, M.; Kirkegaard, K. The NeST Long ncRNA Controls Microbial Susceptibility and Epigenetic Activation of the Interferon-γ Locus. *Cell* **2013**, *152*, 743–754. [CrossRef]
23. Ouyang, J.; Zhu, X.; Chen, Y.; Wei, H.; Chen, Q.; Chi, X.; Qi, B.; Zhang, L.; Zhao, Y.; Gao, G.F.; et al. NRAV, a Long Noncoding RNA, Modulates Antiviral Responses through Suppression of Interferon-Stimulated Gene Transcription. *Cell Host Microbe* **2014**, *16*, 616–626. [CrossRef]
24. Wu, L.; Sun, L.; Hua, Y.; Yang, C.; Teng, Y. Overexpression of long non-coding rna h19 protects lung fibroblasts from lps-induced injury by targeting mir-181a and runx2 via activation of notch and jnk pathways. *J. Cell. Biochem.* **2018**, *120*, 12075.
25. Zhang, H.; Lu, W. LncRNA SNHG12 regulates gastric cancer progression by acting as a molecular sponge of miR-320. *Mol. Med. Rep.* **2018**, *2*, 2743–2749. [CrossRef]

26. Takekoshi, T.; Fang, L.; Paragh, G.; Hwang, S.T. CCR7-expressing B16 melanoma cells downregulate interferon-γ-mediated inflammation and increase lymphangiogenesis in the tumor microenvironment. *Oncogenesis* **2012**, *1*, e9. [CrossRef] [PubMed]
27. Carlos, T.S.; Young, D.; Stertz, S.; Kochs, G.; Randall, R.E. Interferon-induced inhibition of parainfluenza virus type 5; the roles of MxA, PKR and oligo A synthetase/RNase L. *Virology* **2007**, *363*, 166–173. [CrossRef]
28. Kawai, T.; Akira, S. Innate immune recognition of viral infection. *Uirusu* **2006**, *7*, 131–137. [CrossRef] [PubMed]
29. Hiroyuki Oshiumi, M.O.K.F. The TLR3/TICAM-1 Pathway Is Mandatoryfor Innate Immune Responses to Poliovirus Infection. *J. Immunol.* **2011**, *10*, 5320–5327. [CrossRef] [PubMed]
30. Kikkert, M. Innate Immune Evasion by Human Respiratory RNA Viruses. *J. Innate Immun.* **2020**, *12*, 4–20. [CrossRef]
31. Xue, Q.; Liu, H.; Zhu, Z.; Yang, F.; Xue, Q.; Cai, X.; Liu, X.; Zheng, H. Seneca Valley Virus 3C protease negatively regulates the type I interferon pathway by acting as a viral deubiquitinase. *Antivir. Res.* **2018**, *160*, 183–189. [CrossRef] [PubMed]
32. Xue, Q.; Liu, H.; Zhu, Z.; Yang, F.; Ma, L.; Cai, X.; Xue, Q.; Zheng, H. Seneca Valley Virus 3Cpro abrogates the IRF3- and IRF7-mediated innate immune response by degrading IRF3 and IRF7. *Virology* **2018**, *518*, 1–7. [CrossRef] [PubMed]

microorganisms MDPI

Systematic Review

Biomarkers Correlated with Tuberculosis Preventive Treatment Response: A Systematic Review and Meta-Analysis

Haoran Zhang [1,†], Zuyu Sun [2,†], Yi Liu [1], Rongrong Wei [1] and Nanying Che [1,2,*]

1 Biobank of Beijing Chest Hospital, Capital Medical University, Beijing 100054, China
2 Department of Pathology, Beijing Tuberculosis and Thoracic Tumor Research Institute, Beijing Chest Hospital, Capital Medical University, Beijing 100054, China
* Correspondence: chenanying@bjxkyy.cn
† These authors contributed equally to this work.

Abstract: Background: There is a need to identify alternative biomarkers to predict tuberculosis (TB) preventive treatment response because observing the incidence decline renders a long follow-up period. Methods: We searched PubMed, Embase and Web of Science up to 9 February 2023. The biomarker levels during preventive treatment were quantitatively summarized by means of meta-analysis using the random-effect model. Results: Eleven eligible studies, published during 2006–2022, were included in the meta-analysis, with frequently heterogeneous results. Twenty-six biomarkers or testing methods were identified regarding TB preventive treatment monitoring. The summarized standard mean differences of interferon-γ (INF-γ) were -1.44 (95% CI: -1.85, -1.03) among those who completed preventive treatment ($\tau^2 = 0.21$; $I^2 = 95.2\%$, $p < 0.001$) and -0.49 (95% CI: -1.05, 0.06) for those without preventive treatment ($\tau^2 = 0.13$; $I^2 = 82.0\%$, $p < 0.001$), respectively. Subgroup analysis showed that the INF-γ level after treatment decreased significantly from baseline among studies with high TB burden (-0.98, 95% CI: -1.21, -0.75) and among those with a history of Bacillus Calmette–Guérin vaccination (-0.87, 95% CI: -1.10, -0.63). Conclusions: Our results suggested that decreased INF-γ was observed among those who completed preventive treatment but not in those without preventive treatment. Further studies are warranted to explore its value in preventive treatment monitoring due to limited available data and extensive between-study heterogeneity.

Keywords: tuberculosis; preventive treatment; biomarkers; monitoring; meta-analysis

Citation: Zhang, H.; Sun, Z.; Liu, Y.; Wei, R.; Che, N. Biomarkers Correlated with Tuberculosis Preventive Treatment Response: A Systematic Review and Meta-Analysis. *Microorganisms* **2023**, *11*, 743. https://doi.org/10.3390/microorganisms11030743

Academic Editors: Shengxi Chen and Fabio Zicker

Received: 20 February 2023
Revised: 7 March 2023
Accepted: 11 March 2023
Published: 14 March 2023

1. Introduction

The World Health Organization (WHO) recommended tuberculosis (TB) preventive treatment as a key intervention strategy [1]. However, in the absence of an effective vaccine, TB preventive treatment is currently a difficulty in global TB control. The implementation of preventive treatment is beset with numerous obstacles due to its lack of objective treatment indication and effect evaluation index, which seriously affects the realization of the global END TB strategy. According to the 2022 WHO TB report, the global number of people provided with TB preventive treatment in 2021 was 3.5 million [2]. In addition, the population of latent tuberculosis infection (LTBI) in China is hundreds of millions [3]. In order to control the TB epidemic in China, developing suitable strategies for LTBI management is of utmost importance. In addition to preventing infection, the scale-up of preventive treatment coverage for the high-risk groups and evaluation of the therapeutic efficacy of LTBI must be addressed [4,5]. The evaluation of efficacy is essential to ensure that the vast number of people initiated on TB preventive treatment can benefit more. The efficacy of currently available preventive treatment ranges from 60 to 90% [6]. Direct observation of the decline in incidence is the standard method used to evaluate the protective effect of the preventive treatment. However, this usually requires a long follow-up period and huge resource costs. From the viewpoint of clinical and public health needs, instant and sensitive biomarkers or tests for monitoring the performance and predicting the outcomes of TB preventive treatment are warranted.

For adult pulmonary TB patients, the WHO recommends sputum smear microscopy and/or culture conversion at the end of the intensive phase of treatment as methods of monitoring treatment response [2,7,8]. Although sputum testing also could be used in evaluating successful TB preventive treatment, there is no instant and sensitive advantage in practice [9]. Additionally, no current WHO guidelines are recommended for the assessment of TB preventive treatment efficacy. A systematic review summarized the available evidence of active TB treatment monitoring biomarkers [10], but no systematic review has evaluated changes in biomarker levels with respect to TB preventive treatment. The requirement for biomarkers in the evaluation of TB preventive treatment stems from the critical aspect: the long and variable natural history of Mycobacterium tuberculosis (MTB) [11–13]. Timely efficacy evaluation will guarantee the optimization of TB control strategies and technical pathways to some extent. The main purpose of conducting a centered biomarker investigation is to concisely assess the effectiveness of preventive treatment and thereafter to prevent the development of active diseases [14]. In addition, new drugs, vaccines and other therapies will be required to realize the goal of END TB worldwide [4]. The important role of biomarkers also helps to accelerate intervention development by offering surrogate endpoints. Some studies were conducted to explore the potential biomarkers or assays for predicting clinical outcomes [9,15–19]. The types of specimens (e.g., blood and urine), the methods of testing (e.g., ELISA and transcriptomic sequencing) and the targeted biomarkers (e.g., gene expression signatures and cytokines) were heterogeneous [20,21]. Effective supporting evidence needs to overcome the study heterogeneity, especially by appropriately differentiating between clinical heterogeneity such as preventive treatment regimen [15,16,22,23]. No previous systematic review has addressed this. However, knowledge about specific biomarkers or assays that might represent promising options to optimize preventive treatment monitoring is still largely limited. The most valuable biomarkers are given great expectations of being directly involved in pathogenesis or protection and for which changes early during the preventive treatment stage can be related to the pharmacology of the intervention. This dynamic of the response and its relation to short- and long-term outcomes should be further evaluated by means of treatment trials. In this sense, a systematic summary of the related evidence on biomarkers during TB preventive treatment will benefit the constitution of efficacy indicators. More importantly, candidate biomarkers need to verify their influence on the delivery of routine care. Therefore, to better guide the recommendations for the use of biomarkers for TB preventive treatment response, this systematic review aims to summarize the evidence for biomarkers that is related to TB preventive treatment by means of systematic review and meta-analysis.

2. Materials and Methods

2.1. Methods

The updated Preferred Reporting Items for Systematic Reviews and Meta-analyses (PRISMA 2020, 27-item checklist) guideline was used in reporting our findings [24] (Supplementary Table S1). The study was registered in PROSPERO (CRD42023393104). No patients or members of the public were directly involved in this research study.

2.2. Search Strategy

Electronic databases PubMed, Web of Science and the Embase were searched to obtain articles addressing the biomarkers that correlated with TB preventive treatment response from database inception to 9 February 2023. The searches included combinations of key blocks of terms involving medical subject heading terms and text words: "tuberculosis infection", "latent tuberculosis infection", "LTBI", "latent tuberculosis" and "latent TB" to represent the exposed population; "biomarkers", "biomarker", "markers" and "marker" to indicate the outcomes index; and combined them with terms related to preventive treatment, such as "prophylaxis", "prevention", "prophylactic treatment" or "prophylactic therapy". The search strategies are detailed in Supplementary Table S2. The references list of relevant systematic reviews and eligible studies was manually examined to identify additional literature.

2.3. Eligibility Criteria

The criteria for inclusion in the meta-analyses were as follows: original articles investigating the longitudinal changes in biomarker levels during TB preventive treatment irrespective of study design; participants should initially test MTB-positive and then receive a monitored course of TB infection preventive treatment; and level of biomarkers should be measured at two or more time points. The decision to include only articles published in English was due to language capabilities of study team. If the study was reported in duplication, the version first published or which provided more detailed information was included. Review articles, animal studies, case reports, commentaries/editorials, mathematical modeling studies, conference abstracts and studies addressing active pulmonary TB treatment or exploring diagnostic biomarkers were excluded. We also excluded studies in which required data was unavailable, there was no full text, or there was no response to the request for data from authors. To minimize the potential bias caused by too small a sample size, articles with a sample size of fewer than 10 participants were excluded from the meta-analyses. In addition, to improve validity of data, non-peer-reviewed articles in preprint databases were excluded.

2.4. Study Selection and Data Extraction

All publications identified from the search strategy were imported into the reference management database EndNote (version X9, Clarivate^{TM}, Philadelphia, PA, USA). After duplications were removed, studies were screened in two stages: first by title and abstract and then by full-text article. Two researchers (ZHR and SZY) independently screened each title, abstract, full text and data extraction, with discrepancies resolved by consensus with a third researcher (LY). All full texts were checked against eligibility criteria (WRR).

During the extraction process, a predetermined proforma in Microsoft Excel Version 16.54 was used. All key extracted data were reviewed and quality-checked at the end of the data extraction phase by the same two researchers. For the quantitative assessment, we only extracted data on biomarkers when their quantitative level changes were reported by five or more studies. The levels of biomarkers and measures of spread data were extracted directly from the texts or tables when available. If not available, the data were extracted directly from available figures. In addition, for each included study, extracted data on study characteristics comprised of study site, study design, first author and published year, and study population. Participant data comprised age, sex, history of prior TB disease and anti-TB treatment, Bacillus Calmette–Guérin (BCG) vaccination and HIV-infected or acquired immune deficiency syndrome. Preventive treatment-related data included regimen of preventive treatment, number of participants receiving preventive treatment with serial test results and intervals between follow-ups. Outcome-related data comprised assay type, method and numbers of measurement, biomarkers levels of baseline and follow-up and follow-up period.

2.5. Assessment of Quality and Risk of Bias

Risk of bias was assessed using the Quality Assessment of Diagnostic Accuracy Score 2 (QUADAS-2). This tool consists of four domains: patient selection, index test, reference standard, and flow and timing. Each domain was evaluated using a set of guiding questions (Supplementary Table S3). Items were scored as "high concern," "low concern," or "unclear concern." The overall risk of bias was evaluated as "high risk" for studies with more than one area of high concern, "low risk" for those with two or more areas of low concern and no high risk and "unclear risk" for those with three or more areas of unclear concern and no high risk. Two reviewers (LY and WRR) appraised the risk of bias in the results of all studies that met inclusion criteria independently, with discrepancies resolved by discussion with a third reviewer (ZHR).

2.6. Outcomes

The prespecified outcome in the meta-analysis was association of dynamic changes in biomarkers levels with respect to TB preventive treatment. In addition to declining the incidence of active TB disease, infection clearance could be used for estimating for efficacy of preventive treatment. However, due to lack of golden standard for defining MTB infection, current infection testing could not absolutely reflect the real infection status. Therefore, we only indirectly evaluated the response to TB preventive treatment by comparison across groups. For instance, we compared the biomarker levels in different time points between different subgroups: preventive treatment group and untreated controls; infection testing positive group, negative groups or healthy control group; and comparison of baseline and post-treatment levels in the same population group.

Biomarker outcomes were required to collect more than one time point, at least with results at baseline and after preventive treatment. If more than two time points were reported, we extracted baseline results and the last testing results after preventive treatment. If more than two biomarkers in one study were reported, data for each study would be extracted. If two or more measurement tools were used in one study, corresponding results would be extracted and described, respectively.

2.7. Data Synthesis and Analysis

The extracted data were first transformed and standardized as mean and standard deviation (SD) values. The Box–Cox (BC) method [17,24] was applied to estimate the sample mean and SD from studies that reported the median accompanied by first and third quartiles. Standardized mean difference (SMD) was calculated as the effect size for each biomarker and summarized it using the random-effects model based on the Hartung–Knapp–Sidik–Jonkman method after adjustment to Hedges' g [25,26]. SMD was applied because different testing methods would be used to measure the same outcomes. Confidence intervals (CIs) were converted to SDs. When measures of variation were missing for mean differences within each treatment arm of a given study but a test of difference between treatment arms was reported, we converted F-statistics, t-statistics and *p*-values to standard errors and SDs. Here, t was taken as the square root of F, and it was assumed that the SDs of the mean differences in each treatment arm were equal. However, if the study showed no test of difference between groups, we used the highest SD recorded in the same meta-analysis for each treatment arm instead of the study's own data. At least two studies were required for each meta-analysis [27]. When studies included more than two intervention groups, we excluded irrelevant groups or combined relevant groups as recommended in the *Cochrane Handbook* in order to avoid arbitrary decisions [28]. For example, a regimen of isoniazid (INH) and rifapentine for 3 months (3HP) was compared with a regimen of INH for 6 months (6H) and 9H regimen group. Using the formula provided for combining two treatment groups (*Cochrane Handbook* 6.5.2.10, 23.3.4), when we aimed to target 3HP regimen as a subgroup, we calculated the combined mean difference and SD for 6H regimen and 9H regimen groups as a single control group.

As different biomarkers or assays were included, we performed statistical pooling restricted to biomarkers or assays, which had two or more studies that quantitatively presented the data of measures at different follow-up time points. We investigated heterogeneity by performing subgroup analyses on the following variables: TB burden of country of enrollment, age of participants (years), sex, history of BCG vaccination and sample size. Heterogeneity was assessed using the Cochrane Q test and quantified as I^2 values and τ^2 (the restricted maximum likelihood estimator was used to estimate this between-study variance) [10,29]. Heterogeneity was considered significant if the *p*-value of Cochran's Q test was <0.10 or if the I^2 statistic was $\geq 50\%$ [30]. Publication bias for one specific outcome rather than studies was shown using a funnel plot. Small-study effect was assessed by calculating Egger's test score and Begg's rank correlation analysis [31,32]. Meta-analyses were carried out using STATA Meta-Analysis (V2.0, Biostat, Englewood, NJ, USA). A two-sided $p < 0.05$ was considered to be statistically significant.

3. Results

3.1. Study Identification and Selection

As shown in Figure 1, a total of 5718 articles were obtained by database searches using different combinations of key terms. After removing duplicates, 5325 records were screened by title and abstract, of which 121 full texts were retrieved for detailed evaluation. Supplementary Table S4 shows a list of excluded studies with reasons. Of the excluded 105 articles which underwent full-text screening, 59 were excluded due to no information on TB preventive treatment, 18 due to mathematics modeling studies, 9 due to no serial data on TB preventive treatment monitoring, 5 due to no biomarkers being tested, 10 due to no data on treatment monitoring, 3 due to the required data being unavailable and 1 due to a sample < 10. Finally, 11 studies were eligible for data extraction and quantitative analysis [33–43] and are listed in Supplementary Table S5.

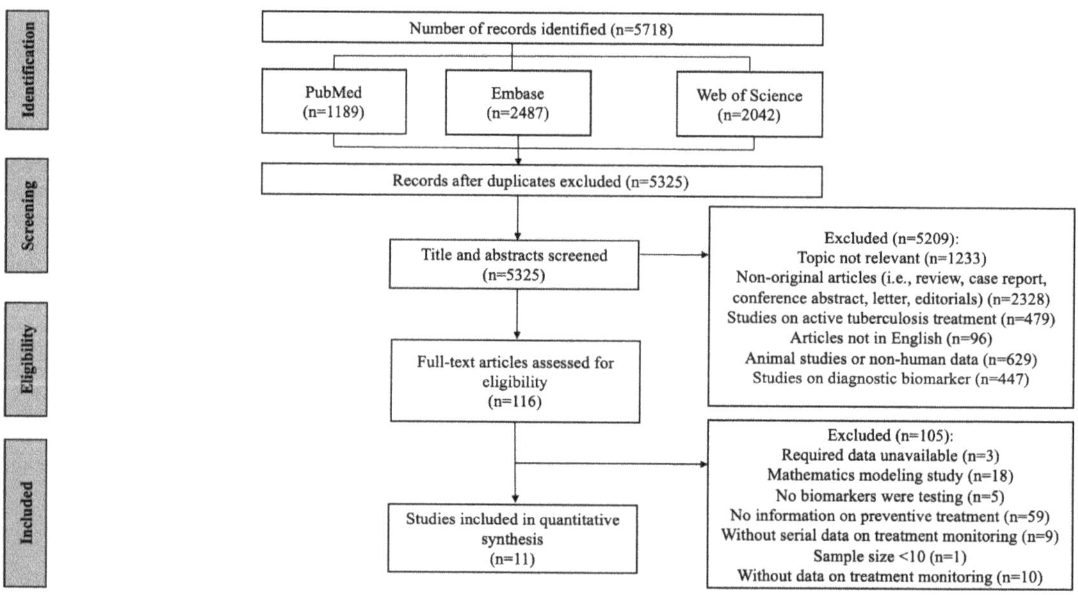

Figure 1. Flow chart of the included study identification.

3.2. Quality and Risk of Bias Assessment

The QUADAS-2 assessments are summarized in Supplementary Table S3 and Figure 2. When considering the four main categories of the QUADAS-2 tool, only two studies showed an overall low risk of bias. Specifically, the risk of bias for patient selection was high for studies that used a case-control study design ($n = 4$). Regarding preventive treatment monitoring reference standards, the results which were interpreted without knowledge of the reference standard received a "high risk of bias" ($n = 3$). Most studies did not report whether the reference standard was blinded while interpreting the results of the index ($n = 6$). Finally, the "flow and timing" of the study were generally at a "low risk of bias" as most samples for testing were either processed immediately or frozen.

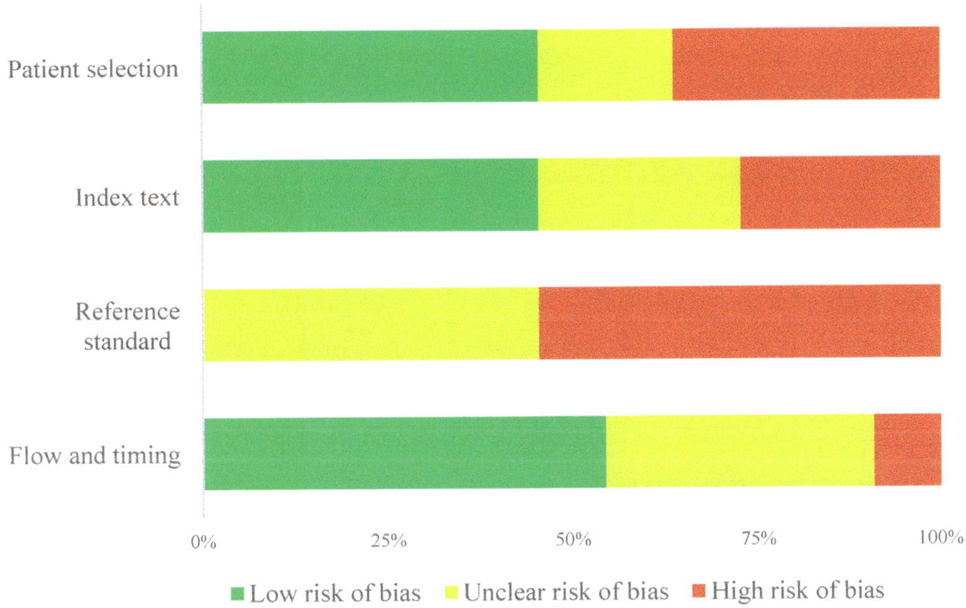

Figure 2. Summary of the QUADAS-2 risk of bias assessment.

3.3. Characteristics of Studies Included in the Study

Table 1 summarizes the characteristics of the included studies. The publication of the studies occurred between 2006 and 2022, and the sample size varied from 11 to 2618. Six studies were conducted in Asia [33–38], three in Europe and two in Africa [39–43]. Six studies were from high TB burden countries [33,34,37–40], two from middle TB burden countries [35,36] and three from low TB burden countries [41–43]. Five studies were of a randomized controlled trial (RCT) design [33,34,37,38,40], three were prospective studies [35,39,42], two were case-control studies [41,43], and one was a cross-sectional study [36]. Four studies were conducted in close-contact subjects with TB patients [36,39,40,42], six in healthy adults [33–35,37,38,43] and one in both adults and students [41]. Two studies indicated study participants had a prior history of TB [37,43]. Seven studies showed participants had previously received the BCG [33,36,39–43]. No studies included participants living with HIV.

Across all included studies, different biomarkers or testing methods were identified (n = 26). Of all the biomarkers, interferon-γ (IFN-γ) was the most frequently analyzed biomarker for TB preventive treatment monitoring [33,34,36,40,43] (n = 5). Studies that examined well-established diagnostics, such as QuantiFERON-TB Gold Plus (QFT-Plus), were not excluded as no previous systematic reviews have characterized their capabilities in TB preventive treatment monitoring.

Table 1. Characteristics of studies included in the meta-analysis.

First Author (Year)	Country (TB Burden)	Study Design	Study Participant	Age (Years)	Male (%)	Sample Size	Prior History of TB (%)	BCG Vaccination (%)	Preventive Treatment Regimen	Assay Name	Outcome at Baseline (Cutoff Value)	Outcome at Follow-Up (Cutoff Value)	Follow-Up Period
Ying Du (2022) [40]	China (high)	RCT	Rural residents	61.8 (mean)	60.0	40	0	-	Rifapentine plus isoniazid 600 mg for 6 weeks	QFT	INF-γ (0.35 IU/mL): 1.26 (0.95–2.39) C4BPA: 108.5 S100A9: 122.5	INF-γ (0.35 IU/mL): 0.06 (0.00–0.13) C4BPA: 74.1 S100A9: 69.8	24 months
Mulugeta Belay (2021) [33]	Ethiopia (high)	Nested prospective study	Household contacts	-	52.0	284	0	40.0	6 months of 300 mg isoniazid and 50 mg pyridoxine daily	Label-free quantitative protein mass spectrometry	M tuberculosis complex, n (%) DNA: 41 (95.0); QFT-Plus: 17 (40.0)	M tuberculosis complex DNA, n (%): 23 (53.0) QFT-PLUS: 13 (30.0)	6 months
Xuefang Cao (2021) [39]	China (high)	RCT	Rural residents	61 (50–69)	68.3	95	0	Treated participants: 53, untreated controls: 28	6 weeks of twice-weekly rifapentine (RPT) plus INH	Digital polymerase chain reaction	IL-1α: 120.8 (−5.7–421.6) IL-8: 12242.9 (1103.1–37141.7) IFN-γ (0.35 IU/mL): 121.9 (−182.5, 1170.2) IL-2: 162.9 (34.2, 886.9) IL-5: 63.5 (27.5, 220.8) IL-13: 4.9 (−0.6, 15.5) IL-17a: 25.4 (4.5, 73.1)	IL-1α: 88.43 (−12.6, 323.2) IL-8: 4604.9 (−618.3, 16283.7) IFN-γ (0.35 IU/mL): −10.7 (−283.9, 777.0) IL-2: 52.7 (4.5, 267.3) IL-5: 65.5 (22.2, 153.4) IL-13: 2.0 (−0.6, 7.3) IL-17a: 27.2 (−7.3, 51.4)	1 week after treatment
Ock-Hwa Kim (2020) [35]	Korea (medium)	Prospective study	Adults	47.6 ± 11.4	34.1	44	0	-	4 months of isoniazid and rifampin or up to 6 months of rifampin	QFT	IFN-γ (0.35 IU/mL): 3.395 QFT-Plus TB1: 3.060 IU/mL (0.35 IU/mL) QFT-Plus TB2: 2.880 IU/mL (0.35 IU/mL) QFT-Plus: 100%	IFN-γ (0.35 IU/mL): 6.804 QFT-Plus TB1: 2.905 IU/mL (0.35 IU/mL) QFT-Plus TB2: 3.880 IU/mL (0.35 IU/mL) QFT-Plus: 93.2	6 months
Henan Xin (2020) [37]	China (high)	RCT	Rural residents	50–70	55.0	Group A: 910, Group B:890, Group C: 818	1.0		Group A: 8 weeks of once-weekly rifapentine (RPT) plus isoniazid (INH); group B: 6 weeks of twice-weekly RPT plus INH; group C: untreated controls	Elisa kit for human cytokine	IFN-γ (0.35 IU/mL): Group A: 1.5 (0.7–3.5) Group B: 1.3 (0.6–3.5) Group C: 1.4 (0.7–3.2)	IFN-γ (0.35 IU/mL): Group A: 0.8 (0.3–2.4) Group B: 0.8 (0.4–2.3) Group C: 0.8 (0.3–2.2)	24 months

Table 1. *Cont.*

First Author (Year)	Country (TB Burden)	Study Design	Study Participant	Age (Years)	Male (%)	Sample Size	Prior History of TB (%)	BCG Vaccination (%)	Preventive Treatment Regimen	Assay Name	Outcome at Baseline (Cutoff Value)	Outcome at Follow-Up (Cutoff Value)	Follow-Up Period
Haoran Zhang (2020) [38]	China (high)	RCT	Rural residents	69 (65–73)	50.0	63	-	-	8-week regimen of once-weekly RPT plus INH, 6 weeks of twice-weekly RPT plus INH	QFT	IFN-γ (0.35 IU/mL): 16.5 (11.9–23.0)	IFN-γ (0.35 IU/mL): 13.0 (10.7–14.5)	1 week after treatment
Elisa Petruccioli (2018) [43]	Italy (low)	Case-control	HIV-uninfected patients	38 (24–50)	54.0	46	Microbiologically confirmed: 68.0, clinical diagnosis: 32.0	59.0	INH for 6 months or INH and rifampicin (RIF) for 3 months	QFT-Plus	TB1 or TB2, n (%): TB1 or TB2: 46 (100.0) TB1 and TB2 43 (93.0) Only TB1 1 (.0) Only TB2 2 (4.0) TB1 44 (96.0) TB2 45 (98.0)	TB1 or TB2, n (%): TB1 or TB2: 40 (87.0) TB1 and TB2 37 (80.0) Only TB1 1 (2.0) Only TB2 2 (4.0) TB1 38 (83.0) TB2 39 (85.0)	6 months
Irene Andia Biraro (2015) [34]	Sub-Saharan Africa (high)	RCT	Household contacts	24 (median)	37.0	24	-	IPT group: 17 (71.0), no IPT group: 11 (46.0)	Isoniazid plus pyridoxine daily for 6 months	QFT	IFN-γ (pg/mL): 648 (259.6–2605.4) TNF-α: 6.8 (1–74.6) IL-2: 291.2 (80.6–444.4) IL-5: 1 (1–1) IL-13: 2.1 (1–10.6) IL-10:1 (1–1) IL-17a: 1 (1–1.8) IL-17f 1 (1–1.9) IL-21: 1 (1–1) IL-22: 1 (1–4.9) CFP-10: 2700 (2050–4900) ESAT-6: 3700 (2550–5600)	IFN-γ (pg/mL): 541.1 (290.8, 791.3) IL-2:193.9 (108.5, 279.3) TNF-α: 55.6 (−6.3, 117.6) IL-5:1.1 (0.8, 1.4) IL-13: 5.1 (0.9, 9.4) IL-10: 3.3 (−0.8, 7.6) IL-17a: 1.5 (0.7, 2.3) IL-17f: 10.2 (−6.1, 26.6) IL-21: 123.8 (−133.9, 381.6) IL-22: 6.1 (−0.2, 12.6) CFP-10: 3.4 (3.3, 3.5) ESAT-6: 3.5 (3.4, 3.7)	6 months
SW Lee (2012) [36]	Korea (medium)	Cross-sectional	Close-contact soldiers	21 (20–24)	100.0	26	-	76.9	INH and rifampicin daily for 3 months	QFT	IFN-γ (0.35 IU/mL): 3.6 ± 3.4	IFN-γ (0.35 IU/mL): 0.8 ± 1.1	3 months

Table 1. *Cont.*

First Author (Year)	Country (TB Burden)	Study Design	Study Participant	Age (Years)	Male (%)	Sample Size	Prior History of TB (%)	BCG Vaccination (%)	Preventive Treatment Regimen	Assay Name	Outcome at Baseline (Cutoff Value)	Outcome at Follow-Up (Cutoff Value)	Follow-Up Period
Delia Goletti (2007) [42]	Italy (low)	Prospective study	Close contacts	No past exposure: 31, past exposure: 52	33.0	33	-	No INH, no past exposure: 2 (33), past exposure: 1 (20), INH, no past exposure: 11 (46), past exposure: 1 (11)	INH for 6 months	QFT	PHA: 14.7 ± 2.7 PPD: 17.6 ± 2.8 IFN-γ (cutoff value: NA: 17.5 ± 2.7 RD1 proteins: 12.5 ± 2.6 RD1 peptides: 9.2 ± 1.7	PHA: 14.8 ± 2.8 PPD: 14.1 ± 2.5 IFN-γ (cutoff value: NA: 5.2 ± 1.3 RD1 proteins: 2.6 ± 0.9 RD1 peptides: 0.9 ± 0.1	6 months
Katie Ewer (2006) [41]	UK (low)	Case-control	Adults, students	Adults: 47.0 (31–61), students 14.0 (11–15)	Adults: 18.0, students: 58.0	Adults: 11, students: 38	-	Adults: 9 (82.0), students: 30 (79.0)	3-month course of rifampin and isoniazid	Elisa based on RD1 selected peptide and proteins	Adults: RD1 147 (93 to 234), ESAT-6 18 (4 to 96), CFP-10 39 (9 to 171). Students: RD1 247 (173 to 354), ESAT-6 43 (20 to 93), CFP-10 57 (25 to 130)	Adults: 0. Students: RD1 72 (39 to 132), ESAT-6 51 (29 to 89), CFP-10 5 (0 to 74)	18 months

BCG, Bacillus Calmette–Guerin. INH, isoniazid. INF-γ, interferon-γ. QFT, QuantiFERON TB Gold In-Tube. RCT, randomized controlled trial. SMD, standard mean difference. TB, tuberculosis.

3.4. Biomarker Levels before and after TB Preventive Treatment

For biomarkers where there were two or more studies that numerically presented the dynamic levels at different follow-up time points, then meta-analysis was further performed. For INF-γ, the summarized SMD was found to be −1.44 (95% CI: −1.85, −1.03) for those who completed preventive treatment (τ^2 = 0.21; I^2 = 95.2%, $p < 0.001$) and −0.49 (95% CI: −1.05, 0.06) for those without preventive treatment (τ^2 = 0.13; I^2 = 82.0%, $p < 0.001$) (Figure 3A). With respect to QFT-Plus testing, which reflected the levels of TB1 and TB2 tube antigens, we summarized their status after treatment, such as reversion or conversion of MTB infection, based on the reported results of included studies. The summarized rate of reversion was observed to be 9.3% (4.3–14.3%) after preventive treatment (τ^2 = 0; I^2 = 0%, p = 0.613) (Figure 3B). In addition, four cytokine proteins, IL-2, IL-5, IL-13 and IL-17a, were analyzed using quantitative synthesis (Figure 3C). The results of this meta-analysis found that the CI for these four biomarkers crossed the null and therefore did not reach statistical significance.

As shown in Table 2, despite the heterogeneity between the included studies, the summarized biomarker levels before and after TB preventive treatment were found to be significantly influenced by the TB burden of the country of enrollment, age of participants, male gender, history of BCG vaccination and sample size ($p < 0.001$) in the stratified analyses. In addition to the age of participants, male gender and sample size, the changed level was found to be particularly significant in the studies with a high TB burden (−0.98; 95% CI: −1.21, −0.75) (τ^2 = 0.04; I^2 = 92.1%, $p < 0.001$) and among participants with BCG vaccination history (−0.87; 95% CI: −1.10, −0.63) (τ^2 = 0.06; I^2 = 92.1%, $p < 0.001$).

No evident publication bias was observed when all 11 studies were evaluated using Begg's rank correlation analysis (p = 0.216) and Egger's weighted regression analysis (p = 0.389), but the funnel plot seemed to be asymmetric (Supplementary Figure S1).

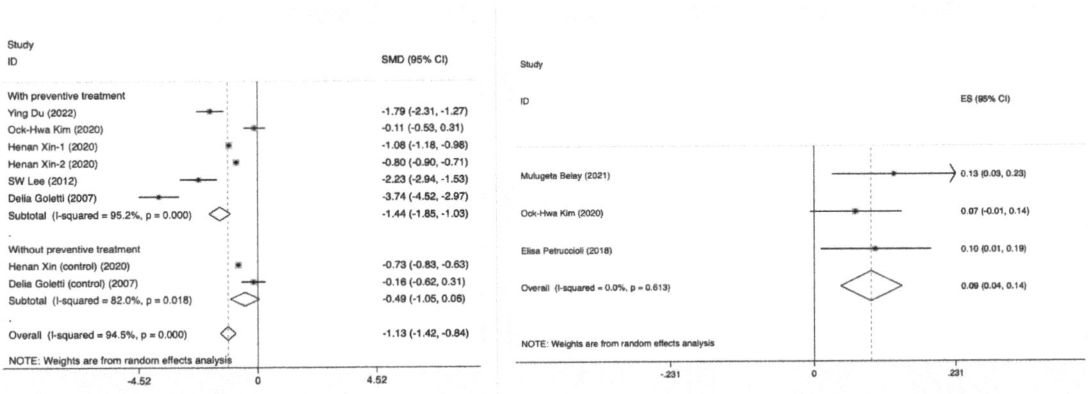

(A) (B)

Figure 3. *Cont.*

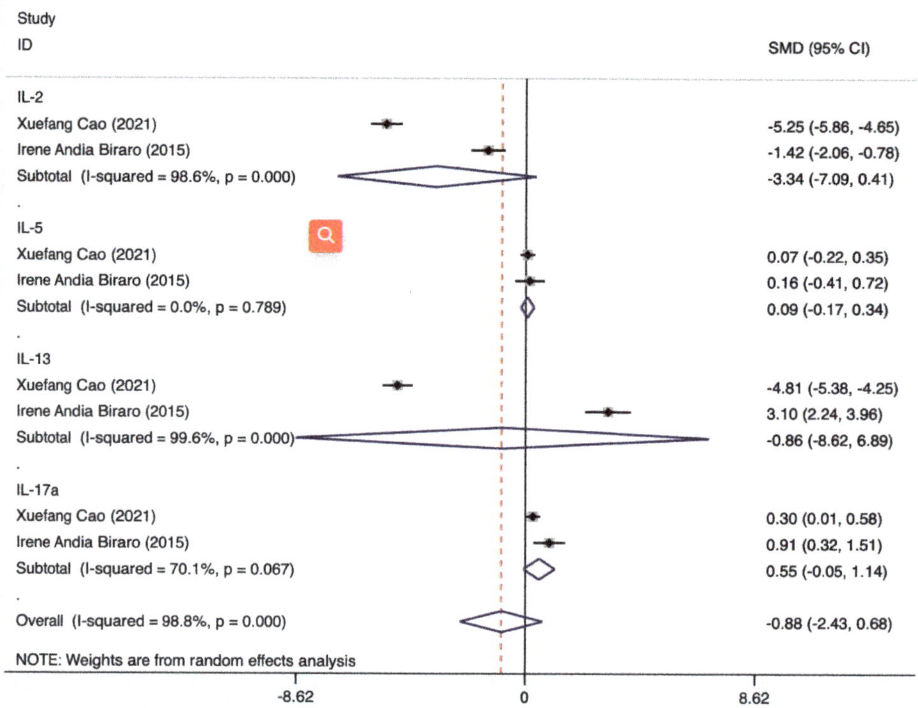

(C)

Figure 3. Forest plot of meta-analysis on summarized SMD of INF-γ (**A**) [35–37,40,42], summarized rate of reversion QFT-Plus testing (**B**) [33,35,43] and summarized SMD of 4 cytokine proteins (**C**) [34,39]. CI, confidence interval. INF-γ, interferon-γ. QFT-Plus, QuantiFERON-TB Gold Plus. SMD, standard mean difference.

Table 2. Stratified analysis of summarized INF-γ levels before and after TB preventive treatment.

	No. of Studies †	Pooled SMD (95% CI)	Heterogeneity		
			I^2	p	τ^2
TB burden of country of enrollment §					
High	4	−0.98 (−1.21, −0.75)	92.1	<0.001	0.04
Medium	2	−1.15 (−3.23, 0.93)	96.1	<0.001	2.17
Low	2	−1.94 (−5.45, 1.57)	98.3	<0.001	6.31
Age of participants (years)					
<60	4	−1.53 (−3.06, −0.01)	96.6	<0.001	2.31
≥60	4	−0.98 (−1.21, −0.75)	92.1	<0.001	0.04
Male (%)					
<50	3	−1.13 (−1.42, −0.84)	97.2	<0.001	2.50
≥50	5	−1.10 (−1.35, −0.85)	92.3	<0.001	0.06
History of BCG vaccination					
No	5	−0.87 (−1.10, −0.63)	92.1	<0.001	0.06
Yes	3	−2.03 (−4.18, 0.12)	97.1	<0.001	3.49
Sample size					
<50	5	−1.58 (−2.79, −0.37)	95.9	<0.001	1.82
≥50	3	−0.87 (−1.08, −0.67)	92.3	<0.001	0.03

§ Based on 2022 data. † If the included studies had multiple treatment or control groups, the pooled analysis was undertaken based on each comparison. Therefore, the sum would be not the total of included studies. BCG, Bacillus Calmette–Guerin. CI, confidence interval. SMD, standard mean difference. TB, tuberculosis.

4. Discussion

To our knowledge, this is the first meta-analysis aiming to evaluate biomarkers that correlated with TB preventive treatment response. The findings of the included studies were found to be frequently heterogeneous. INF-γ, QFT-Plus and cytokines, including IL-2, IL-5, IL-13 and IL-17a, were commonly used for evaluating responses to TB preventive treatment. The summarized changes in INF-γ levels with respect to TB preventive treatment were found to be −1.44 (95% CI: −1.85, −1.03) for those who completed preventive treatment ($I^2 = 95.2\%$, $p < 0.001$). In addition, the summarized reversion rate of QFT-Plus was observed to be 10.0% (1.0–19.0%) after preventive treatment ($I^2 = 0\%$, $p = 0.613$). However, due to the limited included studies, whether the identified biomarkers or testing method in our study could be used to monitor TB preventive treatment efficacy in practice should be further explored.

Globally, the treatment monitoring of LTBI is an essential part of TB control. In recent years, increasing numbers of trial studies have reported the efficacy of different TB preventive treatment regimens across various populations [44–47]. However, previous meta-analysis has rarely examined the association of biomarkers with TB preventive treatment response. Some systematic reviews only assessed the biomarkers that correlated with active pulmonary TB treatment responses [48]. We chose to focus on the performance of potential biomarkers in preventive treatment because there is currently a lack of biomarkers and assays for clinicians to evaluate the effectiveness and characteristics of preventive treatment. Therefore, to date, the association of accessible biomarkers with preventive treatment response is still equivocal, leaving a large knowledge gap on this topic. In our meta-analysis, we found that the level of INF-γ decreased along with preventive treatment. However, a number of studies, even previous meta-analyses, have shown that dynamic changes in INF-γ levels were not associated with preventive treatment [37,49,50]. Inconsistently, our results indicated that INF-γ levels decreased significantly among participants with preventive treatment (−1.44, 95% CI: −1.85, −1.03) but not among those without preventive treatment (−0.49, 95% CI: −1.05, 0.06). One of the possible explanations is that we directly used the levels of blood INF-γ rather than testing the reversion indicator. A testing method such as QuantiFERON-TB Gold In-Tube (QFT) could be a potential reason because a simple "wobble" around the cut-point due to imperfect test reproducibility could induce misclassification [51,52]. More importantly, the underlying mechanism of TB testing reversion is still unclear [50] and needs further exploration. The declined levels of INF-γ after TB preventive treatment may help with clinical decision-making by identifying people who respond favorably to treatment, though further analyses are needed to characterize how this differs between those who respond to treatment and those who do not respond to treatment or are lost at follow-up. Vice versa, the ability to detect the up- or down-regulation of INF-γ may allow for the simpler and earlier identification of people who respond both favorably and unfavorably to treatment. Although the subgroup analyses showed consistent results to some extent, the number of included studies in these subgroups was low.

Additionally, QFT-Plus was not excluded from our analysis because its performance has been scarcely evaluated in TB preventive treatment by previous meta-analyses. QFT-Plus, as a new generation of QFT, was developed with two TB-specific antigen tubes (TB1 and TB2). TB1 tube induces a specific CD4+ T-cell response, and TB2 was designed to induce IFN-γ production by both CD4+ and CD8+ T cells. It has been reported that QFT-Plus has increased sensitivity compared to QFT with some capability of identifying recent infection among contacts [53,54]. Therefore, we anticipated a promising performance of QFT-Plus in monitoring the TB preventive treatment efficacy. A previous study involving 6H or rifampicin for 3 months conducted in Italy showed that TB preventive treatment significantly decreased INF-γ levels in response to the antigen present in QFT-Plus tests in LTBI and TB-active patients [43]. Unfortunately, the summarized QFT-Plus results in the present meta-analysis showed a lower rate of reversion (9.0%) compared to previous serial testing results (13.0%) without preventive treatment but were significantly higher than QFT (5.9%). Based on only three studies, we could not obtain more reliable estimates of the associations between QFT-Plus and preventive treatment responses. As for the policy implications, the evidence to guide practitioners and clinicians on the monitoring efficacy of interventions continues to be constrained by limited confirmatory biomarkers. There is no recommendation for TB preventive treatment evaluation in the guidelines from a national or regional level or even from the WHO.

A lack of good-quality evidence remains a barrier to more conclusive findings. While only two included studies were at low risk of bias, as shown in Supplementary Table S3, imprecision may reduce our certainty in the outcomes, further highlighting the need for well-designed trials. Our pooled results were somewhat clinically meaningful but might have been overestimated owing to the small-study effect [55]. The assessment of publication bias was restricted by the limited number of studies for some comparisons; therefore, meta-regression was not performed in our study. Although no evident heterogeneity was observed between the included studies, as indicated by the τ^2 and I^2 estimates, potential publication bias was found using the visual inspection of asymmetry in the funnel plot, indicating uncertainty around the observed effect estimates. As suggested by our subgroup analyses, it might be partly explained by the high TB burden of the country. In the present meta-analysis, studies from countries with a high TB burden account for a moderate proportion. In such high-transmission settings, repeated exposure to MTB or re-infection might occur more often. However, we have no clue to speculate on various IFN-γ responses to treatment in populations from different regions. Another concern is that no history of BCG vaccination was associated with declined IFN-γ levels after treatment. This was consistent with previous results that BCG vaccination did not affect QFT performance [56]. In addition, some other factors such as age, sex or sample size might be associated with the heterogeneity. This highlights the need for updating and describing the study design more in detail.

There are some limitations in this study. First, only articles published in English were identified in our study, thus a potential language bias might exist. Second, we extracted and analyzed the rawest available data in each study where possible, standardized these data using SMD and then performed subgroup analyses to validate the findings. Despite these precautions, some degree of imprecision was still possible in the pooled effect sizes driven by variations in the aggregate data. Accessing individual participant data could considerably improve the precision, which we strongly recommend in future research. Third, because not all necessary information could be obtained from all the studies included, related stratifications (e.g., by prior history of TB) could not be made. Fourth, few included studies and small sample sizes limited data analysis and generation of our results. Each biomarker identified in this study has only been evaluated in a limited number of studies, precluding us from meta-analyzing the dynamic change of these markers throughout treatment. Additional well-conducted studies may be required to identify and validate surrogate biomarkers for the field to advance.

In conclusion, our results show low certainty in supporting the use of INF-γ and QFT-Plus for evaluating TB preventive treatment responses. These preliminary data may help inform future studies to investigate these biomarkers in a more rigorous and standardized manner. In the era of prevention-centered care and to give strong recommendations and better guidance for clinical operations, further studies are warranted to determine the potential applicability of accessible biomarkers to field trials; only in this way can their particular role in monitoring and even diagnosis be served.

Supplementary Materials: The following supporting information can be downloaded at: https://www.mdpi.com/article/10.3390/microorganisms11030743/s1.

Author Contributions: H.Z. and Z.S. performed the literature search, study identification and data extraction. H.Z. performed statistical analysis and drafted the manuscript. Y.L. and R.W. participated in the data extraction and quality control. N.C. designed the study and contributed to the manuscript revision and interpretation of results and conclusions. All authors have read and agreed to the published version of the manuscript.

Funding: This work was supported by grants from the National Natural Science Foundation of China (Grant Number: 82072381) and the Beijing Municipal Science and Technology Project (Grant Number: Z191100006619079).

Data Availability Statement: This study is registered at https://www.crd.york.ac.uk/prospero/#myprospero (accessed on 30 January 2023) with identifier CRD42023393104. The corresponding authors can provide, upon request, data used for analyses and analytic codes and any other materials used in the review after applying necessary measures to guarantee that no individual is identified or identifiable.

Conflicts of Interest: The authors declare no conflict of interest.

References

1. Behr, M.A.; Edelstein, P.H.; Ramakrishnan, L. Is Mycobacterium tuberculosis infection life long? *BMJ* **2019**, *367*, l5770. [CrossRef] [PubMed]
2. WHO. *Global Tuberculosis Report 2022*; WHO: Geneva, Switzerland, 2022.
3. Houben, R.M.; Dodd, P.J. The Global Burden of Latent Tuberculosis Infection: A Re-estimation Using Mathematical Modelling. *PLoS Med.* **2016**, *13*, e1002152. [CrossRef] [PubMed]
4. Cui, X.; Gao, L.; Cao, B. Management of latent tuberculosis infection in China: Exploring solutions suitable for high-burden countries. *Int. J. Infect. Dis.* **2020**, *92*, S37–S40. [CrossRef]
5. Surie, D.; Interrante, J.D.; Pathmanathan, I.; Patel, M.R.; Anyalechi, G.; Cavanaugh, J.S.; Kirking, H.L. Policies, practices and barriers to implementing tuberculosis preventive treatment-35 countries, 2017. *Int. J. Tuberc. Lung Dis.* **2019**, *23*, 1308–1313. [CrossRef]
6. Getahun, H.; Matteelli, A.; Chaisson, R.E.; Raviglione, M. Latent Mycobacterium tuberculosis infection. *N. Engl. J. Med.* **2015**, *372*, 2127–2135. [CrossRef]
7. WHO. *Consolidated Guidelines on Tuberculosis: Tuberculosis Preventive Treatment: Module 1: Prevention; WHO: Geneva, Switzerland, 2020.*
8. WHO. *Treatment of Tuberculosis: Guidelines*, 4th ed.; WHO Guidelines Approved by the Guidelines Review Committee; WHO: Geneva, Switzerland, 2010.
9. Goletti, D.; Lindestam Arlehamn, C.S.; Scriba, T.J.; Anthony, R.; Cirillo, D.M.; Alonzi, T.; Denkinger, C.M.; Cobelens, F. Can we predict tuberculosis cure? What tools are available? *Eur. Respir. J.* **2018**, *52*, 1801089. [CrossRef]
10. Higgins, J.P.; Thompson, S.G.; Deeks, J.J.; Altman, D.G. Measuring inconsistency in meta-analyses. *BMJ* **2003**, *327*, 557–560. [CrossRef]
11. Rao, S.P.S.; Alonso, S.; Rand, L.; Dick, T.; Pethe, K. The protonmotive force is required for maintaining ATP homeostasis and viability of hypoxic, nonreplicating Mycobacterium tuberculosis. *Proc. Natl. Acad. Sci. USA* **2008**, *105*, 11945–11950. [CrossRef]
12. Schnappinger, D.; Ehrt, S. A broader spectrum of tuberculosis. *Nat. Med.* **2016**, *22*, 1076–1077. [CrossRef]
13. Scriba, T.J.; Penn-Nicholson, A.; Shankar, S.; Hraha, T.; Thompson, E.G.; Sterling, D.; Nemes, E.; Darboe, F.; Suliman, S.; Amon, L.M.; et al. Sequential inflammatory processes define human progression from M. tuberculosis infection to tuberculosis disease. *PLoS Pathog.* **2017**, *13*, e1006687. [CrossRef]
14. Fiore-Gartland, A.; Carpp, L.N.; Naidoo, K.; Thompson, E.; Zak, D.E.; Self, S.; Churchyard, G.; Walzl, G.; Penn-Nicholson, A.; Scriba, T.J.; et al. Considerations for biomarker-targeted intervention strategies for tuberculosis disease prevention. *Tuberculosis* **2018**, *109*, 61–68. [CrossRef]
15. Bruins, W.S.; van Leth, F. Effect of secondary preventive therapy on recurrence of tuberculosis in HIV-infected individuals: A systematic review. *Infect. Dis.* **2017**, *49*, 161–169. [CrossRef]
16. Spyridis, N.P.; Spyridis, P.G.; Gelesme, A.; Sypsa, V.; Valianatou, M.; Metsou, F.; Gourgiotis, D.; Tsolia, M.N. The effectiveness of a 9-month regimen of isoniazid alone versus 3- and 4-month regimens of isoniazid plus rifampin for treatment of latent tuberculosis infection in children: Results of an 11-year randomized study. *Clin. Infect. Dis.* **2007**, *45*, 715–722. [CrossRef]
17. Page, M.J.; McKenzie, J.E.; Bossuyt, P.M.; Boutron, I.; Hoffmann, T.C.; Mulrow, C.D.; Shamseer, L.; Tetzlaff, J.M.; Akl, E.A.; Brennan, S.E.; et al. The PRISMA 2020 statement: An updated guideline for reporting systematic reviews. *BMJ* **2021**, *372*, n71. [CrossRef]
18. Adetifa, I.M.; Ota, M.O.C.; Jeffries, D.J.; Lugos, M.D.; Hammond, A.S.; Battersby, N.J.; Owiafe, P.K.; Donkor, S.D.; Antonio, M.; Ibanga, H.B.; et al. Interferon-γ ELISPOT as a biomarker of treatment efficacy in latent tuberculosis infection: A clinical trial. *Am. J. Respir. Crit. Care Med.* **2013**, *187*, 439–445. [CrossRef]
19. Higuchi, K.; Harada, N.; Mori, T. Interferon-gamma responses after isoniazid chemotherapy for latent tuberculosis. *Respirology* **2008**, *13*, 468–472. [CrossRef]
20. Wallis, R.S.; Doherty, T.M.; Onyebujoh, P.; Vahedi, M.; Laang, H.; Olesen, O.; Parida, S.; Zumla, A. Biomarkers for tuberculosis disease activity, cure, and relapse. *Lancet Infect. Dis.* **2009**, *9*, 162–172. [CrossRef]

21. Biomarkers and surrogate endpoints: Preferred definitions and conceptual framework. *Clin. Pharmacol. Ther.* **2001**, *69*, 89–95. [CrossRef]
22. Sterling, T.R.; Villarino, M.E.; Borisov, A.S.; Shang, N.; Gordin, F.; Bliven-Sizemore, E.; Hackman, J.; Hamilton, C.D.; Menzies, D.; Kerrigan, A.; et al. Three months of rifapentine and isoniazid for latent tuberculosis infection. *N. Engl. J. Med.* **2011**, *365*, 2155–2166. [CrossRef]
23. Danel, C.; Moh, R.; Gabillard, D.; Badje, A.; Le Carrou, J.; Ouassa, T.; Ouattara, E.; Anzian, A.; Ntakpé, J.-B.; Minga, A.; et al. A Trial of Early Antiretrovirals and Isoniazid Preventive Therapy in Africa. *N. Engl. J. Med.* **2015**, *373*, 808–822. [CrossRef]
24. McGrath, S.; Zhao, X.; Steele, R.; Thombs, B.D.; Benedetti, A. Estimating the sample mean and standard deviation from commonly reported quantiles in meta-analysis. *Stat. Methods Med. Res.* **2020**, *29*, 2520–2537. [CrossRef] [PubMed]
25. IntHout, J.; Ioannidis, J.P.A.; Borm, G.F. The Hartung-Knapp-Sidik-Jonkman method for random effects meta-analysis is straightforward and considerably outperforms the standard DerSimonian-Laird method. *BMC Med. Res. Methodol.* **2014**, *14*, 25. [CrossRef]
26. Lin, L.; Aloe, A.M. Evaluation of various estimators for standardized mean difference in meta-analysis. *Stat. Med.* **2021**, *40*, 403–426. [CrossRef]
27. Avery, N.; McNeilage, A.G.; Stanaway, F.; Ashton-James, C.E.; Blyth, F.M.; Martin, R.; Gholamrezaei, A.; Glare, P. Efficacy of interventions to reduce long term opioid treatment for chronic non-cancer pain: Systematic review and meta-analysis. *BMJ* **2022**, *377*, e066375. [CrossRef]
28. Higgins, J.P.; Eldridge, S.; Li, T. Including variants on randomized trials. In *Cochrane Handbook for Systematic Reviews of Interventions*; John Wiley & Sons: Chichester, UK, 2019; pp. 569–593. [CrossRef]
29. Veroniki, A.; Jackson, D.; Viechtbauer, W.; Bender, R.; Knapp, G.; Kuss, O. Recommendations for quantifying the uncertainty in the summary intervention effect and estimating the between-study heterogeneity variance in random–effects meta-analysis. *Cochrane Methods* **2015**, 25–27.
30. Higgins, J.P.T.; Thompson, S.G. Quantifying heterogeneity in a meta-analysis. *Stat. Med.* **2002**, *21*, 1539–1558. [CrossRef]
31. Egger, M.; Davey Smith, G.; Schneider, M.; Minder, C. Bias in meta-analysis detected by a simple, graphical test. *BMJ* **1997**, *315*, 629–634. [CrossRef]
32. Duval, S.; Tweedie, R. Trim and fill: A simple funnel-plot-based method of testing and adjusting for publication bias in meta-analysis. *Biometrics* **2000**, *56*, 455–463. [CrossRef]
33. Belay, M.; Tulu, B.; Younis, S.; Jolliffe, D.A.; Tayachew, D.; Manwandu, H.; Abozen, T.; Tirfie, E.A.; Tegegn, M.; Zewude, A.; et al. Detection of Mycobacterium tuberculosis complex DNA in CD34-positive peripheral blood mononuclear cells of asymptomatic tuberculosis contacts: An observational study. *Lancet Microbe.* **2021**, *2*, E267–E275. [CrossRef]
34. Biraro, I.A.; Egesa, M.; Kimuda, S.; Smith, S.G.; Toulza, F.; Levin, J.; Joloba, M.; Katamba, A.; Cose, S.; Dockrell, H.M.; et al. Effect of isoniazid preventive therapy on immune responses to mycobacterium tuberculosis: An open label randomised, controlled, exploratory study. *BMC Infect. Dis.* **2015**, *15*, 438. [CrossRef]
35. Kim, O.H.; Jo, K.W.; Park, S.; Jo, Y.H.; Kim, M.N.; Sung, H.; Shim, T.S. Comparison of the change in QuantiFERON-TB Gold Plus and QuantiFERON-TB Gold In-Tube results after preventive therapy for latent tuberculosis infection. *PLoS ONE* **2020**, *15*, e0234700. [CrossRef]
36. Lee, S.W.; Lee, S.H.; Yim, J.J. Serial interferon-gamma release assays after chemoprophylaxis in a tuberculosis outbreak cohort. *Infection* **2012**, *40*, 431–435. [CrossRef]
37. Xin, H.; Cao, X.; Zhang, H.; Liu, J.; Pan, S.; Li, X.; Guan, L.; Shen, F.; Liu, Z.; Wang, D.; et al. Dynamic changes of interferon gamma release assay results with latent tuberculosis infection treatment. *Clin. Microbiol. Infect.* **2020**, *26*, e1551–e1555. [CrossRef]
38. Zhang, H.R.; Cao, X.F.; Xin, H.N.; Liu, J.M.; Pan, S.G.; Guan, L.; Shen, F.; Liu, Z.S.; Wang, D.K.; Guan, X.L.; et al. Serum level of IL-1ra was associated with the treatment of latent tuberculosis infection in a Chinese population. *Bmc Infect. Dis.* **2020**, *20*, 330. [CrossRef]
39. Cao, X.F.; Xin, H.N.; Zhang, H.R.; Liu, J.M.; Pan, S.G.; Du, Y.; Feng, B.X.; Quan, Z.S.; Guan, L.; Shen, F.; et al. The Association Between Mycobacteria-Specific Antigen-Induced Cytokines and Host Response to Latent Tuberculosis Infection Treatment in a Chinese Population. *Front. Microbiol.* **2021**, *12*, 716900. [CrossRef]
40. Du, Y.; Xin, H.A.; Cao, X.F.; Liu, Z.S.; He, Y.J.; Zhang, B.; Yan, J.X.; Wang, D.K.; Guan, L.; Shen, F.; et al. Association Between Plasma Exosomes S100A9/C4BPA and Latent Tuberculosis Infection Treatment: Proteomic Analysis Based on a Randomized Controlled Study. *Front. Microbiol.* **2022**, *13*, 934716. [CrossRef]
41. Ewer, K.; Millington, K.A.; Deeks, J.J.; Alvarez, L.; Bryant, G.; Lalvani, A. Dynamic antigen-specific point-source exposure to T-cell responses after Mycobacterium tuberculosis. *Am. J. Respir. Crit. Care Med.* **2006**, *174*, 831–839. [CrossRef]
42. Goletti, D.; Parracino, M.P.; Butera, O.; Bizzoni, F.; Casetti, R.; Dainotto, D.; Anzidei, G.; Nisii, C.; Ippolito, G.; Poccia, F.; et al. Isoniazid prophylaxis differently modulates T-cell responses to RD1-epitopes in contacts recently exposed to Mycobacterium tuberculosis: A pilot study. *Respir. Res.* **2007**, *8*, 5. [CrossRef]
43. Petruccioli, E.; Chiacchio, T.; Vanini, V.; Cuzzi, G.; Codecasa, L.R.; Ferrarese, M.; Schinina, V.; Palmieri, F.; Ippolito, G.; Goletti, D. Effect of therapy on Quantiferon-Plus response in patients with active and latent tuberculosis infection. *Sci. Rep.* **2018**, *8*, 15626. [CrossRef]
44. Stout, J.E.; Sterling, T.R.; Horsburgh, C.R. One Month of Rifapentine plus Isoniazid to Prevent HIV-Related Tuberculosis. *N. Engl. J. Med.* **2019**, *381*, e23. [CrossRef]

45. Villarino, M.E.; Scott, N.A.; Weis, S.E.; Weiner, M.; Conde, M.B.; Jones, B.; Nachman, S.; Oliveira, R.; Moro, R.N.; Shang, N.; et al. Treatment for preventing tuberculosis in children and adolescents: A randomized clinical trial of a 3-month, 12-dose regimen of a combination of rifapentine and isoniazid. *JAMA Pediatr.* **2015**, *169*, 247–255. [CrossRef] [PubMed]
46. Gupta, A.; Montepiedra, G.; Aaron, L.; Theron, G.; McCarthy, K.; Bradford, S.; Chipato, T.; Vhembo, T.; Stranix-Chibanda, L.; Onyango-Makumbi, C.; et al. Isoniazid Preventive Therapy in HIV-Infected Pregnant and Postpartum Women. *N. Engl. J. Med.* **2019**, *381*, 1333–1346. [CrossRef] [PubMed]
47. Diallo, T.; Adjobimey, M.; Ruslami, R.; Trajman, A.; Sow, O.; Obeng Baah, J.; Marks, G.B.; Long, R.; Elwood, K.; Zielinski, D.; et al. Safety and Side Effects of Rifampin versus Isoniazid in Children. *N. Engl. J. Med.* **2018**, *379*, 454–463. [CrossRef] [PubMed]
48. Zimmer, A.J.; Lainati, F.; Aguilera Vasquez, N.; Chedid, C.; McGrath, S.; Benedetti, A.; MacLean, E.; Ruhwald, M.; Denkinger, C.M.; Kohli, M. Biomarkers That Correlate with Active Pulmonary Tuberculosis Treatment Response: A Systematic Review and Meta-analysis. *J. Clin. Microbiol.* **2022**, *60*, e0185921. [CrossRef]
49. Zhang, H.; Xin, H.; Wang, D.; Pan, S.; Liu, Z.; Cao, X.; Wang, J.; Li, X.; Feng, B.; Li, M.; et al. Serial testing of Mycobacterium tuberculosis infection in Chinese village doctors by QuantiFERON-TB Gold Plus, QuantiFERON-TB Gold in-Tube and T-SPOT.TB. *J. Infect.* **2019**, *78*, 305–310. [CrossRef] [PubMed]
50. Zhang, H.; Xin, H.; Li, X.; Li, H.; Li, M.; Feng, B.; Gao, L. Reversion of QuantiFERON-TB Gold In-Tube test in individuals with and without prophylactic treatment for latent tuberculosis infection: A systematic review and meta-analysis. *J. Infect.* **2018**, *77*, 276–282. [CrossRef]
51. Nemes, E.; Rozot, V.; Geldenhuys, H.; Bilek, N.; Mabwe, S.; Abrahams, D.; Makhethe, L.; Erasmus, M.; Keyser, A.; Toefy, A.; et al. Optimization and Interpretation of Serial QuantiFERON Testing to Measure Acquisition of Mycobacterium tuberculosis Infection. *Am. J. Respir. Crit. Care Med.* **2017**, *196*, 638–648. [CrossRef]
52. Rafiza, S.; Rampal, K.G. Serial testing of Malaysian health care workers with QuantiFERON®-TB Gold In-Tube. *Int. J. Tuberc. Lung Dis.* **2012**, *16*, 163–168. [CrossRef]
53. Rozot, V.; Patrizia, A.; Vigano, S.; Mazza-Stalder, J.; Idrizi, E.; Day, C.L.; Perreau, M.; Lazor-Blanchet, C.; Ohmiti, K.; Goletti, D.; et al. Combined use of Mycobacterium tuberculosis-specific CD4 and CD8 T-cell responses is a powerful diagnostic tool of active tuberculosis. *Clin. Infect. Dis.* **2015**, *60*, 432–437. [CrossRef] [PubMed]
54. Petruccioli, E.; Chiacchio, T.; Pepponi, I.; Vanini, V.; Urso, R.; Cuzzi, G.; Barcellini, L.; Cirillo, D.M.; Palmieri, F.; Ippolito, G.; et al. First characterization of the CD4 and CD8 T-cell responses to QuantiFERON-TB Plus. *J. Infect.* **2016**, *73*, 588–597. [CrossRef]
55. Dechartres, A.; Trinquart, L.; Boutron, I.; Ravaud, P. Influence of trial sample size on treatment effect estimates: Meta-epidemiological study. *BMJ* **2013**, *346*, f2304. [CrossRef] [PubMed]
56. Gao, L.; Lu, W.; Bai, L.; Wang, X.; Xu, J.; Catanzaro, A.; Cárdenas, V.; Li, X.; Yang, Y.; Du, J.; et al. Latent tuberculosis infection in rural China: Baseline results of a population-based, multicentre, prospective cohort study. *Lancet Infect. Dis.* **2015**, *15*, 310–319. [CrossRef] [PubMed]

microorganisms

Review

Dynamics of Hepatitis B Virus Covalently Closed Circular DNA: A Mini-Review

Jie-Li Hu * and Ai-Long Huang *

Key Laboratory of Molecular Biology on Infectious Diseases, Ministry of Education, Chongqing Medical University, Chongqing 400016, China
* Correspondence: 102564@cqmu.edu.cn (J.-L.H.); ahuang@cqmu.edu.cn (A.-L.H.); Tel./Fax: +86-(023)-68486780 (J.-L.H.); +86-(023)-68485112 (A.-L.H.)

Abstract: Eradication of cccDNA is an ideal goal of chronic hepatitis B (CHB) therapy. Understanding the changes in the cccDNA pool during therapy provides a basis for developing CHB treatment strategies. On the other hand, the shift in the balance of the cccDNA pool following therapies allowed researchers to investigate the dynamics of cccDNA. Central to the description of cccDNA dynamics is a parameter called cccDNA half-life. CccDNA half-life is not an intrinsic property of cccDNA molecules, but a description of an observed phenomenon characterized by cccDNA pool decline. Since cccDNA has to be in the nuclei of host cells to function, the half-life of cccDNA is determined by the state and destiny of the host cells. The major factors that drive cccDNA decay include noncytopathic effects and hepatocyte turnover (death and division). In some cases, the determining factor is not the half-life of cccDNA itself, but rather the half-life of the hepatocyte. The main purpose of this review is to analyze the major factors affecting cccDNA half-life and determine the areas requiring further study. In addition, the discrepancy in cccDNA half-life between short-term and long-term nucleot(s)ide analog (NUC) therapy was reported. Hypotheses were proposed to explain the multi-phasic decline of cccDNA during NUC therapy, and a framework based on cccDNA dynamics was suggested for the consideration of various anti-HBV strategies.

Keywords: hepatitis B virus; cccDNA; half-life; nucleot(s)ide analogues; hepatocyte; turnover

Citation: Hu, J.-L.; Huang, A.-L. Dynamics of Hepatitis B Virus Covalently Closed Circular DNA: A Mini-Review. *Microorganisms* **2023**, *11*, 600. https://doi.org/10.3390/microorganisms11030600

Academic Editors: Shengxi Chen and Fabio Zicker

Received: 17 January 2023
Revised: 23 February 2023
Accepted: 24 February 2023
Published: 27 February 2023

1. Introduction

Covalently closed circular DNA (cccDNA) is the first viral product after hepatitis B virus (HBV) infection of hepatocytes. It serves as a stable repository of HBV genetic information in liver cells and represents the most impenetrable barrier to a functional cure for chronic hepatitis B (CHB) [1–3]. CccDNA is converted from its precursor, relaxed circular DNA (rcDNA) (Figure 1). With great effort, in almost two decades, researchers have largely characterized the host factors required for cccDNA formation, intermediates formed during the conversion and the functional regulation mechanisms of cccDNA minichromosomes (see reviews [4–6]). There is a consensus that the functional cure of CHB requires eradication or persistent suppression of cccDNA [2,7–10]. However, current therapies based on the nucleot(s)ide analog (NUC) or/and pegylated-interferon-α (PEG-IFNα) have limited efficacy in achieving a functional cure for CHB, necessitating the development of new therapeutic strategies [11–16].

It is well known that serum HBV virions are in a highly dynamic equilibrium; approximately 10^{11} viruses decay and are produced daily [17]. Similarly, the cccDNA pool in hepatocytes is in a dynamic equilibrium [18]. Since any treatment to cure CHB must reduce the number of cccDNA molecules in the livers, it is important to understand the dynamic changes in cccDNA quantity during therapy. In fact, cccDNA dynamics can only be revealed by treatments that shift the equilibrium in the cccDNA pool. Both NUC and PEG-IFNα can reduce cccDNA levels either indirectly by suppressing cccDNA synthesis or directly by degrading cccDNA. The decline of cccDNA during therapy denotes a shift in the dynamic balance between cccDNA synthesis and decay. The dynamics of the cccDNA pool can be investigated by exploring this shift. However, cccDNA dynamics have not been extensively discussed to date. In this mini-review, we focused on the property of cccDNA half-life, a critical parameter to describe the dynamics of cccDNA, and the factors that affect it. In addition, the discrepancy in cccDNA half-life between short-term and long-term nucleot(s)ide analog (NUC) therapy was analyzed, and a framework based on cccDNA dynamics was suggested for the consideration of various anti-HBV strategies.

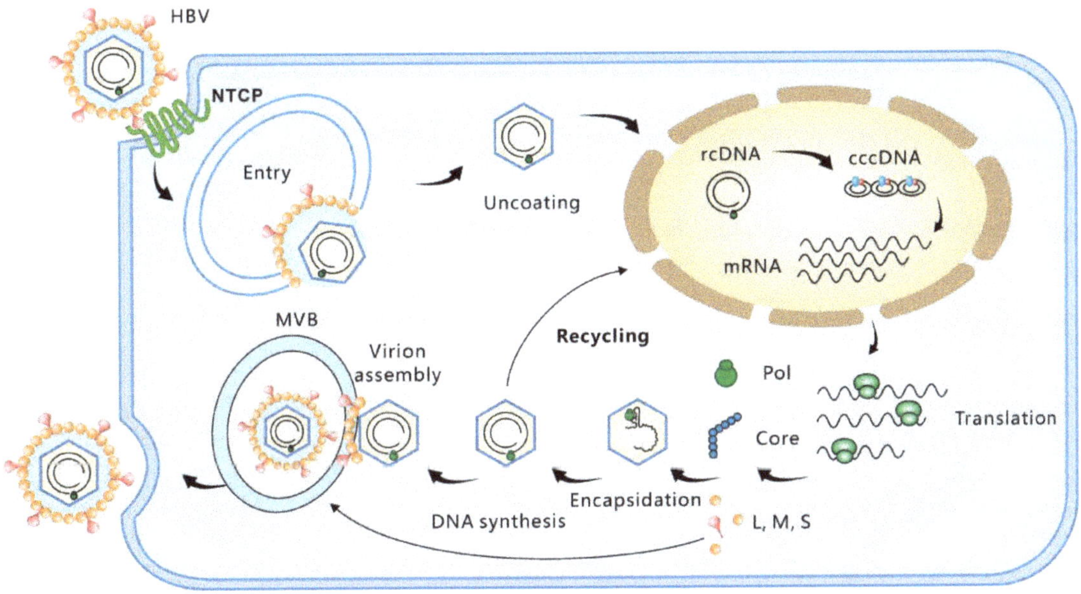

Figure 1. Replication cycle of HBV. The viruses infect hepatocytes by interacting with the receptor sodium taurocholate cotransporting polypeptide (NTCP). Relaxed circular DNA (rcDNA) is released into the nuclei after nucleocapsid uncoating. With the help of the host factors, rcDNA is converted into cccDNA, which serves as the template of viral RNA transcription. The productive nucleocapsid containing progeny rcDNA either secretes through the multivesicular body (MVB) pathway or recycles to the nuclei.

2. The Term 'cccDNA Half-Life' Does Not Refer to an Intrinsic Property of cccDNA

The half-life of pure cccDNA independent of its host cells is practically meaningless, since cccDNA has to be in the nuclei of host cells in order to function. It is the destiny and environment of the host cells that govern the 'half-life' of cccDNA. The cccDNA disappear when the host cells die, and decrease when the host cells divide. CccDNA can also be degraded in noncytopathic ways. The cccDNA pool may exhibit very different half-lives under different conditions. In chimpanzees recovering from an acute HBV infection, cccDNA can decline with a half-life as short as 3 days [19]. In contrast, patients receiving long-term NUC therapy exhibited a very slow cccDNA decline, with a half-life of as long as 26 months [20]. Apparently, 'cccDNA half-life' involves the quantified amplitudes of the cccDNA decline phenomenon, which vary considerably depending on the situation. Therefore, this term does not refer to the intrinsic nature of cccDNA, and that the values represent the observed kinetics of cccDNA decay as a pool that would change depending on the circumstances. In this sense, the term 'apparent cccDNA half-life', as proposed by Boettler et al., is more accurate than 'cccDNA half-life' [18].

3. Major Factors That Affect cccDNA Half-Life

Given that cccDNA is located in the nuclei of the host cells and performs its functions, the state of the host cells will exert a significant influence on the dynamics of cccDNA. Therefore, factors affecting the situations of host cells affect cccDNA half-life. Noncytopathic effects and hepatocyte turnover are the major factors impacting cccDNA half-life.

3.1. Noncytopathic Effects Contribute to cccDNA Degradation in Acute Infection

Chisari et al. first reported that HBV-specific cytotoxic T lymphocytes could abolish HBV gene expression and replication in the liver of transgenic mice via noncytopathic cytokines [21]. Based on these findings, they postulated that this antiviral process might be primarily responsible for the viral clearance during human HBV infection rather than the destruction of infected cells. To test this hypothesis, they observed the clearance of HBV in acutely infected chimpanzees. In these animals, a significant decrease in HBV replicative intermediates and cccDNA from the liver was observed long before the peak of T cell infiltration and most of the liver disease. Xia et al. also found that HBV-specific T cells inhibit HBV replication and reduce cccDNA in infected cells without the direct contact required for cytolysis [22]. The findings support the presence of a noncytopathic mechanism contributing to the clearance of the virus [23]. Mathematical modeling was later performed to evaluate the extent to which cytopathic and noncytopathic T cell effector functions contribute to the resolution of HBV infection in three acutely infected chimpanzees. CccDNA demonstrated a rapid decay in the first phase of the decline, with a half-life of 3 days. If cell death and cell division were the only mechanisms responsible for clearance of HBV infection, such a short half-life would imply the destruction and regeneration of approximately 11 livers. In contrast, simulation incorporating the cytokine effects indicated significantly less hepatocyte death and regeneration (1.4–2.8 livers) [19]. Studies using woodchuck models also supported that both noncytopathic effects and hepatocyte death were responsible for the elimination of cccDNA during recovery from transient infections [24]. Cytokines associated with this noncytopathic effect include interferon-gamma (IFNγ), tumor necrosis factor-alpha (TNFα), and IFNα/β [22,25]. Later studies showed that IFNα accelerates the decay of DHBV cccDNA in culture [26] and induces the specific degradation of the nuclear HBV cccDNA without hepatotoxicity by up-regulating APOBEC3A and APOBEC3B cytidine deaminases [22,27].

3.2. Hepatocyte Turnover Is a Major Factor Promoting cccDNA Decay

The liver is a solid organ with a high regenerative capacity to ensure that the liver-to-bodyweight ratio is always at 100% of what is required for body homeostasis [28]. Hepatocyte turnover is regulated by two closely related events, namely hepatocyte death and division. As hepatocytes die, the number of hepatocytes must be replenished to maintain liver mass and new hepatocytes are produced. Typically, a cell death results in a new cell being produced, and vice versa. It is believed that cccDNA is degraded or eliminated when the host cell dies and cannot infect another hepatocyte.

3.2.1. Evidence That Hepatocyte Turnover Promotes cccDNA Decay

The influence of hepatocyte turnover on cccDNA levels (or infected hepatocytes) was first demonstrated by Mason et al. [29]. An NUC, 2′-deoxycarbocyclic guanosine (2′-CDG), was found to significantly reduce DHBV cccDNA levels (with a half-life of approximately 2 weeks) in the livers of ducks congenitally infected with DHBV. 5-bromo-2′-deoxyuridine (BrdU) labelling showed that hepatocytes proliferation increased by 10-fold after 2 weeks of therapy. This result suggests that the inhibition of viral replication and acceleration of hepatocyte turnover with 2′-CDG therapy caused a rapid clearance of infected hepatocytes. In contrast, 5-fluoro-2′,3′-dideoxy-3′-thiacytidine (524 W), another potent inhibitor of DHBV DNA synthesis that did not expedite the turnover of hepatocytes in ducks, led to a strong inhibition of virus production but a slower rate of decline in the number of infected hepatocytes and cccDNA (with a half-life of more than 6 weeks) than that with 2′-CDG.

Addison et al. analyzed the decline of DHBV cccDNA in ducks treated with a combination of lamivudine (LAM) and a dideoxyguanosine prodrug for 5 months [30]. The cccDNA pools in three ducks demonstrated an exponential decline, with half-lives ranging from 35 to 57 days. Liver sections stained with the cell division marker, proliferating cell nuclear antigen (PCNA), showed that animals with a continuous cccDNA loss had a significantly higher number of PCNA-positive nuclei than those whose cccDNA levels had plateaued.

Mason et al. also treated WHV-infected primary hepatocyte cultures with [1-(2-fluoro-5-methyl-β-L-arabinofuranosyl) uracil] (L-FMAU), a reverse transcriptase inhibitor. They found that although L-FMAU caused a 200-fold suppression in viral DNA replication, no significant loss in cccDNA was observed in the infected hepatocytes during the 40-day treatment [31]. In chronically infected woodchucks, L-FMAU treatment for 30 weeks reduced cccDNA levels to between 1.2 and 5.4% of pretreatment levels, suggesting a half-life of 33 to 50 days. This reduction could be explained by an infected-cell death rate of 1.3 to 2.1% per day, which is in reasonable agreement with the PCNA staining results.

Outstanding examples exhibiting the influence of hepatocyte death on cccDNA is from the experimental therapy of HBV infection by T cells that were engineered to express HBV-specific chimeric antigen receptors (CARs) or T cell receptors (TCRs). Protzer's lab reported that expression of CARs directed against the HBV surface proteins enables human T cells to kill HBV-infected human hepatocytes and to eliminate viral cccDNA in vitro and in vivo [32,33]. T cells stably expressing high-affinity HBV envelope- or core-specific TCRs also effectively control HBV infection in humanized mice by specifically clearing infected hepatocytes without damaging noninfected cells [34].

3.2.2. Effect of Cell Division on cccDNA

In nonproliferating primary cultures of woodchuck hepatocyte infected with woodchuck hepatitis virus (WHV), treatment of L-FMAU did not facilitate loss of cccDNA. No significant loss of cccDNA from the treated cultures was seen between days 8 and 40 post-infection, indicating a cccDNA half-life of at least 32 days [31]. Ko et al. tested cccDNA half-life in an HBV-infected cell line (HepG2-NTCP-K7) in the presence of 2.5% DMSO, which induces cell cycle arrest. Treating cells with entecavir (ETV) reduced cccDNA levels by 48% at day 45 post-infection, suggesting a cccDNA half-life of approximately 40 days in these nondividing cells [35].

During hepatocyte division, the HBV cccDNA minichromosome could be distributed unequally or even lost during mitosis since it is not a cellular chromosome equipped with centromere structures. Chong et al. examined the dynamic changes in HBV cccDNA in different cellular growth stages of a stably HBV-producing cell line, 1.3ES2 [36]. They found that the amount of cccDNA decreased dramatically in the cells during their exponential proliferation, and cccDNA could be removed when proliferating cells were subjected to long term of lamivudine (3TC) treatment. The half-life of cccDNA in the exponentially proliferating cells was approximately 5 days. Once the cells had grown to confluence, the half-life of cccDNA was approximately 9 days, but 20% of the remaining cccDNA was retained stably inside the cells for more than 30 days [36]. To quantify the impact of cell division on cccDNA loss, Tu et al. passaged cells every 3 days to induce mitosis of HBV-infected HepG2-NTCP and HepaRG-NTCP cells. Then, the number of cccDNA copies was measured by precise PCR assays and the number of HBV-expressing cells was monitored over time using reporter viruses. They observed that cccDNA levels undergo a 5-fold decrease after each round of mitosis, which is the exact rate predicted by mathematical models assuming a complete loss of cccDNA in daughter cells [37].

In contrast, some other studies did not observe significant cccDNA loss during cell division. Li et al. determined the distribution of HBV cccDNA in continuously passaged HepAD38 cells using a fluorescence imaging in situ hybridization (FISH)-based assay. The findings showed that nuclear HBV DNA symmetrically distribute to daughter cells. It is worth noting that the detected nuclear DNA was not necessarily cccDNA, since the probes used were not cccDNA-specific [38]. Dandri et al. evaluate whether nuclear cccDNA becomes unstable during cell division in cultures of primary hepatocytes isolated from a WHV chronically infected woodchuck [39]. The cells were treated with epidermal growth factor (EGF) for 24 days to induce proliferation in the presence of adefovir (ADV). They found that cccDNA signals had the same intensity in ADV-treated plates despite the fact that a 50% increase in cell number (70% vs. 105%) did occur when EGF was added. Cautions should be taken while interpreting this result, because the cells seemed to proliferate at a marginal rate during a long period.

To determine the effect of cell proliferation on cccDNA in vivo, Lutgehetmann et al. transplanted primary tupaia hepatocytes (PTHs) chronically infected with woolly monkey HBV (WM-HBV) from chimeric mice into the urokinase-type plasminogen activator (uPA)/severe combined immunodeficiency (SCID)/beige (USB) mouse model. Transplantation of WM-HBV-infected hepatocytes led to an average of 3.8 PTH doublings within 80 days. Remarkably, a median 2-log decline of cccDNA per cell determined during PTH proliferation was due to both dilution of the cccDNA pool among daughter cells and a 0.5-log loss of intrahepatic cccDNA loads [40]. The effect of hepatocyte division on the HBV cccDNA pool was also observed in humanized mice. Primary human hepatocytes (PHH) from HBV-infected humanized mice were serially transplanted into naïve recipients. The findings demonstrated that human hepatocyte division, triggers substantial cccDNA loss even without the involvement of cytolytic mechanisms. CccDNA copies per PHH decreased by 2.4 log during cell division (from day 3 post-transplantation until day 30), and the total cccDNA amounts per liver reduced by 0.7 log during this period [41]. In contrast, Hayashi et al. demonstrated that in HBV-infected human-liver-chimeric mice (PXB-mice), the total cccDNA content did not change during liver repopulation after entecavir treatment. An explanation for the inconsistence is that the experimental procedure used by Hayashi et al. is different from that used by Allweiss et al. In PXB-mice, observation started 8 weeks after transplantation and 4 weeks after HBV infection. Allweiss et al. adopted a different procedure in which human liver cells from USB mice already infected with HBV were transplanted to naïve recipients and observation started 3 days after transplantation. Probably, this procedure maximized the proliferation rate of the cells and thus presented its effect on cccDNA more readily.

3.2.3. Hepatocyte Turnover Rate

Knowing the lifespan or turnover rate of hepatocytes is essential in exploring the dynamics of the cccDNA pool due to the close relationship between cell death and the change in the cccDNA pool. In 1961, Richard determined the lifespan of rat liver cells by tracing tritiated thymidine (H^3-thymidine) incorporated into cell nuclei [42]. He found that 0.22% of hepatic nuclei were labelled following a single injection of H^3-thy-midine in normal adult rats, suggesting that a liver would renew in 450 days. Of note, it was estimated that in normal liver approximately 41% of labeled cells had died or divided by 60 days, and the remaining labeled cells had lifespans varying up to at least 6 additional months [42] (Figure 2A). This implies a great heterogeneity of hepatocytes lifespan. For fatty and cirrhotic liver, the lifespan of hepatocytes was considerably less than that of normal liver (26 days vs. 450 days). Again, those hepatocytes exhibited heterogeneity in lifespan. Although 98% of labelled hepatocytes died or divided in 60 days, the remaining 2% of cells did not go further change for at least several months. Magami et al. used H^3-thymidine to trace the life of hepatocytes in mice [43]. They found that the proportion of labelled hepatocytes decreased from 33.7% to 16.6%, 10.9% and 7.9% after 100, 200 and 300 days (Figure 2A). This indicates an overall hepatocyte half-life of approximately 100 days, but the heterogeneity is apparent at individual cell level, given that 7.9% of the cells survived longer than 300 days [43].

The concept of the balance between hepatocyte death and production provides a basis for estimating hepatocyte death rate by monitoring newly produced hepatocytes over time. Recently, He et al. assessed hepatocyte proliferation in a mouse model using Ki67-induced expression of the fluorescent protein [44]. Their dual recombinase-based method, called Protracer (proliferation tracer, Figure 2B), enabled the continuous recording of the proliferation events of entire cell populations over time in multiple organs, including the liver. Although their primary goal was to characterize the cell source of hepatocytes during homeostasis and regeneration, the study provided valuable information to estimate the turnover of hepatocytes. One set of data revealed that from week 2 to week 12 after tamoxifen induction, a gradual increase in the number of newly produced hepatocytes was observed, as demonstrated by the GFP expression in the cells induced by Ki67 expression. At week 12, over 60% of the total population were GFP-positive cells (Figure 2C). This implies that more than 60% of the hepatocytes died during the course of the 10 weeks, as approximately 60% of the hepatocytes were newly produced at this time. Given that some GFP-positive hepatocytes must have already died, 10 weeks may be a low estimate of the half-life of mouse hepatocytes in homeostasis. This estimate closes to the half-life of hepatocytes (100 days) inferred from the H^3-thymidine labelling experiments. In addition, different regions of the liver lobule contribute differentially to hepatocyte turnover, and zone 2 is the primary source of new hepatocytes during homeostasis and regeneration (Figure 2D) [44,45].

The half-life of normal human hepatocytes does not permit experimental determination in vivo by lineage or H^3-thymidine tracing. However, indirect estimation of normal human hepatocyte turnover can be made by immunohistochemical detection of PCNA. PCNA is present in the cell nuclei throughout the cell cycle but binds tightly to chromatin at the peak of the S-phase. Immunofluorescent studies have shown that in cells only PCNA associated with DNA replication sites (S-phase-specific PCNA) can be detected. In contrast, non-S-phase nuclear PCNA, which is present in lower amounts and is not physically associated with DNA replication sites, is likely lost or undetectable by conventional immunocytochemical methods [46–48]. In methanol-fixed normal human liver, the PCNA-labeling index was from 0.05% to 0.78% for hepatocytes [49,50]. This translates to 0.15 to 2.34% of hepatocyte death per day, assuming that S-phase lasts for approximately 8 h and that PCNA staining therefore only reflects approximately one-third of the cells in S-phase over a 24 h period [51]. Another proliferation marker, Ki67, can also be used to estimate hepatocyte turnover rate. Farinati et al. found that 0.2% of the hepatocytes were Ki6-positive in the periportal area (zone 1) in the samples from patients with hepatitis B, translating to a 0.6% of hepatocyte death per day. However, this might be a low estimate since the majority of proliferating hepatocytes were found in zone 2 of mouse lobules [44,45].

In summary, normal hepatocytes of rat and mice exhibit a heterogeneity in half-life. The majority of hepatocytes have a half-life of 50–100 days, but a minority of hepatocytes can live longer than 300 days. The death rate of normal human hepatocytes is estimated to be approximately 0.5% per day, equivalent to a half-life of 100 days. However, details of half-life heterogeneity of human hepatocytes need further investigation.

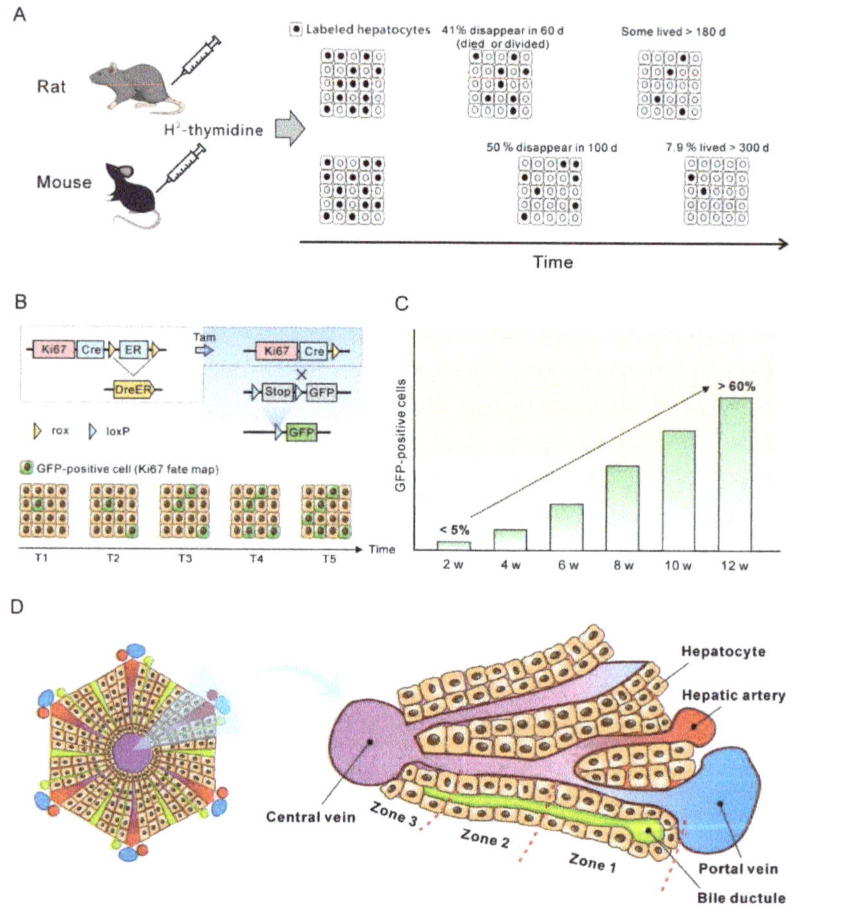

Figure 2. Monitoring the lifespan and proliferation of hepatocytes. (**A**) H^3-thymidine was injected into rat and mouse models to label hepatocytes and monitor the changes in the labeled cells [42,43]. (**B**) Ki67 expression induced a constitutive expression of GFP which was used to monitor the fate of liver cells. (**C**) From week 2 to week 12, the number of hepatocytes with Ki67-expression-induced GFP expression increased to more than 50%, which represent newly produced hepatocytes. The figure is a modification of a figure from reference [45]. (**D**) Structure of the liver lobule. Hepatocytes in the liver lobule are organized into three zones and those in zones 2 and 1 have higher proliferation rates than those in zone 3 during homeostasis.

4. HBV Genome Recycling and cccDNA Replenishment

Hepatoma cell lines, such as HepG2, transfected stably or transiently with replication-capable HBV genome support cccDNA formation [52–54]. These cccDNA must be completely converted from intracellular recycling rcDNA, since these cell lines do not support de novo HBV infection. In a HepG2 cell line expressing sodium taurocholate cotransporting polypeptide (NTCP) (HepG2-NTCP-K7), transduction of L-HBsAg-deficient, but replication-competent overlength HBV genome (HBV1.3L-) via an adenoviral vector (Ad-HBV1.3L-) also resulted in cccDNA. Because cccDNA formation in this system can only result from nuclear import of capsids, this demonstrates that the intracellular recycling pathway of rcDNA-containing capsids is a driver of cccDNA replenishment [35]. Recently, Kostyushev et al. found that CRISPR-Cas9-mediated inactivation of HBV cccDNA is not sufficient for curing infection, because HBV replication rapidly rebounds in their system due to formation of HBV cccDNA from recycling rcDNA [55].

However, in cells supporting HBV infection, intracellular recycling of rcDNA seems not to play a primary role in establishing and sustaining the cccDNA pool. In HepG2-NTCP A3 cells infected with wild-type HBV or HBV that do not encode core protein (ΔHBc HBV) from its genome, cccDNA levels were comparable over 9 weeks of infection [56]. ΔHBc HBV do not produce rcDNA after infection because of no core protein, suggesting that intracellular recycling of rcDNA is not critical in sustaining the cccDNA pool in this model. Volz et al. treated humanized uPA/SCID mice infected with HBV with the entry inhibitor, myrcludex B (or bulevirtide), from week 3 to week 9 post-infection. They found that intrahepatic cccDNA loads did not differ significantly between these mice and mice that were sacrificed 3 weeks post-infection [57]. This result indicates that intracellular recycling of rcDNA does not increase the amount of cccDNA in those cells in which infection has already been established. Possibly, there is a mechanism for HBV to limit cccDNA in a low copy number in primary human hepatocytes [52]. This number might be easily fulfilled by de novo infection and maintained thereafter in quiescent cells, letting it unnecessary to recycle rcDNA. Probably, recycling acts as a 'salvage' pathway to rescue an occasional loss of cccDNA (e.g., by the activity of nucleases) rather than a strong 'driver' for counteracting ongoing degradation [56].

5. cccDNA Decay Dominates the Second Phase of Serum HBV Decline during Short-Term NUC Therapy

A two-phase decline in serum HBV has repeatedly been observed during short-term NUC therapy. Tsiang et al. reported a biphasic clearance of serum HBV in a cohort of 10 chronic hepatitis B patients receiving adefovir (ADV) for 12 weeks [58]. The initial, fast phase of viral load decline reflects the clearance of HBV particles from the serum, with an average half-life of 1.1 days. The second, slower phase of viral load decline has a mean half-life of 18 days and is believed to closely mirror the rate-limiting process of infected cell loss. Typically, the two phases' inflection points were found between weeks 1 and 2. Wolters et al. also recorded a biphasic viral load decline in 10 CHB patients receiving entecavir (ETV) for 28 days. The half-life of serum HBV decline in the first phase was 16 h, and 10.7 days in the second phase [59], with inflection points occurring within 1 week of therapy. The bi-phasic decline of serum HBV was also reported in patients receiving core protein allosteric modulators. In a cohort of 35 patients treated with RG7907 for 28 days, HBV DNA declined fast in the first phase showing a half-life of 17 ± 6 h, and declined slowly in the second phase with a half-life of 6 ± 0.8 days [60].

The decay of the cccDNA pool is the most probable mechanism for the second-phase viral decline, since a model without cccDNA decay would predict a viral load plateau instead. Assuming that the cccDNA pool remained constant during therapy, this pool would produce an equal amount of new viruses at any time point, but at a slower rate (inhibiting efficiency) than in the natural steady state. For example, if a patient had a total serum virus load of 2×10^{11} and thus produced 10^{11} viruses per day before therapy (equal to the number of viruses lost per day), 10^9 viruses would be produced daily under a 99% inhibiting efficiency during therapy. Serum HBV levels enter the fast decline phase first, declining to 2×10^9 after approximately 7 days (7 half-lives). Assuming a constant cccDNA pool, 10^9 viruses would also be produced at this time. This would be in accordance with the amount of serum HBV that was eliminated each day (half of 2×10^9), predicting that the serum viral load would not decrease further. However, the experimental observations contradicted this hypothesis. Of note, a different inhibiting efficiency could only affect the first phase. For example, it would take 3.3 days for an agent with a 90% inhibiting efficiency (allowing the production of 10^{10} new viruses per day) to reduce the total viral load from 2×10^{11} to 2×10^{10}, after which it would plateau.

Therefore, decreases in cccDNA or hepatocyte turnover should be involved in the second phase of decline in serum viral load during short-term therapy of NUC. The first, fast phase of viral load decline results from the short half-life of serum HBV (1 day). The subsequent decline in serum viral load must be primarily caused by the decline of cccDNA due to the daily production of new viruses approaching the daily decay of viruses. As depicted in Figure 3, the amount of serum virus decayed per day (decay rate), which is determined by the current serum viral load, decreases with time passes. However, the daily amount of newly produced virus also declines during NUC treatment, which is determined by the cccDNA pool size and inhibiting potency of the drugs used. The daily decay of virus declines faster than the daily production of new virus. Before the transition point, the decline of serum viral load is primarily determined by the decrease in serum virus levels. After the transition point, the amount of decayed viruses per day is roughly equal to the number of newly produced viruses. Hence, the viral load is primarily determined by the decline in production rate.

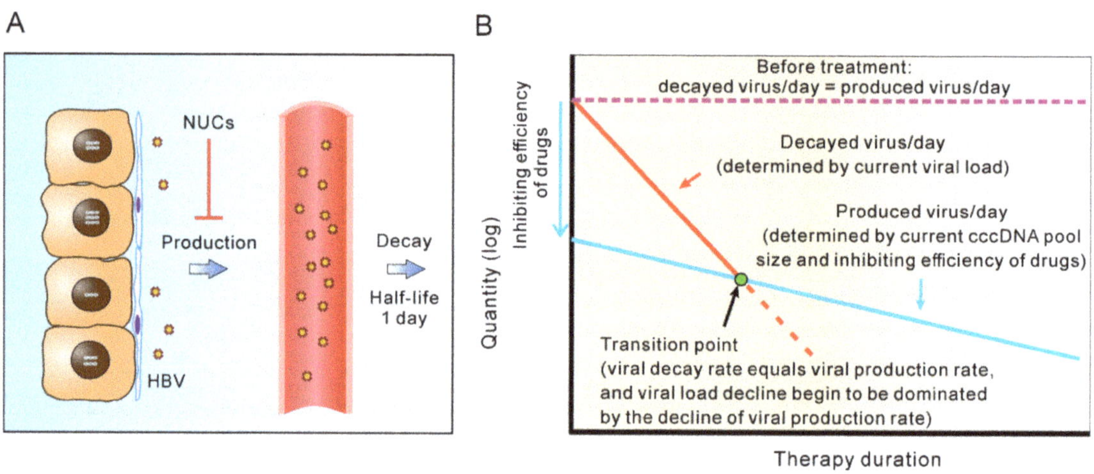

Figure 3. cccDNA decline is responsible for the second phase of serum viral load decrease during NUC therapy. (**A**) Serum HBV has a half-life of one day [17], and new viruses derived from the transcription of cccDNA replenish serum viral load during homeostasis prior to treatment. NUC treatment alters the balances in serum viral load and cccDNA pool size by inhibiting the production of new viruses [58]. (**B**) The serum HBV decay rate (amount of viruses that decay each day) is determined by the current viral load, and the production rate of new viruses is determined by the cccDNA pool size. Both rates decrease over the course of NUC treatment because the total viral load and the cccDNA pool size both decline. However, the decay rate declines faster than the production rate since serum HBV has a much shorter half-life than cccDNA (or infected hepatocyte). At the transition point, the viral decay rate equals the production rate. Subsequently, the production rate (or cccDNA) shows a faster decline than the serum viral load.

Assuming that the decline of cccDNA is the primary cause of the second phase of decline in serum viral load, the cccDNA half-life would correspond to approximately 6–18 days [58–61] (Table 1). This value is comparable to the result (16 days) obtained by Nowark et al. using a different method [17]. Notably, the researchers appeared to explain the decline in production rate completely by the death of infected cells, and they deduced the half-life of hepatocytes on this basis. We argue that this observation is better to explain this by cccDNA decline, which involves cell death, cell division and noncytopathic effects. The half-lives of infected cells might have been overestimated, perhaps by a factor of less than two.

Table 1. cccDNA half-life during short-term NUC therapy.

Therapy	Sample Size	Observation Period	Half-Life of Serum Virion	Half-Life of Infected Cells *	Reference
LAM	45	28 days	1 day	16 days (10–100 days)	[17]
ADV	13	12 weeks	1.1 days	18 days (11–30 days)	[58]
LAM	21	4 weeks	17 h	7 days	[61]
ETV	10	4 weeks	16 h	10.7 days	[59]

* Half-life of infected cells in the original publications should be half-life of cccDNA.

6. A Significant Variation in cccDNA Half-Life Was Observed during Different Terms of NUC Therapy

Direct analyses of cccDNA in the liver have also been performed in patients receiving long-term therapy. However, varying cccDNA decline rates have been observed throughout the course of the therapy. Data on cccDNA decline from 10 previous studies were collected (Table 2). These studies provided data on the cccDNA of paired liver biopsies (pre- and post-treatment) from the same patients, allowing the assessment of the cccDNA half-life. After 48 weeks of NUC therapy, cccDNA declined by 0.7 log for LAM therapy (n = 146), 0.8 log for ADV (n = 22) and 0.9 log for ETV (n = 159) or 1.0 log (another study, n = 40). Table 2 also includes studies that used a combination therapy or had a shorter observation period (12 weeks). The average cccDNA decline within 48 weeks is usually less than 1 log, corresponding to a half-life of more than 14 weeks.

Table 2. cccDNA half-life during long-term therapy.

Therapy	Sample Size (Patients)	Observation Period	cccDNA Reduction	cccDNA Half-Life	Reference
ADV	22	48 weeks	0.8 log		[62]
PEG-IFNα-2b + ADV	26	48 weeks	2.4 log		[63]
PEG-IFNα-2a + ADV	40	48 weeks	1.03 log (HBeAg (+)) 0.44 log (HBeAg (−))		[64]
ETV or LAM	305	48 weeks	0.9 log (ETV) 0.7 log (LAM)		[65]
ETV	40	48 weeks	1 log		[66]
NUC	117	52 weeks	0.93 log		[67]
ADV	15	12 weeks	0.65 log		[68]
ETV or ADV	54	60 months	1.56 log		[69]
TDF	27	6.9 years		8.6 months (HBeAg+) 26.2 months (HBeAg−)	[20]
NUC	43	72–145 months		1.03 log in the 1st year, 2.94 log in 2–10 year	[70]

A recent study indirectly analyzed cccDNA turnover by monitoring the composition of HBV RNA from different viral quasi-species during NUC therapy [71]. They found that LAM^R mutations emerged and increased from undetectable to 40–90% within 16–28 weeks in serum HBV RNA from telbivudine-treated patients experiencing virological breakthrough. From these results, the cccDNA half-life for the majority of patients was inferred to be <12 weeks. However, these cccDNA half-lives might be overestimated (shorter than reality), since the composition of LAM^R would be influenced by the sizes of cccDNA pools, which most likely increased during the breakthrough.

Three studies were performed with a long-term observation period of at least 60 months. One of the three studies used three sequential samples from each patient while they were receiving NUC treatment (baseline, 1 year and a long-term point), which provided useful information for our analysis. This study reported a 1.03 log cccDNA reduction within the first year, equivalent to a half-life of 14.1 weeks, which is close to the half-life calculated above. Within the next 9 years, the same cohort experienced a further 1.94 log cccDNA reduction, which corresponds to a half-life of 76 weeks (this value was an underestimate because the cccDNA levels were below the detection limit in half of the long-term samples). This half-life was approximately 5 times longer than during the first year of treatment. Furthermore, HIV and HBV co-infected patients who received NUC treatment for an average of 65.8 months displayed half-lives of 36 weeks (HBe+) and 112 weeks (HBe−), both of which were significantly longer than the half-life during the first year of therapy. In addition, another study that tracked cccDNA changes during 60 months of NUC treatment reported a 1.56 log cccDNA reduction. Assuming that the cccDNA decreased by 0.8 log in the 1st year, the cccDNA reduction over the next 48 months would be 0.76 log, equivalent to a half-life of 82 weeks. Collectively, these findings demonstrated that during NUC treatment, cccDNA declines with varying half-lives, ranging from 1–2 weeks within the first 12 weeks, to 14 weeks within the first year, to 36–112 weeks within 2–10 years. Collectively, cccDNA declined in a multi-phasic model during long-term NUC therapy (Figure 4A).

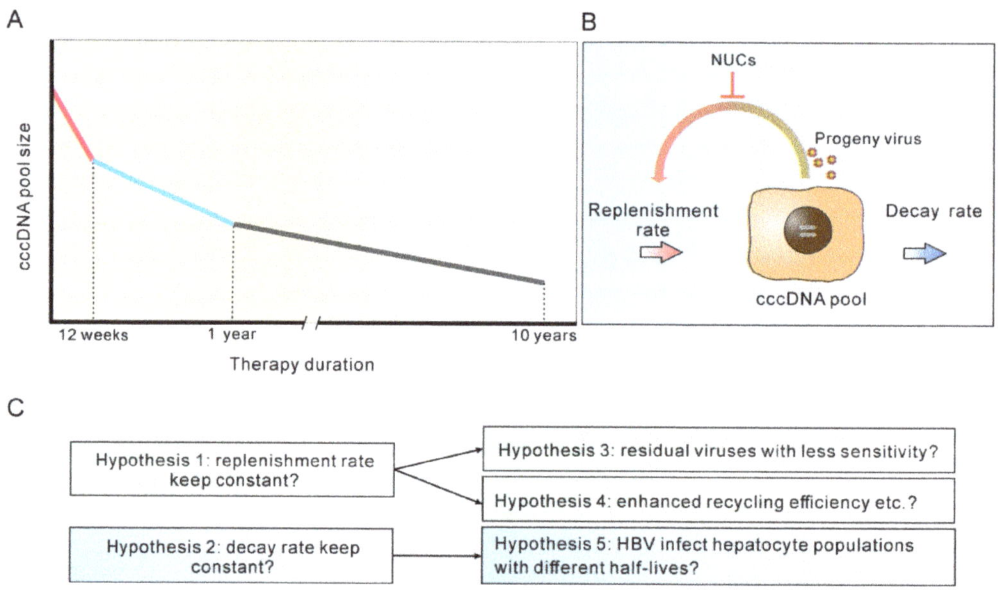

Figure 4. Explanations for the multi-phasic decline of cccDNA during long-term NUC therapy. (**A**) cccDNA declined in a multi-phasic model during long-term NUC therapy ([70], Tables 1 and 2). (**B**) NUC shifts the balance of the cccDNA pool by reducing replenishment. (**C**) There are several hypotheses for the multi-phasic decline of cccDNA during long-term NUC therapy.

7. Possible Explanations for the Multi-Phasic Decline of cccDNA during NUC Therapy

At the steady pretreatment state, the daily newly synthesized cccDNA must equal the daily decayed cccDNA in order to maintain the cccDNA pool size (Figure 4B). The rate of new cccDNA formation is mainly determined by the quantity of newly produced viruses and core particles, which lead to new infection or intracellular recycling [35,52]. The cccDNA decay rate primarily depends on the hepatocyte turnover rate and noncytopathic effects that degrade cccDNA. Assuming that during NUC treatment, new virus production is suppressed with a constant efficiency at any given time point (Hypothesis 1), and that the cccDNA decay rate is kept constant (Hypothesis 2), the cccDNA pool will continue to decline and eventually disappear (Figure 4C).

However, the fact that the cccDNA pool size declines in a multi-phasic manner and rarely disappears opposes one or both of the hypotheses. Rejection of Hypothesis 1 implies that the inhibition efficiency of NUC decreases (Hypothesis 3), or the conversion efficiency of newly produced viruses into cccDNA increases with therapy duration (Hypothesis 4).

In both circumstances, the efficiency of cccDNA synthesis would be relatively higher in latter stage of therapy than in earlier stage. Hypothesis 3 may be plausible in the case of residual cccDNA producing HBV variants with lower sensitivity to the NUC being used. Comparing the sensitivity of residual variants with that of initial viruses would help to clarify this issue. Hypothesis 4 requires mechanisms such as enhanced intracellular recycling efficiency of HBV DNA-containing capsid particles into nuclei to form cccDNA. However, this possibility seems remote given that intracellular recycling of DNA-containing nucleocapsids is not essential for the maintenance of HBV cccDNA [56].

If Hypothesis 2 is rejected, it would imply that cccDNA decays at different rates depending on the stage of therapy. For example, after a year of therapy, cccDNA would decay more slowly than it would have earlier. This suggests HBV-infected hepatocytes would have different half-lives, since cccDNA half-life is primarily determined by the half-life of hepatocytes (assuming NUC does not affect noncytopathic effects) (Hypothesis 5). In this scenario, cccDNA residing in hepatocytes with a short half-life would be eliminated at a faster rate, leading to a relatively larger slope in the declining curve. In contrast, cccDNA residing in hepatocytes with a long half-life would be eliminated slowly, resulting in a smaller slope. Previous studies indicate that hepatocyte half-life of rat and mouse are heterogenous and a small part of cells lived much longer than other cells [42,43]. Recent studies in mice have confirmed that hepatocytes in distinct liver lobule zones exhibit different proliferation activity [44,45]. Human hepatocytes can also be divided into different clusters according to the transcriptional profiles, as shown by single-cell sequencing [72,73]. Although the turnover rates of these diverse hepatocytes remain unknown, it is possible that a small part of hepatocytes, such as those in rat and mouse, have a long half-life. Given that the proportion of HBV-infected hepatocytes, both in acute and chronic infection, can be or close to 100% in some patients [19,74–77], viruses have a big chance to infect those long-lived hepatocytes. Testing this and the other four hypotheses would greatly advance our knowledge of the forces controlling the cccDNA dynamics during therapy (Figure 4C).

8. Considering CHB Treatments in Light of cccDNA Dynamics

The balance between cccDNA replenishment (synthesis) and decay during homeostasis provides an alternative framework for considering the effects of various therapies. An effective treatment could either slow down the cccDNA replenishment, speed up cccDNA decay, or exert both effects (Figure 5). NUCs inhibit viral DNA synthesis, which reduces the amount of newly productive viruses and slows the synthesis of downstream cccDNA. Capsid protein assembly modulators (CpAM) interfere with pregenomic RNA (pgRNA) encapsidation into the capsid [78–81], which also reduces viral production and cccDNA replenishment. Entry inhibitors of various types, including peptides, monoclonal antibodies and compounds [82–84], suppress cccDNA replenishment by preventing viruses from entering the cells. None of these treatments affect cccDNA decay, if there are no unintended effects.

SiRNA, anti-sense oligonucleotide (ASO) and locked nucleic acid (LNA) suppress new virus production and cccDNA replenishment by destroying virus RNA directly or suppress cccDNA transcription by reducing HBx level [85,86]. On the other hand, the expression of the surface antigen decreases as a result of preS2/S RNA degradation [85,87–90]. This may alleviate immune repression brought on by high antigen load and restore the HBV-specific immune response. Indeed, experiments in mouse models demonstrated that siRNA treatment following the administration of a therapeutic vaccine improved HBV-specific immunity and facilitated virus clearance [91]. In clinical trials, an ASO (bepirovirsen or GSK3228836) caused a transient and self-resolved ALT elevation in some patients who experienced a significant HBsAg reduction, indicating that immune reconstruction may occur in these patients [92,93]. Apparently, this immune reconstitution accelerated hepatocyte turnover through cytotoxic effects. Nucleic acid polymers (NAP) can also lower HBsAg load by inhibiting HBsAg release from the cells, and thus might help to restore immune response [94]. PEG-IFNα affects both cccDNA replenishment and decay by simultaneously inhibiting cccDNA transcription and degrading cccDNA [26,27,95]. Endonucleases such as CRISPR/cas9, ZFNs and TALENs affect cccDNA decay by directly cleaving/editing cccDNA molecules [96–98]. Immune modulators, such as Toll-like receptor agonists [99,100], PD-L1 antibodies [101–103], and therapeutic vaccines [104–109], were designed to restore innate immunity or HBV-specific immune response, which is believed to both restrict viral replication and accelerate cccDNA decay.

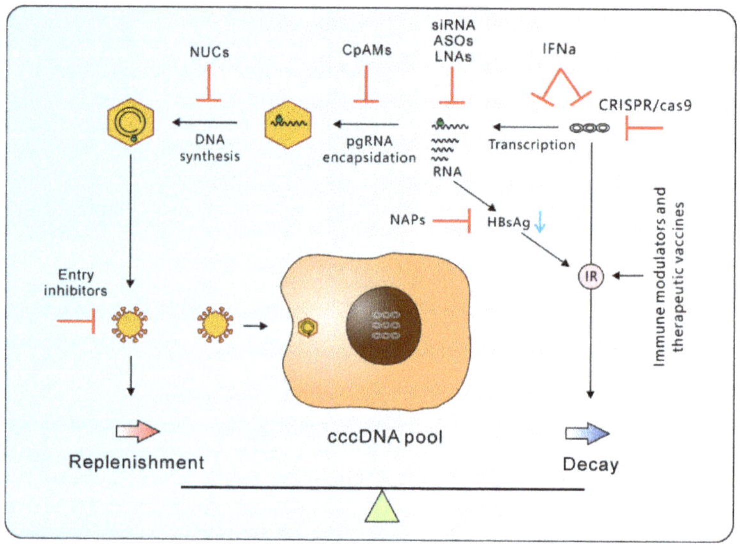

Figure 5. A framework for considering various anti-HBV strategies. Effective therapies shift the balance of the cccDNA pool by either blocking replenishment, accelerating cccDNA decay or both. NUCs, CpAMs, entry inhibitors, siRNA/ASO/LNA and PEG-IFNα all can inhibit cccDNA replenishment indirectly by suppressing steps upstream of cccDNA synthesis. SiRNA/ASO/LNA, NAPs, PEG-IFNα and immune modulators affect cccDNA decay directly or indirectly by facilitating immune restoration.

The cccDNA half-life required for the elimination of cccDNA within a finite course of treatment (e.g., 1 year) can be calculated. Assuming that a liver contains 10^{10} copies of cccDNA in total (approximately 0.1 copy/cell) before treatment, and the patient receives an ideal therapy that can completely block the production of progeny viruses (so that any cccDNA replenishment can be ignored), the cccDNA half-life should be 10 days. This value is within the range of some estimates derived from studies of short-term NUC therapy (Table 1). However, as was mentioned in Section 5, the cccDNA pool does not decline at a constant rate under long-term NUC therapy. This raises another hypothesis (Hypothesis 5), which describes the heterogeneity of the cccDNA pool's half-life due to the varying half-lives of hepatocyte populations. If this hypothesis is true, a much longer course of treatment should be expected for the complete elimination of cccDNA from the liver, even if high-efficiency agents are used to inhibiting viral production but no effect on cccDNA decay. Under this scenario, a therapy that can affect the cccDNA decay rate is highly desired.

Although it would be ideal for increasing cccDNA decay solely through noncytopathic effects, creating such methods might not be feasible. CRISIPR/cas9 strategies targeting cccDNA were successful in in vitro and animal models [110–112]. However, the delivery and editing efficiency and safety issues associated with off-targeting have yet to be solved for clinical usage. IFNα was reported to degrade cccDNA but with many side effects and is effective for a minority of patients. A recent study fused IFNα with a PD-L1 antibody and found the anti-PDL1-IFNα heterodimer preferentially targeted the liver and could overcome HBV-induced immune tolerance to an HBsAg vaccine in a mouse model [113]. This sophisticated approach used the PD-L1 antibody to target IFNα to HBV-infected cells with high PD-L1 expression and block immune checkpoints simultaneously. This approach offers a promising translatable therapeutic strategy for the functional cure of CHB. The first-in-class orally available cccDNA destabilizer, ccc-R08, was recently discovered [114]. Although its clinical use might be problematic because of safety concerns, it provides a proof of concept for developing noncytopathic compounds accelerating cccDNA decay. Strategies that involve initially lowering HBsAg levels first using siRNA, ASO or other compounds, followed by therapeutic vaccine and immune modulators treatment might be worth exploring [91]. If these strategies can restore innate and HBV-specific immunity, a functional cure for CHB would be achievable with an accelerated cccDNA decay rate.

9. Conclusions

The genetic materials of HBV, such as HBV DNA, HBV RNA and cccDNA, are distributed in different compartments of the host and exhibit different half-lives. CccDNA in the nucleus decay much slower than viral particles in serum, resulting in a two-phase decline in serum viral load during short-term NUC therapy. Current evidence indicates that cccDNA also declined in a multi-phasic manner during long-term NUC therapy. Considering that cccDNA half-life is mainly determined by the hepatocyte turnover rate in the immune-inactive stage, cccDNA may be distributed across hepatocyte populations with different half-lives. If this hypothesis is true, strategies should be developed to accelerate cccDNA decay in order to reduce cccDNA levels within a set treatment period. Direct antiviral agents (DAA) such as NUC can be used to inhibit cccDNA replenishment, and new agents such as CpAM would support this position even more. However, there is still no efficient treatment promoting cccDNA decay on the market. Strategies are currently being developed to reestablish innate and HBV-specific immune responses.

Author Contributions: Conceptualization, J.-L.H. and A.-L.H. writing—original draft preparation, J.-L.H.; writing—review and editing, J.-L.H. and A.-L.H.; visualization, J.-L.H.; funding acquisition, J.-L.H. and A.-L.H. All authors have read and agreed to the published version of the manuscript.

Funding: This work was funded by the National Key R&D Program of China (2022YFA1303600), the Natural Science Foundation of Chongqing (cstc2021jcyj-msxmX0298, cstc2020jcyj-msxmX0764 and cstc2020jscx-dxwtBX0022), the 111 Project (No. D20028), the Key Laboratory of Molecular Biology on Infectious Diseases, the Ministry of Education, Chongqing Medical University (No. 202104), and the CQMU Program for Youth Innovation in Future Medicine (No. W0049).

Data Availability Statement: No new data were used in this manuscript.

Conflicts of Interest: The authors of this study declared that they do not have any conflict of interest.

List of Abbreviations

cccDNA, covalently closed circular DNA; HBV, hepatitis B Virus; CHB, chronic hepatitis B; NUC, nucleot(s)ide analogue; PEG-IFNα, pegylated-interferon-α; IFNγ, interferon γ; TNFα, tumor necrosis factor α; IFNα/β, interferon α or β; 2′-CDG, 2′-deoxycarbocyclic guanosine; BrdU, 5-bromo-2′-deoxyuridine; LAM, lamivudine; 524W, 5-fluoro-2′,3′-dideoxy-3′-thiacytidine; PTHs, primary tupaia hepatocytes; L-FMAU, [1-(2-fluoro-5-methyl-β-L-arabinofuranosyl) uracil]; WM-HBV, woolly monkey HBV; uPA, urokinase-type plasminogen activator; SCID, severe combined immunodeficiency; PCNA, proliferating cell nuclear antigen; ADV, adefovir; ETV, entecavir; HBe, HBV e antigen; CpAM, capsid protein assembly modulators; pgRNA, pregenomic RNA; ASO, anti-sense oligonucleotide; LNA, locked nucleic acid; NAP, nucleic acid polymers; ZFNs, zinc finger endonucleases; TALENs, transcription activator-like effector nucleases; PD-L1, programmed death ligand 1.

References

1. Lok, A.S.; Zoulim, F.; Dusheiko, G.; Ghany, M.G. Hepatitis B cure: From discovery to regulatory approval. *Hepatology* **2017**, *66*, 1296–1313. [CrossRef]
2. Ning, Q.; Wu, D.; Wang, G.-Q.; Ren, H.; Gao, Z.-L.; Hu, P.; Han, M.-F.; Wang, Y.; Zhang, W.-H.; Lu, F.-M.; et al. Roadmap to functional cure of chronic hepatitis B: An expert consensus. *J. Viral Hepat.* **2019**, *26*, 1146–1155. [CrossRef]
3. Wong, G.L.H.; Gane, E.; Lok, A.S.F. How to achieve functional cure of HBV: Stopping NUCs, adding interferon or new drug de-velopment? *J. Hepatol.* **2022**, *76*, 1249–1262. [CrossRef] [PubMed]
4. Wei, L.; Ploss, A. Mechanism of Hepatitis B Virus cccDNA Formation. *Viruses* **2021**, *13*, 1463. [CrossRef]
5. Xia, Y.; Guo, H. Hepatitis B virus cccDNA: Formation, regulation and therapeutic potential. *Antivir. Res.* **2020**, *180*, 104824. [CrossRef]
6. Hong, X.; Kim, E.S.; Guo, H. Epigenetic regulation of hepatitis B virus covalently closed circular DNA: Implications for epigenetic therapy against chronic hepatitis B. *Hepatology* **2017**, *66*, 2066–2077. [CrossRef]
7. Block, T.M.; Alter, H.; Brown, N.; Brownstein, A.; Brosgart, C.; Chang, K.-M.; Chen, P.-J.; Cohen, C.; El-Serag, H.; Feld, J.; et al. Research priorities for the discovery of a cure for chronic hepatitis B: Report of a workshop. *Antivir. Res.* **2018**, *150*, 93–100. [CrossRef]
8. Martinez, M.G.; Boyd, A.; Combe, E.; Testoni, B.; Zoulim, F. Covalently closed circular DNA: The ultimate therapeutic target for curing HBV infections. *J. Hepatol.* **2021**, *75*, 706–717. [CrossRef] [PubMed]
9. Lim, S.G.; Baumert, T.F.; Boni, C.; Gane, E.; Levrero, M.; Lok, A.S.; Maini, M.K.; Terrault, N.A.; Zoulim, F. The scientific basis of combination therapy for chronic hepatitis B functional cure. *Nat. Rev. Gastroenterol. Hepatol.* **2023**, 1–16. [CrossRef]
10. Dusheiko, G.; Agarwal, K.; Maini, M.K. New Approaches to Chronic Hepatitis B. *N. Engl. J. Med.* **2023**, *388*, 55–69. [CrossRef] [PubMed]
11. Degasperi, E.; Anolli, M.P.; Lampertico, P. Towards a Functional Cure for Hepatitis B Virus: A 2022 Update on New Antiviral Strategies. *Viruses* **2022**, *14*, 2404. [CrossRef]
12. Zoulim, F. Inhibition of hepatitis B virus gene expression: A step towards functional cure. *J. Hepatol.* **2018**, *68*, 386–388. [CrossRef] [PubMed]
13. Mak, L.-Y.; Cheung, K.-S.; Fung, J.; Seto, W.-K.; Yuen, M.-F. New strategies for the treatment of chronic hepatitis B. *Trends Mol. Med.* **2022**, *28*, 742–757. [CrossRef]
14. Fanning, G.C.; Zoulim, F.; Hou, J.; Bertoletti, A. Therapeutic strategies for hepatitis B virus infection: Towards a cure. *Nat. Rev. Drug Discov.* **2019**, *18*, 827–844. [CrossRef] [PubMed]
15. Naggie, S.; Lok, A.S. New Therapeutics for Hepatitis B: The Road to Cure. *Annu. Rev. Med.* **2021**, *72*, 93–105. [CrossRef] [PubMed]
16. Ma, Z.; Zhang, E.; Gao, S.; Xiong, Y.; Lu, M. Toward a Functional Cure for Hepatitis B: The Rationale and Challenges for Therapeutic Targeting of the B Cell Immune Response. *Front. Immunol.* **2019**, *10*, 2308. [CrossRef]
17. Nowak, M.A.; Bonhoeffer, S.; Hill, A.M.; Boehme, R.; Thomas, H.C.; McDade, H. Viral dynamics in hepatitis B virus infection. *Proc. Natl. Acad. Sci. USA* **1996**, *93*, 4398–4402. [CrossRef]
18. Boettler, T.; Gill, U.S.; Allweiss, L.; Pollicino, T.; Tavis, J.E.; Zoulim, F. Assessing immunological and virological responses in the liver: Implications for the cure of chronic hepatitis B virus infection. *JHEP Rep.* **2022**, *4*, 100480. [CrossRef]
19. Murray, J.M.; Wieland, S.F.; Purcell, R.H.; Chisari, F.V. Dynamics of hepatitis B virus clearance in chimpanzees. *Proc. Natl. Acad. Sci. USA* **2005**, *102*, 17780–17785. [CrossRef]
20. Boyd, A.; Lacombe, K.; Lavocat, F.; Maylin, S.; Miailhes, P.; Lascoux-Combe, C.; Delaugerre, C.; Girard, P.M.; Zoulim, F. Decay of ccc-DNA marks per-sistence of intrahepatic viral DNA synthesis under tenofovir in HIV-HBV co-infected patients. *J. Hepatol.* **2016**, *65*, 683–691. [CrossRef]
21. Guidotti, L.G.; Ishikawa, T.; Hobbs, M.V.; Matzke, B.; Schreiber, R.; Chisari, F.V. Intracellular Inactivation of the Hepatitis B Virus by Cytotoxic T Lymphocytes. *Immunity* **1996**, *4*, 25–36. [CrossRef] [PubMed]
22. Xia, Y.; Stadler, D.; Lucifora, J.; Reisinger, F.; Webb, D.; Hösel, M.; Michler, T.; Wisskirchen, K.; Cheng, X.; Zhang, K.; et al. Interferon-gamma and Tumor Necrosis Factor-alpha Produced by T Cells Reduce the HBV Persistence Form, cccDNA, Without Cytolysis. *Gastroenterology* **2016**, *150*, 194–205. [CrossRef] [PubMed]
23. Guidotti, L.G.; Rochford, R.; Chung, J.; Shapiro, M.; Purcell, R.; Chisari, F.V. Viral Clearance Without Destruction of Infected Cells During Acute HBV Infection. *Science* **1999**, *284*, 825–829. [CrossRef] [PubMed]
24. Mason, W.S.; Xu, C.; Low, H.C.; Saputelli, J.; Aldrich, C.E.; Scougall, C.; Grosse, A.; Colonno, R.; Litwin, S.; Jilbert, A.R. The amount of hepatocyte turnover that oc-curred during resolution of transient hepadnavirus infections was lower when virus replication was inhibited with entecavir. *J. Virol.* **2009**, *83*, 1778–1789. [CrossRef] [PubMed]
25. Guidotti, L.G.; Ando, K.; Hobbs, M.V.; Ishikawa, T.; Runkel, L.; Schreiber, R.D.; Chisari, F.V. Cytotoxic T lymphocytes inhibit hepatitis B virus gene expression by a noncytolytic mechanism in transgenic mice. *Proc. Natl. Acad. Sci. USA* **1994**, *91*, 3764–3768. [CrossRef] [PubMed]
26. Liu, F.; Campagna, M.; Qi, Y.; Zhao, X.; Guo, F.; Xu, C.; Li, S.; Li, W.; Block, T.M.; Chang, J.; et al. Alpha-Interferon Suppresses Hepadnavirus Transcription by Altering Epigenetic Modification of cccDNA Minichromosomes. *PLoS Pathog.* **2013**, *9*, e1003613. [CrossRef]
27. Lucifora, J.; Xia, Y.; Reisinger, F.; Zhang, K.; Stadler, D.; Cheng, X.; Sprinzl, M.F.; Koppensteiner, H.; Makowska, Z.; Volz, T.; et al. Specific and Nonhepatotoxic Degradation of Nuclear Hepatitis B Virus cccDNA. *Science* **2014**, *343*, 1221–1228. [CrossRef]

28. Michalopoulos, G.K.; Bhushan, B. Liver regeneration: Biological and pathological mechanisms and implications. *Nat. Rev. Gastroenterol. Hepatol.* **2021**, *18*, 40–55. [CrossRef]
29. Fourel, I.; Cullen, J.M.; Saputelli, J.; Aldrich, C.E.; Schaffer, P.; Averett, D.R.; Pugh, J.; Mason, W.S. Evidence that hepatocyte turnover is required for rapid clearance of duck hepatitis B virus during antiviral therapy of chronically infected ducks. *J. Virol.* **1994**, *68*, 8321–8330. [CrossRef]
30. Addison, W.R.; Walters, K.A.; Wong, W.W.; Wilson, J.S.; Madej, D.; Jewell, L.D.; Tyrrell, D.L. Half-life of the duck hepatitis B virus co-valently closed circular DNA pool in vivo following inhibition of viral replication. *J. Virol.* **2002**, *76*, 6356–6363. [CrossRef]
31. Zhu, Y.; Yamamoto, T.; Cullen, J.; Saputelli, J.; Aldrich, C.E.; Miller, D.S.; Litwin, S.; Furman, P.A.; Jilbert, A.; Mason, W.S. Kinetics of Hepadnavirus Loss from the Liver during Inhibition of Viral DNA Synthesis. *J. Virol.* **2001**, *75*, 311–322. [CrossRef] [PubMed]
32. Bohne, F.; Chmielewski, M.; Ebert, G.; Wiegmann, K.; Kürschner, T.; Schulze, A.; Urban, S.; Krönke, M.; Abken, H.; Protzer, U. T Cells Redirected Against Hepatitis B Virus Surface Proteins Eliminate Infected Hepatocytes. *Gastroenterology* **2008**, *134*, 239–247. [CrossRef] [PubMed]
33. Krebs, K.; Böttinger, N.; Huang, L.; Chmielewski, M.; Arzberger, S.; Gasteiger, G.; Jäger, C.; Schmitt, E.; Bohne, F.; Aichler, M.; et al. T Cells Expressing a Chimeric Antigen Receptor That Binds Hepatitis B Virus Envelope Proteins Control Virus Replication in Mice. *Gastroenterology* **2013**, *145*, 456–465. [CrossRef] [PubMed]
34. Wisskirchen, K.; Kah, J.; Malo, A.; Asen, T.; Volz, T.; Allweiss, L.; Wettengel, J.M.; Lütgehetmann, M.; Urban, S.; Bauer, T.; et al. T cell receptor grafting allows virological control of hepatitis B virus infection. *J. Clin. Investig.* **2019**, *129*, 2932–2945. [CrossRef] [PubMed]
35. Ko, C.; Chakraborty, A.; Chou, W.-M.; Hasreiter, J.; Wettengel, J.M.; Stadler, D.; Bester, R.; Asen, T.; Zhang, K.; Wisskirchen, K.; et al. Hepatitis B virus genome recycling and de novo secondary infection events maintain stable cccDNA levels. *J. Hepatol.* **2018**, *69*, 1231–1241. [CrossRef] [PubMed]
36. Chong, C.L.; Chen, M.L.; Wu, Y.C.; Tsai, K.N.; Huang, C.C.; Hu, C.P.; Jeng, K.S.; Chou, Y.C.; Chang, C. Dynamics of HBV cccDNA expression and tran-scription in different cell growth phase. *J. Biomed. Sci.* **2011**, *18*, 96. [CrossRef] [PubMed]
37. Tu, T.; Zehnder, B.; Wettengel, J.M.; Zhang, H.; Coulter, S.; Ho, V.; Douglas, M.W.; Protzer, U.; George, J.; Urban, S. Mitosis of hepatitis B virus-infected cells in vitro results in uninfected daughter cells. *JHEP Rep.* **2022**, *4*, 100514. [CrossRef]
38. Li, M.; Sohn, J.A.; Seeger, C. Distribution of Hepatitis B Virus Nuclear DNA. *J. Virol.* **2018**, *92*, e01391-17. [CrossRef]
39. Dandri, M.; Burda, M.R.; Will, H.; Petersen, J. Increased hepatocyte turnover and inhibition of woodchuck hepatitis B virus repli-cation by adefovir in vitro do not lead to reduction of the closed circular DNA. *Hepatology* **2000**, *32*, 139–146. [CrossRef]
40. Lutgehetmann, M.; Volz, T.; Köpke, A.; Broja, T.; Tigges, E.; Lohse, A.W.; Fuchs, E.; Murray, J.M.; Petersen, J.; Dandri, M. In vivo proliferation of hepadnavirus-infected hepatocytes induces loss of covalently closed circular DNA in mice. *Hepatology* **2010**, *52*, 16–24. [CrossRef]
41. Allweiss, L.; Volz, T.; Giersch, K.; Kah, J.; Raffa, G.; Petersen, J.; Lohse, A.W.; Beninati, C.; Pollicino, T.; Urban, S.; et al. Proliferation of primary human hepatocytes and prevention of hepatitis B virus reinfection efficiently deplete nuclear cccDNA in vivo. *Gut* **2018**, *67*, 542–552. [CrossRef]
42. Macdonald, R.A. "Lifespan" of liver cells. Autoradio-graphic study using tritiated thymidine in normal, cirrhotic, and partially hepatectomized rats. *Arch. Intern. Med.* **1961**, *107*, 335–343. [CrossRef] [PubMed]
43. Magami, Y.; Azuma, T.; Inokuchi, H.; Kokuno, S.; Moriyasu, F.; Kawai, K.; Hattori, T. Cell proliferation and renewal of normal hepatocytes and bile duct cells in adult mouse liver. *Liver Int.* **2002**, *22*, 419–425. [CrossRef] [PubMed]
44. He, L.; Pu, W.; Liu, X.; Zhang, Z.; Han, M.; Li, Y.; Huang, X.; Han, X.; Li, Y.; Liu, K.; et al. Proliferation tracing reveals regional hepatocyte generation in liver homeostasis and repair. *Science* **2021**, *371*, eabc4346. [CrossRef] [PubMed]
45. Wei, Y.; Wang, Y.G.; Jia, Y.; Li, L.; Yoon, J.; Zhang, S.; Wang, Z.; Zhang, Y.; Zhu, M.; Sharma, T.; et al. Liver homeostasis is maintained by midlobular zone 2 hepatocytes. *Science* **2021**, *371*, eabb1625. [CrossRef] [PubMed]
46. Celis, J.E.; Celis, A. Cell cycle-dependent variations in the distribution of the nuclear protein cyclin proliferating cell nuclear antigen in cultured cells: Subdivision of S phase. *Proc. Natl. Acad. Sci. USA* **1985**, *82*, 3262–3266. [CrossRef] [PubMed]
47. Bravo, R.; Macdonald-Bravo, H. Changes in the nuclear distribution of cyclin (PCNA) but not its synthesis depend on DNA replication. *EMBO J.* **1985**, *4*, 655–661. [CrossRef]
48. Bravo, R.; Macdonald-Bravo, H. Existence of two populations of cyclin/proliferating cell nuclear antigen during the cell cycle: Association with DNA replication sites. *J. Cell Biol.* **1987**, *105*, 1549–1554. [CrossRef]
49. Mancini, R.; Marucci, L.; Benedetti, A.; Jezequel, A.-M.; Orlandi, F. Immunohistochemical analysis of S-phase cells in normal human and rat liver by PC10 monoclonal antibody. *Liver Int.* **1994**, *14*, 57–64. [CrossRef]
50. Nakajima, T.; Kagawa, K.; Ueda, K.; Ohkawara, T.; Kimura, H.; Kakusui, M.; Deguchi, T.; Okanoue, T.; Kashima, K.; Ashihara, T. Evaluation of hepatic proliferative activity in chronic liver diseases and hepatocellular carcinomas by proliferating cell nuclear antigen (PCNA) immunohisto-chemical staining of methanol-fixed tissues. *J. Gastroenterol.* **1994**, *29*, 450–454. [CrossRef]
51. Mason, W.S.; Jilbert, A.R.; Litwin, S. Hepatitis B Virus DNA Integration and Clonal Expansion of Hepatocytes in the Chronically Infected Liver. *Viruses* **2021**, *13*, 210. [CrossRef] [PubMed]
52. Köck, J.; Rösler, C.; Zhang, J.-J.; Blum, H.E.; Nassal, M.; Thoma, C. Generation of Covalently Closed Circular DNA of Hepatitis B Viruses via Intracellular Recycling Is Regulated in a Virus Specific Manner. *PLoS Pathog.* **2010**, *6*, e1001082. [CrossRef] [PubMed]

53. Guo, H.; Jiang, D.; Zhou, T.; Cuconati, A.; Block, T.M.; Guo, J.-T. Characterization of the Intracellular Deproteinized Relaxed Circular DNA of Hepatitis B Virus: An Intermediate of Covalently Closed Circular DNA Formation. *J. Virol.* **2007**, *81*, 12472–12484. [CrossRef] [PubMed]
54. Gao, W.; Hu, J. Formation of Hepatitis B Virus Covalently Closed Circular DNA: Removal of Genome-Linked Protein. *J. Virol.* **2007**, *81*, 6164–6174. [CrossRef] [PubMed]
55. Kostyushev, D.; Kostyusheva, A.; Brezgin, S.; Ponomareva, N.; Zakirova, N.F.; Egorshina, A.; Yanvarev, D.V.; Bayurova, E.; Sudina, A.; Goptar, I.; et al. Depleting hepatitis B virus relaxed circular DNA is necessary for resolution of infection by CRISPR-Cas9. *Mol. Ther. Nucleic Acids* **2023**, *31*, 482–493. [CrossRef]
56. Tu, T.; Zehnder, B.; Qu, B.; Urban, S. De novo synthesis of hepatitis B virus nucleocapsids is dispensable for the maintenance and transcriptional regulation of cccDNA. *JHEP Rep.* **2021**, *3*, 100195. [CrossRef]
57. Volz, T.; Allweiss, L.; Ben Ḿbarek, M.; Warlich, M.; Lohse, A.W.; Pollok, J.M.; Alexandrov, A.; Urban, S.; Petersen, J.; Lütgehetmann, M.; et al. The entry inhibitor Myrcludex-B efficiently blocks intrahepatic virus spreading in humanized mice previously infected with hepatitis B virus. *J. Hepatol.* **2013**, *58*, 861–867. [CrossRef]
58. Tsiang, M.; Rooney, J.F.; Toole, J.J.; Gibbs, C.S. Biphasic clearance kinetics of hepatitis B virus from patients during adefovir dipivoxil therapy. *Hepatology* **1999**, *29*, 1863–1869. [CrossRef]
59. Wolters, L.M.; E Hansen, B.; Niesters, H.G.; DeHertogh, D.; de Man, R.A. Viral dynamics during and after entecavir therapy in patients with chronic hepatitis B. *J. Hepatol.* **2002**, *37*, 137–144. [CrossRef]
60. Goncalves, A.; Lemenuel-Diot, A.; Cosson, V.; Jin, Y.; Feng, S.; Bo, Q.; Guedj, J. What drives the dynamics of HBV RNA during treatment? *J. Viral. Hepat.* **2021**, *28*, 383–392. [CrossRef]
61. Wolters, L.; Hansen, B.; Niesters, H.; Zeuzem, S.; Schalm, S.; De, M.R.; De Man, R.A. Viral dynamics in chronic hepatitis B patients during lamivudine therapy. *Liver Int.* **2002**, *22*, 121–126. [CrossRef] [PubMed]
62. Werle-Lapostolle, B.; Bowden, S.; Locarnini, S.; Wursthorn, K.; Petersen, J.; Lau, G.; Trepo, C.; Marcellin, P.; Goodman, Z.; Delaney, W.E., IV; et al. Persistence of cccDNA during the natural history of chronic hepatitis B and decline during adefovir dipivoxil therapy. *Gastroenterology* **2004**, *126*, 1750–1758. [CrossRef]
63. Wursthorn, K.; Lutgehetmann, M.; Dandri, M.; Volz, T.; Buggisch, P.; Zollner, B.; Longerich, T.; Schirmacher, P.; Metzler, F.; Zankel, M.; et al. Peginterferon alpha-2b plus adefovir induce strong cccDNA decline and HBsAg reduction in patients with chronic hepatitis B. *Hepatology* **2006**, *44*, 675–684. [CrossRef]
64. Takkenberg, B.; Terpstra, V.; Zaaijer, H.; Weegink, C.; Dijkgraaf, M.; Jansen, P.; Beld, M.; Reesink, H. Intrahepatic response markers in chronic hepatitis B patients treated with peginterferon alpha-2a and adefovir. *J. Gastroenterol. Hepatol.* **2011**, *26*, 1527–1535. [CrossRef] [PubMed]
65. Bowden, S.; Locarnini, S.; Chang, T.-T.; Chao, Y.-C.; Han, K.-H.; Gish, R.G.; de Man, R.A.; Yu, M.; Llamoso, C.; Tang, H. Covalently closed-circular hepatitis B virus DNA reduction with entecavir or lamivudine. *World J. Gastroenterol.* **2015**, *21*, 4644–4651. [CrossRef] [PubMed]
66. Wong, D.K.-H.; Yuen, M.-F.; Ngai, V.W.-S.; Fung, J.; Lai, C.-L. One-Year Entecavir or Lamivudine Therapy Results in Reduction of Hepatitis B Virus Intrahepatic Covalently Closed Circular DNA Levels. *Antivir. Ther.* **2006**, *11*, 909–916. [CrossRef] [PubMed]
67. Wong, D.K.-H.; Seto, W.-K.; Fung, J.; Ip, P.; Huang, F.-Y.; Lai, C.-L.; Yuen, M.-F. Reduction of Hepatitis B Surface Antigen and Covalently Closed Circular DNA by Nucleos(t)ide Analogues of Different Potency. *Clin. Gastroenterol. Hepatol.* **2013**, *11*, 1004–1010.e1. [CrossRef] [PubMed]
68. Zheng, Q.; Zhu, Y.Y.; Chen, J.; Liu, Y.R.; You, J.; Dong, J.; Zeng, D.W.; Gao, L.Y.; Chen, L.H.; Jiang, J.J. Decline in intrahepatic cccDNA and increase in immune cell reactivity after 12 weeks of antiviral treatment were associated with HBeAg loss. *J. Viral Hepat.* **2014**, *21*, 909–916. [CrossRef]
69. Wang, Y.; Liu, Y.; Liao, H.; Deng, Z.; Bian, D.; Ren, Y.; Yu, G.; Jiang, Y.; Bai, L.; Liu, S.; et al. Serum HBV DNA plus RNA reflecting cccDNA level before and during NAs treatment in HBeAg positive CHB patients. *Int. J. Med. Sci.* **2022**, *19*, 858–866. [CrossRef]
70. Lai, C.-L.; Wong, D.; Ip, P.; Kopaniszen, M.; Seto, W.-K.; Fung, J.; Huang, F.-Y.; Lee, B.; Cullaro, G.; Chong, C.K.; et al. Reduction of covalently closed circular DNA with long-term nucleos(t)ide analogue treatment in chronic hepatitis B. *J. Hepatol.* **2017**, *66*, 275–281. [CrossRef]
71. Huang, Q.; Zhou, B.; Cai, D.; Zong, Y.; Wu, Y.; Liu, S.; Mercier, A.; Guo, H.; Hou, J.; Colonno, R.; et al. Rapid Turnover of Hepatitis B Virus Covalently Closed Circular DNA Indicated by Monitoring Emergence and Reversion of Signature-Mutation in Treated Chronic Hepatitis B Patients. *Hepatology* **2021**, *73*, 41–52. [CrossRef] [PubMed]
72. MacParland, S.A.; Liu, J.C.; Ma, X.-Z.; Innes, B.T.; Bartczak, A.M.; Gage, B.K.; Manuel, J.; Khuu, N.; Echeverri, J.; Linares, I.; et al. Single cell RNA sequencing of human liver reveals distinct intrahepatic macrophage populations. *Nat. Commun.* **2018**, *9*, 4383. [CrossRef]
73. Payen, V.L.; Lavergne, A.; Sarika, N.A.; Colonval, M.; Karim, L.; Deckers, M.; Najimi, M.; Coppieters, W.; Charloteaux, B.; Sokal, E.M.; et al. Single-cell RNA sequencing of human liver reveals hepatic stellate cell heterogeneity. *JHEP Rep.* **2021**, *3*, 100278. [CrossRef] [PubMed]
74. Balagopal, A.; Grudda, T.; Ribeiro, R.M.; Saad, Y.S.; Hwang, H.S.; Quinn, J.; Murphy, M.; Ward, K.; Sterling, R.K.; Zhang, Y.; et al. Single hepatocytes show persistence and transcriptional inactivity of hepatitis B. *J. Clin. Investig.* **2020**, *5*. [CrossRef] [PubMed]
75. Balagopal, A.; Hwang, H.S.; Grudda, T.; Quinn, J.; Sterling, R.K.; Sulkowski, M.S.; Thio, C.L. Single Hepatocyte Hepatitis B Virus Transcriptional Landscape in HIV Coinfection. *J. Infect. Dis.* **2020**, *221*, 1462–1469. [CrossRef]

76. Wieland, S.F.; Spangenberg, H.C.; Thimme, R.; Purcell, R.H.; Chisari, F.V. Expansion and contraction of the hepatitis B virus tran-scriptional template in infected chimpanzees. *Proc. Natl. Acad. Sci. USA* **2004**, *101*, 2129–2134. [CrossRef] [PubMed]
77. Thimme, R.; Wieland, S.; Steiger, C.; Ghrayeb, J.; Reimann, K.A.; Purcell, R.H.; Chisari, F.V. CD8$^+$ T Cells Mediate Viral Clearance and Disease Pathogenesis during Acute Hepatitis B Virus Infection. *J. Virol.* **2003**, *77*, 68–76. [CrossRef] [PubMed]
78. Nijampatnam, B.; Liotta, D.C. Recent advances in the development of HBV capsid assembly modulators. *Curr. Opin. Chem. Biol.* **2019**, *50*, 73–79. [CrossRef]
79. Zoulim, F.; Zlotnick, A.; Buchholz, S.; Donaldson, E.; Fry, J.; Gaggar, A.; Hu, J.; Kann, M.; Lenz, O.; Lin, K.; et al. Nomenclature of HBV core protein-targeting antivirals. *Nat. Rev. Gastroenterol. Hepatol.* **2022**, *19*, 748–750. [CrossRef]
80. Taverniti, V.; Ligat, G.; Debing, Y.; Kum, D.B.; Baumert, T.F.; Verrier, E.R. Capsid Assembly Modulators as Antiviral Agents against HBV: Molecular Mechanisms and Clinical Perspectives. *J. Clin. Med.* **2022**, *11*, 1349. [CrossRef]
81. Viswanathan, U.; Mani, N.; Hu, Z.; Ban, H.; Du, Y.; Hu, J.; Chang, J.; Guo, J.-T. Targeting the multifunctional HBV core protein as a potential cure for chronic hepatitis B. *Antivir. Res.* **2020**, *182*, 104917. [CrossRef] [PubMed]
82. Roca Suarez, A.A.; Testoni, B.; Zoulim, F. HBV 2021: New therapeutic strategies against an old foe. *Liver Int.* **2021**, *41* (Suppl. 1), 15–23. [CrossRef] [PubMed]
83. Tu, T.; Urban, S. Virus entry and its inhibition to prevent and treat hepatitis B and hepatitis D virus infections. *Curr. Opin. Virol.* **2018**, *30*, 68–79. [CrossRef] [PubMed]
84. Fung, S.; Choi, H.S.J.; Gehring, A.; Janssen, H.L.A. Getting to HBV cure: The promising paths forward. *Hepatology* **2022**, *76*, 233–250. [CrossRef] [PubMed]
85. Hui, R.W.-H.; Mak, L.-Y.; Seto, W.-K.; Yuen, M.-F. RNA interference as a novel treatment strategy for chronic hepatitis B infection. *Clin. Mol. Hepatol.* **2022**, *28*, 408–424. [CrossRef]
86. Allweiss, L.; Giersch, K.; Pirosu, A.; Volz, T.; Muench, R.C.; Beran, R.K.; Urban, S.; Javanbakht, H.; Fletcher, S.P.; Lütgehetmann, M.; et al. Therapeutic shutdown of HBV transcripts promotes reappearance of the SMC5/6 complex and silencing of the viral genome in vivo. *Gut* **2022**, *71*, 372–381. [CrossRef]
87. Van den Berg, F.; Limani, S.W.; Mnyandu, N.; Maepa, M.B.; Ely, A.; Arbuthnot, P. Advances with RNAi-Based Therapy for Hepatitis B Virus Infection. *Viruses* **2020**, *12*, 851. [CrossRef]
88. Yuen, M.; Schiefke, I.; Yoon, J.; Ahn, S.H.; Heo, J.; Kim, J.H.; Chan, H.L.Y.; Yoon, K.T.; Klinker, H.; Manns, M.; et al. RNA Interference Therapy With ARC-520 Results in Prolonged Hepatitis B Surface Antigen Response in Patients with Chronic Hepatitis B Infection. *Hepatology* **2020**, *72*, 19–31. [CrossRef]
89. Yuen, M.-F.; Wong, D.K.-H.; Schluep, T.; Lai, C.-L.; Ferrari, C.; Locarnini, S.; Lo, R.C.-L.; Gish, R.G.; Hamilton, J.; Wooddell, C.I.; et al. Long-term serological, virological and histological responses to RNA inhibition by ARC-520 in Chinese chronic hepatitis B patients on entecavir treatment. *Gut* **2022**, *71*, 789–797. [CrossRef]
90. Yuen, M.-F.; Locarnini, S.; Lim, T.H.; Strasser, S.I.; Sievert, W.; Cheng, W.; Thompson, A.J.; Given, B.D.; Schluep, T.; Hamilton, J.; et al. Combination treatments including the small-interfering RNA JNJ-3989 induce rapid and sometimes prolonged viral responses in patients with CHB. *J. Hepatol.* **2022**, *77*, 1287–1298. [CrossRef]
91. Michler, T.; Kosinska, A.D.; Festag, J.; Bunse, T.; Su, J.; Ringelhan, M.; Imhof, H.; Grimm, D.; Steiger, K.; Mogler, C.; et al. Knockdown of Virus Antigen Expression Increases Therapeutic Vaccine Efficacy in High-Titer Hepatitis B Virus Carrier Mice. *Gastroenterology* **2020**, *158*, 1762–1775.e9. [CrossRef] [PubMed]
92. Yuen, M.-F.; Heo, J.; Jang, J.-W.; Yoon, J.-H.; Kweon, Y.-O.; Park, S.-J.; Tami, Y.; You, S.; Yates, P.; Tao, Y.; et al. Safety, tolerability and antiviral activity of the antisense oligonucleotide bepirovirsen in patients with chronic hepatitis B: A phase 2 randomized controlled trial. *Nat. Med.* **2021**, *27*, 1725–1734. [CrossRef]
93. Yuen, M.-F.; Lim, S.-G.; Plesniak, R.; Tsuji, K.; Janssen, H.L.; Pojoga, C.; Gadano, A.; Popescu, C.P.; Stepanova, T.; Asselah, T.; et al. Efficacy and Safety of Bepirovirsen in Chronic Hepatitis B Infection. *N. Engl. J. Med.* **2022**, *387*, 1957–1968. [CrossRef] [PubMed]
94. Bazinet, M.; Pântea, V.; Placinta, G.; Moscalu, I.; Cebotarescu, V.; Cojuhari, L.; Jimbei, P.; Iarovoi, L.; Smesnoi, V.; Musteata, T.; et al. Safety and Efficacy of 48 Weeks REP 2139 or REP 2165, Tenofovir Disoproxil, and Pegylated Interferon Alfa-2a in Patients with Chronic HBV Infection Naive to Nu-cleos(t)ide Therapy. *Gastroenterology* **2020**, *158*, 2180–2194. [CrossRef]
95. Belloni, L.; Allweiss, L.; Guerrieri, F.; Pediconi, N.; Volz, T.; Pollicino, T.; Petersen, J.; Raimondo, G.; Dandri, M.; Levrero, M. IFN-alpha inhibits HBV transcription and replication in cell culture and in humanized mice by targeting the epigenetic regulation of the nuclear cccDNA minichromo-some. *J. Clin. Investig.* **2012**, *122*, 529–537. [CrossRef] [PubMed]
96. Martinez, M.G.; Combe, E.; Inchauspe, A.; Mangeot, P.E.; Delberghe, E.; Chapus, F.; Neveu, G.; Alam, A.; Carter, K.; Testoni, B.; et al. CRISPR-Cas9 Targeting of Hepatitis B Virus Covalently Closed Circular DNA Generates Transcriptionally Active Episomal Variants. *Mbio* **2022**, *13*, e02888-21. [CrossRef]
97. Lin, H.; Li, G.; Peng, X.; Deng, A.; Ye, L.; Shi, L.; Wang, T.; He, J. The Use of CRISPR/Cas9 as a Tool to Study Human Infectious Viruses. *Front. Cell. Infect. Microbiol.* **2021**, *11*, 590989. [CrossRef]
98. Yang, Y.-C.; Yang, H.-C. Recent Progress and Future Prospective in HBV Cure by CRISPR/Cas. *Viruses* **2021**, *14*, 4. [CrossRef]
99. Amin, O.E.; Colbeck, E.J.; Daffis, S.; Khan, S.; Ramakrishnan, D.; Pattabiraman, D.; Chu, R.; Micolochick Steuer, H.; Lehar, S.; Peiser, L.; et al. Therapeutic Potential of TLR8 Agonist GS-9688 (Selgantolimod) in Chronic Hepatitis B: Remodeling of Antiviral and Regulatory Mediators. *Hepatology* **2021**, *74*, 55–71. [CrossRef]
100. Kayesh, M.E.H.; Kohara, M.; Tsukiyama-Kohara, K. Toll-Like Receptor Response to Hepatitis B Virus Infection and Potential of TLR Agonists as Immunomodulators for Treating Chronic Hepatitis B: An Overview. *Int. J. Mol. Sci.* **2021**, *22*, 10462. [CrossRef]

101. Balsitis, S.; Gali, V.; Mason, P.J.; Chaniewski, S.; Levine, S.M.; Wichroski, M.J.; Feulner, M.; Song, Y.; Granaldi, K.; Loy, J.K.; et al. Safety and efficacy of anti-PD-L1 therapy in the woodchuck model of HBV infection. *PLoS ONE* **2018**, *13*, e0190058. [CrossRef] [PubMed]
102. Ferrando-Martinez, S.; Huang, K.; Bennett, A.S.; Sterba, P.; Yu, L.; Suzich, J.A.; Janssen, H.L.; Robbins, S.H. HBeAg seroconversion is associated with a more effective PD-L1 blockade during chronic hepatitis B infection. *JHEP Rep.* **2019**, *1*, 170–178. [CrossRef]
103. Feray, C.; Lopez-Labrador, F.X. Is PD-1 blockade a potential therapy for HBV? *JHEP Rep.* **2019**, *1*, 142–144. [CrossRef]
104. Whitacre, D.C.; Peters, C.J.; Sureau, C.; Nio, K.; Li, F.; Su, L.; Jones, J.E.; Isogawa, M.; Sallberg, M.; Frelin, L.; et al. Designing a therapeutic hepatitis B vaccine to circumvent immune tolerance. *Hum. Vaccines Immunother.* **2020**, *16*, 251–268. [CrossRef]
105. Lim, S.G.; Agcaoili, J.; De Souza, N.N.A.; Chan, E. Therapeutic vaccination for chronic hepatitis B: A systematic review and me-ta-analysis. *J. Viral. Hepat.* **2019**, *26*, 803–817. [CrossRef] [PubMed]
106. Cargill, T.; Barnes, E. Therapeutic vaccination for treatment of chronic hepatitis B. *Clin. Exp. Immunol.* **2021**, *205*, 106–118. [CrossRef]
107. Stasi, C.; Silvestri, C.; Voller, F. Hepatitis B vaccination and immunotherapies: An update. *Clin. Exp. Vaccine Res.* **2020**, *9*, 1–7. [CrossRef] [PubMed]
108. Kosinska, A.D.; Moeed, A.; Kallin, N.; Festag, J.; Su, J.; Steiger, K.; Michel, M.L.; Protzer, U.; Knolle, P.A. Synergy of therapeutic heterologous prime-boost hepatitis B vaccination with CpG-application to improve immune control of persistent HBV infection. *Sci. Rep.* **2019**, *9*, 10808. [CrossRef]
109. Zhang, T.-Y.; Guo, X.-R.; Wu, Y.; Kang, X.-Z.; Zheng, Q.-B.; Qi, R.-Y.; Chen, B.-B.; Lan, Y.; Wei, M.; Wang, S.-J.; et al. A unique B cell epitope-based particulate vaccine shows effective suppression of hepatitis B surface antigen in mice. *Gut* **2020**, *69*, 343–354. [CrossRef]
110. Lin, S.R.; Yang, H.C.; Kuo, Y.T.; Liu, C.J.; Yang, T.Y.; Sung, K.C.; Lin, Y.Y.; Wang, H.Y.; Wang, C.C.; Shen, Y.C.; et al. The CRISPR/Cas9 System Facilitates Clearance of the Intrahepatic HBV Templates In Vivo. *Mol. Ther. Nucleic Acids* **2014**, *3*, e186. [CrossRef]
111. Seeger, C.; Sohn, J.A. Targeting Hepatitis B Virus With CRISPR/Cas. *Mol. Ther. Nucleic Acids* **2014**, *3*, e216. [CrossRef] [PubMed]
112. Yang, Y.-C.; Chen, Y.-H.; Kao, J.-H.; Ching, C.; Liu, I.-J.; Wang, C.-C.; Tsai, C.-H.; Wu, F.-Y.; Liu, C.-J.; Chen, P.-J.; et al. Permanent Inactivation of HBV Genomes by CRISPR/Cas9-Mediated Non-cleavage Base Editing. *Mol. Ther. Nucleic Acids* **2020**, *20*, 480–490. [CrossRef] [PubMed]
113. Meng, C.Y.; Sun, S.; Liang, Y.; Xu, H.; Zhang, C.; Zhang, M.; Wang, F.S.; Fu, Y.-X.; Peng, H. Engineered anti-PDL1 with IFNalpha targets both im-munoinhibitory and activating signals in the liver to break HBV immune tolerance. *Gut* **2022**. [CrossRef]
114. Wang, L.; Zhu, Q.; Zhang, J.D.; Zhang, Y.; Ni, X.; Xiang, K.; Jiang, J.; Li, B.; Yu, Y.; Hu, H.; et al. Discovery of a first-in-class orally available HBV cccDNA inhibitor. *J. Hepatol.* **2022**. [CrossRef] [PubMed]

microorganisms

Review

Current Methods for the Detection of Antibodies of Varicella-Zoster Virus: A Review

Dequan Pan, Wei Wang * and Tong Cheng *

State Key Laboratory of Molecular Vaccinology and Molecular Diagnostics, National Institute of Diagnostics and Vaccine Development in Infectious Diseases, School of Life Sciences, School of Public Health, Xiamen University, Xiamen 361102, China
* Correspondence: lukewang@xmu.edu.cn (W.W.); tcheng@xmu.edu.cn (T.C.)

Abstract: Infection with the varicella-zoster virus (VZV) causes chickenpox and shingles, which lead to significant morbidity and mortality globally. The detection of serum VZV-specific antibodies is important for the clinical diagnosis and sero-epidemiological research of VZV infection, and for assessing the effect of VZV vaccine immunization. Over recent decades, a variety of methods for VZV antibody detection have been developed. This review summarizes and compares the current methods for detecting VZV antibodies, and discussed future directions for this field.

Keywords: varicella-zoster virus (VZV); antibody detection methods; FAMA; ELISA

Citation: Pan, D.; Wang, W.; Cheng, T. Current Methods for the Detection of Antibodies of Varicella-Zoster Virus: A Review. *Microorganisms* **2023**, *11*, 519. https://doi.org/10.3390/microorganisms11020519

Academic Editors: Shengxi Chen and Fabio Zicker

Received: 18 January 2023
Revised: 10 February 2023
Accepted: 16 February 2023
Published: 17 February 2023

1. Introduction

Varicella-zoster virus (VZV) is a highly contagious alpha-herpesvirus that infects more than 90% of people worldwide [1,2]. Chickenpox (varicella) is the outcome of primary infection with VZV and mainly affects children. As per the report published by the World Health Organization (WHO) in 2014, the minimum of the annual global disease burden of chickenpox was estimated to be 140 million cases, of which 4.2 million have severe complications leading to hospitalization and death [3]. Although usually a mild and self-limiting illness, chickenpox poses a greater risk of severe disease to pregnant women, neonates, VZV-seronegative adults, and immunocompromised individuals [4]. The reactivation of latent VZV causes shingles (herpes zoster), which occurs mainly in people ≥50 years of age and is usually associated with intense neuralgia [5]. It was estimated that about one third of individuals who have had chickenpox will develop shingles [6]. The incidence and severity of shingles increases with age or immunosuppression [7]. Recent studies have shown that COVID-19 vaccination may also increase the risk of VZV reactivation and thus potentially increase the incidence of shingles, especially in the elderly population [8–10]. To date, there is still no specific cure for VZV-induced diseases. Vaccination is among the most cost-effective ways for preventing chickenpox and shingles. The most widely used chickenpox vaccines consist of the Oka strain of live-attenuated VZV (vOka), and there are currently two kinds of shingles vaccines on the market, including a live-attenuated vaccine also based on vOka (ZOSTAVAX®, Merck Sharp & Dohme, Rahway, NJ, USA) and a recombinant subunit vaccine based on VZV glycoprotein E (Shingrix®, GlaxoSmithKline, Brentford, Middlesex, UK) [6,11,12]. However, the use of these vaccines is not universal and thus a large proportion of the global population has not been vaccinated against VZV. Consequently, VZV infection is still prevalent and accounts for a significant disease burden worldwide.

Laboratory testing is essential for the diagnosis and surveillance of VZV-induced diseases. Virus isolation was once the "gold standard" for the diagnosis of VZV infection [13,14]. However, this method is time-consuming and not readily accessible, and has thus been replaced by polymerase chain reaction (PCR) for the detection of viral DNA and direct fluorescence assays (DFA) for the detection of viral proteins [15,16]. Most recently, a recombinase-aided amplification-lateral flow system (RAA-LF) has been used for the rapid detection of VZV DNA, which is easier to use and requires no equipment [17]. PCR is currently regarded as the most sensitive and reliable method for VZV detection [14]. Meanwhile, serological assays, which detect VZV-specific antibodies, also represent reliable diagnostic tools in the detection of VZV infection, and have been implemented as complementary approaches to PCR. Furthermore, since the quantitative detection of anti-VZV antibody is necessary for measuring the infection history and evaluating the immune status against VZV in the population, serological assays have been widely used in epidemiological investigations of VZV and evaluations of immune responses to VZV vaccination. These assays include the fluorescent-antibody-to-membrane-antigen (FAMA) test, the complement fixation (CF) test, the immune adherence hemagglutination assay (IAHA), the latex agglutination (LA) test, the radioimmunoassay (RIA), the neutralization assay, the chemiluminescent immunoassay (CLIA), the enzyme-linked immunosorbent assay (ELISA), the immunofluorescence assay (IFA), the time-resolved fluorescence immunoassay (TRFIA), and the lateral flow immunochromatographic assay (LFIA). Among these tests, FAMA and ELISA are the most widely used. On the other hand, the CF test has low sensitivity, while RIA uses harmful radioactive materials; therefore, neither method has been widely used in recent years.

Herein, we review the existing serological assays for the detection of anti-VZV antibodies and compare their advantages and limitations. This paper could help clinicians and technicians to choose the appropriate serological method for diagnosing VZV infection or assessing the efficacy of VZV vaccines.

2. Methods for the Detection of Anti-VZV Antibodies

2.1. FAMA

The FAMA test, which was initially developed by Williams et al. [18], is the most extensively validated assay and is considered the "gold standard" for VZV antibody detection [19,20]. This method determines the presence of antibodies specific to viral proteins that distribute on the surface of VZV-infected cells, which correlate with protection from disease. Different human cells, including HFF [21], MRC-5 [22], Vero [23] and Raji [24], have been used to perform the FAMA test in previous studies. According to the standard FAMA procedure (schematic shown in Figure 1A), cells in culture are infected with VZV for 48–72 h and harvested by trypsin digestion until 70–90% of them show a cytopathic effect. Then, the infected cells are resuspended in PBS and incubated with serial dilutions of human sera to bind with VZV-specific antibodies. Following washing, cells are incubated with a fluorescein-conjugated anti-human secondary antibody (specific for IgG, IgM, or IgM and IgG). After a second washing step, the cells are transferred into small wells on glass slides and incubated for certain time to allow the cells to attach. Finally, cells are sealed with 90% glycerin and a cover glass before observation under a fluorescence microscope [18]. When the viral proteins (e.g., VZV glycoproteins) distributed on the surface of VZV-infected cells bind to their specific antibodies in the serum samples, ring-like fluorescence patterns are typically observed. The highest dilution that can still cause a positive ring-like fluorescence reaction is taken as the FAMA titer of the serum sample. Given the false-positive results caused by the non-specific reactivity of undiluted sera, serum samples with a titer of $\geq$1:2 or 1:4 are generally considered positive [18,25,26]. Several studies have shown that healthy children can be protected from chickenpox infection when they have serum FAMA titers of $\geq$1:4 for VZV [27,28].

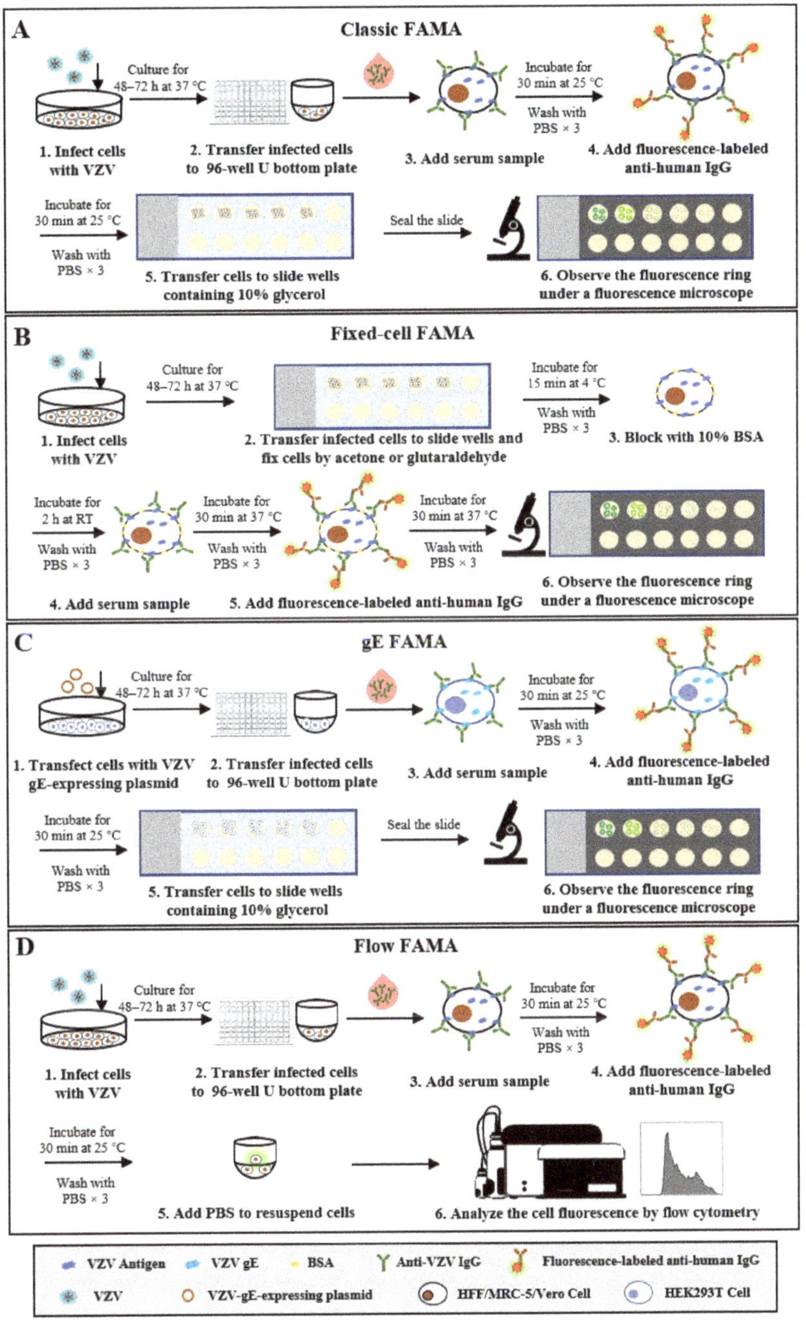

Figure 1. Schematic representation of the basic protocols for four types of FAMA that detect anti-VZV antibodies. (**A**) Classic FAMA using live VZV-infected cells to capture anti-VZV antibodies. (**B**) Fixed-cell FAMA using chemically fixed VZV-infected cells to capture anti-VZV antibodies. (**C**) gE-FAMA using VZV-gE-expressing live cells to capture anti-VZV-gE antibodies. (**D**) Flow FAMA using flow cytometry to analyze the fluorescence-labeled VZV-infected cells.

The FAMA test can be performed using chemically fixed VZV-infected cells. The fixed-cell FAMA was developed from the classic FAMA to increase throughput and efficiency and has been used to study human immunity to VZV [29–32]. In this modified FAMA (schematic shown in Figure 1B), VZV-infected cells are first fixed on the slide with cold acetone [31,32] or glutaraldehyde [29,30], while the other steps and the cutoff values (1:2 or 1:4) are similar to or the same as those in the classic live-cell FAMA test. The acetone fixation increases membrane permeability and allows antibody access to the cell cytoplasm and binding to other viral proteins besides membrane antigens [33]. In comparison, glutaraldehyde fixation does not change the permeability of the cell membrane and detects only the membrane antigen, but changes the natural conformation of the antigen, resulting in reduced sensitivity to a certain extent [19]. The fixed-cell FAMA has several advantages over the classic FAMA, e.g., the slides with attached cells can be prepared in large batches in advance and stored in a freezer for a long time, which enables testing at any time and reduces hands-on time. In addition, the used FAMA slides can be kept for re-reading, which creates the possibility of them being inspected by the drug administration. However, the specificity of fixed-cell FAMA could be challenged, since the procedure of fixed-cell FAMA is similar to that of indirect fluorescent antibody test (IFAT), which also uses fixed VZV-infected cells, and previous studies have documented that serum samples from children weakly cross-reacted between VZV and herpes simplex virus (HSV) in IFAT [19,34,35]. Nonetheless, the influence of cell fixation on the specificity of FAMA remains controversial and needs to be further clarified.

To avoid the use of infectious viruses, one study has reported a simple and safer FAMA using HEK293T cells transfected with a plasmid encoding VZV glycoprotein E (gE) to replace VZV-infected cells, which is called the gE FAMA (schematic shown in Figure 1C) [36]. The gE FAMA exhibited a similar staining effect to classic FAMA, and the gE-FAMA titers were closely correlated with the gp-ELISA data. However, this assay only detects antibodies against VZV gE, and the abundance of expressed gE on plasmid-transfected cells is different from that of VZV-infected cells; the cutoff value of gE FAMA may be different from that of classic FAMA, and remains to be determined.

All the above-mentioned FAMA assays rely on experienced technicians making result judgments under fluorescence microscopy, which not only leads to subjective bias but also limits the throughput of the detection. Some studies have reported a flow cytometry-adapted FAMA (flow FAMA; schematic shown in Figure 1D), in which flow cytometry is used instead of examination under a microscope to analyze the fluorescence-labeled cells [37–39]. The positive cutoff value determined by flow cytometry analysis can make the judgment of FAMA more objective, and the automated measurement can also reduce the complexity of operations and increase the detection throughput of FAMA. In a study involving 62 human serum samples, the detection accuracy of flow FAMA was 90.32% compared with that of standard FAMA [38]. In the flow FAMA, the matter of how to select the appropriate cutoff value remains a key problem. In addition, a flow cytometer is required for the flow FAMA, which limits the application of this method.

Taken together, the classic live-cell FAMA test is regarded as the "gold standard" to detect anti-VZV antibodies because of its high sensitivity and specificity. However, the standard FAMA procedure is semi-quantitative, low-throughput, and labor-intensive, and requires a subjective evaluation by trained, experienced technicians, which limits its widespread use. Fixed-cell FAMA and gE FAMA are modified from the classic live-cell FAMA to improve the throughput and safety, but both of them have their own defects, including possibly lower specificity, a lack of validation, or ambiguous cutoff values (shown in Table 1).

2.2. ELISA

ELISA is one of the most common antibody detection methods and has been widely used for the quantitative detection of anti-VZV antibodies for epidemiological investigations of VZV infection and for efficacy evaluation of the varicella and zoster vaccines [25,40,41]. Indirect ELISA is the most commonly used type. According to the procedure of indirect ELISA (schematic shown in Figure 2A), VZV antigens are coated on 96-well polystyrene ELISA plates and subsequently blocked with bovine serum albumin or normal goat serum. Diluted serum samples are then added into the plate wells. After incubation and washing, peroxidase- or alkaline-phosphatase-conjugated anti-human antibodies are added to detect the captured anti-VZV antibodies. After another washing step, the substrate solution is added for a chromogenic reaction, which is then terminated with excess acid or base. Finally, the optical density (OD) or absorbance value of appropriate wavelength is measured quantitatively using a spectrophotometer. Within a certain range, the OD or absorbance value is proportional to the number of binding antibodies on the plate, so the standard curve can be drawn through the detection of standard products and achieve quantitative detection of anti-VZV antibodies [42–44].

Table 1. Comparison of four types of FAMA tests for detecting anti-VZV antibodies.

Type	Characteristics	Advantages	Limitations
Classic FAMA	Using live VZV-infected cells	• High sensitivity and specificity. • Gold standard for VZV antibody detection.	• Tedious procedure. • Subjective interpretation.
Fixed-cell FAMA	Using chemically fixed VZV-infected cells	• High sensitivity. • FAMA slides can be stored for a long time. • Reduced hands-on time.	• Tedious procedure. • Subjective interpretation. • Supposedly low specificity due to possible cross reaction with anti-HSV antibodies.
gE FAMA	Using live cells expressing VZV gE	• High sensitivity and specificity. • No contact with infectious VZV during operation.	• Tedious procedure. • Subjective interpretation. • Lacking further validation. • Unclear cutoff value.
Flow FAMA	Flow cytometry-adapted	• Objective and automated measurement.	• Special equipment is required. • Unclear cutoff value.

Nowadays, there are several commercial VZV antibody ELISA kits available [19,45]. They are mostly indirect ELISAs and use either whole VZV-infected cell lysate (WC ELISA) [25,44] or purified glycoprotein (gp-ELISA) [46,47] as the antigen to capture anti-VZV antibodies. Different ELISA kits are calibrated according to the first international standard for varicella-zoster immunoglobulin, and results of <50 mIU/mL are considered negative, but the cutoff value varies [25,41,48]. The majority of these commercial ELISAs are designed to measure antibody levels after natural infection and are found to be insufficiently sensitive to measure antibody responses to chickenpox vaccination [19,49,50]. One study compared four commercial ELISAs and showed that their sensitivity ranged from 60.4% to 91.8%, values that are low compared with those of the FAMA test [25]. To address the issue of sensitivity, Merck has developed an in-house, highly sensitive, and specific gp-ELISA that uses lentil lectin-purified VZV glycoproteins, including gE, gB, and gH, from VZV-infected cells as the antigen [51]. The Merck gp-ELISA has been used extensively to evaluate antibody responses in children immunized with the Varivax Oka vaccine and a titer of 5 gp-ELISA units/mL (equivalent to 10 mIU/mL, by the international reference standard) was found to be associated with a high degree of protection against breakthrough infection during seven follow-up years [52,53]. However, the Merck gp-ELISA is not commercially available and is restricted to only a few specialist testing centers. In general, the gp-ELISA has high sensitivity and specificity, and a high consistency with FAMA, and is thus considered to be the most likely alternative to FAMA [54]. In addition, GSK also developed an in-house ELISA to detect VZV gE antibody (gE-ELISA) and applied it to evaluating the immunogenicity of a herpes zoster subunit vaccine Shingrix [55].

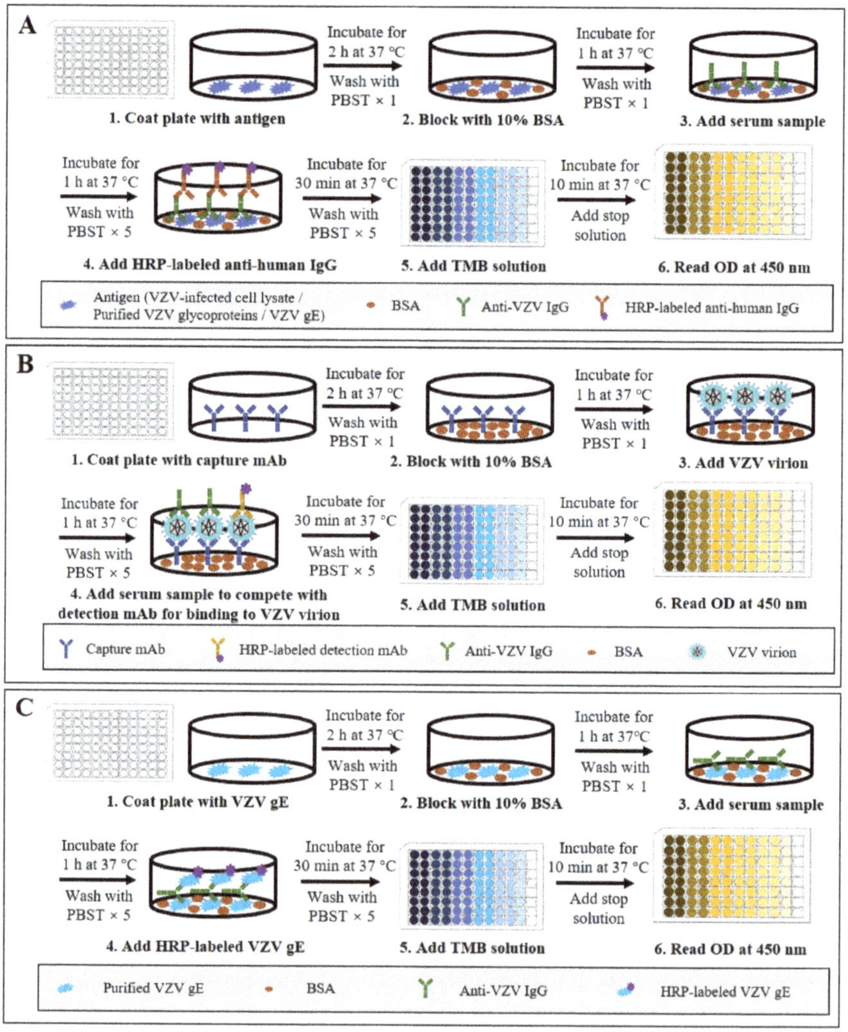

Figure 2. Schematic representation of the basic protocols for three types of ELISA that detect anti-VZV antibodies. (**A**) Indirect ELISA using VZV-infected cell lysate, purified VZV glycoproteins, or VZV gE as the coating antigen. (**B**) Double antibody sandwich competitive ELISA using anti-ORF9 antibody as the detection antibody and HRP-labeled anti-gE antibody as the detection antibody. (**C**) Double gE antigen sandwich ELISA using purified VZV gE as the coating antigen and HRP-labeled gE as the detection antigen.

In addition to indirect ELISA, some researchers have developed a competitive ELISA and a gE double-antigen sandwich ELISA for VZV antibody detection [56,57]. Firstly, for the competitive ELISA (schematic shown in Figure 2B), an anti-ORF9 antibody is used to capture VZV particles, and then an HRP-labeled anti-gE antibody is used to compete with serum anti-VZV antibodies for virus binding. Using the following formula: PI (%) = 100 × [1 − (positive serum OD450/negative reference serum OD450)], the blocking rate can be calculated to evaluate the VZV antibody levels in serum samples. The study showed that the competitive sandwich ELISA had a sensitivity of 95.6%, a specificity of 99.77%, and coincidence of 97.61% compared to the FAMA test. Secondly, the gE double-antigen sandwich ELISA (schematic shown in Figure 2C) is modified from the gE-based indirect ELISA by replacing the enzyme-conjugated anti-human secondary antibodies with HRP-labeled recombinant gE protein. The study showed that the gE double-antigen sandwich ELISA had sensitivity of 95.08% and specificity of 100% compared to the FAMA test. The findings in these studies suggest that the use of antibody competition or a dual-gE-antigen sandwich could increase the sensitivity and specificity of ELISA for VZV antibody detection. However, to date, there are no reports on the further application and validation of these two methods.

Taken together, to date, many types of ELISAs have been developed for anti-VZV antibody detection (summarized in Table 2). Compared to the FAMA test, these ELISAs are quantitative, easy to use, and high throughput, but are considered to be less sensitive, except for the Merck gp-ELISA, which is regarded as sensitive enough to be an optimal alternative reference assay to FAMA. Recent advances have improved the performance of several ELISAs, achieving sensitivity and specificity similar to FAMA, but these new methods still lack further validation or else are not commercially available.

Table 2. Comparison of ELISAs for detecting anti-VZV antibodies.

Type	Characteristics	Advantages	Limitations
WC-ELISA	Using whole lysates of VZV-infected cells as antigens	• Commercially available. • Antibodies against all VZV antigens can be detected.	• Not sensitive enough to measure antibody responses to chickenpox vaccination.
gp-ELISA	Using purified VZV glycoproteins as antigens	• Commercially available. • Higher sensitivity and specificity than WC-ELISA.	• Not sensitive enough to measure antibody responses to chickenpox vaccination (expect the Merck gp-ELISA). • High cost of glycoprotein purification.
gE-ELISA	Using purified VZV gE as an antigen	• Higher sensitivity and specificity than WC-ELISA.	• Not commercially available. • Only test for anti-gE antibodies.
Double antibody sandwich competitive ELISA	Capture antibody: anti-ORF9 antibody Detection antibody: HRP-labeled anti-gE antibody	• Comparable sensitivity and specificity to FAMA	• Not commercially available. • Lacking further validation.
Double gE antigen sandwich ELISA	Using purified VZV gE as the coating antigen and HRP-labeled gE as the detection antigen.	• Comparable sensitivity and specificity to FAMA	• Not commercially available. • Lacking further validation.

2.3. Neutralization Assay

The neutralization assay measures the titers of neutralizing antibodies that confer protection from VZV infections. However, the sensitivity of the earliest neutralization assay was low, making it difficult to detect anti-VZV neutralizing antibodies in individuals many years after infection [58]. Some studies have reported that the addition of guinea pig complement and anti-immunoglobulin antibodies can make the sensitivity of the neutralization assay for VZV 2 to 16 fold and 7 to 100 fold higher, respectively, and thus the enhanced neutralization assay is more frequently used for the evaluation of antibody responses to VZV infection, compared to the original method [59–63].

According to the procedure of the complement-enhanced neutralization assay for VZV (schematic shown in Figure 3) [59,64,65], which is modified from the universal plaque-reduction neutralization test (PRNT), hundreds of PFUs of cell-free VZV are mixed with diluted heat-inactivated serum samples (at 56 °C for 30 min) and guinea pig complement, and co-incubated at 37 °C for 1 h before being added into the cultured cells (e.g., MRC-5). After incubation for 5–7 days, the number of virus plaques is directly counted under an inverted microscope, and the highest dilutions of serum that result in ≥50% reduction in plaque counts are defined as the neutralization titers. In this experiment, cells can be stained with dyes such as crystal violet to make the plaques easier to observe. The neutralization assay requires viral plaque formation and takes about one week to complete. To reduce the testing time, an indirect immunoperoxidase assay (IPA) was used to stain VZV-infected cells and shortened the test period to 72 h [66,67]. Another study established a neutralization test basing on an enzyme-linked immunosorbent spot (Elispot) assay with VZV-gK protein as the detection target, and shortened the test period to 36 h [68].

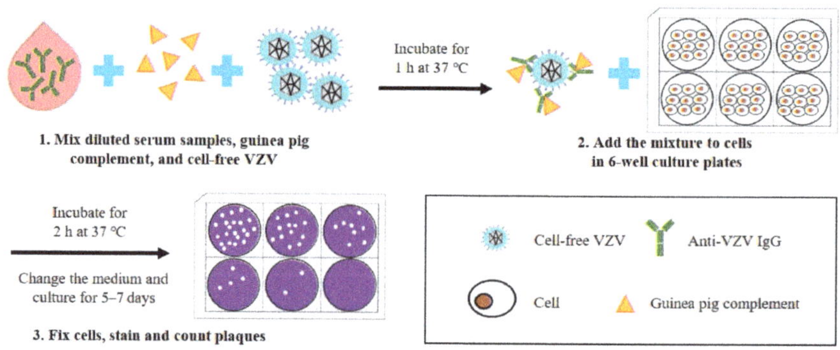

Figure 3. Schematic diagram depicting the basic steps of a neutralization assay for the detection of anti-VZV antibodies. (1) Diluted serum samples, guinea pig complement, and cell-free VZV are mixed and incubated for 1 h at 37 °C. (2) The mixture is added to cells in 6-well culture plates and incubated for 2 h at 37 °C, and then the culture medium is replaced. (3) After culturing for 5–7 days, the cells are fixed and stained to visualize the plaques, and the neutralization titer of serum samples can be calculated according to the number of plaques.

Taken together, neutralization assays can directly detect the presence of neutralizing antibodies to various types of viruses in sera, and have been widely used to determine virus infection and evaluate the protective efficacy of vaccines. However, this conventional method has a low sensitivity when detecting anti-VZV neutralizing antibodies, possibly because VZV has a highly cell-associated nature and grows to low titers in culture, thus affecting the interaction between serum antibodies and cell-free VZV particles. Furthermore, the test period of the neutralization assay is relatively long, and the operation is labor-intensive and low-throughput. Given all these limitations, the neutralization assay for VZV is not commonly used nowadays. Nonetheless, neutralization assays in combination with high-sensitivity immunodetection methods (e.g., Elispot) have, in recent years, shown promise in achieving high-throughput quantitative analysis of anti-VZV neutralizing antibodies.

2.4. IFA

IFAs for the detection of anti-VZV antibodies include the anti-complement immunofluorescence (ACIF) assay [69] and the IFAT [70]. Since the procedure of IFAT is similar to that of FAMA using fixed cells, and detects not only viral glycoproteins on the surface of infected cells but other VZV antigens within them, they are generally regarded as the same immunoassay. For the ACIF assay, the complement is mixed with diluted serum samples before incubation with chemically fixed VZV-infected cells, and then the bound complement is detected with fluorescence-conjugated anti-C3. Positive and negative samples can be confirmed by comparing the fluorescence of infected and uninfected cells, and, like the FAMA test, the highest dilution that causes a positive fluorescence reaction is regarded as the antibody titer of the serum sample [69]. For the IFAT, the only difference is that fluorescence-labeled anti-human secondary antibodies are directly used for detection without the aid of a complement. Although it has been reported that IFA is more sensitive than FAMA, its specificity seems problematic since a cross-reaction was found with other herpesviruses, such as HSV [71]. The same doubt about specificity also exists in FAMA using fixed cells. Nevertheless, the specificity of these immunoassays can be improved by adjusting the experimental methods and materials. For example, Sauerbrei et al. used fixed VZV-infected A549 cells in IFA for VZV antibody detection (schematic shown in Figure 4), which showed high specificity without cross-reaction with anti-HSV antibodies and was 100% consistent with FAMA [31].

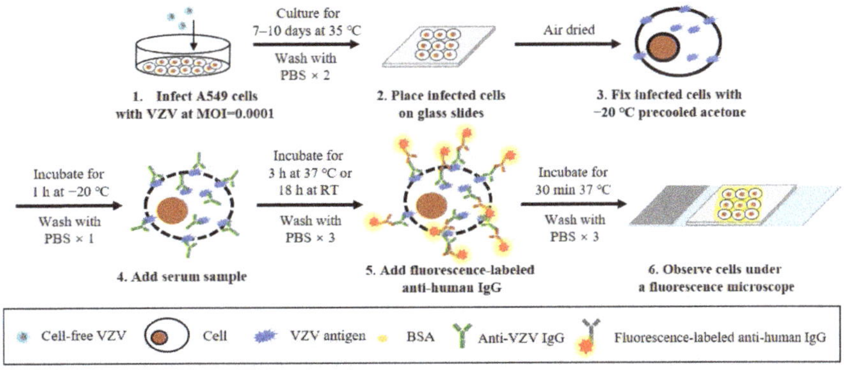

Figure 4. Schematic diagram of the protocol of a reported IFAT for the detection of anti-VZV antibodies. (1) Human lung carcinoma cells (A549) are infected with VZV at MOI = 0.0001 and cultured for 7–10 days at 35 °C. (2) After washing, the infected cells are placed on glass slides and air-dried. (3) Cells are fixed with precooled acetone for 1 h at −20 °C. (4) After washing, diluted serum samples are added and incubated for 3 h at 37 °C or for 18 h at room temperature (RT). (5) After washing, fluorescence-labeled anti-human IgG is added and incubated for 30 min at 37 °C. (6) After washing, the stained cells are observed under a fluorescence microscope.

2.5. TRFIA

TRFIA, which is also called dissociation enhanced-lanthanide fluorescence immunoassay (DELFIA), was first used to detect anti-VZV antibodies in 2006 [72]. As shown in Figure 5, the procedure of TRFIA is similar to that of ELISA. Firstly, purified VZV antigens are coated on DELFIA microtiter plates, and, after washing, the plates are incubated with diluted sera. Then, the plates are incubated with europium (EU)-labeled anti-human IgG conjugate as a secondary antibody to form EU-labeled antibody-antigen immune complexes, and the DELFIA enhancement solution is added to enhance the fluorescence signal of EU3+. Finally, the fluorescence signal is captured by a DELFIA plate reader, and the concentration of VZV antibodies in the serum samples is further calculated according to the standard curve. The standard curves can be determined with international standard VZV antibodies, so the results of TRFIA can be expressed in the international standard unit (mIU/mL). Some researchers consider 150 mIU/mL to be the suitable cutoff value for TRFIA, since antibody concentrations of >150 mIU/mL seem to provide a protective effect, while others suggest that 130 mIU/mL is sufficient to distinguish between uninfected and infected individuals with TRFIA [73,74].

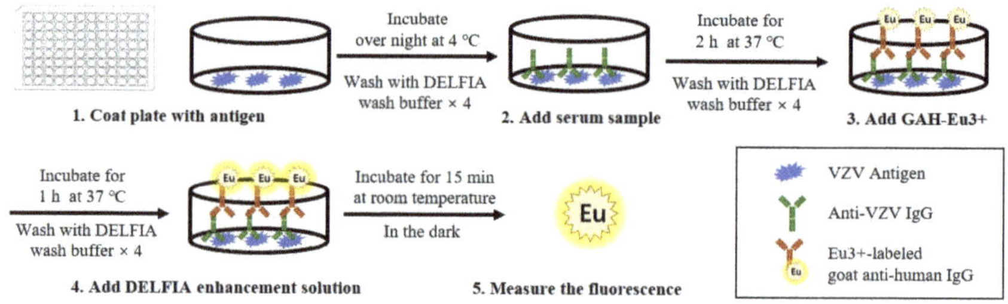

Figure 5. Schematic diagram depicting the basic steps of TRFIA for the detection of anti-VZV antibodies. (1) DELFIA microtiter plates are coated with VZV antigen and incubated overnight at 4 °C. (2) After washing, diluted serum samples are added to the plates and incubated for 2 h at 37 °C. (3) After washing, europium (Eu3+)-labeled goat anti-human IgG (GAH-Eu3+) is added and incubated for 1 h at 37 °C. (4) After washing, DELFIA enhancement solution is added and incubated for 15 min at room temperature in the dark. (5) The plate is read using a DELFIA 1234 reader to measure the fluorescence intensity of EU3+.

TRFIA uses EU3+ as the fluorescent probe; it has a long decay time, and antibody concentrations can be measured after other fluorescent substances with short half-lives are decayed, thus eliminating the interference of non-specific fluorescence. Coupled with the narrow emission light crest of lanthanum fluorescence, the background of TRFIA is further reduced. TRFIA has sensitivity and specificity equivalent to those of the Merck gp-ELISA. Furthermore, compared with FAMA, this method is easy to use and has a relatively short testing time. However, TRFIA requires special equipment and is only used in a few specialist testing centers.

2.6. IAHA

IAHA was once commonly used to detect anti-VZV antibodies for the evaluation of the immune status of the population against VZV and the immune effect of chickenpox vaccination [54]. IAHA is performed by mixing VZV antigens, serum, a complement, and human type O red blood cells (RBCs), and a positive reaction is indicated by the agglutination of RBCs, which is mediated by their surface C3 receptors [75]. Since the sensitivity of IAHA is low, its application has great limitations.

LA is a modified version of IAHA. According to the procedure of LA (schematic shown in Figure 6), serially diluted serum samples are added to the synthetic latex particles coated with VZV antigen (e.g., VZV gE), and the test samples are determined as positive by the observation of the agglutination reaction [27]. Serum with a titer of $\geq$1:2 by LA is considered positive [76]. LA is convenient and fast to operate, requires no special equipment, and is commercially available. The sensitivity of LA is almost as good as that of FAMA and is better than that of a commercial ELISA. However, false-positive results may occur in the detection process since LA cannot distinguish between IgG and IgM, and it is difficult to automate and difficult to use on a large scale.

2.7. CLIA

CLIA has also been used to detect anti-VZV antibodies. According to the procedure of CLIA (schematic shown in Figure 7), magnetic particles are first coated with purified VZV antigens (e.g., glycoproteins). After incubation with diluted serum samples, secondary anti-human IgA/IgM/IgG antibodies conjugated with isoluminol or acridinium are used to detect the captured VZV-specific antibodies. Next, chemiluminescent detection reagents are added to produce a signal, and the relative light units (RLU) are measured using a full-automatic chemical luminescence immune analyzer and converted to the antibody concentration according to the standard curve [77,78]. The reported CLIA used cutoff values of 150 mIU/mL [79] and 100 mIU/mL [80]. It has been reported that women with CLIA values of <100 mIU/mL are more likely to develop varicella than those with values of >100 mIU/mL. Thus, a value of 100 mIU/mL may distinguish women who are susceptible to chickenpox infection from those who are protected from exposure. However, while emphasizing the importance of this value, some international guidelines note that the CLIA cutoff may vary depending on vaccination status, race, or age.

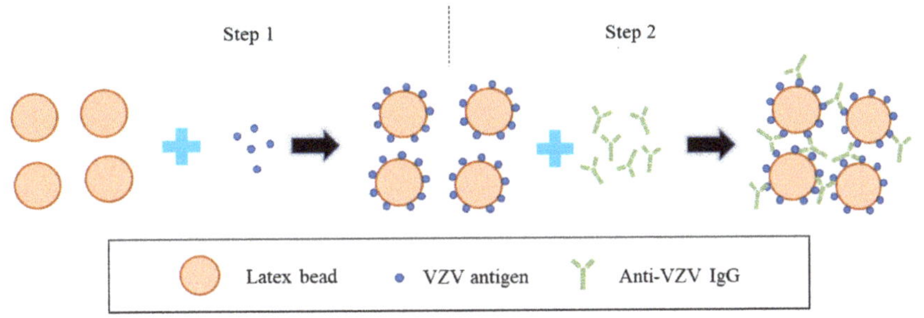

Figure 6. Schematic diagram depicting the basic steps of LA for the detection of anti-VZV antibodies. (1) The latex beads are coated with VZV antigen (e.g., purified VZV gE). (2) Serially diluted serum samples are added to the synthetic latex beads coated with VZV antigen, and the test samples are determined as positive by the observation of the agglutination reaction.

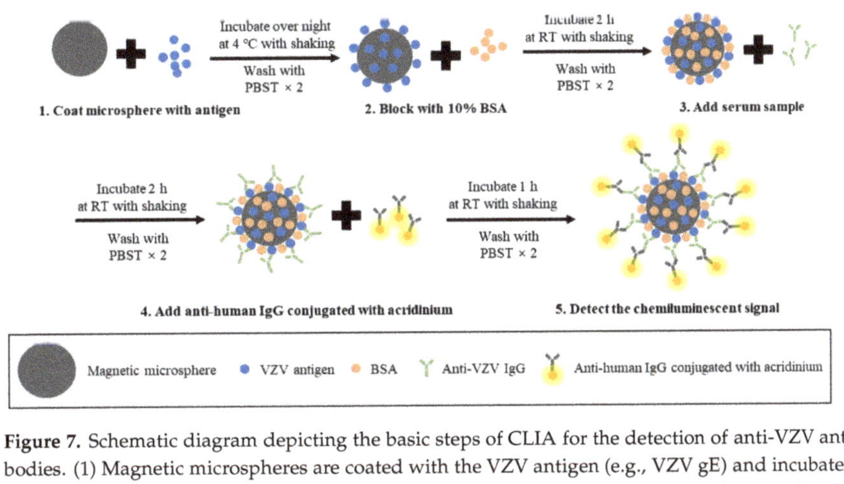

Figure 7. Schematic diagram depicting the basic steps of CLIA for the detection of anti-VZV antibodies. (1) Magnetic microspheres are coated with the VZV antigen (e.g., VZV gE) and incubated overnight at 4 °C with shaking. (2) After washing, microspheres are blocked with 10% BSA and incubated for 2 h at 37 °C with shaking. (3) After washing, diluted serum samples are added to the coated microspheres and incubated for 2 h at 37 °C with shaking. (4) After washing, anti-human IgG conjugated with acridinium is added and incubated for 1 h at room temperature (RT) with shaking. (5) After washing, the magnetic microspheres are resuspended with PBS and the chemiluminescent signal is detected with a chemical luminescence immune analyzer.

The CLIA is easy to operate, can achieve automated detection, and has a commercial kit. However, the sensitivity of CLIA still needs to be improved. Recently, a VZV gE-CLIA has shown better sensitivity and specificity than the gp-ELISA (Abcam, Cambridge, MA, USA) [78], but further application and validation of this method are still required.

2.8. LFIA

LFIA, a simple point-of-care testing (POCT) based on antigen and antibody immune responses, has been used for the detection of VZV antibodies [81]. The reported VZV LIFA is a paper system that uses truncated VZV gE protein as the capture antigen and consists of a substrate, nitrocellulose membrane, sample pad, binding pad, test line, and control line [82,83]. The protocol for this assay is shown in Figure 8. First, 15 to 20 µL of serum or whole blood is added to the sample port, filtered through the blood separation membrane, and absorbed by the test strip. Then, three drops of chase buffer are added to allow migration of colloidal gold-conjugated goat anti-human IgG and the sample onto the test line. The mixture contacts the test line and the control line in turn, and anti-VZV gE antibodies present in the sample will be captured by the VZV gE protein in the test line, resulting in the visualization of the test line along with the control line after 15 to 20 min [81]. A comparative study between Viro VZV IgG LFIA and the Diamedix VZV IgG ELISA showed that VZV LFIA was more sensitive than ELISA, while having comparable specificity [81]. LFIA provides a simple, inexpensive, and rapid method for VZV antibody detection without equipment and sample pretreatment. However, this detection method cannot achieve the absolute quantitative detection of antibodies.

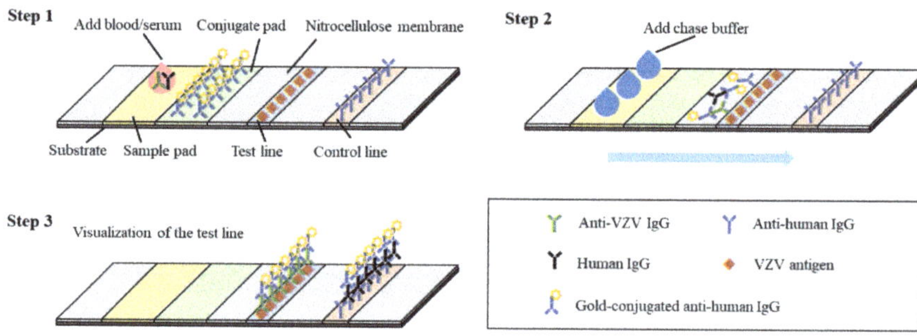

Figure 8. The fabricated LFIA strip for the detection of anti-VZV antibodies comprises of a base nitrocellulose membrane, a sample loading pad, a conjugate pad with immobilized gold-conjugated anti-human IgG, a test line with pre-absorbed VZV antigen (e.g., VZV gE) and a control line that serves to confirm if the strip is working. The basic steps of LFIA are as follows: (1) Add 15 to 20 µL of whole blood or serum to the sample pad of the strip. (2) Add three drops of chase buffer to allow migration of gold-antibody complexes onto the test line and the control line. (3) Anti-VZV antibodies in the sample are captured by the VZV antigen in the test line, and the presence of both a control line and a test line is used to define a VZV-positive test result.

3. Concluding Remarks and Perspectives

In recent decades, a variety of serological methods for detecting anti-VZV antibodies have been established for the clinical auxiliary diagnosis of VZV infection; these methods have facilitated VZV-related epidemiological studies and vaccine studies, as well as risk assessments of healthcare workers. FAMA is the most widely recognized and commonly used method due to its relatively high sensitivity and specificity. However, standard live-cell FAMA is labor-intensive with low throughput, and is susceptible to subjective judgment. ELISA is the most accessible method, but commercial ELISA kits may yield false-positive or false-negative results and are not reliable in evaluating serum conversion after vaccination for chickenpox, which generates lower antibody levels compared to the wild-type VZV infection. Neutralization tests can directly evaluate the immune protection effect, but have a low sensitivity for detecting anti-VZV antibody. IFA and LA are very sensitive but have limitations in determining positive reactions, which make them less than ideal for both chickenpox susceptibility screening and detecting seroconversion to VZV vaccines. TRFIA and CLIA showed good specificity and sensitivity but require further validation.

LFIA is an easy-to-use POCT without the requirement for experienced technicians and equipment, but it only provides qualitative or semi-quantitative results.

The existing VZV serological tests mainly detect IgG antibodies, but the analysis of different antibody subtypes is also of value in the diagnosis of VZV infection. For example, it has been reported that the IgG3 subtype was the main subtype in the recovery period of chickenpox, and the IgGl subtype was the main subtype in the recovery period of shingles [84]. Meanwhile, detection of IgGl and IgG2a represents the activation of Th2- and Th1-type immunity, respectively, and would be helpful in elucidating the immune-protective mechanisms of VZV vaccines [85]. Furthermore, the detection of IgM-and IgA-anti-VZV antibodies is also helpful for the diagnosis of VZV infection in immunocompromised individuals [86]. Although FAMA, ELISA, IFA, TRFIA, CLIA, and LFIA can theoretically be applied to analyze different subtypes of anti-VZV antibodies, the relevant research is still somewhat lacking.

While current methods to detect anti-VZV antibodies are numerous and relatively mature, researchers can still strive to make improvements to obtain better detection efficiency and measurement accuracy, as well as greater convenience. Some potential directions for future development are as follows:

(1) Biosensor techniques, such as lateral flow assays and electrochemical assays, are evolving rapidly, and have shown promising prospects in making inexpensive, easy-to-use diagnostic tools (e.g., POCT devices) for quicker sensitive and specific detection of anti-VZV antibodies. These biosensor-based assays would not require elaborate instrumentation and/or a laboratory set-up and would therefore be more accessible to researchers, clinicians, and the general public, thus better meeting the needs of large-scale population screening. They could be useful complements to conventional laboratory tests.

(2) For national institutions or testing centers involved in disease control and prevention, it is better to combine high-precision detection equipment with different assay strategies to develop standardized methods for the automated measurement of anti-VZV antibodies, to not only achieve high sensitivity, accuracy, and reliability but also to enable high-throughput, objective, and stable testing.

(3) It is hoped that the methods developed in the future will be able to detect different subtypes of anti-VZV antibodies with similarly high sensitivity, which would be beneficial for clinical diagnosis, increasing our understanding of the role of antibody responses in the prevention of VZV infection, and may provide insights into ways to improve the effectiveness of VZV vaccines.

Author Contributions: The manuscript was written by D.P. and edited by W.W. and T.C. All authors have read and agreed to the published version of the manuscript.

Funding: This work was funded by grants from the National Natural Science Foundation of China (Nos. 82171833 and 81871648).

Institutional Review Board Statement: Not applicable.

Informed Consent Statement: Not applicable.

Data Availability Statement: Not applicable.

Conflicts of Interest: The authors declare no conflict of interest.

References

1. Gershon, A.A.; Breuer, J.; Cohen, J.I.; Cohrs, R.J.; Gershon, M.D.; Gilden, D.; Grose, C.; Hambleton, S.; Kennedy, P.G.; Oxman, M.N.; et al. Varicella zoster virus infection. *Nat. Rev. Dis. Primers* **2015**, *1*, 15016. [CrossRef] [PubMed]
2. Arvin, A.M.; Gilden, D. Varicella-zoster virus. In *Fields Virology*, 6th ed.; Lippincott Williams & Wilkins: Philadelphia, PA, USA, 2013; pp. 2015–2057.
3. World Health Organization. Varicella and herpes zoster vaccines: WHO position paper, June 2014. *Relev. Epidemiol. Hebd.* **2014**, *89*, 265–287.
4. Heininger, U.; Seward, J.F. Varicella. *Lancet* **2006**, *368*, 1365–1376. [CrossRef] [PubMed]

5. Cohen, J.I. Clinical practice: Herpes zoster. *N. Engl. J. Med.* **2013**, *369*, 255–263. [CrossRef] [PubMed]
6. Harpaz, R.; Ortega-Sanchez, I.R.; Seward, J.F. Prevention of herpes zoster: Recommendations of the Advisory Committee on Immunization Practices (ACIP). *Morb. Mortal. Wkly. Rep. Recomm. Rep.* **2008**, *57*, 1–30.
7. Sampathkumar, P.; Drage, L.A.; Martin, D.P. Herpes zoster (shingles) and postherpetic neuralgia. *Mayo Clin. Proc.* **2009**, *84*, 274–280. [CrossRef]
8. Crooke, S.N.; Ovsyannikova, I.G.; Poland, G.A.; Kennedy, R.B. Immunosenescence and human vaccine immune responses. *Immun. Ageing* **2019**, *16*, 25. [CrossRef]
9. Kechagias, K.; Giannos, P. Varicella Zoster Virus Reactivation Following COVID-19 Vaccination: A Systematic Review of Case Reports. *Vaccines* **2021**, *9*, 1013.
10. Psichogiou, M.; Samarkos, M.; Mikos, N.; Hatzakis, A. Reactivation of varicella zoster virus after vaccination for SARS-CoV-2. *Vaccines* **2021**, *9*, 572. [CrossRef]
11. Gershon, A.; Gershon, M.; Shapiro, E.D. Live Attenuated Varicella Vaccine: Prevention of Varicella and of Zoster. *J. Infect. Dis.* **2021**, *224*, S387–S397. [CrossRef]
12. Harbecke, R.; Cohen, J.; Oxman, M.N. Herpes Zoster Vaccines. *J. Infect. Dis.* **2021**, *224*, S429–S442. [CrossRef] [PubMed]
13. Kennedy, P.G.E.; Gershon, A.A. Clinical Features of Varicella-Zoster Virus Infection. *Viruses* **2018**, *10*, 609. [CrossRef] [PubMed]
14. Guarner, J.; Del Rio, C.; Malani, P.N. Monkeypox in 2022-What Clinicians Need to Know. *JAMA* **2022**, *328*, 139–140. [CrossRef] [PubMed]
15. McCollum, A.M.; Damon, I.K. Human monkeypox. *Clin. Infect. Dis.* **2014**, *58*, 260–267. [CrossRef]
16. MacNeil, A.; Reynolds, M.G.; Carroll, D.S.; Karem, K.; Braden, Z.; Lash, R.; Moundeli, A.; Mombouli, J.-V.; Jumaan, A.O.; Schmid, D.S.; et al. Monkeypox or varicella? Lessons from a rash outbreak investigation in the Republic of the Congo. *Am. J. Trop. Med. Hyg.* **2009**, *80*, 503–507. [CrossRef]
17. Bienes, K.M.; Mao, L.; Selekon, B.; Gonofio, E.; Nakoune, E.; Wong, G.; Berthet, N. Rapid Detection of the Varicella-Zoster Virus Using a Recombinase-Aided Amplification-Lateral Flow System. *Diagnostics* **2022**, *12*, 2957. [CrossRef]
18. Williams, V.; Gershon, A.; Brunell, P.A. Serologic response to varicella-zoster membrane antigens measured by indirect immunofluorescence. *J. Infect. Dis.* **1974**, *130*, 669–672. [CrossRef]
19. Breuer, J.; Schmid, D.S.; Gershon, A.A. Use and limitations of varicella-zoster virus-specific serological testing to evaluate breakthrough disease in vaccinees and to screen for susceptibility to varicella. *J. Infect. Dis.* **2008**, *197*, S147–S151. [CrossRef]
20. Gershon, A.A.; Gershon, M.D. Perspectives on vaccines against varicella-zoster virus infections. In *Varicella-Zoster Virus*; Springer: Berlin/Heidelberg, Germany, 2010; pp. 359–372.
21. Landry, M.L.; Cohen, S.; Mayo, D.; Fong, C.; Andiman, W.A. Comparison of fluorescent-antibody-to-membrane-antigen test, indirect immunofluorescence assay, and a commercial enzyme-linked immunosorbent assay for determination of antibody to varicella-zoster virus. *J. Clin. Microbiol.* **1987**, *25*, 832–835. [CrossRef]
22. Keller, P.M.; Neff, B.J.; Ellis, R.W. Three major glycoprotein genes of varicella-zoster virus whose products have neutralization epitopes. *J. Virol.* **1984**, *52*, 293–297. [CrossRef]
23. Baba, K.; Yoshida, M.; Tawa, A.; Yabuuchi, H.; Maeda, K.; Takahashi, M. A simplified immunofluorescence technique for antibody to varicella-zoster membrane antigen (FAMA). *Biken J.* **1984**, *27*, 23–29. [PubMed]
24. Iltis, J.; Castellano, G.; Gerber, P.; Le, C.; Vujcic, L.; Quinnan, G.V., Jr. Comparison of the Raji cell line fluorescent antibody to membrane antigen test and the enzyme-linked immunosorbent assay for determination of immunity to varicella-zoster virus. *J. Clin. Microbiol.* **1982**, *16*, 878–884. [CrossRef] [PubMed]
25. de Ory, F.; Echevarría, J.M.; Kafatos, G.; Anastassopoulou, C.; Andrews, N.; Backhouse, J.; Berbers, G.; Bruckova, B.; Cohen, D.I.; de Melker, H.; et al. European seroepidemiology network 2: Standardisation of assays for seroepidemiology of varicella zoster virus. *J. Clin. Virol.* **2006**, *36*, 111–118. [CrossRef] [PubMed]
26. Maple, P.; Haedicke, J.; Quinlivan, M.; Steinberg, S.; Gershon, A.; Brown, K.; Breuer, J. The differences in short-and long-term varicella-zoster virus (VZV) immunoglobulin G levels following varicella vaccination of healthcare workers measured by VZV fluorescent-antibody-to-membrane-antigen assay (FAMA), VZV time-resolved fluorescence immunoassay and a VZV purified glycoprotein enzyme immunoassay. *Epidemiol. Infect.* **2016**, *144*, 2345–2353.
27. Gershon, A.A.; LaRussa, P.; Steinberg, S. Detection of antibodies to varicella-zoster virus using a latex agglutination assay. *Clin. Diagn. Virol.* **1994**, *2*, 271–277. [CrossRef] [PubMed]
28. Kim, Y.H.; Hwang, J.Y.; Shim, H.M.; Lee, E.; Park, S.; Park, H. Evaluation of a commercial glycoprotein enzyme-linked immunosorbent assay for measuring vaccine immunity to varicella. *Yonsei Med. J.* **2014**, *55*, 459–466. [CrossRef]
29. Zaia, J.A.; Oxman, M.N. Antibody to varicella-zoster virus-induced membrane antigen: Immunofluorescence assay using monodisperse glutaraldehyde-fixed target cells. *J. Infect. Dis.* **1977**, *136*, 519–530. [CrossRef]
30. Wutzler, P.; Färber, I.; Wagenpfeil, S.; Bisanz, H.; Tischer, A. Seroprevalence of varicella-zoster virus in the German population. *Vaccine* **2001**, *20*, 121–124. [CrossRef]
31. Sauerbrei, A.; Färber, I.; Brandstädt, A.; Schacke, M.; Wutzler, P. Immunofluorescence test for sensitive detection of varicella-zoster virus-specific IgG: An alternative to fluorescent antibody to membrane antigen test. *J. Virol. Methods* **2004**, *119*, 25–30. [CrossRef]
32. Han, S.; Kang, K.; Huh, D.; Lee, H.; Kim, J.; Kang, J.; Ma, S.H. Seroepidemiology of varicella-zoster virus in Korean adolescents and adults using fluorescent antibody to membrane antigen test. *Epidemiol. Infect.* **2015**, *143*, 1643–1650. [CrossRef]

33. Larussa, P.; Steinberg, S.; Waithe, E.; Hanna, B.; Holzman, R. Comparison of five assays for antibody to varicella-zoster virus and the fluorescent-antibody-to-membrane-antigen test. *J. Clin. Microbiol.* **1987**, *25*, 2059–2062. [CrossRef] [PubMed]
34. Cradock-Watson, J.; Ridehalgh, M.K.; Bourne, M.S. Specific immunoglobulin responses after varicella and herpes zoster. *Epidemiol. Infect.* **1979**, *82*, 319–336. [CrossRef] [PubMed]
35. Kitamura, K.; Namazue, J.; Campo-Vera, H.; Ogino, T.; Yamanishi, K. Induction of neutralizing antibody against varicella-zoster virus (VZV) by VZV gp3 and cross-reactivity between VZV gp3 and herpes simplex viruses gB. *Virology* **1986**, *149*, 74–82. [CrossRef] [PubMed]
36. Park, R.; Hwang, J.; Lee, K.; Namkoong, S.; Choi, S.; Park, S.; Park, H.; Park, J. Measurement of antibodies to varicella-zoster virus using a virus-free fluorescent-antibody-to-membrane-antigen (FAMA) test. *J. Microbiol. Biotechnol.* **2015**, *25*, 268–273. [CrossRef] [PubMed]
37. Litwin, V.; Sandor, M.; Grose, C. Cell surface expression of the varicella-zoster virus glycoproteins and Fc receptor. *Virology* **1990**, *178*, 263–272. [CrossRef] [PubMed]
38. Lafer, M.M.; Weckx, L.; de Moraes-Pinto, M.; Garretson, A.; Steinberg, S.; Gershon, A.; LaRussa, P.S. Comparative study of the standard fluorescent antibody to membrane antigen (FAMA) assay and a flow cytometry-adapted FAMA assay to assess immunity to varicella-zoster virus. *Clin. Vaccine Immunol.* **2011**, *18*, 1194–1197. [CrossRef]
39. Tourtelot, E.; Quataert, S.; Glantz, J.C.; Perlis, L.; Muthukrishnan, G.; Mosmann, T. Women who received varicella vaccine versus natural infection have different long-term T cell immunity but similar antibody levels. *Vaccine* **2020**, *38*, 1581–1585. [CrossRef]
40. Smith-Norowitz, T.A.; Saadia, T.A.; Norowitz, K.B.; Joks, R.; Durkin, H.G.; Kohlhoff, S. Negative IgG varicella zoster virus antibody status: Immune responses pre and post re-immunization. *Infect. Dis. Ther.* **2018**, *7*, 175–181. [CrossRef]
41. Gillard, P.; Povey, M.; Carryn, S. Clinically-versus serologically-identified varicella: A hidden infection burden. A ten-year follow-up from a randomized study in varicella-endemic countries. *Hum. Vaccines Immunother.* **2021**, *17*, 3747–3756. [CrossRef]
42. Chenghua, P.; Xiaolin, L.; Peixia, P.; Guiqiu, W.; Ming, D.; Yongji, W. Detection of VZV IgG by indirect ELISA. *Chin. J. Biol.* **1995**, *8*, 34–36.
43. Enders, G. Serodiagnosis of Varicella-Zoster virus infection in pregnancy and standardization of the ELISA IgG and IgM antibody tests. *Dev. Biol. Stand.* **1982**, *52*, 221–236. [PubMed]
44. Ono, E.; Lafer, M.M.; Weckx, L.Y.; Granato, C.; Moraes-Pinto, M.I.D. A simple and cheaper in house varicella zoster virus antibody indirect ELISA. *Rev. Do Inst. De Med. Trop. De São Paulo* **2004**, *46*, 165–168. [CrossRef] [PubMed]
45. Sauerbrei, A.; Schäfler, A.; Hofmann, J.; Schacke, M.; Gruhn, B.; Wutzler, P. Evaluation of three commercial varicella-zoster virus IgG enzyme-linked immunosorbent assays in comparison to the fluorescent-antibody-to-membrane-antigen test. *Clin. Vaccine Immunol.* **2012**, *19*, 1261–1268. [CrossRef] [PubMed]
46. Sauerbrei, A.; Wutzler, P. Serological detection of varicella-zoster virus-specific immunoglobulin G by an enzyme-linked immunosorbent assay using glycoprotein antigen. *J. Clin. Microbiol.* **2006**, *44*, 3094–3097. [CrossRef] [PubMed]
47. Sauerbrei, A.; Wutzler, P. Serological detection of specific IgG to varicella-zoster virus by novel ELISA based on viral glycoprotein antigen. *Clin. Lab.* **2009**, *55*, 1–7.
48. Heininger, U.; Desgrandchamps, D.; Schaad, U.B. Seroprevalence of varicella-zoster virus IgG antibodies in Swiss children during the first 16 months of age. *Vaccine* **2006**, *24*, 3258–3260. [CrossRef] [PubMed]
49. Ceroni, A.; Sibani, S.; Baiker, A.; Pothineni, V.R.; Bailer, S.M.; LaBaer, J.; Haas, J.; Campbell, C.J. Systematic analysis of the IgG antibody immune response against varicella zoster virus (VZV) using a self-assembled protein microarray. *Mol. Biosyst.* **2010**, *6*, 1604–1610. [CrossRef]
50. Rolando, L.; Schneider, W.J.; Steinberg, S.; Low, S.; Stiles, J.; Gomez, L.; Gershon, A.A.; Brown, A.E. Effect of varicella-zoster virus (VZV) fluorescent-antibody-to-membrane-antigen (FAMA) testing on sensitivity of determining VZV immunity in healthcare workers and on furlough days. *Infect. Control. Hosp. Epidemiol.* **2010**, *31*, 972–974. [CrossRef]
51. Wasmuth, E.H.; Miller, W.J. Sensitive enzyme-linked immunosorbent assay for antibody to varicella-zoster virus using purified VZV glycoprotein antigen. *J. Med. Virol.* **1990**, *32*, 189–193. [CrossRef]
52. Li, S.; CHAN IV, A.S.; Matthews, H.; Heyse, J.F.; Chan, C.Y.; Kuter, B.J.; Kaplan, K.M.; Vessey, S.R.; Sadoff, J.C. Inverse relationship between six week postvaccination varicella antibody response to vaccine and likelihood of long term breakthrough infection. *Pediatr. Infect. Dis. J.* **2002**, *21*, 337–342. [CrossRef]
53. Mitra, M.; Faridi, M.; Ghosh, A.; Shah, N.; Shah, R.; Chaterjee, S.; Narang, M.; Bhattacharya, N.; Bhat, G.; Choudhury, H.; et al. Safety and immunogenicity of single dose live attenuated varicella vaccine (VR 795 Oka strain) in healthy Indian children: A randomized controlled study. *Hum. Vaccines Immunother.* **2015**, *11*, 443–449. [CrossRef] [PubMed]
54. Otani, N.; Tanaka, M.; Maeda, K.; Gomi, Y.; Nakajima, K.; Tanimura, S.; Takesue, Y.; Shima, M.; Okuno, T. Varicella zoster virus antibody detection: A comparison of four commonly used techniques. *J. Infect. Chemother.* **2016**, *22*, 225–228. [CrossRef]
55. Muchtar, E.; Koehler, A.B.; Johnson, M.J.; Rabe, K.G.; Ding, W.; Call, T.G.; Leis, J.F.; Kenderian, S.S.; Hayman, S.R.; Wang, Y.; et al. Humoral and cellular immune responses to recombinant herpes zoster vaccine in patients with chronic lymphocytic leukemia and monoclonal B cell lymphocytosis. *Am. J. Hematol.* **2022**, *97*, 90–98. [CrossRef] [PubMed]
56. Liu, J.; Ye, X.; Jia, J.; Zhu, R.; Wang, L.; Chen, C.; Yang, L.; Wang, Y.; Wang, W.; Ye, J.; et al. Serological evaluation of immunity to the varicella-zoster virus based on a novel competitive enzyme-linked immunosorbent assay. *Sci. Rep.* **2016**, *6*, 20577. [CrossRef] [PubMed]

57. Liu, J.; Chen, C.; Zhu, R.; Ye, X.; Jia, J.; Yang, L.; Wang, Y.; Wang, W.; Ye, J.; Li, Y.; et al. Evaluation of immunity to varicella zoster virus with a novel double antigen sandwich enzyme-linked immunosorbent assay. *Appl. Microbiol. Biotechnol.* **2016**, *100*, 9321–9329. [CrossRef]
58. Caunt, A.E.; Shaw, D.G. Neutralization tests with varicella-zoster virus. *Epidemiol. Infect.* **1969**, *67*, 343–352. [CrossRef]
59. Grose, C.; Edmond, B.J.; Brunell, P.A. Complement-enhanced neutralizing antibody response to varicella-zoster virus. *J. Infect. Dis.* **1979**, *139*, 432–437. [CrossRef]
60. Asano, Y.; Takahashi, M. Clinical and serologic testing of a live varicella vaccine and two-year follow-up for immunity of the vaccinated children. *Pediatrics* **1977**, *60*, 810–814. [CrossRef]
61. Asano, Y.; Albrecht, P.; Stagno, S.; Takahashi, M. Potentiation of neutralization of varicella-zoster virus by antibody to immunoglobulin. *J. Infect. Dis.* **1982**, *146*, 524–529. [CrossRef]
62. Krah, D.L.; Provost, P.J.; Ellis, R.W. Combined use of complement and anti-immunoglobulin in an enhanced neutralization assay for antibodies to varicella-zoster virus. *J. Virol. Methods* **1995**, *53*, 176–187. [CrossRef]
63. Asano, Y.; Nakayama, H.; Yazaki, T.; Ito, S.; Isomura, S. Protective efficacy of vaccination in children in four episodes of natural varicella and zoster in the ward. *Pediatrics* **1977**, *59*, 8–12. [PubMed]
64. Schmidt, N.J.; Lennette, E.H. Neutralizing antibody responses to varicella-zoster virus. *Infect. Immun.* **1975**, *12*, 606–613. [CrossRef] [PubMed]
65. Asano, Y.; Takahashi, M. Studies on neutralization of varicella-zoster virus and serological follow-up of cases of varicella and zoster. *Biken J.* **1978**, *21*, 15–23. [PubMed]
66. Gerna, G.; Chambers, R.W. Varicella-zoster plaque assay and plaque reduction neutralization test by the immunoperoxidase technique. *J. Clin. Microbiol.* **1976**, *4*, 437–442. [CrossRef]
67. Schmidtmayerova, H.; Mayer, V.; Zachar, V. Focus assay for varicella-zoster virus in human embryo cells stained with immunoperoxidase method. *Acta Virol.* **1986**, *30*, 468–474.
68. Chen, L.; Liu, J.; Wang, W.; Ye, J.; Wen, L.; Zhao, Q.; Zhu, H.; Cheng, T.; Xia, N. Development of a varicella-zoster virus neutralization assay using a glycoprotein K antibody enzyme-linked immunosorbent spot assay. *J. Virol. Methods* **2014**, *200*, 10–14. [CrossRef]
69. Shigeta, S.; Baba, M.; Ogata, M.; Iijima, S.; Murai, C. Anticomplement Immunofluorescence for the Titration of Antibody to Varicella-Zoster Virus. *Microbiol. Immunol.* **1981**, *25*, 295–303. [CrossRef]
70. Sauerbrei, A.; Eichhorn, U.; Schacke, M.; Wutzler, P. Laboratory diagnosis of herpes zoster. *J. Clin. Virol.* **1999**, *14*, 31–36. [CrossRef]
71. Krah, D.L. Assays for antibodies to varicella-zoster virus. *Infect. Dis. Clin. N. Am.* **1996**, *10*, 507–527. [CrossRef]
72. Maple, P.; Gray, J.; Breuer, J.; Kafatos, G.; Parker, S.; Brown, D. Performance of a time-resolved fluorescence immunoassay for measuring varicella-zoster virus immunoglobulin G levels in adults and comparison with commercial enzyme immunoassays and Merck glycoprotein enzyme immunoassay. *Clin. Vaccine Immunol.* **2006**, *13*, 214–218. [CrossRef]
73. McDonald, E.M.; de Kock, J.; Ram, F.S. Antivirals for management of herpes zoster including ophthalmicus: A systematic review of high-quality randomized controlled trials. *Antivir. Ther.* **2012**, *17*, 255–264. [CrossRef] [PubMed]
74. Chris Maple, P.; Gunn, A.; Sellwood, J.; Brown, D.; Gray, J.J. Comparison of fifteen commercial assays for detecting Varicella Zoster virus IgG with reference to a time resolved fluorescence immunoassay (TRFIA) and the performance of two commercial assays for screening sera from immunocompromised individuals. *J. Virol. Methods* **2009**, *155*, 143–149. [CrossRef] [PubMed]
75. Gershon, A.A.; Kalter, Z.G.; Steinberg, S. Detection of antibody to varicella-zoster virus by immune adherence hemagglutination. *Exp. Biol. Med.* **1976**, *151*, 762–765. [CrossRef] [PubMed]
76. Steinberg, S.P.; Gershon, A. Measurement of antibodies to varicella-zoster virus by using a latex agglutination test. *J. Clin. Microbiol.* **1991**, *29*, 1527–1529. [CrossRef] [PubMed]
77. Tafuri, S.; Gallone, M.; Cappelli, M.; Gallone, M.; Larocca, A.; Germinario, C. A seroprevalence survey on varicella among adults in the vaccination era in Apulia (Italy). *Vaccine* **2014**, *32*, 6544–6547. [CrossRef] [PubMed]
78. Kombe Kombe, A.J.; Xie, J.; Zahid, A.; Ma, H.; Xu, G.; Deng, Y.; Nsole Biteghe, F.A.; Mohammed, A.; Dan, Z.; Yang, Y.; et al. Detection of Circulating VZV-Glycoprotein E-Specific Antibodies by Chemiluminescent Immunoassay (CLIA) for Varicella–Zoster Diagnosis. *Pathogens* **2022**, *11*, 66. [CrossRef] [PubMed]
79. Maple, P.; Rathod, P.; Smit, E.; Gray, J.; Brown, D.; Boxall, E.H. Comparison of the performance of the LIAISON VZV-IgG and VIDAS automated enzyme linked fluorescent immunoassays with reference to a VZV-IgG time-resolved fluorescence immunoassay and implications of choice of cut-off for LIAISON assay. *J. Clin. Virol.* **2009**, *44*, 9–14. [CrossRef]
80. Boxall, E.; Maple, P.; Rathod, P.; Smit, E. Follow-up of pregnant women exposed to chicken pox: An audit of relationship between level of antibody and development of chicken pox. *Eur. J. Clin. Microbiol. Infect. Dis.* **2011**, *30*, 1193–1200. [CrossRef]
81. Vafai, N.; Self, K.; Sheffield, B.; Hojvat, S.; Kusi-Appiah, A.; Vaughan, P.; Cowan, E.; Vafai, A. Rapid, sensitive, and specific lateral-flow immunochromatographic point-of-care device for detection of varicella-zoster virus immunoglobulin G antibodies in fingerstick blood. *J. Immunol. Methods* **2023**, *514*, 113429. [CrossRef]
82. Boehringer, H.R.; O'Farrell, B.J. Lateral flow assays in infectious disease diagnosis. *Clin. Chem.* **2022**, *68*, 52–58. [CrossRef]
83. Laderman, E.I.; Whitworth, E.; Dumaual, E.; Jones, M.; Hudak, A.; Hogrefe, W.; Carney, J.; Groen, J. Rapid, sensitive, and specific lateral-flow immunochromatographic point-of-care device for detection of herpes simplex virus type 2-specific immunoglobulin G antibodies in serum and whole blood. *Clin. Vaccine Immunol.* **2008**, *15*, 159–163. [CrossRef] [PubMed]

84. Møyner, K.; Michaelsen, T.E. IgG subclass distribution among antibodies to varicella-zoster virus in human varicella/zoster immunoglobulin preparations and the corresponding donor plasma. *J. Biol. Stand.* **1988**, *16*, 157–164. [CrossRef] [PubMed]
85. Nisonoff, A.; Hopper, J.; Spring, S.B. Allotypes of rabbit, human, and mouse immunoglobulins. In *The Antibody Molecule*; Academic Press: New York, NY, USA, 1975; pp. 346–406. [CrossRef]
86. Gross, G.; Schöfer, H.; Wassilew, S.; Friese, K.; Timm, A.; Guthoff, R.; Pau, H.; Malin, J.; Wutzler, P.; Doerr, H.W. Herpes zoster guideline of the German Dermatology Society (DDG). *J. Clin. Virol.* **2003**, *26*, 277–289, discussion 291–273. [CrossRef] [PubMed]

microorganisms

MDPI

Review

Drug Reaction with Eosinophilia and Systemic Symptoms (DRESS): Focus on the Pathophysiological and Diagnostic Role of Viruses

Giuseppe A. Ramirez [1,2], Marco Ripa [2,3], Samuele Burastero [1], Giovanni Benanti [1,2], Diego Bagnasco [4], Serena Nannipieri [1,2], Roberta Monardo [2,3], Giacomo Ponta [2,3], Chiara Asperti [1,2], Maria Bernadette Cilona [1,2], Antonella Castagna [2,3], Lorenzo Dagna [1,2,†] and Mona-Rita Yacoub [1,2,*,†]

[1] Unit of Immunology, Rheumatology, Allergy and Rare Diseases, IRCCS Ospedale San Raffaele, 20132 Milan, Italy
[2] Faculty of Medicine, Università Vita-Salute San Raffaele, 20132 Milan, Italy
[3] Unit of Infectious Diseases, IRCCS Ospedale San Raffaele, 20132 Milan, Italy
[4] IRCCS Policlinico San Martino, Department of Internal Medicine (DIMI), University of Genoa, 16132 Genoa, Italy
* Correspondence: yacoub.monarita@hsr.it
† These authors contributed equally to this work.

Abstract: Drug reaction with eosinophilia and systemic symptoms (DRESS) is a heterogeneous, multiorgan and potentially life-threatening drug-hypersensitivity reaction (DHR) that occurs several days or weeks after drug initiation or discontinuation. DHRs constitute an emerging issue for public health, due to population aging, growing multi-organ morbidity, and subsequent enhanced drug prescriptions. DRESS has more consistently been associated with anticonvulsants, allopurinol and antibiotics, such as sulphonamides and vancomycin, although new drugs are increasingly reported as culprit agents. Reactivation of latent infectious agents such as viruses (especially Herpesviridae) plays a key role in prompting and sustaining aberrant T-cell and eosinophil responses to drugs and pathogens, ultimately causing organ damage. However, the boundaries of the impact of viral agents in the pathophysiology of DRESS are still ill-defined. Along with growing awareness of the multifaceted aspects of immune perturbation caused by severe acute respiratory syndrome coronavirus 2 (SARS-CoV-2) during the ongoing SARS-CoV-2-related disease (COVID-19) pandemic, novel interest has been sparked towards DRESS and the potential interactions among antiviral and anti-drug inflammatory responses. In this review, we summarised the most recent evidence on pathophysiological mechanisms, diagnostic approaches, and clinical management of DRESS with the aim of increasing awareness on this syndrome and possibly suggesting clues for future research in this field.

Keywords: DRESS; virus; eosinophils; reaction; T-cells; herpesvirus; viral reactivation

Citation: Ramirez, G.A.; Ripa, M.; Burastero, S.; Benanti, G.; Bagnasco, D.; Nannipieri, S.; Monardo, R.; Ponta, G.; Asperti, C.; Cilona, M.B.; et al. Drug Reaction with Eosinophilia and Systemic Symptoms (DRESS): Focus on the Pathophysiological and Diagnostic Role of Viruses. *Microorganisms* **2023**, *11*, 346. https://doi.org/10.3390/microorganisms11020346

Academic Editors: Shengxi Chen and Fabio Zicker

Received: 31 October 2022
Revised: 13 January 2023
Accepted: 16 January 2023
Published: 30 January 2023

1. Introduction

Adverse drug reactions constitute an emerging issue for public health due to their increasing incidence over time, at least in Westernised countries [1,2]. Population aging along with the growing prevalence of dementia and multi-organ morbidity are associated with frequent institutionalisation and enhanced drug prescribing and may account for this trend [3]. Hospitalisation-related adverse drug reactions are particularly relevant from an epidemiological and economic standpoint and can occur in up to 30% of patients entering ordinary wards [4]. Drug-related hypersensitivity reactions (DHR) constitute a subgroup of adverse drug reactions, occurring with an incidence rate of 80 cases/1000 person-years in in-patient settings [5]. DHRs pose major challenges to the management of hospitalised patients, since they occur unpredictably (in contrast with non-immune-mediated adverse drug reactions) and affect patients' ability to receive appropriate treatments for their acute conditions, besides bearing intrinsic morbidity and mortality risks [6]. Multiple pathogenic factors are thought to contribute to the development and maintenance of hypersensitivity reactions to drugs. Aberrant haptenation of drugs in the setting of acute inflammation and/or direct drug-related activation of immune cells might combine with predisposing genetic factors, such as a permissive human leukocyte antigen (HLA) repertoire, to prompt drug sensitisation [7–13]. Microbial factors might also contribute to the development of DHRs by promoting systemic inflammation and affecting immune tolerance due to molecular mimicry among self and microbial antigens. Consistently, the majority of DHRs are associated with the use of antimicrobials [5].

Drug reaction with eosinophilia and systemic symptoms (DRESS, also known as drug-induced hypersensitivity syndrome, DIHS) is a rare but potentially life-threatening delayed-type systemic DHR characterised by elevated blood eosinophil counts along with constitutional symptoms and multi-organ failure [14,15]. Affordable estimates of DRESS epidemiology are to date missing. The current literature suggests that DRESS incidence can range from less than 0.01 cases to 0.7 cases per 1000 hospitalised patients depending on the healthcare system and demographic context [16–19]. Changing epidemiological trends within the same cohort according to variations in drug prescription attitudes and environmental factors have also been reported [20]. Infectious agents including exogenous or latent virus along with endogenous retroviral elements constitute known perturbators of the physiological immune response and are increasingly recognised as cofactors in the onset of DRESS. This review summarises the clinical and pathophysiological evidence addressing the role of microbial cofactors in DRESS up to date.

2. Aetiology and Pathogenesis

DRESS is a systemic disorder sustained by two pairs of fundamental pathophysiological pillars: (1) inciting stimuli, encompassing drugs; and viruses; (2) deranged immune responses including (a) HLA-restricted aberrant T-cell activation and; (b) eosinophilic inflammation (Figure 1).

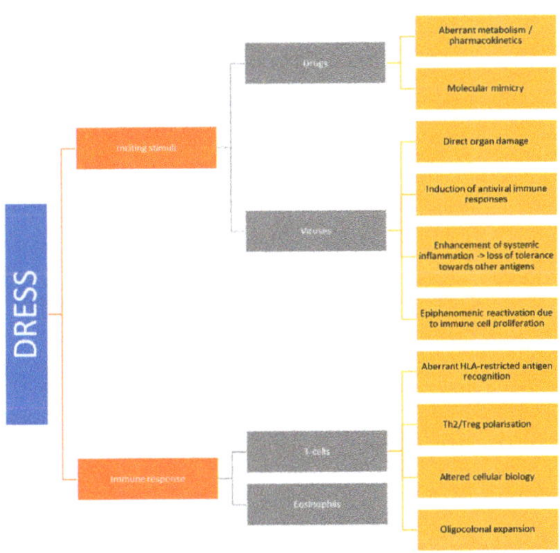

Figure 1. Simplified DRESS aetiopathogenesis. Flow-chart depicting the pathophysiological relationships among the main exogenous and host-related factors involved in the development of DRESS.

2.1. Drugs

A straightforward association with drug exposure is found in 80% of patients, and DRESS onset typically occurs 2–8 weeks after treatment start with the causative drug [21].

Recent drug discontinuation is also associated with DRESS. Numerous drugs have been described as possible triggers of DRESS, but around 75% of cases can be traced back to a high-risk group of drugs [19,22], including anticonvulsants, allopurinol, antibiotics such as vancomycin, minocycline, trimethoprim-sulfamethoxazole and other sulphonamides, anti-tuberculosis agents and antiviral drugs such as nevirapine [23]. Shorter lag times for DRESS onset are observed when antibiotics or iodinated contrast media are implicated [24,25].

However, other drugs have also been reported in association with DRESS onset. Some of these drugs are in more widespread use, such as anti-inflammatory drugs (e.g., non-steroidal anti-inflammatory drugs, NSAIDs, and paracetamol) or antipsychotic drugs. Special attention should be given to special populations such as patients with cancer, rheumatic diseases and chronic viral infection. These patients bear disease-specific risk factors for developing DRESS, including concomitant polytherapy, concomitant infection, kidney or liver failure, and intrinsic immune dysfunction [23]. Furthermore, novel drugs are continuously introduced in the setting of these disorders, expanding the spectrum of potential DRESS triggers. Anti-cancer targeted therapies such as imatinib [21] or sorafenib [22], immune-modulators such as IL1 or IL6 inhibitors (e.g., anakinra, canakinumab, tocilizumab or hydroxychloroquine) or anti-HCV therapies including Telaprevir and Boceprevir constitute potential examples of emerging culprit drugs for DRESS (Table 1) [26].

Culprit drug cross-reactivity is not conventionally expected in DRESS. However, very limited evidence suggests that sensitisation to glycopeptides and β-lactams might compromise patient ability to eventually receive any members of these drug classes [27]. In addition, patients with DRESS may develop secondary neosensitisation to unrelated chemical compounds concurrently administered during DRESS [28].

From a pathophysiological standpoint, trigger drugs are thought to constitute the main target of the immune response. The strength of association between drugs and DRESS is affected by interindividual and inter-ethnic variations in the HLA repertoire (see below). Distinct HLA variants might in fact segregate with selected ethnicities. Additional inherited factors may promote altered drug metabolism and variably combine with HLA-related factors to contribute to DRESS susceptibility. Polymorphisms in cytochrome P (CYP) 450 and N-acetyltransferase (NAT1, NAT2) may affect drug pharmacokinetics and cause active ingredients of metabolism by-product overload [29]. Examples of the roles of these polymorphisms are constituted by the associations between CYP2C9*3 and severe reactions to phenytoin (in Asian ethnicities) and between variants of the NAT gene and sulphonamides [30,31]. This evidence raises the possibility that non-immunological, non-virological factors impacting drug metabolism may play a role on DRESS.

Besides constituting a target for deranged immune response, culprit drugs might also be involved in other disease mechanisms, including viral reactivation. For example, continuous anticonvulsant therapy has been shown to associate with IgG production decreases. Impaired humoral immunity in turn constitutes a risk factor for viral reactivation (see below).

Table 1. Most frequent and newly reported culprit drugs in DRESS.

Drugs Categories	Drug		Ref.
Urate lowering agents	Allopurinol	High Risk	[32]
	Febuxostat		[33]
Aromatic antiepileptic agents	Carbamazepine	High Risk	
	Phenytoin	High Risk	
	Lamotrigine	High Risk	[14]
	Oxcarbazepine	High Risk	
	Phenobarbital	High Risk	

Table 1. *Cont.*

Drugs Categories	Drug		Ref.
Sulphonamides	Sulfasalazine	High Risk	[14]
	Dapsone	High Risk	
	Trimethoprim-Sulfamethoxazole	High Risk	
	Sulfadiazine	High Risk	
Antibiotics	Vancomycin	High Risk	[23]
	Minocycline	High Risk	
	Piperacillin-Tazobactam		
	Antituberculosis Agents	High Risk	
	Other Penicillins and Cephalosporins		
Antiviral agents	Nevirapine	High Risk	[34]
	Abacavir	High Risk	[35]
	Efavirenz		[36]
	Boceprevir		[37]
	Telaprevir		[38]
Anti-inflammatory drugs	Diclofenac		[23]
	Celecoxib		
	Ibuprofen		
Anti-IL1 antibodies	Anakinra		[39]
	Canakinumab		
Anti IL6 antibodies	Tocilizumab		[39]
Targeted therapies	Imatinib		[21]
	Sorafenib		[40]
	Vismodegib		[41]
	Vemurafenib		[42]
Antipsychotic agents	Fluoxetine		[23]
	Olanzapine		
Anti-coagulant	Rivaroxaban		[43]
Immunomodulators	Hydroxychloroquine		[44]

2.2. Viral Factors

Clinically, viral reactivation occurs up to two weeks after the onset of DRESS symptoms and is associated with a worse prognosis in term of disease duration, relapses, constitutional symptoms and organ involvement [45–47], as compared to patients with no evidence of viral reactivation [47,48]. The pathophysiological meaning of viral reactivation in DRESS and the reciprocal interactions between viral reactivation and inflammation are the subjects of an ongoing debate [49]. Viral reactivation may take part in DRESS pathogenesis in four ways (Figure 1):

1. Viruses may cause direct tissue damage and contribute to the early manifestations of DRESS.
2. In a later phase of the disease course, they can be the target of the immune response [50,51]. In this regard, an "immune-reconstitution like" syndrome may occur

as the result of corticosteroid treatment and/or immunosuppression to control DRESS manifestations [48].

3. Viral reactivation might constitute the epiphenomenon of a wider expansion of virus-harbouring immune cells in the setting of systemic inflammation. In fact, latent human herpesviruses (HHVs) chronically resides in cells of the immune system, including T-lymphocytes and cells of the monocyte/macrophage lineage. Thus, viral reactivation and release could represent an early marker of stimulation of these cell reservoirs following drug-driven expansion, rather than representing a trigger event of DRESS [52].
4. Viruses might promote anti-drug responses and mis-differentiation of antigen-specific lymphocytes by molecular mimicry. In fact, T-lymphocytes previously selected and expanded by viral antigens might eventually be activated by drugs, inducing DRESS (see also below at Section 2.3.2) [53]. Furthermore, challenging EBV-immortalised B-lymphocytes from healthy subjects and from patients with DRESS with DRESS culprit drugs selectively prompts EBV production increases in DRESS subjects [54], suggesting generalised dysfunction of tolerance and pathogen control in both arms of the immune response during DRESS.

Therefore, DRESS pathophysiological unicity may reside in the feed forward loop linking drug-induced triggering of memory lymphocytes followed by additional antigenic stimulation due to viral reactivation.

2.2.1. Human Herpesviruses (HHVs)

A typical feature of DIHS/DRESS is the reactivation of latent HHVs, namely HHV-6, HHV-7, Epstein–Barr virus (EBV), and Cytomegalovirus (CMV) [22,49,55–66]. Herpes viruses are known to promote the reactivation of other viruses upon reactivating themselves with a peculiar reactivation sequency, as originally reported by Kano 2006 [63] and recently reviewed by Anci 2021 [58]. As suggested by the literature, herpesviruses have differential reactivation kinetics, which can also be observed in the same subjects. The first viruses to reactivate are HHV-6 and EBV, followed by HHV-7 and finally CMV, in the same order that occurs in graft versus host disease (GVHD).

Sequential herpesvirus reactivation can also account for the long-lasting clinical picture of DIHS/DRESS and the occurrence of delayed organ complications even after discontinuation of culprit drugs [46,49,56,57,63,67]. This evidence also suggests that viral reactivation itself is presumably not involved in the onset of DRESS but may be a crucial factor determining the prolonged clinical course of this condition [29,58,65,68]. Furthermore, the characteristic 20%-mortality risk of DRESS/DIHS mortality is significantly affected by CMV reactivation besides older age, hepatic and kidney involvement, while EBV reactivation is most often observed in patients with milder DRESS presentations [29,55,69].

HHV-6 positivity also appears to be associated with a more severe disease course and a later onset following drug exposure than in the case of HHV-6-negative DRESS [48,51,58]. HHV-6 is the most frequent HHV to be associated with DRESS, and its reactivation typically occurs during the course of DRESS and up to 2–3 weeks after DIHS/DRESS rash onset [46,49,67]. HHV-6 infects the vast majority of the general population during infancy and has been shown to be able to be chromosomally integrated into host DNA both in the general population and in the setting of DIHS/DRESS [29,70–72]. Conversely, DIHS/DRESS has rarely been reported in patients during primary viral infection [73,74]. HHV-6 reactivation is normally a transitory event; nevertheless, in some cases, the virus can be detected also several weeks after the onset of DRESS, leading generally to the recurrence of skin rash [49,75,76]. During the course of DRESS, HHV-6 DNA can be found in the skin, lymph nodes, kidney, and liver, along with detection of HHV-6-derived microRNAs in serum and circulating mononuclear cells, suggesting the potential concurrent role of reactivated HHV-6 in the development of DRESS-related rash, lymphadenopathy, and organ failure [49,64,77–79]. The detection of HHV-6 DNA was also associated with symptom flare-ups, and the increase in HHV-6 DNA levels correlated well with the severities of the inflammatory responses [49,57,70,79,80].

HHV-7 has also been demonstrated to reactivate in patients with DRESS, albeit its clinical impact is not fully elucidated. Notably, two separate prospective studies systemically evaluated the proportion of DRESS patients with HHV-7 reactivation. In a study by Picard et al., 32% of patients demonstrated HHV-7 reactivation (compared to 45% of subjects with HHV-6 reactivation) [54]. On the other hand, in another study by Chen et al., only 1/23 patients experienced HHV-7 reactivation [66]. As the studies were performed in different geographical settings, it is possible that this striking difference in the proportion of patients with HHV-7 reactivation could be more related to epidemiological factors than to actual pathophysiological mechanisms.

Among other human herpesviruses, herpes simplex virus (HSV) reactivation has rarely been reported, and usually occurs early during the course of the disease followed by a rapid reduction of HSV-DNA titres [58,81]. Few reports of complications due to reactivation of HSV or VZV in DIHS/DRESS have been published [81,82]. By contrast CMV reactivation can induce some of the late-onset complications of DIHS/DRESS [49,59,83], which can present up to two months after the onset of DIHS/DRESS and even culminate with death. This occurs especially in the case of evident CMV disease, whose manifestations can be hepatitis, pneumonia, gastroenteritis, and skin and gastrointestinal ulcers. Indeed, the higher mortality described in patients with DIHS/DRESS with CMV reactivation may, at least in part, be related to viral end-organ disease, which occurs more commonly in the case of CMV reactivation compared to other viruses. Nonetheless, the association between CMV reactivation and mortality in critically ill patients has been described in multiple studies (recently summarized by Lachance et al. [84] and Schildermans et al. [85]), even though the underlying physiopathological mechanisms are not entirely understood. Interestingly, studies analysing the use of antivirals for prophylaxis of CMV reactivation in critically ill patients failed to show a benefit of this intervention [86–88], and, therefore, this approach should be discouraged in patients with DRESS, while treatment of end-organ disease should be promptly instated.

2.2.2. Severe Acute Respiratory Syndrome Coronavirus 2 (SARS-CoV-2)

DIHS/DRESS has been described among patients with SARS-CoV-2-related disease (COVID-19). These patients are particularly prone to classical risk factors for DIHS/DRESS, as they are frequently affected by multiple comorbidities and exposed to several drugs potentially associated with this syndrome. Few isolated cases of DIHS/DRESS in patients with COVID-19 have been described in the literature [89–98], but the actual incidence of this condition in patients with SARS-CoV-2 pneumonia is still to be fully explored. In a retrospective cohort analysing 9330 patients hospitalised with COVID-19 from a US healthcare system [99] between January 2020 and May 2021, six cases of DRESS syndrome were identified, corresponding to an incidence of 6.43 per 10,000 patients. The most likely culprit drugs were antibiotics, which were administered to all patients who developed DRESS (specifically vancomycin in 6/6, cefepime in 4/6, and meropenem in 1/6 patients). Nevertheless, all antibiotics were prescribed empirically, without microbiological evidence of a secondary bacterial infection. Interestingly, all patients in the study had markedly high eosinophilia (>3.00 $\times$ 10^6 cells/L), and no deaths were reported. In another cohort study [20], five cases of DRESS syndrome were identified among 2721 patients admitted with COVID-19 between February 2020 and March 2021. Notably, all cases were identified during the first wave (February to May 2020), with a corresponding incidence rate of 0.17/100 patient-months (compared to 0.0005/100 patient-months recorded in the previous 3 years in the same institution). Hydroxychloroquine (prescribed in 4/5 cases) and β-lactam antibiotics (administered to 4/5 patients) were considered the most probable culprit drugs, even though all patients received multiple drugs that could, at least, be considered to have a possible causative role. The authors argued that the differences in incidence of DRESS syndrome between separate COVID-19 waves could have been attributable to the evolution in the management of COVID-19, as patients were less exposed to potential culprit drugs such as hydroxychloroquine or lopinavir/ritonavir (and possibly antibiotics) and more frequently received corticosteroids, which could have dampened the immunological mechanisms leading to DRESS syndrome. Indeed, hydroxychloroquine is an emerging potential culprit drug in the setting of DRESS [14,44,90,94,100] (Table 1).

Interestingly, some authors postulated a possible role of SARS-CoV-2 infection in the pathogenesis of DRESS syndrome. The cytokine storm seen in patients with COVID-19 and maculopapular drug rashes was shown to promote the activation of monocytes/macrophages and a robust cytotoxic CD8+ T-cell response. This immunological profile was seen, to a lesser extent, in non-COVID-19 patients with DRESS syndrome, but not in patients with other maculopapular drug rashes [101]. Specifically, COVID-19 and DRESS patients were shown to share an upregulation of several inflammatory cytokines, namely IL-6, TNF, IL-8, IFN-γ, CXCL9, CXCL10 and CXCL11, accompanied by an increase in IL-4 and IL-5 (representing a type 2 response) and proteins associated with eosinophil chemotaxis and immune suppressive phenotype. Therefore, it is possible to speculate that the T-cell hyperactivation and systemic cytokine storm seen in COVID-19 patients may be a predisposing factor for delayed drug hypersensitivity reactions [102], also given the absence of SARS-CoV-2 RNA in skin biopsies from patients affected by maculopapular skin rashes. Moreover, the reactivation of HHV-6, EBV and CMV has been described in patients with COVID-19 [103–105], highlighting a possible shared pathophysiological mechanism.

In conclusion, there is currently a paucity of data regarding the possible relationship between SARS-CoV-2 infection and DRESS, from both the clinical and biological points of view. While some authors described a higher incidence of DRESS in COVID-19 patients compared to historical cohorts, it is plausible that this finding may be related to the presence of several risk factors in these patients, namely a widespread use of antibiotics and other potential culprit drugs. Studies including a similar population (matched for demographic characteristics, comorbidities, and drug prescription) could shed light on this matter. Based on the data currently available, physicians should maintain a high index of suspicion and promptly discontinue potential culprit drugs in patients with COVID-19 with suspected DRESS, even in the absence of an intrinsically higher risk in this population.

2.2.3. Other Viruses

DIHS/DRESS has been described also in association with viruses that do not belong to the Herpesviridae family. In a single case, Coxsackie B4 was reported in a patient who developed a fulminant type 1 diabetes mellitus during DIHS/DRESS induced by carbamazepine. A serological panel was requested and a rise in anti-Coxsackie B4 immunoglobulin titre from <1:4 to 1:64 was observed. However, specific anti-HHV-6 IgG increased by 64-fold, too. Thus, it remains difficult to establish a clear pathogenetic relationship between DRESS and Coxsackie virus [106].

In another report, influenza virus was associated with DIHS/DRESS. A 35-year-old woman with rheumatoid arthritis developed DRESS-related symptoms 6 weeks after starting sulfasalazine. She was tested for serology and antigens of different viruses, and only influenza A and B turned out positive [107]. This is the only DIHS/DRESS case related to influenza virus currently described in the literature.

Another case report describes a young man who developed DIHS/DRESS syndrome and was diagnosed with chikungunya fever (chikungunya IgM titre 1:80 with a reference range of 1:10). As in the previous case, the patient had been treated with sulfasalazine for joint pain in the previous months, making it difficult to link chikungunya virus to DRESS syndrome [108]. Moreover, chikungunya virus-infected patients can develop mucocutaneous changes that may mimic the clinical presentation of DIHS/DRESS, thus making a differential diagnosis more difficult [109,110]. Notably, the macular hyperpigmentation of the nose and cheeks that sometimes follows chikungunya infection (Chik sign) initially appears as a maculopapular exanthem [111] resembling DRESS rash. In both scenarios, activation of skin-resident memory T-cells may account for anti-infectious responses and hypersensitivity reactions, such as DRESS [112]. The pathogenesis of the later hyperpigmentation of chikungunya is unclear, even though some reports in the literature hint at increased intraepidermal melanin dispersion or retention, triggered by chikungunya fever [113].

Finally, another virus that is often cited in relation to DIHS/DRESS is HIV. However, the use of antiretroviral drugs seems to be the trigger, while the virus itself is likely a bystander. An interesting case series describes six patients who developed DIHS/DRESS under treatment with raltegravir for HIV infection; five of these patients were of African ethnicity and four of them possessed the HLA-B*53:01 allele, thus suggesting a possible genetic predisposition for the development of DIHS/DRESS when exposed to raltegravir [114]. Other antiretroviral drugs that have been linked to DIHS/DRESS are nevirapine [34,115] and abacavir in patients who expressed the HLA-B*57-01 allele [35], even though with the latter drug, a hypersensitivity reaction occurs without haematological abnormalities or internal organ involvement. Lastly, a South African case series reported six patients coinfected with HIV and tuberculosis who developed DRESS syndrome after starting rifampicin [116], but even in this case, the link between DIHS/DRESS and HIV was weak, given that antitubercular drugs are often associated with cutaneous adverse drug reactions [117].

2.2.4. Immunological Mechanisms of Virus Reactivation

Multiple factors coincide with virus reactivation in the setting of DRESS. T-cells and antibodies act synergically against virus dissemination by preventing viral reactivation from a latent stage and by preventing the spread of reactivating lytic virus, respectively. Therefore, DIHS/DRESS may develop at the crossroads between transient humoral adaptive immune dysfunction with decreased B-cell counts and antibody secretion, reactivated HHV and expansion of drug-specific T-cells (see, for instance, Aihara 2003 [118] and Kano 2004 [119]). An early decrease in total IgG levels (mostly observed during the acute phase of DRESS) might also corroborate the clinical suspicion of DRESS and might facilitate HHV-6 reactivation [49,56]. Indeed, viral reactivation may potentially induce a secondary immune response with subsequent increase in the levels of specific anti-HHV-6 IgG, mostly observed in later stages of DRESS. It is also possible that the causative drug may induce a state of immunosuppression, subsequently allowing HHV reactivation [50].

One of the reservoirs of human latent HHV-6 infection is represented by mono/myeloid cells, which appear particularly prone to HHV-6 spreading in patients with DRESS. Specifically, circulating CD11b + CD13 + CD14 − CD16^{high} mono/myeloid precursors rise in the early stage of the disease course. In addition, these cells express high levels of OX40L, promoting interactions with their lymphocytic counterpart, which in turn expresses supranormal levels of the cognate receptor CD134 (=OX40) following systemic activation [49,70,120]. Strikingly, CD134 is also a cell-specific receptor for HHV-6 [49,70]. Furthermore, DRESS cases following immune checkpoint inhibitor exposure are increasingly reported [121,122]. Circulating CD11b + CD13 + CD14 − CD16^{high} mono/myeloid precursor cells harbouring HHV-6 also express a skin-homing molecule, CD194 (=CCR4), and are responsive to high mobility group box (HMGB-1). In the skin and in the blood of patients with DIHS/DRESS, high levels of HMGB-1 have been found. Taken together, these data suggest that HHV-6 reactivation might initiate in the skin [22] and consists in monocytes/macrophages latently infected by HHV-6 reactivating during the early phase of DIHS/DRESS, leading to increased viral loads, and subsequent infection of CD4+ T cells via CD134 [49]. This mechanism might account for the preferential involvement of the skin in the clinical spectrum of DRESS. Consistently, the expression of HHV-6 cellular receptors in skin lesions soon after onset positively correlates with DIHS severity [49,70].

Patients with DIHS/DRESS can present high levels of plasmacytoid dendritic cells (pDCs) in the affected skin regions, but, on the contrary, low levels of pDCs in the peripheral blood. Interferon α (IFNα), produced by pDCs, inhibits viral infection and connects innate and adaptative antiviral immunity. In fact, IFNa triggers the antiviral response of myeloid dendritic cells (mDCs), T-cells and natural killer cells and also the maturation of B-cells in order to promote IgG production for antiviral response. When pDCs migrate from the circulation to the skin, the number of pDCs in blood is reduced, possibly resulting in reduced antiviral responses [123].

The cytokine milieu can also affect viral reactivation. Interestingly, a G-CSF-, MIP-1α-, TNF-α-, IL-8-, IL-10-, IL-12p40-, and IL-15-enhanced profile as observed in DRESS has been shown to be associated with CMV reactivation, and higher eotaxin, IL-10, and G-CSF levels accompanied with lower IL12p40 levels at baseline might be useful for predicting the development of CMV disease [124]. Patients at risk of CMV reactivation can be identified by surveillance of these cytokine/chemokine levels prior to and after beginning immunosuppressive therapy. This may help in preventing morbidity and mortality.

2.3. T-Cell Responses

The role of T-cells in DRESS is clinically supported by evidence of positive patch testing and of activation of drug-specific CD4+ and CD8+ T-lymphocytes in patients with DRESS [125,126]. Pharmacogenomic and functional data (see below) point to a prominent role of CD8+ T cells in mediating anti-drug and anti-viral responses along with non-typical support for eosinophil recruitment. Nonetheless, evidence of antigen-specific CD4+ T-cell activation and expansion in DRESS has also consistently been reported [49,125,126]. Drug-reacting cells typically produce large amounts of cytokines potentially associated with a broad spectrum of inflammatory phenotypes that include IL-4, IL-5, IL-13, IFN-γ, and TNF-α [60]. Multiple aspects of T-cell biology may contribute to pathophysiological mechanisms underlying DRESS [22]. Recent evidence indicates that genetically determined dysfunctions in the control of apoptosis and proliferation might contribute to susceptibility to severe cutaneous adverse reactions (including DRESS) in populations of European descendent. Nicoletti et al. performed a recent meta-analysis of two genome-wide association studies (GWAS) on patients with phenotypically defined carbamazepine-serious cutaneous adverse reaction (CBZ-SCAR) and carbamazepine-drug-induced liver injury (CBZ-DILI). They found that an uncommon variant in the ALK gene conferred a supranormal risk of CBZ-SCAR. Indeed, the ALK gene is a receptor tyrosine kinase found in numerous tissues, being involved in cellular proliferation and cell death. This evidence could suggest that the expression of this gene variant may have a relevant role both in T-cell function (as far as proliferation is concerned) and keratinocyte biology (by affecting mechanisms of cell death). These findings also suggest that cellular homeostasis, besides immune-specific functionality, might be altered in T-cells in the setting of DRESS [127].

2.3.1. HLA

HLA-restricted antigen-specific recognition followed by cellular activation constitutes the hallmark of T-cell-mediated responses. HLA is a complex of genes mapping to chromosome 6p21.3 in humans and encoding cell-surface proteins responsible for several activities of the immune system, including self-non-self-recognition and presentation of antigen on the membranes of specialized cells. HLA is highly polymorphic in the human population, and associations between the risk of developing DRESS and several HLA genetic variants have been reported [56,128,129]. These associations are usually drug-specific, possibly implicating that some HLA molecules are able to interact with a specific drug in a more efficacious way to activate T lymphocytes [130]. Although both CD4+ and CD8+ T lymphocytes can be activated by drug exposure in DRESS [125,126], the class I HLA profile shows a stronger epidemiological association with DRESS than the class II HLA profile [55]. Besides the association with the risk of becoming sensitised to selected drugs, HLA is also linked to susceptibility to infection and chronicisation of viral infection. A summary of most frequent DRESS-related HLA variants and their effects on viral infection is reported in Table 2.

Table 2. Selected HLA haplotypes associated with DRESS.

HLA	Viral Infection	Effects on Viral Infection	Drugs	Population	OR (95% C.I.) Ref.
HLA-A*24:02			Lamotrigine	Spanish	34.5 (2.03–209.71) [131]
HLA-A*31:01			Carbamazepine	Han Chinese	12.9 (3.7–45.3) [132]
				Japanese	[133]
				European	24.1 (9.6–60.3) [134]
				North African	32.0 (2.6–389.2) [133]
			Lamotrigine	Korean	11.43 (1.95–59.77) § [135]
HLA-A*32:01			Vancomycin	European	[129]
HLA-A*33:03			Allopurinol	Korean	25.2 (5.2–121.8) [136]
HLA-B*13:01			Dapsone	Han Chinese	[137]
				Thai	60.75 (7.44–496.18) [138]
				Taiwanese, Malaysian	49.64 (5.89–418.13) [137]
			Sulfasalazine	Han Chinese	11.16 (1.98–62.85) [139]
			Sulphamethoxazole	Asian	61 (21.5–175) [140]
HLA-B14:02			Nevirapine	Caucasian	
HLA-B* 51:01			Carbamazepine	Han Chinese	4.6 (2.0–10.5) [132]
			Phenytoin	Thai	5.2 (1.2–22.7) [141]
HLA-B*53:01	HIV		Raltegravir	African	[114]
HLA-B*56:02			Phenytoin	Australian Aboriginal	[142]
HLA-B*58:01			Allopurinol	Han Chinese, Thai, Japanese, Korean, European	580.3 (34.4–9780.9) § [10]
			Carbamazepine	Asian	7.55 (1.20–47.58) [143]
	CMV	Increased reactivation risk			
HLA-B*15:13			Phenytoin	Malaysian	59.0 (2.5–1395.7) [144]
HLA-C*03:02			Allopurinol	Korean	135.7 (15.6–1177.8) [136]
HLA-C*04:01			Nevirapine	Malawian	2.6 (1.1–2.6) § [145]
HLA-DRB1*15:01			IL-1 and IL-6 inhibitors	European patients with AOSD or JIA	40.8 (5.3–316) [39]
	EBV	Coreceptor to EBV infection on B cells			[146]

§ Odds ratios available only for DRESS + Stevens-Johnson's Syndrome + Toxic epidermal necrolysis cumulated in the original paper; AOSD—Adult-onset Still's disease; IJA—Juvenile idiopathic arthritis.

Secondary to the prominent pathogenic role of HLA in DRESS and other DHRs, HLA testing has high specificity and negative predictive value for predicting the occurrence of such reactions in patients exposed to known DHR triggers, suggesting its potential clinical use [19]. However, implementation of HLA genotyping into routine clinical practice is mostly affected by the number needed to test (NNT) in order to prevent one case of DHR. In turn, NNT is affected by DHR incidence and HLA frequency in a given population. HLA-B*57:01 screening is part of routine clinical practice for candidates for abacavir, due to the relatively high frequency of abacavir hypersensitivity syndrome in patients treated with abacavir, at least in Caucasians. Similarly, screening for HLA-B*15:02 and HLA-B*58:01 has a low NNT for carbamazepine-related SJS/TEN and for allopurinol-related DHR, respectively, in Asian populations, due to the high frequency of these alleles in these populations. Conversely, some drug regulatory agencies recommend HLA-A*31:01 genotyping for non-Asian patients due to receive carbamazepine. Due to the low incidence of DRESS, NNT estimates for HLA testing might vary significantly among studies [147,148]. Konvinse et al. [129] have estimated an NNT of 75 for the HLA-drug pair HLA-B*32:01–vancomycin in European populations, supporting its potential use routinely. However, given that the population of the European Union is 447 million, the annual hospitalisation rate approximately 1/10 [149], that 2% of hospitalised patients are usually exposed to vancomycin [150,151], and that more than 40% of them receive vancomycin for 2 weeks or more [152], more than 4000 DRESS diagnoses due to HLA-B*32:01 should be expected yearly in the European Union, which largely exceeds the annual rate of total drug hypersensitivity reactions reported in the EudraVigilance tool (n = 383 for the year 2022) [151]. Consistently, HLA-B*32:01 testing is currently not included among recommended tests by drug regulatory authorities and pharmacogenetics working groups [150].

2.3.2. Molecular Mechanisms of T-Cell Activation and Aberrant HLA/TcR Interactions

Viral and pharmacological triggers can disrupt physiological HLA–T cell receptor (TcR) interactions through multiple mechanisms (Figure 2), which may also co-occur in the same subject [153]. A first set of mechanisms are supposed to alter peptide presentation by haptenation of self-molecules or by modification of HLA steric properties.

The hapten-carrier model constitutes the simplest pathophysiological mechanism accounting for drug hypersensitivity. In this setting, the culprit drug activates T-cells after binding intracellular proteins, which are subsequently processed and presented by antigen-presenting cells [154]. Drug binding to HLA might also induce conformational changes causing a shift in HLA affinity to self-peptides, which in turn promotes autoreactive responses [155].

Drugs might also bind HLA or the TcR by non-covalent direct pharmacological interaction (p-i concept), prompting T-cell activation [156]. This mechanism has been described with several drugs classically involved in DRESS, such as carbamazepine [157] and allopurinol [10] For a more comprehensive review on the p-i model, the reader is referred to Pichler 2019 [158]. The p-i concept seems relatively more suitable to explain severe reactions to drugs, such as those observed in DRESS, since it is compatible with the activation of different T-cell clonotypes, rather than the one or few expected in the case of the hapten-specific activation model [158]. This scenario is reminiscent of an alloreactive activation, similar to that observed in GvHD reactions [159]. The paucity of HLA variants capable of being engaged in this dangerous liaison with the drug (or its metabolites) could contribute to explain the low frequency of this condition. This consideration needs to be kept in mind in order to explain why only a subgroup of individuals bearing a high-risk HLA allele do actually develop DRESS following intake of the culprit drug [53].

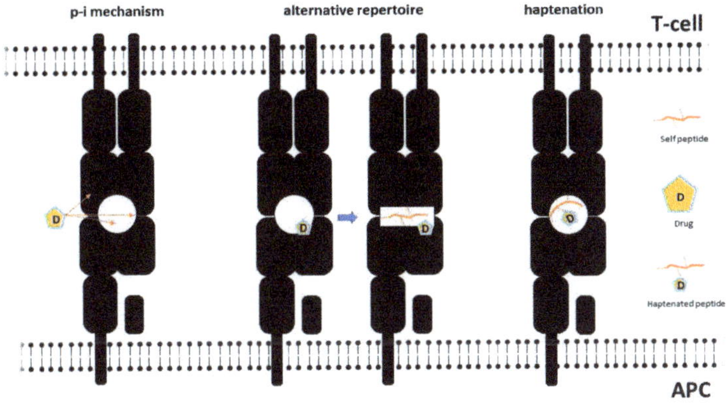

Figure 2. Pathogenic mechanisms for HLA-restricted, drug-induced activation of T-cells in DRESS. In this simplified depiction of HLA-restricted antigen presenting cell (APC)–T-cell interactions, the three main hypothesised mechanisms accounting for drug-induced T-cell activation in DRESS are represented. Drugs might directly interfere with HLA–T-cell receptor interactions, causing T-cell activation without the need for self peptides (p-i mechanism, **left** side). In this setting, drugs might activate the T-cell receptor through allosteric mechanisms or by binding HLA either inside or outside the peptide binding groove. Drugs bound to the peptide groove might cause conformational changes enabling self-peptides to be accommodated within HLA and presented to T-cells, promoting self-reactive responses (alternative peptide repertoire hypothesis, central section). Drugs can also bind self molecules through conventional hapten-carrier models (**right** side).

On the other hand, heterologous reactivity of TcR (heterologous immunity) is a well-recognised mechanism accounting for the ability of the relatively limited human repertoire of T-cell clonotypes to respond to a broad variety of pathogens, even after first antigen exposure, and also for natural autoreactivity towards drugs [160,161]. Promiscuous T-cell activation might also lead to cross-reactivity among drugs sharing a similar chemical structure [162] as well as among viral and self-peptides [163], possibly accounting for hypersensitivity in drug-naïve subjects [164]. Viral factors might also play a role in shifting the immune response towards selected T-cell clones prone to drug-induced activation through recurrent reactivations [48,54,63,165]. Consistent with this model, a study by Yerly et al. [166] showed that self-peptides able to bind permissive HLA variants (e.g., HLA-B*57:01 for abacavir hypersensitivity) may show sequence similarity to (herpes) viral peptides, which can in turn promote T-cell activation. In another report, HLA-B*57:01-restricted HIV-specific T-cells proliferated in response to HLA B*57:01-expressing cells in vitro only in the presence of abacavir [167].

Little is known about the role of heterologous immunity in DRESS (Figure 3). Picard et al. showed that drug exposure prompts expansion of CD8+ T-cells sharing the same TcR repertoire of EBV-specific cytotoxic T-cells in DRESS patients [54]. Niu et al. also showed that EBV-specific CD8+ responses correlate with CD8+ plasticity and DRESS severity, suggesting that repeat challenge by viral factors might continuously renew a pool of autoreactive CD8+ T-cells in predisposed individuals [168]. Heterologous immune mechanisms might also account for potential class sensitisation to multiple drugs in the setting of DRESS [27].

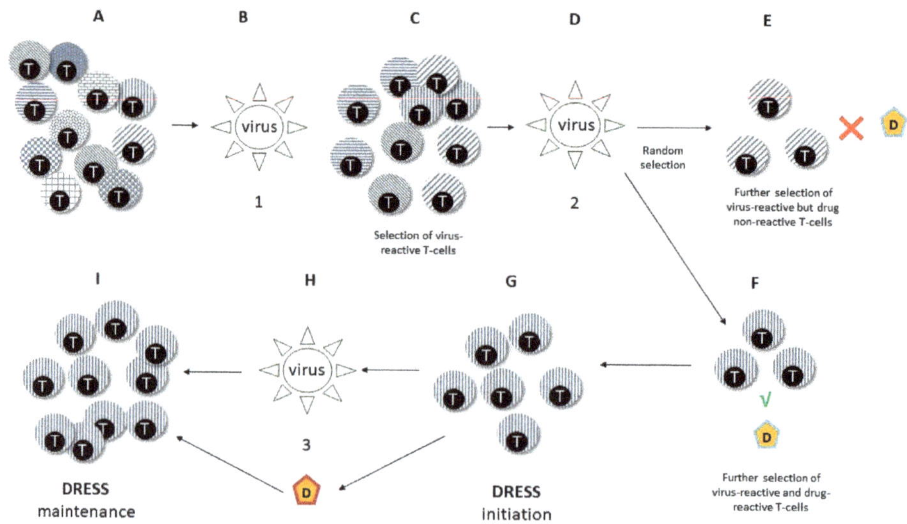

Figure 3. Potential mechanisms of heterologous immunity in DRESS. This figure depicts features of heterologous immunity with a potential pathogenic role in DRESS. Viral infections or reactivations (1–3) prompt selective pressure on a heterogeneous pool of T-cells (**A**). Therefore, after exposure to viruses (**B**), virus-reactive T-cells expand (**C**) and are readily available for eventual viral encounters or reactivations (**D**). Among virus-reactive T-cells, subpopulations harbouring T-cell receptors devoid of the ability to be activated by potential drug allergens might be selected (**E**), preventing the occurrence of hypersensitivity. In other cases (**F**), either occurring in distinct subjects or in the same subjects during distinct phases of life, virus–drug cross-reactive T-lymphocytes might be selected by viral stimulation. When challenged with culprit drugs, these cells might initiate hypersensitivity reactions, possibly including DRESS (**G**). Eventually, re-challenge with re-activating viruses (**H, top**) or chemically related drugs (**H, bottom**) might promote DRESS progression and/or persistence (**I**).

2.3.3. T-Cell Polarisation and Functionality

Aberrant antigen processing and T-cell activation in DRESS encompass alterations in T-cell polarisation and functional specialisation. In fact, patients with DRESS are characterised by oligoclonal expansion of lymphocytes expressing defined subsets of TcR [169]. In addition, patients show a Th2/Treg-skewed phenotype characterised by the possible coexistence of defective control of viral stimuli (that is, reactivated viruses) and enhanced eosinophil proliferation and organ infiltration leading to tissue damage. Supranormal expression of CD134 on circulating CD4+ cells has been detected in patients with DRESS and might contribute to the promotion of Th2 responses [120], besides directly facilitating viral spreading (see above). Enhanced systemic expression of the CD194 (CCR4) ligand *Thymus and Activation-Regulated Chemokine* (TARC) constitutes another hallmark of the acute phase of DRESS. Mechanistically, elevated TARC levels might promote tissue infiltration by Th2 and Treg along with HHV-6 reservoir cells such as mono/myeloid precursors [22,170], favouring HHV-6 spreading and replication, through immunosuppressive responses, and eosinophil-driven tissue damage [57,58,171]. Consistently, patients with predominant HHV-6 reactivation present TARC levels significantly higher than those without HHV-6 reactivation, and in the acute stage of DIHS, these levels correlate with disease activity [22,57,58,171].

In contrast to the acute phase of the disease, where IL10-producing classical monocytes are increased, patrolling monocytes are mobilised and release high amounts of IL6. This in turn induces a drift towards Th17-dominated responses [172], while inhibiting Treg cells. Treg cells collected from patients with DRESS in the late phase of the disease show impaired ability to inhibit their effector counterpart in comparison to Tregs from healthy subjects and Tregs collected from patients with early-stage DRESS [173]. In this scenario, late reactivation of other herpesviruses usually occurs.

2.4. Eosinophils

Deranged eosinophil inflammation is a hallmark of DRESS and contributes to organ damage. Eosinophils take part in the early-phase response against microbial threats, and their defensive capacities are impaired in eosinophil-driven diseases [174–177]. Besides performing direct viral clearance tasks as granulocytes, eosinophils also contribute to shaping the downstream inflammatory response to a Th2 profile. In this setting, eosinophils may be stimulated by IL5 released from type II innate-like lymphoid cells (ILC2s) following alarmin release from infected/damaged tissues and in turn promote CD4+ T-cell polarisation towards a Th2 profile by enhancing ILC2 activation and T-cell maturation by releasing IL4 [178]. Consistently, activation-prone ILC2s have been shown to increase in the blood and skin lesions of patients with DRESS along with elevated circulating levels of the alarmin thymic stromal lymphopoietin (TSLP), of the alarmin receptor ST2 (which binds IL33, a potent stimulator of ILC2) and of IL5 [179]. Besides primary eosinophil activation following organ damage, IL5-dependent systemic eosinophilic responses may be sustained by T-cell activation (see above) [180]. Elevated levels of other Th2-associated chemokines, including TARC and macrophage-derived chemokine (CCL22), have also been described in patients with DIHS/DRESS [120]. Persistently high eosinophil counts are thought to correlate linearly with the development of organ damage [181] and are associated with an increased likelihood of eosinophil infiltration of non-physiological eosinophil-homing tissues such as the skin, the liver, the myocardium, and peripheral or central nervous fibres [178,182]. Consistently, these tissues are an integral part of the clinical spectrum of DRESS.

2.5. Pathophysiological Basis of DRESS Clinical Manifestations

Multiple aspects of DRESS pathophysiology remain obscure. Nonetheless, available evidence globally suggests that viral reactivation, aberrant drug metabolism, and drug–receptor interaction might prime T-cells to activation besides contributing to part of the early tissue/organ damage. T-cell stimulation may then lead to reactivation of viral genomes harboured by leukocytes, which would eventually further stimulate the immune response to control the spread of actively replicating viral particles. Aberrant differentiation of antiviral T-cell precursors, due to heterologous immune mechanisms, might also enhance drug hypersensitivity and promote delayed-type eosinophil responses, exacerbating organ damage. Persisting viral replication after drug clearance might account for slowly resolving symptoms and potential long-term sequelae as observed in DRESS [29,51].

3. Clinical Presentation and Laboratory Findings: When to Suspect DRESS

Multiple organs and tissues can be affected by DRESS syndrome. Systemic findings sorted by declining frequency include cutaneous, lymphatic, haematological, and hepatic manifestations, followed by renal, pulmonary, and cardiac involvement. Severe, atypical cases of DRESS may show neurologic, gastrointestinal, and endocrine dysfunction.

3.1. Systemic and Laboratory Findings

Systemic symptoms constitute a hallmark of DRESS. Fever develops in 90% of subjects, but body temperature rarely (7% of cases) exceeds 38.5 °C. Lymph-node enlargement may be detected in up to 60% of patients with DRESS. In the absence of peripheral lymphadenopathy, pathological lymph nodes may be relatively more frequently detected in the mediastinum [183]. Activation of the reticuloendothelial system also causes leucocytosis, with white blood cell count exceeding 10,000 cells/μL in the vast majority of patients. Lymphocytosis and detection of atypical lymphocytes (with increased cellular volume due to expanded cytoplasm along with irregularly shaped nuclei) are also common. Elevated eosinophil count is a defining feature of DRESS, and up to 80% of subjects show hypereosinophilia (that is, eosinophil count exceeding 1500 cells/μL). Neutrophilia and monocytosis might also be part of the blood cell count profile of patients with DRESS, while alterations in platelet count have been less frequently reported [15,184,185].

Abnormal erythrocyte morphologies (AEMs) were studied in a cohort of 215 patients: 32 had AEMs (14%). AEMs were more frequent among patients with DRESS than in patients with other skin manifestations. This phenomenon may be due to DRESS-related perturbation of haematopoiesis. In fact, in DRESS, toxic eosinophilic granule proteins are released and could affect bone marrow. The most frequent AEMs found in DRESS patients are poikilocytosis (48% of patients with AEM), polychromasia (48%), burr cells (33%), ovalocytes (33%) and others [186]. Recently, a study found that levels of TNF- α in blood samples could be useful biomarkers to detect HHV-6 infection. Indeed, its levels were higher in the reactivation group and decreased together with C-reactive protein and lactate dehydrogenase after infection resolution.

3.2. *Cutaneous Manifestations*

Cutaneous eruptions are the most common clinical finding in DRESS. Symmetrical maculopapular eruption involving either the trunk or the extremities is present in 15% of patients [187]. According to the RegiSCAR prospective study [15], polymorphous maculopapular rash is the most common presentation (85%) and encompasses findings such as purpura, infiltrated plaques, blisters, and exfoliative dermatitis. Facial oedema is observed in 70% of patients with skin manifestations. Some patients can develop an exfoliative dermatitis. The rash extent is > 50% of the body surface in approximately 70% of patients [15]. Some patients can present mucosal lesions, most commonly oral lesions and cheilitis [188]. Cutaneous eruption may last more than 2 weeks.

3.3. *Internal Organ Involvement*

3.3.1. Liver, Gastrointestinal, and Pancreatic Involvement

Liver is the most common extracutaneous organ involved in DRESS/DIHS and is often completely asymptomatic. Conversely, liver function abnormalities occur in up to 70% of patients. Anicteric hepatitis is more prevalent, but, if icteric hepatitis occurs, the prognosis is usually poorer, with progression to hepatic failure [80,189]. Hepatic necrosis may rarely develop, although more than 10% of cases may progress to death or need for liver transplantation [83,190,191]. Sulphonamides/sulfones pose the highest risk of inducing liver injury in DRESS, followed by antiepileptic drugs and allopurinol. A 2013 retrospective study on 136 patients suggested that antibiotics, especially β-lactams, are the most frequent culprits for liver injury [192]. According to this study, liver injury is more common in DRESS/DIHS than SJS/TEN, and is usually accompanied by renal failure. Indeed, both drugs and herpesviruses are known to cause liver injury. In fact, HHV-6 not only infects lymphocytes but can also show hepatotropism [57,193], and several reports of CMV hepatitis in the context of DRESS/DIHS have been described in the literature [83]. As anticipated, reactivation of HHV-6 has been more commonly observed in patients with severe clinical findings including long-lasting high fever, leucocytosis, renal failure, and severe hepatitis [48]. Viral hepatitis panels are usually negative in DRESS, but when DRESS associates with an underlying viral hepatitis infection, the disease course can be more complicated and severe [194].

Intestinal involvement has been seldom described in association with DRESS. Descamps et al. reported on a 32-year-old man with sulfasalazine-induced DIHS/DRESS and reactivation of HHV-6 associated with colonic infection and subsequent development of Crohn's disease [195]. Intestinal involvement in DRESS has also been linked to CMV reactivation, in light of CMV tropism for the intestinal mucosa. This should be suspected when patients develop trunk and intestinal ulcers. The diagnosis is confirmed by detecting anti-CMV IgM and increased CMV DNA copies in blood samples. As anticipated, CMV reactivation is observed 4–5 weeks from the beginning of the disease and can be associated with concomitant or previous detection of HHV-6 reactivation [196]. Gastrointestinal haemorrhage caused by CMV has an unpredictable course and may frequently result in death. For this reason, early detection of CMV reactivation is necessary for successful management of DIHS/DRESS patients, and early administration of anti-CMV treatment can keep these patients from developing acute symptoms [56,59]. A case of esophagitis in DRESS syndrome also has been described [197].

Very limited evidence exists for pancreatic involvement in DRESS. A 2003 case report described a 40-year-old black woman treated with allopurinol who developed facial oedema, erythroderma and pyrexia, along with pancreatitis and hepatitis. A potential diagnosis of EBV-associated DRESS was hypothesised. However, non-DRESS-related EBV-induced pancreatitis cannot be ruled out, given that EBV infection might independently cause pancreatitis [198]. Some other cases of pancreatic injuries are described, one of them with concomitant development of diabetes mellitus type 1 after carbamazepine treatment and HHV-6 positivity detection [199].

3.3.2. Kidney Involvement

Kidney involvement occurs in 10–35% of patients [14] and usually manifests as acute interstitial nephritis. Acute renal failure occurs in up to 8% of patients, with a minority requiring renal replacement treatments. Patients are usually clinically silent, but some can present with mild haematuria and proteinuria. In blood analysis, elevated blood urea nitrogen and creatinine levels may point towards renal impairment. Eosinophils may be detected on urinalysis. Kidney ultrasound is usually negative [200]. In most cases, there is only mild renal impairment, which usually resolves after withdrawal of the offending drug. Renal involvement in DRESS is more common after allopurinol treatment, followed by carbamazepine and dapsone [201]. Indeed, allopurinol is generally associated with acute kidney injury (AKI), and renal biopsy typically yields an acute interstitial nephritis (AIN). Rarely, allopurinol can elicit renal vasculitis and glomerulonephritis [202]. On the other hand, renal failure in CBZ-DIHS/DRESS is considered to be attributable to acute interstitial nephritis. Acute interstitial nephritis is typically reversible after withdrawal of the causative agent. Viral agents can directly affect the development of renal involvement in DRESS. Hagiya et al. described a case report of granulomatous tubulointerstitial nephritis accompanying the proliferation of HHV-6 in tubular epithelial cells, demonstrating a possible association between reactivation of HHV-6 directly in the renal tissue with tubulointerstitial nephritis and renal dysfunction [203].

3.3.3. Heart and Muscle Involvement

The myocardium represents a preferential site for eosinophilic infiltration. Consistently, heart involvement in DRESS/DIHS syndrome has been described, and myocarditis represents one of the most relevant prognostic factors in DIHS/DRESS patients [29,46,49,59,67,82,83]. Heart disease is described at onset of disease or after approximately 40 days. Most frequent symptoms are tachycardia, chest pain, dyspnoea, and hypotension, although some patients are completely asymptomatic. According to the literature, ampicillin and minocycline are more frequently responsible for this manifestation [194].

Chest radiography shows cardiomegaly and/or pleural effusion. ECG shows ST-T non-specific abnormalities and sometimes arrythmias. Echocardiography shows significant ejection fraction reduction. Moreover, elevation of creatine kinase and troponin T is usually detected [200]. Two forms of myocarditis are recognized in DRESS syndrome: hypersensitivity myocarditis and acute necrotizing eosinophilic myocarditis (ANEM). The first is usually self-limiting. ANEM is associated with >50% mortality and a median survival of 3 to 4 days [194]. Echocardiography in patients with ANEM shows increased wall thickness, severe biventricular failure, and a pericardial effusion. For both clinical entities, endomyocardial biopsy is important for a definite diagnosis, and it helps in differential diagnosis from other myocarditis. Eosinophilic and mixed lymphohistiocytic infiltrate without necrosis is a typical histological finding of eosinophilic myocarditis. ANEM is defined as eosinophilic and lymphocytic infiltrate with associated myocyte necrosis. Moreover, it is largely known that HHV-6 can cause myocarditis. The hypersensitive response to drug metabolites and the reactivation of a virus such as HHV-6 may be the requisite "immune alteration" in certain individuals that leads to the severe damage of myocardial tissue by eosinophilic degranulation in this disease [204].

3.3.4. Lung Involvement

As well as other organs, the lungs may also be involved in DRESS syndrome. Indeed, pulmonary manifestations may be among the first evidence of the syndrome, anticipating even skin manifestations. According to a review of the literature, pulmonary symptoms might be found in up to 72% of patients on hospital admission [183]. Symptoms of pulmonary involvement are generally dyspnoea, cough and pleurisy [194,200,205]. The most common pulmonary findings include infiltrating lesions of an interstitial nature, pneumonia (50%), and pleural effusion (22.7%) [25,205]. In some cases, pulmonary nodulations have been reported. Very severe manifestations of DRESS syndrome in the lungs can also lead to the development of acute respiratory distress syndrome (ARDS) with acute respiratory failure (31% of cases). The most important risk factors for the development of severe pulmonary manifestations of DRESS accompanied by ARDS seem to be an onset latency shorter than 30 days and age less than or equal to 60 years [183].

3.4. Nervous System Involvement

Neurological manifestations of DRESS might involve the central and peripheral nervous system. Encephalitis and meningoencephalitis usually present in 2–4 weeks from disease diagnosis and encompass symptoms such as coma, seizures, headache, and speech disturbance. Evidence of HHV-6 DNA in the cerebrospinal fluid points to a probabile role of HHV-6 in this setting [206]. Moreover, a possible link between DIHS/DRESS with reactivation of EBV and the development of autoimmune limbic encephalitis has been described [207]. Electroencephalography shows diffuse slow waves with an occasional solitary spike and waves in the left frontal and temporal leads without periodic patterns. MRI shows bilateral hyperintensity of the grey matter involving the amygdala, medial temporal lobe, insula, and cingulate gyrus. Peripheral involvement is anecdotal [208].

3.5. Other Manifestations

An increased prevalence of autoimmune diseases has consistently been observed in DRESS/DIHS survivors. While viral triggers such as EBV and HHV-6 are known potential risk factors for the development of autoimmune diseases such as type 1 diabetes mellitus and autoimmune hypothyroidism [69,209] in the general population, increased lymphocyte counts along with late-phase hypergammaglobulinemia (in contrast to the early phase of DRESS: see above), low levels of interleukin (IL)-2 and IL-4 at DRESS onset and severe liver involvement might synergise with persistent reactivation of EBV and HHV-6 in enhancing post-DRESS autoimmunity [55,210]. Unstable CD8+ T cell repertoires, possibly due to viral stimuli, might also selectively associate with autoimmunity [168].

4. Diagnosis

4.1. Diagnostic Approach

DRESS should be highly suspected in patients who have recently started new treatments and present with cutaneous eruptions, fever, hypereosinophilia, and alterations in organ function tests [56,211]. Assessment of the causative drug and of the starting time of therapy is one of the first steps of the diagnostic approach when DHR is suspected.

Initial laboratory investigations are aimed at confirming DRESS diagnosis and evaluating the degree of severity of organ involvement. Laboratory tests include complete blood count with peripheral blood smear for evaluation of eosinophilia (>700/µL), leucocytosis, and the presence of atypical lymphocytes. Significant liver function test abnormalities in more than two measurements are suggestive of liver involvement. Kidney function test abnormalities, proteinuria >1 g/day, or haematuria are suggestive of renal involvement. Cardiac enzymes such as troponins, creatine kinase-MB, and NT-proBNP and pancreatic enzymes such as amylase and lipase should be measured as clinically appropriate [21]. Comprehensive serial laboratory investigations are recommended during the follow-up [21,56].

Reactivation of HHV-6 or other herpesviruses could be assessed by serology or viral genome testing by polymerase chain reaction (PCR) in blood or other tissues [29,56,70,171,190]. There is no universal consensus on the methods to assess viral reactivation, and heterogeneity among different laboratories is the rule [212]. Screening for acute viral hepatitis (e.g., anti-hepatitis A, anti-hepatitis B surface antigen, anti-hepatitis B core antigen IgM, or hepatitis C viral RNA) could be performed to exclude alternative diagnoses in patients with abnormal liver enzymes. The Monospot test is often used as a stand-alone evaluation of infection, despite its low clinical value [213]. Additional tests such as blood cultures or anti-chlamydia, anti-mycoplasma or antinuclear antibodies could be considered for differential diagnosis with other infectious or autoimmune diseases [21,48,54,56,212]. Imaging with ultrasound, computed tomography, echocardiography and cardiac magnetic resonance can be performed to assess the severity of organ involvement. Growing evidence supports the use of ECG and echocardiograms to screen for cardiac manifestations of DRESS, which may have a fulminant course [214].

Patch testing may be useful to ascertain culprit drugs. Recent studies report patch test positivity in 30–60% of patients with DRESS, especially in those cases caused by carbamazepine, β-blockers and PPI administration. Negative results were reported when testing for allopurinol and sulfasalazine [215]. Intradermal testing should be used only in exceptional cases, as there is a risk of reaction recurrence. Drug challenge is contraindicated, although it may be useful in the setting of multiple drug treatment for HIV infection or tuberculosis [211]. In addition, the lymphocyte transformation test (LTT) measures T-cell proliferation, following in vitro exposure to the causative drug [216]. For diagnostic purposes, this assay, which is not available to the routine clinical lab, is best performed in the recovery phase of DRESS. Positive LTT is expected to be found in half DRESS cases [216]. Consistently, LTT sensitivity and specificity are quite high(73% and 82%, respectively).

Skin biopsy may provide further evidence supporting DRESS diagnosis, although no histopathological finding is pathognomonic. Histopathologic examination may help to rule out other diagnoses such as exanthematous drug eruptions, acute generalized exanthematous pustulosis (AGEP), and Stevens–Johnson's syndrome/toxic epidermal necrolysis (SJS/TEN). The main histopathological findings encompass dyskeratosis (53–97% prevalence), interface vacuolization (74–91% prevalence), spongiosis (40–78% prevalence), perivascular lymphocytic and dermal eosinophil infiltrates (prevalence 20–80%) [217]. Wide areas of keratinocyte necrosis are found in severe cases [217]. Biopsy of other commonly involved organs (lymph nodes, kidney, liver, and heart) are not routinely performed due to highly nonspecific inflammatory patterns.

4.2. *Differential Diagnosis*

Due to the heterogeneity of its clinical presentation, DRESS syndrome can be misdiagnosed. The main differential diagnoses include acute generalized exanthematous pustulosis (AGEP), exanthematous drug eruptions and Stevens–Johnson syndrome/toxic epidermal necrolysis (SJS/TEN). SJS/TEN shares with DRESS causative drugs, fever, haematological abnormalities and hepatic involvement. While SJS/TEN generally presents a latency period ranging from a few days to 3 weeks, DRESS is a late severe reaction to pharmacological exposure taking at least 2 weeks to show symptoms. On the other hand, AGEP is usually characterised by rapid onset of skin reaction. Referring to hepatic involvement, while more frequent in DISH/DRESS, noteless elevations up to three times the normal value in serum aminotransferase can be detected in half TEN patients (10% of TEN patients can also develop full-blown hepatitis) [218]. Exfoliative dermatitis differs from DRESS in terms of cutaneous and other systemic features. In particular, AGEP rarely associates with kidney involvement and, when present, it is usually self-limited. In addition, visceral involvement in AGEP has rarely been described. In SJS/TEN, a serum urea level >10 mmol/l is a poor prognosis marker [219]. In terms of skin manifestations, DIHS is characterized from the very beginning by a maculopapular rash, often accompanied with oedema on the face and limbs. These features can later evolve to erythroderma or exfoliative dermatitis. In contrast with SJS/TEN, there is no haemorrhagic mucocutaneous involvement. Moreover, HHV-6 re-activation is commonly observed in patients with DIHS/DRESS, along with atypical lymphocytes, whereas it is rarely found in SJS/TEN [220]. AGEP is characterized by an erythematous rash with pustulosis (non-follicular, sterile pustules < 5 mm in diameter) accompanied by fever and neutrophilia and develops typically within 48 h to 3 weeks after drug ingestion. Skin lesions typically observed in AGEP are usually self-limited desquamative pustules accompanied by erythema, with typical facial and intertriginous distribution/pattern and mucosal involvement. Skin biopsies demonstrate intraepidermal pustules with oedema of the papillary dermis and perivascular infiltrates of neutrophils and eosinophils [221,222] (Table 3).

The diagnosis of DRESS may also be particularly challenging in the case of prominent lung involvement because respiratory symptoms associated with peripheral eosinophilia and rash can be found in different pathologies, both infectious and non-infectious (neoplasm, drug, allergic, autoimmune). Specifically autoimmune diseases, acute eosinophilic pneumonia, eosinophilic granulomatosis with polyangiitis [223], idiopathic hyper-eosinophilic syndrome and systemic lupus erythematosus [224], are particularly difficult to distinguish from DRESS. As for infectious causes, all viral, bacterial, parasitic, and fungal pathogens can mimic the symptomatology. The presence or absence of a certain symptom or sign, and a careful anamnesis, help in making a differential diagnosis. In HIV-immunocompromised individuals, a possible cause of DRESS can be found in patients taking raltegravir, whose manifestations characteristically occur in the lungs [225]. The involvement of other districts in addition to the lungs helps to make a differential diagnosis, e.g., renal co-involvement makes one consider pulmonary-renal syndrome (Goodpasture's), hepatic involvement, and hepatopulmonary amebiasis (Entamoeba histolytica) [183,226]. As the differential diagnosis is quite complex, several studies have shown that at least 50% of patients with DRESS syndrome, given misdiagnosis, were initially treated with antibiotics on suspicion of infection [205,227]. The first step, which is particularly important, is to rule out an infectious aetiology, since corticosteroids, as the mainstay of therapy for DRESS, might instead promote infection and, therefore, be contraindicated. Another condition commonly misdiagnosed with DRESS is lymphoma, for which DRESS is misdiagnosed in up to 75% of cases [14,227]. The presence of characteristic interstitial lung lesions, fever, and dyspnoea may point toward a diagnosis of acute eosinophilic pneumonia, which certainly needs to be differentially diagnosed. In the case of suspected eosinophilic pulmonary pathology, although not a practice that can necessarily be used to make a diagnosis of DRESS, it is useful to perform bronchoscopy in order to collect broncho-lavage (BAL) specimens. Indeed, in the case of differential diagnosis with acute eosinophilic pneumonia (AEP), there is evidence of eosinophilia in BAL samples, in the absence of peripheral eosinophilia, which on the other hand is abundantly present in DRESS. AEP laboratory samples usually show neutrophilic leucocytosis without hypereosinophilia. In the case of eosinophilic pneumonias (EPs), the manifestations may also be secondary to exposure to toxins and drugs. A recent literature review identified 196 cases of drug-induced EP over a 27-year period, with a higher prevalence of AEP than the chronic form. In this case, eosinophilia on peripheral blood was elevated, with mean values from 1232 to 1490 cells/μL in acute and chronic forms, respectively. Compared with the more common forms of EP, drug-induced forms of EP have eosinophilic leucocytosis on blood, as opposed to the far more common neutrophilic form. However, in both the drug-induced and non-drug-induced forms, eosinophilia on BAL is present [228]. Although EPs generally present with low blood eosinophilia, those secondary to drug exposure may instead exhibit hypereosinophilia. In these cases, the differential diagnosis usually relies on cell count in BAL samples, which will show increased eosinophils in almost all cases of EP and lower cellularity in cases of DRESS.

Table 3. Differential skin manifestations among DRESS and other diseases.

Syndrome	Rash Features	Timing of Onset	Disease Extent	Systemic Manifestations	Other than Skin Involvement	Blood Analysis Findings	Histopathological Findings
DIHS/DRESS	Maculopapular exanthem Erythroderma Facial oedema	2–8 weeks	Generalised Mucosal involvement Rare	Fever	Hepatitis Lymphadenopathy Pneumonitis Nephritis	Eosinophilia, atypical lymphocytes, leucocytosis Abnormal liver and renal function tests	Subtle, vacuolar interface dermatitis, with scattered, dyskeratotic keratinocytes along the dermo-epidermal junction zone
AGEP	Generalised erythema Pustules Erythroderma	<3 days	Generalised usually with skin fold and facial localisation, Mucosal involvement rare	Higher fever (>38 °C)	Rare	Leucocytosis with neutrophilia (>7000/mm^3)	Intraepidermal pustules with oedema of the papillary dermis and perivascular infiltrates of neutrophils and eosinophils
SJS/TEN	Dusky red, coalescent macular exanthem Atypical target lesions Bullous lesions Epidermal necrosis Nikolsky sign	4–21 days	Disseminated Mucosal involvement rarely absent (stomatitis, conjunctivitis)	Fever Photophobia Sore throat, Dysphagia	Pneumonitis	Lymphopenia	Necrosis of keratinocytes Epidermis shedding Absent inflammatory infiltrate

4.3. Diagnostic and Prognostic Scoring Systems

Several scoring systems have been suggested over the years to guide the diagnosis of DRESS. Currently three major scoring systems are available, including Bocquet's published in 2006, the Japanese Consensus Group Severe Cutaneous Adverse Reactions (J-SCAR) in 2006, and the European Registry of Severe Cutaneous Adverse Reactions (RegiSCAR) in 2007. Relevant items from the three scores are reported in Table 4. The RegiSCAR scoring system is more frequently used and includes seven clinical characteristics. Each characteristic is scored up to a total score in order to identify the diagnosis as possible, probable, or definite [15]. Recently introduced Spanish guidelines for DRESS suggest the use of RegiSCAR criteria for clinical diagnosis [23]. Indeed, they provide specific laboratory investigations that are necessary to guide the diagnostic process. In a comparative retrospective analysis conducted by Kim et al., Bocquet's criteria resulted in being the easiest to be applied in clinical practice [229], although they did not provide specific parameters to potentially guide the differential diagnostic process. A recent review by Cardones et al. compared the abovementioned scoring systems. Interestingly, the Japanese consensus group expanded Bocquet's criteria, adding more detailed requirements based on clinical manifestations. Differently from Bocquet's and RegiSCAR criteria, the Japanese consensus group also included HHV-6 reactivation among the relevant items for the diagnosis. Nonetheless, DIHS diagnosis through the J-SCAR criteria is largely consistent with probable/definite DRESS diagnosis according to the RegiSCAR algorithm. This finding confirms that DRESS and DIHS are part of the same heterogeneous disease [230]. One of the major advantages of using the RegiSCAR diagnostic criteria is the possibility to simultaneously differentiate DRESS from other, similar conditions such as acute cutaneous lupus and connective tissue diseases through autoantibody testing or skin biopsy. Ruling out infections through screening for viral acute hepatitis, blood culture, and testing for atypical bacteria such as Chlamydia and Mycoplasma is also part of the RegiSCAR algorithm [15,231].

Table 4. Comparison between three scoring systems.

Bocquet et al., 1996	J-SCAR, 2006	RegiSCAR, 2007
Cutaneous drug eruption	Fever	Fever > 38.5 °C
Systemic involvement: lymphadenopathy $\geq$ 2 cm; liver involvement (transaminase twice the upper limit); kidney involvement (e.g., interstitial nephritis); lung and cardiac involvement (e.g., interstitial pneumonitis or myocarditis)	Latency time of 3 weeks from drug exposure to the onset of cutaneous manifestations	Enlarged lymph nodes in $\geq$2 lymph node stations
Hematologic alterations: eosinophilia $\geq 1.5 \times 10^9$/L; presence of atypical lymphocytes	Persistence of the eruption $\geq$ 2 weeks after drug interruption HHV-6 reactivation at PCR or serology tests	Eosinophilia > 700/μL Thrombocytopenia Atypical lymphocytes Skin involvement (rash extended for > 50% of body surface area, biopsy) Organ involvement (one or $\geq$ two organs involved) Resolution in $\geq$15 days $\geq$3 negative laboratory investigations including ANA screening, serological screening for HAV/HBV/HCV, blood cultures, tests for Chlamydia and Mycoplasma to exclude other diseases

Despite evidence suggesting the use of RegiSCAR as a complementary diagnostic tool, some limitations might be highlighted. For example, prolonged resolution time (>15 days) is one of the diagnostic criteria but has little usefulness for early diagnosis. Furthermore, assessment of viral reactivation is not included in the scoring system, even though it is considered as a marker of severe disease progression [23,230]. Interestingly, there is increasing effort in identifying CMV disease as a potential prognostic factor for disease severity in patients with DiHS/DRESS. Mizukawa et al. conducted a retrospective analysis of patients with DiHS/DRESS aimed at developing a composite score based on demographic data, medical history and clinical data to monitor severity at different stages of disease, predict patient prognosis, and stratify the risk of CMV disease and complications. Although the study was conducted on a limited number of patients, the composite scoring system was useful in predicting complications related to CMV and to guide medical decisions for early intervention in those patients considered at high risk of CMV reactivation [232].

5. Management

The management of patients with DRESS is generally in-patient, although in milder cases (patients without visceral involvement) outpatient management may be considered with clinical and laboratory monitoring every 48 h [23]. The first necessary measure in the case of DRESS is discontinuing potential culprit drugs, along with supportive treatments such as fluid integration and antipyretics. Empirical use of NSAIDs and antibiotics in the acute phase is not recommended, because it could trigger DRESS exacerbations [194].

Addition of systemic steroids is usually necessary [55,233], although evidence from controlled trials is lacking. The rationale for corticosteroid use is the anti-inflammatory and immunosuppressive effect through inhibition of activated cytotoxic T-cells and cytokine production [234]. Immunosuppression poses the risk of favouring viral reactivation, especially of late-phase viruses such as CMV. Therefore, the timing and aggressiveness of immunosuppressive treatments (including corticosteroids) should be evaluated on a case by case basis with careful risk weighting [235]. Nonetheless, timely initiation of corticosteroid treatment is crucial to break the core pathogenic loop of the disease, that is, the reciprocal stimulation of viral replication and T-cell responses [233]. Early aggressive corticosteroid treatment appear to contrast both T-cell activation and HHV-6 replication, while delayed or low-dose treatments are thought to have limited impact on T-cell behaviour and HHV-6 viremia [46]. Similar to HHV-6, EBV-DNA loads are significantly debulked by systemic corticosteroids. Furthermore, since EBV is a known trigger of autoimmunity, early systemic corticosteroid treatment possibly prompting reduced EBV-DNA loads might have a fundamental role in optimising patient outcomes especially in settings at high-risk for autoimmunity [56]. It is recommended to start with a minimum dosage of 1 mg/kg/day of prednisone or equivalent, with a taper in 3–6 months [235]. In cases where this is not sufficient, pulse intravenous methylprednisolone may be used.

In refractory cases or when steroids are contraindicated, some studies recommend the use of cyclosporine, due its effects on cytotoxic T-lymphocyte activation and on IL5 inhibition [23,236]. The efficacy of intravenous Immunoglobulins (IVIGs) may be due to their general anti-inflammatory effect, protective effect against herpes virus reactivation and compensation for decreased immunoglobulin levels observed in DRESS [237]. IVIGs also constitute an interesting treatment option as a steroid-sparing agent in cases of concomitant infection [237], but they are not indicated in monotherapy due to potential lack of efficacy and increased adverse event rates [194]. Limited data suggest the potential use of immunosuppressants (cyclophosphamide, mycophenolate, rituximab) or plasmapheresis in refractory cases [235].

CMV reactivation might be integral to the pathophysiology of DIHS/DRESS or constitute an unwanted side-effect of immunosuppression [83]. Since CMV reactivation is among the most important risk factors in the prognosis of DIHS/DRESS, caution in the use of corticosteroids is recommended [46]. Indeed, fatal outcomes possibly related to the use of DRESS-related treatments are found almost exclusively in CMV cases. Delayed anti-CMV therapy is associated with a higher risk of adverse outcomes as patients receiving treatment after three or more days from CMV reactivation detection bear a significantly higher risk of death even compared to patients with treatment start after two days from CMV reactivation detection. Anti-CMV therapy may also have a synergic role in minimising the risk of CMV-related and unrelated pathological events, including other herpesvirus-related complications [56]. Some authors suggest that prophylactic treatment of CMV during viraemic stages could prevent progression to CMV-related clinical manifestations in patients with DIHS/DRESS [23]. Antivirals (ganciclovir or valganciclovir) in addition to standard treatment are usually administered with continuous monitoring of viral loads.

Given the possible systemic involvement, multidisciplinary management may be necessary depending on the organs involved. In cases of exfoliative dermatitis, treatment is similar to that of major burns, and management in burn units should be considered [238]. Generally, most patients respond to treatment; however, it must be remembered that this condition has an estimated mortality of up to 20%, while other subjects may have long-term adverse effects [238]. Patients with DRESS may experience relapses of symptoms in the recovery phase. These events are typically associated with corticosteroid dose reduction and viral reactivation. Interestingly, exposure to new drugs during DRESS could represent an underlying mechanism for relapse occurrence with or without subsequent drug sensitization [239,240].

6. Final Remarks

DRESS is a severe multi-organ syndrome characterised by abnormal T-cell and eosinophil responses to drugs along with abnormal control of viral stimuli. Advances in recent years, possibly boosted by the ongoing COVID-19 pandemic and its sequelae for individuals and public health, are increasingly highlighting the role of viruses in modulating the course and severity of DRESS. Large multicentre studies addressing changes in epidemiology and clinical presentation of DRESS among distinct healthcare and microbiological settings are eagerly needed along with deeper mechanistic insights into the pathophysiological basis of DRESS-related immune dysfunction.

Author Contributions: Conceptualization, M.-R.Y., L.D. and G.A.R.; methodology, M.-R.Y. and G.A.R.; validation, M.-R.Y., L.D., A.C. and G.A.R.; data curation, M.R., D.B., G.B., S.B., R.M., G.P., S.N., M.B.C., C.A. and G.A.R.; writing—original draft preparation, M.R., D.B., G.B., S.B., R.M., G.P., S.N., M.B.C., C.A. and G.A.R.; writing—review and editing, G.A.R.; supervision, M.-R.Y., A.C. and L.D. All authors have read and agreed to the published version of the manuscript.

Funding: This research received no external funding.

Data Availability Statement: Not applicable.

Conflicts of Interest: The authors declare no conflict of interest in connection with this paper.

References

1. Zhang, H.; Du, W.; Gnjidic, D.; Chong, S.; Glasgow, N. Trends in Adverse Drug Reaction-Related Hospitalisations over 13 Years in New South Wales, Australia. *Intern. Med. J.* **2019**, *49*, 84–93. [CrossRef]
2. Lee, S.; Hess, E.P.; Lohse, C.; Gilani, W.; Chamberlain, A.M.; Campbell, R.L. Trends, Characteristics, and Incidence of Anaphylaxis in 2001–2010: A Population-Based Study. *J. Allergy Clin. Immunol.* **2017**, *139*, 182–188.e2. [CrossRef]
3. Zaidi, A.S.; Peterson, G.M.; Bereznicki, L.R.E.; Curtain, C.M.; Salahudeen, M.S. Ten-Year Trends in Adverse Drug Reaction-Related Hospitalizations among People with Dementia. *Ther. Adv. Drug Saf.* **2022**, *13*, 20420986221080796. [CrossRef]
4. Classen, D.C.; Pestotnik, S.L.; Evans, R.S.; Lloyd, J.F.; Burke, J.P. Adverse Drug Events in Hospitalized Patients: Excess Length of Stay, Extra Costs, and Attributable Mortality. *JAMA* **1997**, *277*, 301–306. [CrossRef]
5. Park, C.S.; Kim, T.B.; Kim, S.L.; Kim, J.Y.; Yang, K.A.; Bae, Y.J.; Cho, Y.S.; Moon, H.B. The Use of an Electronic Medical Record System for Mandatory Reporting of Drug Hypersensitivity Reactions Has Been Shown to Improve the Management of Patients in the University Hospital in Korea. *Pharmacoepidemiol. Drug Saf.* **2008**, *17*, 919–925. [CrossRef]
6. Thong, B.Y.H.; Tan, T.C. Epidemiology and Risk Factors for Drug Allergy. *Br. J. Clin. Pharmacol.* **2011**, *71*, 684–700. [CrossRef]
7. Chung, W.H.; Hung, S.I.; Hong, H.S.; Hsih, M.S.; Yang, L.C.; Ho, H.C.; Wu, J.Y.; Chen, Y.T. Medical Genetics: A Marker for Stevens-Johnson Syndrome. *Nature* **2004**, *428*, 486. [CrossRef]
8. Lonjou, C.; Thomas, L.; Borot, N.; Ledger, N.; de Toma, C.; LeLouet, H.; Graf, E.; Schumacher, M.; Hovnanian, A.; Mockenhaupt, M.; et al. A Marker for Stevens-Johnson Syndrome . . . : Ethnicity Matters. *Pharmacogenom. J.* **2006**, *6*, 265–268. [CrossRef]

9. Locharernkul, C.; Loplumlert, J.; Limotai, C.; Korkij, W.; Desudchit, T.; Tongkobpetch, S.; Kangwanshiratada, O.; Hirankarn, N.; Suphapeetiporn, K.; Shotelersuk, V. Carbamazepine and Phenytoin Induced Stevens-Johnson Syndrome Is Associated with HLA-B*1502 Allele in Thai Population. *Epilepsia* **2008**, *49*, 2087–2091. [CrossRef]
10. Hung, S.L.; Chung, W.H.; Liou, L.B.; Chu, C.C.; Lin, M.; Huang, H.P.; Lin, Y.L.; Lan, J.L.; Yang, L.C.; Hong, H.S.; et al. HLA-B*5801 Allele as a Genetic Marker for Severe Cutaneous Adverse Reactions Caused by Allopurinol. *Proc. Natl. Acad. Sci. USA* **2005**, *102*, 4134–4139. [CrossRef]
11. Ozeki, T.; Mushiroda, T.; Yowang, A.; Takahashi, A.; Kubo, M.; Shirakata, Y.; Ikezawa, Z.; Iijima, M.; Shiohara, T.; Hashimoto, K.; et al. Genome-Wide Association Study Identifies HLA-A*3101 Allele as a Genetic Risk Factor for Carbamazepine-Induced Cutaneous Adverse Drug Reactions in Japanese Population. *Hum. Mol. Genet.* **2011**, *20*, 1034–1041. [CrossRef]
12. McCormack, M.; Alfirevic, A.; Bourgeois, S.; Farrell, J.J.; Kasperavičiūtė, D.; Carrington, M.; Sills, G.J.; Marson, T.; Jia, X.; de Bakker, P.I.W.; et al. HLA-A*3101 and Carbamazepine-Induced Hypersensitivity Reactions in Europeans. *N. Engl. J. Med.* **2011**, *364*, 1134–1143. [CrossRef]
13. Pichler, W.J. Pharmacological Interaction of Drugs with Antigen-Specific Immune Receptors: The p-i Concept. *Curr. Opin. Allergy Clin. Immunol.* **2002**, *2*, 301–305. [CrossRef]
14. Cacoub, P.; Musette, P.; Descamps, V.; Meyer, O.; Speirs, C.; Finzi, L.; Roujeau, J.C. The DRESS Syndrome: A Literature Review. *Am. J. Med.* **2011**, *124*, 588–597. [CrossRef] [PubMed]
15. Kardaun, S.H.; Sekula, P.; Valeyrie-Allanore, L.; Liss, Y.; Chu, C.Y.; Creamer, D.; Sidoroff, A.; Naldi, L.; Mockenhaupt, M.; Roujeau, J.C. Drug Reaction with Eosinophilia and Systemic Symptoms (DRESS): An Original Multisystem Adverse Drug Reaction. Results from the Prospective RegiSCAR Study. *Br. J. Dermatol.* **2013**, *169*, 1071–1080. [CrossRef] [PubMed]
16. Fiszenson-Albala, F.; Auzerie, V.; Make, E.; Farinotti, R.; Durand-Stocco, C.; Crickx, B.; Descamps, V. A 6-Month Prospective Survey of Cutaneous Drug Reactions in a Hospital Setting. *Br. J. Dermatol.* **2003**, *149*, 1018–1022. [CrossRef] [PubMed]
17. Ramírez, E.; Medrano-Casique, N.; Tong, H.Y.; Bellón, T.; Cabañas, R.; Fiandor, A.; González-Ramos, J.; Herranz, P.; Trigo, E.; Muñoz, M.; et al. Eosinophilic Drug Reactions Detected by a Prospective Pharmacovigilance Programme in a Tertiary Hospital. *Br. J. Clin. Pharmacol.* **2017**, *83*, 400–415. [CrossRef]
18. Muller, P.; Dubreil, P.; Mahé, A.; Lamaury, I.; Salzer, B.; Deloumeaux, J.; Strobel, M. Drug Hypersensitivity Syndrome in a West Indian Population. *Eur. J. Dermatol.* **2003**, *13*, 478–481.
19. Bluestein, S.B.; Yu, R.; Stone, C.; Phillips, E.J. Reporting of Drug Reaction with Eosinophilia and Systemic Symptoms from 2002 to 2019 in the US Food and Drug Administration Adverse Event Reporting System. *J. Allergy Clin. Immunol. Pract.* **2021**, *9*, 3208–3211.e1. [CrossRef]
20. Ramirez, G.A.; Della-Torre, E.; Tresoldi, M.; Scarpellini, P.; Ciceri, F.; Dagna, L.; Yacoub, M.R. Drug Reaction with Eosinophilia and Systemic Symptoms (DRESS) in Patients with COVID-19. *Clin. Microbiol. Infect.* **2021**, *27*, 1190–1192. [CrossRef]
21. Schunkert, E.M.; Divito, S.J. Updates and Insights in the Diagnosis and Management of DRESS Syndrome. *Curr. Dermatol. Rep.* **2021**, *10*, 192–204. [CrossRef] [PubMed]
22. Cho, Y.T.; Yang, C.W.; Chu, C.Y. Drug Reaction with Eosinophilia and Systemic Symptoms (DRESS): An Interplay among Drugs, Viruses, and Immune System. *Int. J. Mol. Sci.* **2017**, *18*, 1243. [CrossRef] [PubMed]
23. Cabañas, R.; Ramírez, E.; Sendagorta, E.; Alamar, R.; Barranco, R.; Blanca-López, N.; Doña, I.; Fernández, J.; Garcia-Nunez, I.; García-Samaniego, J.; et al. Spanish Guidelines for Diagnosis, Management, Treatment, and Prevention of DRESS Syndrome. *J. Investig. Allergol. Clin. Immunol.* **2020**, *30*, 229–253. [CrossRef] [PubMed]
24. Soria, A.; Bernier, C.; Veyrac, G.; Barbaud, A.; Puymirat, E.; Milpied, B. Drug Reaction with Eosinophilia and Systemic Symptoms May Occur within 2 Weeks of Drug Exposure: A Retrospective Study. *J. Am. Acad. Dermatol.* **2020**, *82*, 606–611. [CrossRef] [PubMed]
25. Lee, J.Y.; Lee, S.Y.; Hahm, J.E.; Ha, J.W.; Kim, C.W.; Kim, S.S. Clinical Features of Drug Reaction with Eosinophilia and Systemic Symptoms (DRESS) Syndrome: A Study of 25 Patients in Korea. *Int. J. Dermatol.* **2017**, *56*, 944–951. [CrossRef]
26. Sharifzadeh, S.; Elyasi, S.; Mohammadpour, A.H. Antiviral Induced Drug Reaction with Eosinophilia and Systemic Symptoms (DRESS) Dyndrome: A Literature Review. *J. Pharm. Care* **2020**, *8*, 35–47. [CrossRef]
27. Gangireddy, M.; Sarao, M.S.; Shrimanker, I.; Nookala, V.K. A Fatal Case of Vancomycin Associated Drug Reaction with Eosinophilia and Systemic Symptoms Syndrome in a Septuagenarian. *Cureus* **2019**, *11*, e5015. [CrossRef]
28. Song, J.M.; Jung, Y.E.; Park, J.H.; Kim, M.D.; Cheon, M.S.; Lee, C.I. Neosensitization to Multiple Drugs Following Valproate-Induced Drug Reaction with Eosinophilia and Systemic Symptoms Syndrome. *Psychiatry Investig.* **2017**, *14*, 518–520. [CrossRef]
29. Gentile, I.; Talamo, M.; Borgia, G. Is the Drug-Induced Hypersensitivity Syndrome (DIHS) Due to Human Herpesvirus 6 Infection or to Allergy-Mediated Viral Reactivation? Report of a Case and Literature Review. *BMC Infect. Dis.* **2010**, *10*, 49. [CrossRef]
30. Chung, W.H.; Chang, W.C.; Lee, Y.S.; Wu, Y.Y.; Yang, C.H.; Ho, H.C.; Chen, M.J.; Lin, J.Y.; Hui, R.C.Y.; Ho, J.C.; et al. Genetic Variants Associated with Phenytoin-Related Severe Cutaneous Adverse Reactions. *JAMA* **2014**, *312*, 525–535. [CrossRef]
31. Rieder, M.J.; Shear, N.H.; Kanee, A.; Tang, B.K.; Spielberg, S.P. Prominence of Slow Acetylator Phenotype among Patients with Sulfonamide Hypersensitivity Reactions. *Clin. Pharmacol. Ther.* **1991**, *49*, 13–17. [CrossRef] [PubMed]
32. Markel, A. Allopurinol-Induced DRESS Syndrome. *Isr. Med. Assoc. J.* **2005**, *7*, 656–660. [PubMed]
33. Chou, H.Y.; Chen, C.B.; Cheng, C.Y.; Chen, Y.A.; Ng, C.Y.; Kuo, K.L.; Chen, W.L.; Chen, C.H. Febuxostat-Associated Drug Reaction with Eosinophilia and Systemic Symptoms (DRESS). *J. Clin. Pharm. Ther.* **2015**, *40*, 689–692. [CrossRef] [PubMed]

34. Pott Junior, H.; Gosuen, G.C.; Gales, A.C. DRESS Syndrome Due to Nevirapine Treated with Methylprednisolone. *Case Rep. Med.* **2013**, *2013*, 269501. [CrossRef]
35. Mallal, S.; Phillips, E.; Carosi, G.; Molina, J.-M.; Workman, C.; Tomažič, J.; Jägel-Guedes, E.; Rugina, S.; Kozyrev, O.; Cid, J.F.; et al. HLA-B*5701 Screening for Hypersensitivity to Abacavir. *N. Engl. J. Med.* **2008**, *358*, 568–579. [CrossRef] [PubMed]
36. Borrás-Blasco, J.; Navarro-Ruiz, A.; Borrás, C.; Casterá, E. Adverse Cutaneous Reactions Associated with the Newest Antiretroviral Drugs in Patients with Human Immunodeficiency Virus Infection. *J. Antimicrob. Chemother.* **2008**, *62*, 879–888. [CrossRef]
37. Samain, A.; Duval-Modeste, A.B.; Joly, P.; Leblanc, C.; Massy, N.; Courville, P.; Goria, O.; Riachi, G. First Case of Drug Rash Eosinophilia and Systemic Symptoms Due to Boceprevir. *J. Hepatol.* **2014**, *60*, 891–893. [CrossRef]
38. Roujeau, J.C.; Mockenhaupt, M.; Tahan, S.R.; Henshaw, J.; Martin, E.C.; Harding, M.; van Baelen, B.; Bengtsson, L.; Singhal, P.; Kauffman, R.S.; et al. Telaprevir-Related Dermatitis. *JAMA Dermatol.* **2013**, *149*, 152–158. [CrossRef]
39. Saper, V.E.; Ombrello, M.J.; Tremoulet, A.H.; Montero-Martin, G.; Prahalad, S.; Canna, S.; Shimizu, C.; Deutsch, G.; Tan, S.Y.; Remmers, E.F.; et al. Severe Delayed Hypersensitivity Reactions to IL-1 and IL-6 Inhibitors Link to Common HLA-DRB1*15 Alleles. *Ann. Rheum. Dis.* **2022**, *81*, 406–415. [CrossRef]
40. Kim, D.K.; Lee, S.W.; Nam, H.S.; Jeon, D.S.; Park, N.R.; Nam, Y.H.; Lee, S.K.; Baek, Y.H.; Han, S.Y.; Lee, S.W. A Case of Sorafenib-Induced DRESS Syndrome in Hepatocelluar Carcinoma. *Korean J. Gastroenterol.* **2016**, *67*, 337. [CrossRef]
41. Thomas, C.L.; Arasaratnam, M.; Carlos, G.; Parasyn, A.; Baumgart, K.W.; Fernandez-Penas, P.; Marx, G. Drug Reaction with Eosinophilia and Systemic Symptoms in Metastatic Basal Cell Carcinoma Treated with Vismodegib. *Australas. J. Dermatol.* **2017**, *58*, 69–70. [CrossRef]
42. Peuvrel, L.; Quéreux, G.; Saint-Jean, M.; Brocard, A.; Nguyen, J.M.; Khammari, A.; Knol, A.C.; Varey, E.; Dréno, B. Profile of Vemurafenib-Induced Severe Skin Toxicities. *J. Eur. Acad. Dermatol. Venereol.* **2016**, *30*, 250–257. [CrossRef]
43. Radu, C.; Barnig, C.; de Blay, F. Rivaroxaban-Induced Drug Reaction With Eosinophilia and Systemic Symptoms. *J. Investig. Allergol. Clin. Immunol.* **2016**, *26*, 124–126. [CrossRef] [PubMed]
44. Volpe, A.; Marchetta, A.; Caramaschi, P.; Biasi, D.; Bambara, L.M.; Arcaro, G. Hydroxychloroquine-Induced DRESS Syndrome. *Clin. Rheumatol.* **2008**, *27*, 537–539. [CrossRef] [PubMed]
45. Shiohara, T.; Iijima, M.; Ikezawa, Z.; Hashimoto, K. The Diagnosis of a DRESS Syndrome Has Been Sufficiently Established on the Basis of Typical Clinical Features and Viral Reactivations. *Br. J. Dermatol.* **2007**, *156*, 1083–1084. [CrossRef] [PubMed]
46. Tohyama, M.; Hashimoto, K.; Oda, F.; Namba, C.; Sayama, K. Influence of Corticosteroid Therapy on Viral Reactivation in Drug-Induced Hypersensitivity Syndrome/Drug Reaction with Eosinophilia and Systemic Symptoms. *J. Dermatol.* **2020**, *47*, 476–482. [CrossRef]
47. Descamps, V.; Brunet-Possenti, F. Monitoring of Human Herpesvirus 6 Infection in the Management of Drug Reaction with Eosinophilia and Systemic Symptoms. *Clin. Exp. Dermatol.* **2021**, *46*, 351–352. [CrossRef]
48. Tohyama, M.; Hashimoto, K.; Yasukawa, M.; Kimura, H.; Horikawa, T.; Nakajima, K.; Urano, Y.; Matsumoto, K.; Iijima, M.; Shear, N.H. Association of Human Herpesvirus 6 Reactivation with the Flaring and Severity of Drug-Induced Hypersensitivity Syndrome. *Br. J. Dermatol.* **2007**, *157*, 934–940. [CrossRef]
49. Miyagawa, F.; Asada, H. Current Perspective Regarding the Immunopathogenesis of Drug-Induced Hypersensitivity Syndrome/Drug Reaction with Eosinophilia and Systemic Symptoms (DIHS/DRESS). *Int. J. Mol. Sci.* **2021**, *22*, 2147. [CrossRef]
50. Drago, F.; Cogorno, L.; Broccolo, F.; Ciccarese, G.; Parodi, A. A Fatal Case of DRESS Induced by Strontium Ranelate Associated with HHV-7 Reactivation. *Osteoporos. Int.* **2016**, *27*, 1261–1264. [CrossRef]
51. Ahluwalia, J.; Abuabara, K.; Perman, M.J.; Yan, A.C. Human Herpesvirus 6 Involvement in Paediatric Drug Hypersensitivity Syndrome. *Br. J. Dermatol.* **2015**, *172*, 1090–1095. [CrossRef]
52. Roujeau, J.C.; Dupin, N. Virus Reactivation in Drug Reaction with Eosinophilia and Systemic Symptoms (Dress) Results from a Strong Drug-Specific Immune Response. *J. Allergy Clin. Immunol. Pract.* **2017**, *5*, 811–812. [CrossRef]
53. Stirton, H.; Shear, N.H.; Dodiuk-Gad, R.P. Drug Reaction with Eosinophilia and Systemic Symptoms (DReSS)/Drug-Induced Hypersensitivity Syndrome (DiHS)-Readdressing the DReSS. *Biomedicines* **2022**, *10*, 999. [CrossRef] [PubMed]
54. Picard, D.; Janela, B.; Descamps, V.; D'Incan, M.; Courville, P.; Jacquot, S.; Rogez, S.; Mardivirin, L.; Moins-Teisserenc, H.; Toubert, A.; et al. Drug Reaction with Eosinophilia and Systemic Symptoms (DRESS): A Multiorgan Antiviral T Cell Response. *Sci. Transl. Med.* **2010**, *2*, 46ra62. [CrossRef]
55. Hama, N.; Abe, R.; Gibson, A.; Phillips, E.J. Drug-Induced Hypersensitivity Syndrome (DIHS)/Drug Reaction With Eosinophilia and Systemic Symptoms (DRESS): Clinical Features and Pathogenesis. *J. Allergy Clin. Immunol. Pract.* **2022**, *10*, 1155–1167.e5. [CrossRef] [PubMed]
56. Shiohara, T.; Mizukawa, Y. Drug-Induced Hypersensitivity Syndrome (DiHS)/Drug Reaction with Eosinophilia and Systemic Symptoms (DRESS): An Update in 2019. *Allergol. Int.* **2019**, *68*, 301–308. [CrossRef] [PubMed]
57. Watanabe, H. Recent Advances in Drug-Induced Hypersensitivity Syndrome/Drug Reaction with Eosinophilia and Systemic Symptoms. *J. Immunol. Res.* **2018**, *2018*, 5163129. [CrossRef]
58. Anci, E.; Braun, C.; Marinosci, A.; Rodieux, F.; Midun, E.; Torres, M.J.; Caubet, J.C. Viral Infections and Cutaneous Drug-Related Eruptions. *Front. Pharmacol.* **2021**, *11*, 586407. [CrossRef]
59. Kagoyama, K.; Makino, T.; Ueda, C.; Takegami, Y.; Shimizu, T. Detection of Cytomegalovirus in the Gastric Ulcer of a Patient with Drug-Induced Hypersensitivity Syndrome. *JAAD Case Rep.* **2015**, *1*, 215–218. [CrossRef] [PubMed]

60. Ganeshanandan, L.; Lucas, M. Drug Reaction with Eosinophilia and Systemic Symptoms: A Complex Interplay between Drug, T Cells, and Herpesviridae. *Int. J. Mol. Sci.* **2021**, *22*, 1127. [CrossRef]
61. Kardaun, S.H.; Sidoroff, A.; Valeyrie-Allanore, L.; Halevy, S.; Davidovici, B.B.; Mockenhaupt, M.; Roujeau, J.C. Variability in the Clinical Pattern of Cutaneous Side-Effects of Drugs with Systemic Symptoms: Does a DRESS Syndrome Really Exist? *Br. J. Dermatol.* **2007**, *156*, 609–611. [CrossRef] [PubMed]
62. Seishima, M.; Yamanaka, S.; Fujisawa, T.; Tohyama, M.; Hashimoto, K. Reactivation of Human Herpesvirus (HHV) Family Members Other than HHV-6 in Drug-Induced Hypersensitivity Syndrome. *Br. J. Dermatol.* **2006**, *155*, 344–349. [CrossRef]
63. Kano, Y.; Hiraharas, K.; Sakuma, K.; Shiohara, T. Several Herpesviruses Can Reactivate in a Severe Drug-Induced Multiorgan Reaction in the Same Sequential Order as in Graft-versus-Host Disease. *Br. J. Dermatol.* **2006**, *155*, 301–306. [CrossRef]
64. Suzuki, Y.; Inagi, R.; Aono, T.; Yamanishi, K.; Shiohara, T. Human Herpesvirus 6 Infection as a Risk Factor for the Development of Severe Drug-Induced Hypersensitivity Syndrome. *Arch. Dermatol.* **1998**, *134*, 1108–1112. [CrossRef]
65. Ishida, T.; Kano, Y.; Mizukawa, Y.; Shiohara, T. The Dynamics of Herpesvirus Reactivations during and after Severe Drug Eruptions: Their Relation to the Clinical Phenotype and Therapeutic Outcome. *Allergy* **2014**, *69*, 798–805. [CrossRef]
66. Chen, Y.C.; Chiang, H.H.; Cho, Y.T.; Chang, C.Y.; Chen, K.L.; Yang, C.W.; Lee, Y.H.; Chu, C.Y. Human Herpes Virus Reactivations and Dynamic Cytokine Profiles in Patients with Cutaneous Adverse Drug Reactions—A Prospective Comparative Study. *Allergy* **2015**, *70*, 568–575. [CrossRef]
67. Pritchett, J.C.; Nanau, R.M.; Neuman, M.G. The Link between Hypersensitivity Syndrome Reaction Development and Human Herpes Virus-6 Reactivation. *Int. J. Hepatol.* **2012**, *2012*, 723062. [CrossRef]
68. Shiohara, T.; Takahashi, R.; Kano, Y. Drug-Induced Hypersensitivity Syndrome and Viral Reactivation. *Drug Hypersensit.* **2007**, 251–266. [CrossRef]
69. Chiou, C.C.; Chung, W.H.; Hung, S.I.; Yang, L.C.; Hong, H.S. Fulminant Type 1 Diabetes Mellitus Caused by Drug Hypersensitivity Syndrome with Human Herpesvirus 6 Infection. *J. Am. Acad. Dermatol.* **2006**, *54* (Suppl. S2), S14–S17. [CrossRef] [PubMed]
70. Lee, E.S.; Kiuchi, Y.; Inomata, N.; Sueki, H. Increased Expression of Human Herpes Virus 6 Receptor CD134/OX40 in Skin Lesions of Patients with Drug-Induced Hypersensitivity Syndrome/Drug Reaction with Eosinophilia and Systemic Symptoms. *J. Dermatol.* **2022**. [CrossRef]
71. Hall, C.; Caserta, M.T.; Schnabel, K.; Shelley, L.; Marino, A.S.; Carnahan, J.A.; Yoo, C.; Lofthus, G.K.; McDermott, M.P. Chromosomal Integration of Human Herpesvirus 6 Is the Major Mode of Congenital Human Herpesvirus 6 Infection. *Pediatrics* **2008**, *122*, 513–520. [CrossRef]
72. Watanabe, H.; Daibata, M.; Tohyama, M.; Batchelor, J.; Hashimoto, K.; Iijima, M. Chromosomal Integration of Human Herpesvirus 6 DNA in Anticonvulsant Hypersensitivity Syndrome. *Br. J. Dermatol.* **2008**, *158*, 640–642. [CrossRef]
73. Nanishi, E.; Hoshina, T.; Ohga, S.; Nishio, H.; Hara, T. Drug Reaction with Eosinophilia and Systemic Symptoms during Primary Epstein-Barr Virus Infection. *J. Microbiol. Immunol. Infect.* **2015**, *48*, 109–112. [CrossRef]
74. Aouam, K.; Bel Hadj Ali, H.; Youssef, M.; Chaabane, A.; Amri, M.; Boughattas, N.A.; Zili, J.E. Carbamazepine-Induced DRESS and HHV6 Primary Infection: The Importance of Skin Tests. *Epilepsia* **2008**, *49*, 1630–1633. [CrossRef]
75. Miyagawa, F.; Nakamura, Y.; Ommori, R.; Miyashita, K.; Iioka, H.; Miyashita, N.; Nishikawa, M.; Himuro, Y.; Ogawa, K.; Asada, H. Predominant Contribution of CD4 T Cells to Human Herpesvirus 6 (HHV-6) Load in the Peripheral Blood of Patients with Drug-Induced Hypersensitivity Syndrome and Persistent HHV-6 Infection. *Acta Derm. Venereol.* **2018**, *98*, 146–148. [CrossRef]
76. Kanatani, Y.; Miyagawa, F.; Ogawa, K.; Arima, A.; Asada, H. Parallel Changes in Serum Thymus and Activation-Regulated Chemokine Levels in Response to Flare-Ups in Drug-Induced Hypersensitivity Syndrome. *J. Dermatol.* **2020**, *47*, e417–e419. [CrossRef]
77. Saraya, T.; Mikoshiba, M.; Kamiyama, H.; Yoshizumi, M.; Tsuchida, S.; Tsukagoshi, H.; Ishioka, T.; Terada, M.; Tanabe, E.; Tomioka, C.; et al. Evidence for Reactivation of Human Herpesvirus 6 in Generalized Lymphadenopathy in a Patient with Drug-Induced Hypersensitivity Syndrome. *J. Clin. Microbiol.* **2013**, *51*, 1979–1982. [CrossRef]
78. Miyashita, K.; Shobatake, C.; Miyagawa, F.; Kobayashi, N.; Onmori, R.; Yonekawa, S.; Tanabe, K.; Kawate, K.; Morita, K.; Asada, H. Involvement of Human Herpesvirus 6 Infection in Renal Dysfunction Associated with DIHS/DRESS. *Acta Derm. Venereol.* **2016**, *96*, 114–115. [CrossRef]
79. Miyashita, K.; Miyagawa, F.; Nakamura, Y.; Ommori, R.; Azukizawa, H.; Asada, H. Up-Regulation of Human Herpesvirus 6B-Derived MicroRNAs in the Serum of Patients with Drug-Induced Hypersensitivity Syndrome/Drug Reaction with Eosinophilia and Systemic Symptoms. *Acta Derm. Venereol.* **2018**, *98*, 612–613. [CrossRef]
80. Shiohara, T.; Inaoka, M.; Kano, Y. Drug-Induced Hypersensitivity Syndrome(DIHS): A Reaction Induced by a Complex Interplay among Herpesviruses and Antiviral and Antidrug Immune Responses. *Allergol. Int.* **2006**, *55*, 1–8. [CrossRef]
81. Kurata, M.; Shiohara, T. Herpes Simplex Virus Reactivation: Is It Common or Rare in Drug-Induced Hypersensitivity Syndrome/Drug Reaction with Eosinophilia and Systemic Symptoms? *Eur. J. Dermatol.* **2017**, *27*, 658–659. [CrossRef] [PubMed]
82. Tsutsumi, R.; Adachi, K.; Yoshida, Y.; Yamamoto, O. Drug-Induced Hypersensitivity Syndrome in Association with Varicella. *Acta Derm. Venereol.* **2015**, *95*, 503–504. [CrossRef]
83. Wong, Y.J.; Choo, K.J.L.; Soh, J.X.J.; Tan, C.K. Cytomegalovirus (CMV) Hepatitis: An Uncommon Complication of CMV Reactivation in Drug Reaction with Eosinophilia and Systemic Symptoms. *Singap. Med. J.* **2018**, *59*, 112–113. [CrossRef] [PubMed]

84. Lachance, P.; Chen, J.; Featherstone, R.; Sligl, W.I. Association Between Cytomegalovirus Reactivation and Clinical Outcomes in Immunocompetent Critically Ill Patients: A Systematic Review and Meta-Analysis. *Open Forum Infect. Dis.* **2017**, *4*, ofx029. [CrossRef]
85. Schildermans, J.; de Vlieger, G. Cytomegalovirus: A Troll in the ICU? Overview of the Literature and Perspectives for the Future. *Front. Med.* **2020**, *7*, 188. [CrossRef] [PubMed]
86. Papazian, L.; Jaber, S.; Hraiech, S.; Baumstarck, K.; Cayot-Constantin, S.; Aissaoui, N.; Jung, B.; Leone, M.; Souweine, B.; Schwebel, C.; et al. Preemptive Ganciclovir for Mechanically Ventilated Patients with Cytomegalovirus Reactivation. *Ann. Intensive Care* **2021**, *11*, 33. [CrossRef]
87. Limaye, A.P.; Stapleton, R.D.; Peng, L.; Gunn, S.R.; Kimball, L.E.; Hyzy, R.; Exline, M.C.; Files, D.C.; Morris, P.E.; Frankel, S.K.; et al. Effect of Ganciclovir on IL-6 Levels Among Cytomegalovirus-Seropositive Adults With Critical Illness: A Randomized Clinical Trial. *JAMA* **2017**, *318*, 731–740. [CrossRef]
88. Cowley, N.J.; Owen, A.; Shiels, S.C.; Millar, J.; Woolley, R.; Ives, N.; Osman, H.; Moss, P.; Bion, J.F. Safety and Efficacy of Antiviral Therapy for Prevention of Cytomegalovirus Reactivation in Immunocompetent Critically Ill Patients: A Randomized Clinical Trial. *JAMA Intern. Med.* **2017**, *177*, 774–783. [CrossRef]
89. Herman, A.; Matthews, M.; Mairlot, M.; Nobile, L.; Fameree, L.; Jacquet, L.M.; Baeck, M. Drug Reaction with Eosinophilia and Systemic Symptoms Syndrome in a Patient with COVID-19. *J. Eur. Acad. Dermatol. Venereol.* **2020**, *34*, e700–e768. [CrossRef]
90. Grandolfo, M.; Romita, P.; Bonamonte, D.; Cazzato, G.; Hansel, K.; Stingeni, L.; Conforti, C.; Giuffrida, R.; Foti, C. Drug Reaction with Eosinophilia and Systemic Symptoms Syndrome to Hydroxychloroquine, an Old Drug in the Spotlight in the COVID-19 Era. *Dermatol. Ther.* **2020**, *33*, e13499. [CrossRef]
91. Minenna, E.; Chaoul, N.; Rossi, M.P.; Giliberti, L.; Albanesi, M.; Nettis, E.; Foschino Barbaro, M.P.; di Bona, D.; Caiaffa, M.F.; Macchia, L. Sustained Drug-Related Reaction with Eosinophilia and Systemic Symptoms (DRESS) Triggered by Low Molecular Weight Heparins in COVID-19: Management and Precision Diagnosis. *Postepy Dermatol. Alergol.* **2022**, *39*, 816–818. [CrossRef] [PubMed]
92. Brunasso, A.M.G.; Castellaneta, M.; Pontali, E.; Raggi, F.; Massone, C. Follow-up of Skin Lesions during COVID-19: A Description of a DRESS Case. *Arch. Dermatol. Res.* **2021**, *313*, 607–610. [CrossRef] [PubMed]
93. Mata, L.; Torres-Zevallos, H.; Guerreros, A.G. Life-Threatening DRESS Syndrome with Kidney Damage Following Severe COVID-19 in a Patient with Down Syndrome. *BMJ Case Rep.* **2021**, *14*, e241418. [CrossRef]
94. Castro Jiménez, A.; Navarrete Navarrete, N.; Gratacós Gómez, A.R.; Florido López, F.; García Rodríguez, R.; Gómez Torrijos, E. First Case of DRESS Syndrome Caused by Hydroxychloroquine with a Positive Patch Test. *Contact Dermat.* **2021**, *84*, 50–51. [CrossRef]
95. Balconi, S.N.; Lopes, N.T.; Luzzatto, L.; Bonamigo, R.R. Detection of SARS-CoV-2 in a Case of DRESS by Sulfasalazine: Could There Be a Relationship with Clinical Importance? *Int. J. Dermatol.* **2021**, *60*, 125–126. [CrossRef] [PubMed]
96. Schmid-Grendelmeier, P.; Steiger, P.; Naegeli, M.C.; Kolm, I.; Lang, C.C.V.; Maverakis, E.; Brüggen, M.C. Benralizumab for Severe DRESS in Two COVID-19 Patients. *J. Allergy Clin. Immunol. Pract.* **2021**, *9*, 481–483.e2. [CrossRef]
97. Cruz, V.B.; Júnior, L.F.F.F.; Kobal, C.R.; da Silva, N.A. Does Sensitization by SARS-CoV-2 Immune Complexes Trigger DRESS Syndrome? *Braz. J. Infect. Dis.* **2022**, *26*. [CrossRef]
98. Mesli, F.; Dumont, M.; Soria, A.; Groh, M.; Turpin, M.; Voiriot, G.; Rafat, C.; Sallé, D.S.; Gibelin, A.; Desnos, C. Benralizumab: A Potential Tailored Treatment for Life-Threatening DRESS in the COVID-19 Era. *J. Allergy Clin. Immunol. Pract.* **2021**, *9*, 3529–3531.e1. [CrossRef]
99. Cucka, B.; Biglione, B.; Zhou, L.; Phillips, E.J.; Bassir, F.; Samarakoon, U.; Rrapi, R.; Chand, S.; Wang, L.; Alvarez-Arango, S.; et al. Drug Reaction with Eosinophilia and Systemic Symptoms in Patients Hospitalized with COVID-19: A Case Series from a Large US Healthcare System. *Br. J. Dermatol.* **2022**, *187*, 619–622. [CrossRef]
100. Schmutz, J.L.; Barbaud, A.; Tréchot, P. Hydroxychloroquine and Dress. *Ann. Dermatol. Venereol.* **2008**, *135*, 903. [CrossRef]
101. Mitamura, Y.; Schulz, D.; Oro, S.; Li, N.; Kolm, I.; Lang, C.; Ziadlou, R.; Tan, G.; Bodenmiller, B.; Steiger, P.; et al. Cutaneous and Systemic Hyperinflammation Drives Maculopapular Drug Exanthema in Severely Ill COVID-19 Patients. *Allergy* **2022**, *77*, 595–608. [CrossRef]
102. Lang, C.C.V.; Schmid-Grendelmeier, P.; Maverakis, E.; Brüggen, M.C. Reply to "Benralizumab: A Potential Tailored Treatment for Life-Threatening DRESS in the COVID-19 Era". *J. Allergy Clin. Immunol. Pract.* **2021**, *9*, 3531–3532. [CrossRef] [PubMed]
103. Brooks, B.; Tancredi, C.; Song, Y.; Mogus, A.T.; Huang, M.L.W.; Zhu, H.; Phan, T.L.; Zhu, H.; Kadl, A.; Woodfolk, J.; et al. Epstein-Barr Virus and Human Herpesvirus-6 Reactivation in Acute COVID-19 Patients. *Viruses* **2022**, *14*, 1872. [CrossRef] [PubMed]
104. Simonnet, A.; Engelmann, I.; Moreau, A.S.; Garcia, B.; Six, S.; el Kalioubie, A.; Robriquet, L.; Hober, D.; Jourdain, M. High Incidence of Epstein-Barr Virus, Cytomegalovirus, and Human-Herpes Virus-6 Reactivations in Critically Ill Patients with COVID-19. *Infect. Dis. Now* **2021**, *51*, 296–299. [CrossRef] [PubMed]
105. Naendrup, J.H.; Garcia Borrega, J.; Eichenauer, D.A.; Shimabukuro-Vornhagen, A.; Kochanek, M.; Böll, B. Reactivation of EBV and CMV in Severe COVID-19-Epiphenomena or Trigger of Hyperinflammation in Need of Treatment? A Large Case Series of Critically Ill Patients. *J. Intensive Care Med.* **2022**, *37*, 1152–1158. [CrossRef]

106. Takeno, A.; Kanazawa, I.; Morita, M.; Takedani, K.; Miyake, H.; Yamamoto, M.; Nogami, K.; Kaneko, S.; Sugimoto, T. A Case Report of Fulminant Type 1 Diabetes Mellitus Associated with Drug-Induced Hypersensitivity Syndrome in an Elderly Patient with Coxsackie B4 Virus Infection and Human Leukocyte Antigen-A24 Haplotype. *Endocr. J.* **2018**, *65*, 129–132. [CrossRef] [PubMed]
107. Girijala, R.L.; Ramamurthi, A.; Wright, D.; Kwak, Y.; Goldberg, L.H. DRESS Syndrome Associated with Influenza Virus. *Proc. (Bayl. Univ. Med. Cent.)* **2019**, *32*, 277–278. [CrossRef]
108. Sil, A.; Bhattacharjee, M.S.; Chandra, A.; Pramanik, J.D. Sulfasalazine-Induced Drug Reaction with Eosinophilia and Systemic Symptoms (DRESS) with Concomitant Acute Chikungunya Virus Infection: Possible Role of New Viral Trigger. *BMJ Case Rep.* **2021**, *14*, e244063. [CrossRef]
109. Panigrahi, A.; Chakraborty, S.; Sil, A. Chik Sign in Chikungunya Fever. *Infection* **2021**, *49*, 1075–1076. [CrossRef] [PubMed]
110. Sil, A.; Biswas, S.K.; Bhanja, D.B.; Das, S.; Panigrahi, A. Post-Chikungunya Hyperpigmentation. *Postgrad. Med. J.* **2021**, *97*, 59–60. [CrossRef] [PubMed]
111. Srivastava, A. Hyperpigmentation and Chikungunya Fever. *An. Bras. Dermatol.* **2016**, *91*, 860–861. [CrossRef]
112. Trubiano, J.A.; Gordon, C.L.; Castellucci, C.; Christo, S.N.; Park, S.L.; Mouhtouris, E.; Konvinse, K.; Rose, M.; Goh, M.; Boyd, A.S.; et al. Analysis of Skin-Resident Memory T Cells Following Drug Hypersensitivity Reactions. *J. Investig. Dermatol.* **2020**, *140*, 1442–1445.e4. [CrossRef]
113. Chakraborty, U.; Biswas, P.; Chandra, A.; Pal, J.; Ray, A.K. Chik Sign: Post-Chikungunya Hyperpigmentation. *QJM* **2021**, *114*, 137–138. [CrossRef]
114. Thomas, M.; Hopkins, C.; Duffy, E.; Lee, D.; Loulergue, P.; Ripamonti, D.; Ostrov, D.A.; Phillips, E. Association of the HLA-B*53:01 Allele With Drug Reaction With Eosinophilia and Systemic Symptoms (DRESS) Syndrome During Treatment of HIV Infection With Raltegravir. *Clin. Infect. Dis.* **2017**, *64*, 1198–1203. [CrossRef]
115. Gill, S.; Sagar, A.; Shankar, S.; Nair, V. Nevirapine-Induced Rash with Eosinophilia and Systemic Symptoms (DRESS). *Indian J. Pharmacol.* **2013**, *45*, 401–402. [CrossRef]
116. Lehloenya, R.J.; Dlamini, S.; Muloiwa, R.; Kakande, B.; Ngwanya, M.R.; Todd, G.; Dheda, K. Therapeutic Trial of Rifabutin After Rifampicin-Associated DRESS Syndrome in Tuberculosis-Human Immunodeficiency Virus Coinfected Patients. *Open Forum Infect. Dis.* **2016**, *3*, ofw130. [CrossRef]
117. Walsh, S.A.; Creamer, D. Drug Reaction with Eosinophilia and Systemic Symptoms (DRESS): A Clinical Update and Review of Current Thinking. *Clin. Exp. Dermatol.* **2011**, *36*, 6–11. [CrossRef]
118. Aihara, Y.; Ito, S.I.; Kobayashi, Y.; Yamakawa, Y.; Aihara, M.; Yokota, S. Carbamazepine-Induced Hypersensitivity Syndrome Associated with Transient Hypogammaglobulinaemia and Reactivation of Human Herpesvirus 6 Infection Demonstrated by Real-Time Quantitative Polymerase Chain Reaction. *Br. J. Dermatol.* **2003**, *149*, 165–169. [CrossRef]
119. Kano, Y.; Inaoka, M.; Shiohara, T. Association between Anticonvulsant Hypersensitivity Syndrome and Human Herpesvirus 6 Reactivation and Hypogammaglobulinemia. *Arch. Dermatol.* **2004**, *140*, 183–188. [CrossRef]
120. Miyagawa, F.; Nakamura-Nishimura, Y.; Kanatani, Y.; Asada, H. Correlation Between Expression of CD134, a Human Herpesvirus 6 Cellular Receptor, on CD4+ T Cells and Th2-Type Immune Responses in Drug-Induced Hypersensitivity Syndrome/Drug Reaction with Eosinophilia and Systemic Symptoms. *Acta Derm. Venereol.* **2020**, *100*, adv00102. [CrossRef]
121. Lu, J.; Thuraisingam, T.; Chergui, M.; Nguyen, K. Nivolumab-Associated DRESS Syndrome: A Case Report. *JAAD Case Rep.* **2019**, *5*, 216–218. [CrossRef]
122. Ai, L.; Gao, J.; Zhao, S.; Li, Q.; Cui, Y.H.; Liu, Q.; Wu, D.; Wang, Y.; Jin, X.; Ji, Y.; et al. Nivolumab-Associated DRESS in a Genetic Susceptible Individual. *J. Immunother. Cancer* **2021**, *9*, e002879. [CrossRef]
123. Sugita, K.; Tohyama, M.; Watanabe, H.; Otsuka, A.; Nakajima, S.; Iijima, M.; Hashimoto, K.; Tokura, Y.; Miyachi, Y.; Kabashima, K. Fluctuation of Blood and Skin Plasmacytoid Dendritic Cells in Drug-Induced Hypersensitivity Syndrome. *J. Allergy Clin. Immunol.* **2010**, *126*, 408–410. [CrossRef]
124. Mizukawa, Y.; Kimishima, M.; Aoyama, Y.; Shiohara, T. Predictive Biomarkers for Cytomegalovirus Reactivation before and after Immunosuppressive Therapy: A Single-Institution Retrospective Long-Term Analysis of Patients with Drug-Induced Hypersensitivity Syndrome (DiHS)/Drug Reaction with Eosinophilia and Systemic Syndrome (DRESS). *Int. J. Infect. Dis.* **2020**, *100*, 239–246. [CrossRef]
125. Farrell, J.; Lichtenfels, M.; Sullivan, A.; Elliott, E.C.; Alfirevic, A.; Stachulski, A.V.; Pirmohamed, M.; Naisbitt, D.J.; Park, B.K. Activation of Carbamazepine-Responsive T-Cell Clones with Metabolically Inert Halogenated Derivatives. *J. Allergy Clin. Immunol.* **2013**, *132*, 493–495. [CrossRef]
126. Ye, Y.M.; Hur, G.Y.; Kim, S.H.; Ban, G.Y.; Jee, Y.K.; Naisbitt, D.J.; Park, H.S.; Kim, S.H. Drug-Specific CD4+ T-Cell Immune Responses Are Responsible for Antituberculosis Drug-Induced Maculopapular Exanthema and Drug Reaction with Eosinophilia and Systemic Symptoms Syndrome. *Br. J. Dermatol.* **2017**, *176*, 378–386. [CrossRef]
127. Nicoletti, P.; Barrett, S.; McEvoy, L.; Daly, A.K.; Aithal, G.; Lucena, M.I.; Andrade, R.J.; Wadelius, M.; Hallberg, P.; Stephens, C.; et al. Shared Genetic Risk Factors Across Carbamazepine-Induced Hypersensitivity Reactions. *Clin. Pharmacol. Ther.* **2019**, *106*, 1028–1036. [CrossRef]
128. Phillips, E.J.; Chung, W.H.; Mockenhaupt, M.; Roujeau, J.C.; Mallal, S.A. Drug Hypersensitivity: Pharmacogenetics and Clinical Syndromes. *J. Allergy Clin. Immunol.* **2011**, *127* (Suppl. S3), S60–S66. [CrossRef]

129. Konvinse, K.C.; Trubiano, J.A.; Pavlos, R.; James, I.; Shaffer, C.M.; Bejan, C.A.; Schutte, R.J.; Ostrov, D.A.; Pilkinton, M.A.; Rosenbach, M.; et al. HLA-A*32:01 Is Strongly Associated with Vancomycin-Induced Drug Reaction with Eosinophilia and Systemic Symptoms. *J. Allergy Clin. Immunol.* **2019**, *144*, 183–192. [CrossRef]
130. Chen, C.B.; Abe, R.; Pan, R.Y.; Wang, C.W.; Hung, S.I.; Tsai, Y.G.; Chung, W.H. An Updated Review of the Molecular Mechanisms in Drug Hypersensitivity. *J. Immunol. Res.* **2018**, *2018*, 6431694. [CrossRef]
131. Ramírez, E.; Bellón, T.; Tong, H.Y.; Borobia, A.M.; de Abajo, F.J.; Lerma, V.; Moreno Hidalgo, M.A.; Castañer, J.L.; Cabañas, R.; Fiandor, A.; et al. Significant HLA Class I Type Associations with Aromatic Antiepileptic Drug (AED)-Induced SJS/TEN Are Different from Those Found for the Same AED-Induced DRESS in the Spanish Population. *Pharmacol. Res.* **2017**, *115*, 168–178. [CrossRef]
132. Hsiao, Y.H.; Hui, R.C.Y.; Wu, T.; Chang, W.C.; Hsih, M.S.; Yang, C.H.; Ho, H.C.; Chang, Y.G.; Chen, M.J.; Lin, J.Y.; et al. Genotype-Phenotype Association between HLA and Carbamazepine-Induced Hypersensitivity Reactions: Strength and Clinical Correlations. *J. Dermatol. Sci.* **2014**, *73*, 101–109. [CrossRef]
133. Ksouda, K.; Affes, H.; Mahfoudh, N.; Chtourou, L.; Kammoun, A.; Charfi, A.; Chaabane, H.; Medhioub, M.; Sahnoun, Z.; Turki, H.; et al. HLA-A*31:01 and Carbamazepine-Induced DRESS Syndrom in a Sample of North African Population. *Seizure* **2017**, *53*, 42–46. [CrossRef]
134. Genin, E.; Chen, D.P.; Hung, S.I.; Sekula, P.; Schumacher, M.; Chang, P.Y.; Tsai, S.H.; Wu, T.L.; Bellón, T.; Tamouza, R.; et al. HLA-A*31:01 and Different Types of Carbamazepine-Induced Severe Cutaneous Adverse Reactions: An International Study and Meta-Analysis. *Pharm. J.* **2014**, *14*, 281–288. [CrossRef]
135. Kim, B.K.; Jung, J.W.; Kim, T.B.; Chang, Y.S.; Park, H.S.; Moon, J.; Lee, S.T.; Jung, K.H.; Jung, K.Y.; Chu, K.; et al. HLA-A*31:01 and Lamotrigine-Induced Severe Cutaneous Adverse Drug Reactions in a Korean Population. *Ann. Allergy Asthma. Immunol.* **2017**, *118*, 629–630. [CrossRef]
136. Kang, H.R.; Jee, Y.K.; Kim, Y.S.; Lee, C.H.; Jung, J.W.; Kim, S.H.; Park, H.W.; Chang, Y.S.; Jang, I.J.; Cho, S.H.; et al. Positive and Negative Associations of HLA Class i Alleles with Allopurinol-Induced SCARs in Koreans. *Pharm. Genom.* **2011**, *21*, 303–307. [CrossRef]
137. Chen, W.T.; Wang, C.W.; Lu, C.W.; Chen, C.B.; Lee, H.E.; Hung, S.I.; Choon, S.E.; Yang, C.H.; Liu, M.T.; Chen, T.J.; et al. The Function of HLA-B*13:01 Involved in the Pathomechanism of Dapsone-Induced Severe Cutaneous Adverse Reactions. *J. Investig. Dermatol.* **2018**, *138*, 1546–1554. [CrossRef]
138. Zhang, F.-R.; Liu, H.; Irwanto, A.; Fu, X.-A.; Li, Y.; Yu, G.-Q.; Yu, Y.-X.; Chen, M.-F.; Low, H.-Q.; Li, J.-H.; et al. HLA-B*13:01 and the Dapsone Hypersensitivity Syndrome. *N. Engl. J. Med.* **2013**, *369*, 1620–1628. [CrossRef]
139. Yang, F.; Gu, B.; Zhang, L.; Xuan, J.; Luo, H.; Zhou, P.; Zhu, Q.; Yan, S.; Chen, S.A.; Cao, Z.; et al. HLA-B∗13:01 Is Associated with Salazosulfapyridine-Induced Drug Rash with Eosinophilia and Systemic Symptoms in Chinese Han Population. *Pharmacogenomics* **2014**, *15*, 1461–1469. [CrossRef]
140. Wang, C.W.; Tassaneeyakul, W.; Chen, C.B.; Chen, W.T.; Teng, Y.C.; Huang, C.Y.; Sukasem, C.; Lu, C.W.; Lee, Y.S.; Choon, S.E.; et al. Whole Genome Sequencing Identifies Genetic Variants Associated with Co-Trimoxazole Hypersensitivity in Asians. *J. Allergy Clin. Immunol.* **2021**, *147*, 1402–1412. [CrossRef]
141. Tassaneeyakul, W.; Prabmeechai, N.; Sukasem, C.; Kongpan, T.; Konyoung, P.; Chumworathayi, P.; Tiamkao, S.; Khunarkornsiri, U.; Kulkantrakorn, K.; Saksit, N.; et al. Associations between HLA Class I and Cytochrome P450 2C9 Genetic Polymorphisms and Phenytoin-Related Severe Cutaneous Adverse Reactions in a Thai Population. *Pharm. Genom.* **2016**, *26*, 225–234. [CrossRef]
142. Somogyi, A.A.; Barratt, D.T.; Phillips, E.J.; Moore, K.; Ilyas, F.; Gabb, G.M. High and Variable Population Prevalence of HLA-B*56:02 in Indigenous Australians and Relation to Phenytoin-Associated Drug Reaction with Eosinophilia and Systemic Symptoms. *Br. J. Clin. Pharmacol.* **2019**, *85*, 2163–2169. [CrossRef] [PubMed]
143. Sukasem, C.; Chaichan, C.; Nakkrut, T.; Satapornpong, P.; Jaruthamsophon, K.; Jantararoungtong, T.; Koomdee, N.; Sririttha, S.; Medhasi, S.; Oo-Puthinan, S.; et al. Association between HLA-B Alleles and Carbamazepine-Induced Maculopapular Exanthema and Severe Cutaneous Reactions in Thai Patients. *J. Immunol. Res.* **2018**, *2018*, 2780272. [CrossRef]
144. Chang, C.C.; Ng, C.C.; Too, C.L.; Choon, S.E.; Lee, C.K.; Chung, W.H.; Hussein, S.H.; Lim, K.S.; Murad, S. Association of HLA-B*15:13 and HLA-B*15:02 with Phenytoin-Induced Severe Cutaneous Adverse Reactions in a Malay Population. *Pharm. J.* **2017**, *17*, 170–173. [CrossRef]
145. Carr, D.F.; Chaponda, M.; Jorgensen, A.L.; Castro, E.C.; van Oosterhout, J.J.; Khoo, S.H.; Lalloo, D.G.; Heyderman, R.S.; Alfirevic, A.; Pirmohamed, M. Association of Human Leukocyte Antigen Alleles and Nevirapine Hypersensitivity in a Malawian HIV-Infected Population. *Clin. Infect. Dis.* **2013**, *56*, 1330–1339. [CrossRef]
146. Menegatti, J.; Schub, D.; Schäfer, M.; Grässer, F.A.; Ruprecht, K. HLA-DRB1*15:01 Is a Co-Receptor for Epstein–Barr Virus, Linking Genetic and Environmental Risk Factors for Multiple Sclerosis. *Eur. J. Immunol.* **2021**, *51*, 2348–2350. [CrossRef]
147. Manson, L.E.N.; Swen, J.J.; Guchelaar, H.J. Diagnostic Test Criteria for HLA Genotyping to Prevent Drug Hypersensitivity Reactions: A Systematic Review of Actionable HLA Recommendations in CPIC and DPWG Guidelines. *Front. Pharmacol.* **2020**, *11*, 1450. [CrossRef]
148. Hsu, Y.S.O.; Lu, K.L.; Fu, Y.; Wang, C.W.; Lu, C.W.; Lin, Y.F.; Chang, W.C.; Yeh, K.Y.; Hung, S.I.; Chung, W.H.; et al. The Roles of Immunoregulatory Networks in Severe Drug Hypersensitivity. *Front. Immunol.* **2021**, *12*, 597761. [CrossRef]

149. Hospital Discharges and Length of Stay Statistics—Statistics Explained. Available online: https://ec.europa.eu/eurostat/statistics-explained/index.php?title=Hospital_discharges_and_length_of_stay_statistics&oldid=574861#Hospital_discharges (accessed on 16 December 2022).
150. Lai, C.C.; Wang, C.Y.; Chu, C.C.; Tan, C.K.; Lu, C.L.; Lee, Y.L.; Huang, Y.T.; Lee, P.I.; Hsueh, P.R. Correlation between Antimicrobial Consumption and Resistance among Staphylococcus Aureus and Enterococci Causing Healthcare-Associated Infections at a University Hospital in Taiwan from 2000 to 2009. *Eur. J. Clin. Microbiol. Infect. Dis.* **2011**, *30*, 265–271. [CrossRef]
151. Antimicrobial Resistance Surveillance in Europe 2022–2020 Data. Available online: https://www.ecdc.europa.eu/en/publications-data/antimicrobial-resistance-surveillance-europe-2022-2020-data (accessed on 16 December 2022).
152. Junior, M.S.; Correa, L.; Marra, A.R.; Camargo, L.F.A.; Pereira, C.A.P. Analysis of Vancomycin Use and Associated Risk Factors in a University Teaching Hospital: A Prospective Cohort Study. *BMC Infect. Dis.* **2007**, *7*, 88. [CrossRef]
153. Pichler, W.J.; Hausmann, O. Classification of Drug Hypersensitivity into Allergic, p-i, and Pseudo-Allergic Forms. *Int. Arch. Allergy Immunol.* **2016**, *171*, 166–179. [CrossRef] [PubMed]
154. Yun, J.; Adam, J.; Yerly, D.; Pichler, W.J. Human Leukocyte Antigens (HLA) Associated Drug Hypersensitivity: Consequences of Drug Binding to HLA. *Allergy* **2012**, *67*, 1338–1346. [CrossRef]
155. Ostrov, D.A.; Grant, B.J.; Pompeu, Y.A.; Sidney, J.; Harndahl, M.; Southwood, S.; Oseroff, C.; Lu, S.; Jakoncic, J.; de Oliveira, C.A.F.; et al. Drug Hypersensitivity Caused by Alteration of the MHC-Presented Self-Peptide Repertoire. *Proc. Natl. Acad. Sci. USA* **2012**, *109*, 9959–9964. [CrossRef]
156. Pichler, W.J.; Beeler, A.; Keller, M.; Lerch, M.; Posadas, S.; Schmid, D.; Spanou, Z.; Zawodniak, A.; Gerber, B. Pharmacological Interaction of Drugs with Immune Receptors: The p-i Concept. *Allergol. Int.* **2006**, *55*, 17–25. [CrossRef] [PubMed]
157. Wei, C.Y.; Chung, W.H.; Huang, H.W.; Chen, Y.T.; Hung, S.I. Direct Interaction between HLA-B and Carbamazepine Activates T Cells in Patients with Stevens-Johnson Syndrome. *J. Allergy Clin. Immunol.* **2012**, *129*, 1562–1569.e5. [CrossRef]
158. Pichler, W.J. Immune Pathomechanism and Classification of Drug Hypersensitivity. *Allergy* **2019**, *74*, 1457–1471. [CrossRef] [PubMed]
159. Natkunarajah, J.; Watson, K.; Diaz-Cano, S.; Mufti, G.; du Vivier, A.; Creamer, D. Drug Rash with Eosinophilia and Systemic Symptoms and Graft-versus-Host Disease Developing Sequentially in a Patient. *Clin. Exp. Dermatol.* **2009**, *34*, 199–201. [CrossRef] [PubMed]
160. Nakkam, N.; Gibson, A.; Mouhtouris, E.; Konvinse, K.C.; Holmes, N.E.; Chua, K.Y.; Deshpande, P.; Li, D.; Ostrov, D.A.; Trubiano, J.; et al. Cross-Reactivity between Vancomycin, Teicoplanin, and Telavancin in Patients with HLA-A∗32:01-Positive Vancomycin-Induced DRESS Sharing an HLA Class II Haplotype. *J. Allergy Clin. Immunol.* **2021**, *147*, 403–405. [CrossRef]
161. Azoury, M.E.; Filì, L.; Bechara, R.; Scornet, N.; de Chaisemartin, L.; Weaver, R.J.; Claude, N.; Maillere, B.; Parronchi, P.; Joseph, D.; et al. Identification of T-Cell Epitopes from Benzylpenicillin Conjugated to Human Serum Albumin and Implication in Penicillin Allergy. *Allergy* **2018**, *73*, 1662–1672. [CrossRef] [PubMed]
162. Nakkam, N.; Trubiano, J.; Gibson, A.; Phillips, E.J. Considerations for Cross-Reactivity between Vancomycin and Other Glycopeptides. *J. Allergy Clin. Immunol. Pract.* **2021**, *9*, 3233. [CrossRef]
163. Monneaux, F.; Muller, S. Epitope Spreading in Systemic Lupus Erythematosus: Identification of Triggering Peptide Sequences. *Arthr. Rheum* **2002**, *46*, 1430–1438. [CrossRef] [PubMed]
164. Lucas, A.; Lucas, M.; Strhyn, A.; Keane, N.M.; McKinnon, E.; Pavlos, R.; Moran, E.M.; Meyer-Pannwitt, V.; Gaudieri, S.; D'Orsogna, L.; et al. Abacavir-Reactive Memory T Cells Are Present in Drug Naïve Individuals. *PLoS ONE* **2015**, *10*, e0117160. [CrossRef]
165. Sekiguchi, A.; Kashiwagi, T.; Ishida-Yamamoto, A.; Takahashi, H.; Hashimoto, Y.; Kimura, H.; Tohyama, M.; Hashimoto, K.; Iizuka, H. Drug-Induced Hypersensitivity Syndrome Due to Mexiletine Associated with Human Herpes Virus 6 and Cytomegalovirus Reactivation. *J. Dermatol.* **2005**, *32*, 278–281. [CrossRef]
166. Yerly, D.; Pompeu, Y.A.; Schutte, R.J.; Eriksson, K.K.; Strhyn, A.; Bracey, A.W.; Buus, S.; Ostrov, D.A. Structural Elements Recognized by Abacavir-Induced T Cells. *Int. J. Mol. Sci.* **2017**, *18*, 1464. [CrossRef] [PubMed]
167. Almeida, C.A.; van Miert, P.; O'Driscoll, K.; Zoet, Y.M.; Chopra, A.; Witt, C.; John, M.; Claas, F.H.J.; D'Orsogna, L.J. Virus-Specific T-Cell Clonotypes Might Contribute to Drug Hypersensitivity Reactions through Heterologous Immunity. *J. Allergy Clin. Immunol.* **2019**, *144*, 608–611.e4. [CrossRef]
168. Niu, J.; Jia, Q.; Ni, Q.; Yang, Y.; Chen, G.; Yang, X.; Zhai, Z.; Yu, H.; Guan, P.; Lin, R.; et al. Association of CD8(+) T Lymphocyte Repertoire Spreading with the Severity of DRESS Syndrome. *Sci. Rep.* **2015**, *5*, 9913. [CrossRef] [PubMed]
169. Chung, W.H.; Pan, R.Y.; Chu, M.T.; Chin, S.W.; Huang, Y.L.; Wang, W.C.; Chang, J.Y.; Hung, S.I. Oxypurinol-Specific T Cells Possess Preferential TCR Clonotypes and Express Granulysin in Allopurinol-Induced Severe Cutaneous Adverse Reactions. *J. Investig. Dermatol.* **2015**, *135*, 2237–2248. [CrossRef]
170. Kurose, K.; Ohue, Y.; Sato, E.; Yamauchi, A.; Eikawa, S.; Isobe, M.; Nishio, Y.; Uenaka, A.; Oka, M.; Nakayama, E. Increase in Activated Treg in TIL in Lung Cancer and in Vitro Depletion of Treg by ADCC Using an Antihuman CCR4 MAb (KM2760). *J. Thorac. Oncol.* **2015**, *10*, 74–83. [CrossRef] [PubMed]
171. Ogawa, K.; Morito, H.; Hasegawa, A.; Miyagawa, F.; Kobayashi, N.; Watanabe, H.; Sueki, H.; Tohyama, M.; Hashimoto, K.; Kano, Y.; et al. Elevated Serum Thymus and Activation-Regulated Chemokine (TARC/CCL17) Relates to Reactivation of Human Herpesvirus 6 in Drug Reaction with Eosinophilia and Systemic Symptoms (DRESS)/Drug-Induced Hypersensitivity Syndrome (DIHS). *Br. J. Dermatol.* **2014**, *171*, 425–427. [CrossRef] [PubMed]

172. Ushigome, Y.; Mizukawa, Y.; Kimishima, M.; Yamazaki, Y.; Takahashi, R.; Kano, Y.; Shiohara, T. Monocytes Are Involved in the Balance between Regulatory T Cells and Th17 Cells in Severe Drug Eruptions. *Clin. Exp. Allergy* **2018**, *48*, 1453–1463. [CrossRef] [PubMed]
173. Takahashi, R.; Kano, Y.; Yamazaki, Y.; Kimishima, M.; Mizukawa, Y.; Shiohara, T. Defective Regulatory T Cells in Patients with Severe Drug Eruptions: Timing of the Dysfunction Is Associated with the Pathological Phenotype and Outcome. *J. Immunol.* **2009**, *182*, 8071–8079. [CrossRef] [PubMed]
174. Ramirez, G.A.; Cariddi, A.; Noviello, S.; Campochiaro, C.; Canti, V.; Moroni, L.; Yacoub, M.-R.; Baldissera, E.M.; Bozzolo, E.P.; Dagna, L. Real-Life Efficacy and Safety of Mepolizumab for Eosinophilic Granulomatosis with Polyangiitis. *Clin. Immunol. Commun.* **2022**, *2*, 23–29. [CrossRef]
175. Yamaya, M.; Sasaki, H. Rhinovirus and Asthma. *Viral Immunol.* **2003**, *16*, 99–109. [CrossRef] [PubMed]
176. Khatri, S.; Moore, W.; Gibson, P.G.; Leigh, R.; Bourdin, A.; Maspero, J.; Barros, M.; Buhl, R.; Howarth, P.; Albers, F.C.; et al. Assessment of the Long-Term Safety of Mepolizumab and Durability of Clinical Response in Patients with Severe Eosinophilic Asthma. *J. Allergy Clin. Immunol.* **2019**, *143*, 1742–1751.e7. [CrossRef]
177. Sabogal Piñeros, Y.S.; Bal, S.M.; Dijkhuis, A.; Majoor, C.J.; Dierdorp, B.S.; Dekker, T.; Hoefsmit, E.P.; Bonta, P.I.; Picavet, D.; van der Wel, N.N.; et al. Eosinophils Capture Viruses, a Capacity That Is Defective in Asthma. *Allergy* **2019**, *74*, 1898–1909. [CrossRef] [PubMed]
178. Ramirez, G.A.; Yacoub, M.R.; Ripa, M.; Mannina, D.; Cariddi, A.; Saporiti, N.; Ciceri, F.; Castagna, A.; Colombo, G.; Dagna, L. Eosinophils from Physiology to Disease: A Comprehensive Review. *BioMed Res. Int.* **2018**, *2018*, 9095275. [CrossRef] [PubMed]
179. Tsai, Y.G.; Liou, J.H.; Hung, S.I.; Chen, C.B.; Chiu, T.M.; Wang, C.W.; Chung, W.H. Increased Type 2 Innate Lymphoid Cells in Patients with Drug Reaction with Eosinophilia and Systemic Symptoms Syndrome. *J. Investig. Dermatol.* **2019**, *139*, 1722–1731. [CrossRef] [PubMed]
180. Stoeckle, C.; Simon, H.U. CD8(+) T Cells Producing IL-3 and IL-5 in Non-IgE-Mediated Eosinophilic Diseases. *Allergy* **2013**, *68*, 1622–1625. [CrossRef] [PubMed]
181. Lyons, P.A.; Peters, J.E.; Alberici, F.; Liley, J.; Coulson, R.M.R.; Astle, W.; Baldini, C.; Bonatti, F.; Cid, M.C.; Elding, H.; et al. Genome-Wide Association Study of Eosinophilic Granulomatosis with Polyangiitis Reveals Genomic Loci Stratified by ANCA Status. *Nat. Commun.* **2019**, *10*, 5120. [CrossRef] [PubMed]
182. Fulkerson, P.C.; Rothenberg, M.E. Targeting Eosinophils in Allergy, Inflammation and Beyond. *Nat. Rev. Drug Discov.* **2013**, *12*, 117–129. [CrossRef] [PubMed]
183. James, J.; Sammou, Y.M.; Virata, A.R.; Nordin, T.A.; Dumic, I. Drug Reaction with Eosinophilia and Systemic Symptoms (DRESS) Syndrome Secondary to Furosemide: Case Report and Review of Literature. *Am. J. Case Rep.* **2018**, *19*, 163–170. [CrossRef] [PubMed]
184. Hiransuthikul, A.; Rattananupong, T.; Klaewsongkram, J.; Rerknimitr, P.; Pongprutthipan, M.; Ruxrungtham, K. Drug-Induced Hypersensitivity Syndrome/Drug Reaction with Eosinophilia and Systemic Symptoms (DIHS/DRESS): 11 Years Retrospective Study in Thailand. *Allergol. Int.* **2016**, *65*, 432–438. [CrossRef] [PubMed]
185. Behera, S.K.; Das, S.; Xavier, A.S.; Selvarajan, S. DRESS Syndrome: A Detailed Insight. *Hosp. Pract. (1995)* **2018**, *46*, 152–162. [CrossRef] [PubMed]
186. Dorrell BA, D.N.; Whitaker BS, L.F.; Anderson, K.L.; Strowd, L.C. Abnormal Erythrocyte Morphology in Drug Reaction with Eosinophilia and Systemic Symptoms. *J. Am. Acad. Dermatol.* **2018**, *80*, 1159–1160. [CrossRef]
187. Peyrière, H.; Dereure, O.; Breton, H.; Demoly, P.; Cociglio, M.; Blayac, J.P.; Hillaire-Buys, D. Variability in the Clinical Pattern of Cutaneous Side-Effects of Drugs with Systemic Symptoms: Does a DRESS Syndrome Really Exist? *Br. J. Dermatol.* **2006**, *155*, 422–428. [CrossRef] [PubMed]
188. Choudhary, S.; McLeod, M.; Torchia, D.; Romanelli, P. Drug Reaction with Eosinophilia and Systemic Symptoms (DRESS) Syndrome. *J. Clin. Aesthet. Dermatol.* **2013**, *6*, 31–37. [PubMed]
189. Syn, W.K.; Naisbitt, D.J.; Holt, A.P.; Pirmohamed, M.; Mutimer, D.J. Carbamazepine-Induced Acute Liver Failure as Part of the DRESS Syndrome. *Int. J. Clin. Pract.* **2005**, *59*, 988–991. [CrossRef]
190. de Campos, F.P.F.; de Lima, P.P.; Maragno, L.; Watanabe, F.T. Hepatic Necrosis Associated with Drug-Induced Hypersensitivity Syndrome. *Autops. Case Rep.* **2012**, *2*, 5–14. [CrossRef] [PubMed]
191. Subhani, M.; Dong, V.; Connolly, A.; Salisbury, J.; Miquel, R.; Walsh, S.; Pirani, T. Trimethoprim-Induced Drug Reaction with Eosinophilia and Systemic Symptoms (DRESS) Associated with Reactivation of Human Herpes Virus-6 (HHV-6) Leading to Acute Liver Failure. *Clin. Case Rep.* **2020**, *8*, 2568–2573. [CrossRef]
192. Lee, T.; Lee, Y.S.; Yoon, S.Y.; Kim, S.; Bae, Y.J.; Kwon, H.S.; Cho, Y.S.; Moon, H.B.; Kim, T.B. Characteristics of Liver Injury in Drug-Induced Systemic Hypersensitivity Reactions. *J. Am. Acad. Dermatol.* **2013**, *69*, 407–415. [CrossRef]
193. Miyasaka, A.; Kumagai, I.; Masda, T.; Takikawa, Y. A 51-Year-Old Woman with Drug-Induced Hypersensitivity Syndrome Associated with Carbamazepine, Reactivation of Human Herpesvirus 6, and Acute Liver Failure: A Case Report. *Am. J. Case Rep.* **2021**, *22*, e928587. [CrossRef] [PubMed]
194. Husain, Z.; Reddy, B.Y.; Schwartz, R.A. DRESS Syndrome: Part I. Clinical Perspectives. *J. Am. Acad. Dermatol.* **2013**, *68*, 693.e1–693.e14. [CrossRef] [PubMed]

195. Descamps, V.; Gautheret-Dejean, A.; Pelletier, A.L.; Bonnafous, P.; Deschamps, L.; Prusty, B.K. Chronic Persistent HHV-6B Infection after Sulfasalazine-Induced DRESS with Demonstration of HHV-6 Encoded Small Noncoding RNAs (SncRNAs) in Crohn's-like Colitis: Case Report. *Clin. Case Rep.* **2020**, *9*, 841–844. [CrossRef] [PubMed]
196. Asano, Y.; Kagawa, H.; Kano, Y.; Shiohara, T. Cytomegalovirus Disease during Severe Drug Eruptions: Report of 2 Cases and Retrospective Study of 18 Patients with Drug-Induced Hypersensitivity Syndrome. *Arch. Dermatol.* **2009**, *145*, 1030–1036. [CrossRef] [PubMed]
197. Do-Pham, G.; Charachon, A.; Duong, T.A.; Thille, A.W.; Benhaiem, N.; Bagot, M.; Chosidow, O.; Roujeau, J.C.; Wolkenstein, P.; Valeyrie-Allanore, L. Drug Reaction with Eosinophilia and Systemic Symptoms and Severe Involvement of Digestive Tract: Description of Two Cases. *Br. J. Dermatol.* **2011**, *165*, 207–209. [CrossRef] [PubMed]
198. Descamps, V.; Mahe, E.; Houhou, N.; Abramowitz, L.; Rozenberg, F.; Ranger-Rogez, S.; Crickx, B. Drug-Induced Hypersensitivity Syndrome Associated with Epstein–Barr Virus Infection. *Br. J. Dermatol.* **2003**, *148*, 1032–1034. [CrossRef] [PubMed]
199. Sekine, N.; Motokura, T.; Oki, T.; Umeda, Y.; Sasaki, N.; Hayashi, M.; Sato, H.; Fujita, T.; Kaneko, T.; Asano, Y.; et al. Rapid Loss of Insulin Secretion in a Patient with Fulminant Type 1 Diabetes Mellitus and Carbamazepine Hypersensitivity Syndrome. *JAMA* **2001**, *285*, 1153–1154. [CrossRef]
200. Kano, Y.; Ishida, T.; Hirahara, K.; Shiohara, T. Visceral Involvements and Long-Term Sequelae in Drug-Induced Hypersensitivity Syndrome. *Med. Clin. N. Am.* **2010**, *94*, 743–759. [CrossRef] [PubMed]
201. Kano, Y.; Shiohara, T. The Variable Clinical Picture of Drug-Induced Hypersensitivity Syndrome/Drug Rash with Eosinophilia and Systemic Symptoms in Relation to the Eliciting Drug. *Immunol. Allergy Clin. N. Am.* **2009**, *29*, 481–501. [CrossRef] [PubMed]
202. Esposito, A.J.; Murphy, R.C.; Toukatly, M.N.; Amro, O.W.; Kestenbaum, B.R.; Najafian, B. Acute Kidney Injury in Allopurinol-Induced DRESS Syndrome: A Case Report of Concurrent Tubulointerstitial Nephritis and Kidney-Limited Necrotizing Vasculitis. *Clin. Nephrol.* **2017**, *87*, 316–319. [CrossRef] [PubMed]
203. Hagiya, H.; Iwamuro, M.; Tanaka, T.; Hasegawa, K.; Hanayama, Y.; Kimura, M.; Otsuka, F. Reactivation of Human Herpes Virus-6 in the Renal Tissue of a Patient with Drug-Induced Hypersensitivity Syndrome/Drug Rash with Eosinophilia and Systemic Symptoms (DIHS/DRESS). *Intern. Med.* **2016**, *55*, 1769–1774. [CrossRef] [PubMed]
204. Bourgeois, G.P.; Cafardi, J.A.; Groysman, V.; Hughey, L.C. A Review of DRESS-Associated Myocarditis. *J. Am. Acad. Dermatol.* **2012**, *66*, e229–e236. [CrossRef]
205. Taweesedt, P.T.; Nordstrom, C.W.; Stoeckel, J.; Dumic, I. Pulmonary Manifestations of Drug Reaction with Eosinophilia and Systemic Symptoms (DRESS) Syndrome: A Systematic Review. *BioMed Res. Int.* **2019**, *2019*, 7863815. [CrossRef] [PubMed]
206. Masaki, T.; Fukunaga, A.; Tohyama, M.; Koda, Y.; Okuda, S.; Maeda, N.; Kanda, F.; Yasukawa, M.; Hashimoto, K.; Horikawa, T.; et al. Human Herpes Virus 6 Encephalitis in Allopurinol-Induced Hypersensitivity Syndrome. *Acta Derm. Venereol.* **2003**, *83*, 128–131. [CrossRef] [PubMed]
207. Ozisik, L.; Tanriover, M.D.; Saka, E. Autoimmune Limbic Encephalitis and Syndrome of Inappropriate Antidiuretic Hormone Secretion Associated with Lamotrigine-Induced Drug Rash with Eosinophilia and Systemic Symptoms (DRESS) Syndrome. *Intern. Med.* **2016**, *55*, 1393–1396. [CrossRef]
208. Yokote, A.; Tomita, S.; Sawada, H. Sensory Ganglionopathy Associated with Drug-Induced Hypersensitivity Syndrome Caused by Mexiletine. *BMJ Case Rep.* **2018**, *2018*, bcr-2017222540. [CrossRef]
209. Zhu, B.; Wu, J.; Chen, G.; Yang, Y.; Yi, C. Fulminant Type 1 Diabetes Mellitus Caused by Drug Reaction With Eosinophilia and Systemic Symptoms (DRESS): A Case Report and Review of the Literature. *Front. Endocrinol.* **2019**, *10*, 474. [CrossRef]
210. Mizukawa, Y.; Aoyama, Y.; Takahashi, H.; Takahashi, R.; Shiohara, T. Risk of Progression to Autoimmune Disease in Severe Drug Eruption: Risk Factors and the Factor-Guided Stratification. *J. Investig. Dermatol.* **2022**, *142*, 960–968.e9. [CrossRef]
211. Owen, C.E.; Jones, J.M. Recognition and Management of Severe Cutaneous Adverse Drug Reactions (Including Drug Reaction with Eosinophilia and Systemic Symptoms, Stevens-Johnson Syndrome, and Toxic Epidermal Necrolysis). *Med. Clin. N. Am.* **2021**, *105*, 577–597. [CrossRef]
212. Milani-Nejad, N.; Trinidad, J.; Kaffenberger, B.H. Viral Reactivation in Hospitalized Patients with Drug Reaction with Eosinophilia and Systemic Symptoms: A Retrospective Study from a Tertiary Medical Center in the United States. *J. Am. Acad. Dermatol.* **2020**, *83*, 278–279. [CrossRef]
213. Abdelnabi, M.; Elmssary, M.; Sekhon, J.; Benjanuwattra, J. Acute Onset of Fever, Eosinophilia, Rash, Acute Kidney Injury, and a Positive Monospot Test in a Patient on Lamotrigine: DRESS Syndrome. *Lancet* **2022**, *399*, 1902. [CrossRef]
214. Thongsri, T.; Chularojanamontri, L.; Pichler, W.J. Cardiac Involvement in DRESS Syndrome. *Asian Pac. J. Allergy Immunol.* **2017**, *35*, 3–10. [CrossRef]
215. de Groot, A.C. Patch Testing in Drug Reaction with Eosinophilia and Systemic Symptoms (DRESS): A Literature Review. *Contact Dermat.* **2022**, *86*, 443–479. [CrossRef] [PubMed]
216. Pichler, W.J.; Tilch, J. The Lymphocyte Transformation Test in the Diagnosis of Drug Hypersensitivity. *Allergy* **2004**, *59*, 809–820. [CrossRef]
217. Chi, M.H.; Hui, R.C.Y.; Yang, C.H.; Lin, J.Y.; Lin, Y.T.; Ho, H.C.; Chung, W.H.; Kuo, T.T. Histopathological Analysis and Clinical Correlation of Drug Reaction with Eosinophilia and Systemic Symptoms (DRESS). *Br. J. Dermatol.* **2014**, *170*, 866–873. [CrossRef] [PubMed]
218. Revuz, J.; Penso, D.; Roujeau, J.C.; Guillaume, J.C.; Payne, C.R.; Wechsler, J.; Touraine, R. Toxic Epidermal Necrolysis. Clinical Findings and Prognosis Factors in 87 Patients. *Arch. Dermatol.* **1987**, *123*, 1160–1165. [CrossRef] [PubMed]

219. Ye, L.-P.; Zhang, C.; Zhu, Q.-X. The Effect of Intravenous Immunoglobulin Combined with Corticosteroid on the Progression of Stevens-Johnson Syndrome and Toxic Epidermal Necrolysis: A Meta-Analysis. *PloS ONE* **2016**, *11*, e0167120. [CrossRef] [PubMed]
220. Teraki, Y.; Shibuya, M.; Izaki, S. Stevens–Johnson Syndrome and Toxic Epidermal Necrolysis Due to Anticonvulsants Share Certain Clinical and Laboratory Features with Drug-Induced Hypersensitivity Syndrome, despite Differences in Cutaneous Presentations. *Clin. Exp. Dermatol.* **2010**, *35*, 723–728. [CrossRef]
221. Zalewska-Janowska, A.; Spiewak, R.; Kowalski, M.L. Cutaneous Manifestation of Drug Allergy and Hypersensitivity. *Immunol. Allergy Clin. N. Am.* **2017**, *37*, 165–181. [CrossRef]
222. Bircher, A.J.; Scherer, K. Delayed Cutaneous Manifestations of Drug Hypersensitivity. *Med. Clin. N. Am.* **2010**, *94*, 711–725. [CrossRef]
223. Bagnasco, D.; Ferrando, M.; Caminati, M.; Bragantini, A.; Puggioni, F.; Varricchi, G.; Passalacqua, G.; Canonica, G.W. Targeting Interleukin-5 or Interleukin-5Rα: Safety Considerations. *Drug Saf.* **2017**, *40*, 559–570. [CrossRef] [PubMed]
224. Ruiz Matta, J.M.; Domínguez Cherit, J.; Méndez Flores, S. Drug Reaction with Eosinophilia and Systemic Symptoms (DRESS) and Its Relation with Autoimmunity in a Reference Center in Mexico. *An. Bras. Dermatol.* **2017**, *92*, 30–33. [CrossRef] [PubMed]
225. Yee, B.E.; Nguyen, N.H.; Lee, D. Extensive Pulmonary Involvement with Raltegravir-Induced DRESS Syndrome in a Postpartum Woman with HIV. *BMJ Case Rep.* **2014**, *2014*, bcr2013201545. [CrossRef] [PubMed]
226. Simmons, R.P.; Dudzinski, D.M.; Shepard, J.-A.O.; Hurtado, R.M.; Coffey, K.C. Case 16-2019: A 53-Year-Old Man with Cough and Eosinophilia. *N. Engl. J. Med.* **2019**, *380*, 2052–2059. [CrossRef]
227. Lee, H.Y.; Walsh, S.; Creamer, D. Initial Presentation of DRESS: Often Misdiagnosed as Infections. *Arch. Dermatol.* **2012**, *148*, 1085–1087. [CrossRef] [PubMed]
228. Bartal, C.; Sagy, I.; Barski, L. Drug-Induced Eosinophilic Pneumonia: A Review of 196 Case Reports. *Medicine* **2018**, *97*, e9688. [CrossRef]
229. Kim, D.H.; Koh, Y.-I. Comparison of Diagnostic Criteria and Determination of Prognostic Factors for Drug Reaction with Eosinophilia and Systemic Symptoms Syndrome. *Allergy Asthma. Immunol. Res.* **2014**, *6*, 216–221. [CrossRef]
230. Cardones, A.R. Drug Reaction with Eosinophilia and Systemic Symptoms (DRESS) Syndrome. *Clin. Dermatol.* **2020**, *38*, 702–711. [CrossRef]
231. Duong, T.A.; Valeyrie-Allanore, L.; Wolkenstein, P.; Chosidow, O. Severe Cutaneous Adverse Reactions to Drugs. *Lancet* **2017**, *390*, 1996–2011. [CrossRef]
232. Mizukawa, Y.; Hirahara, K.; Kano, Y.; Shiohara, T. Drug-Induced Hypersensitivity Syndrome/Drug Reaction with Eosinophilia and Systemic Symptoms Severity Score: A Useful Tool for Assessing Disease Severity and Predicting Fatal Cytomegalovirus Disease. *J. Am. Acad. Dermatol.* **2019**, *80*, 670–678.e2. [CrossRef]
233. Natkunarajah, J.; Goolamali, S.; Craythorne, E.; Benton, E.; Smith, C.; Morris-Jones, R.; Wendon, J.; Higgins, E.; Creamer, D. Ten Cases of Drug Reaction with Eosinophilia and Systemic Symptoms (DRESS) Treated with Pulsed Intravenous Methylprednisolone. *Eur. J. Dermatol.* **2011**, *21*, 385–391. [CrossRef]
234. Yacoub, M.R.; Berti, A.; Campochiaro, C.; Tombetti, E.; Ramirez, G.A.; Nico, A.; Leo, E.; Fantini, P.; Sabbadini, M.G.; Nettis, E.; et al. Drug Induced Exfoliative Dermatitis: State of the Art. *Clin. Mol. Allergy* **2016**, *14*, 9. [CrossRef]
235. Gottlieb, M.; Figlewicz, M.R.; Rabah, W.; Buddan, D.; Long, B. Drug Reaction with Eosinophilia and Systemic Symptoms: An Emergency Medicine Focused Review. *Am. J. Emerg. Med.* **2022**, *56*, 1–6. [CrossRef] [PubMed]
236. Nguyen, E.; Yanes, D.; Imadojemu, S.; Kroshinsky, D. Evaluation of Cyclosporine for the Treatment of DRESS Syndrome. *JAMA Dermatol.* **2020**, *156*, 704–706. [CrossRef]
237. Della-Torre, E.; Yacoub, M.R.; Pignatti, P.; Della-Torre, F.; Sabbadini, M.G.; Colombo, G.; Tresoldi, M. Optimal Management of DRESS Syndrome in Course of Infectious Endocarditis. *Ann. Allergy Asthma Immunol.* **2013**, *110*, 303–305. [CrossRef] [PubMed]
238. Husain, Z.; Reddy, B.Y.; Schwartz, R.A. DRESS Syndrome: Part II. Management and Therapeutics. *J. Am. Acad. Dermatol.* **2013**, *68*, 709.e1–709.e9. [CrossRef] [PubMed]
239. Picard, D.; Vellar, M.; Janela, B.; Roussel, A.; Joly, P.; Musette, P. Recurrence of Drug-Induced Reactions in DRESS Patients. *J. Eur. Acad. Dermatol. Venereol.* **2015**, *29*, 801–804. [CrossRef] [PubMed]
240. Jörg, L.; Helbling, A.; Yerly, D.; Pichler, W.J. Drug-Related Relapses in Drug Reaction with Eosinophilia and Systemic Symptoms (DRESS). *Clin. Transl. Allergy* **2020**, *10*, 52. [CrossRef] [PubMed]

MDPI
St. Alban-Anlage 66
4052 Basel
Switzerland
www.mdpi.com

Microorganisms Editorial Office
E-mail: microorganisms@mdpi.com
www.mdpi.com/journal/microorganisms

www.ingramcontent.com/pod-product-compliance
Lightning Source LLC
Chambersburg PA
CBHW061805210926
51850CB00006B/616
9783725806683